THE PHYSICS OF VIBRATIONS AND WAVES

Sixth Edition

THE PHYSICS OF VIBRATIONS AND WAVES

Sixth Edition

H. J. Pain
Formerly of Department of Physics,
Imperial College of Science and Technology, London, UK

John Wiley & Sons, Ltd

Copyright © 2005   John Wiley & Sons Ltd, The Atrium, Southern Gate, Chichester,
West Sussex PO19 8SQ, England

Telephone   (+44) 1243 779777

Email (for orders and customer service enquiries): cs-books@wiley.co.uk
Visit our Home Page on www.wileyeurope.com or www.wiley.com

Reprinted with corrections October 2005

Other Wiley Editorial Offices

John Wiley & Sons Inc., 111 River Street, Hoboken, NJ 07030, USA

Jossey-Bass, 989 Market Street, San Francisco, CA 94103-1741, USA

Wiley-VCH Verlag GmbH, Boschstr. 12, D-69469 Weinheim, Germany

John Wiley & Sons Australia Ltd, 33 Park Road, Milton, Queensland 4064, Australia

John Wiley & Sons (Asia) Pte Ltd, 2 Clementi Loop #02-01, Jin Xing Distripark, Singapore 129809

John Wiley & Sons Canada Ltd, 22 Worcester Road, Etobicoke, Ontario, Canada M9W 1L1

Wiley also publishes its books in a variety of electronic formats. Some content that appears in
print may not be available in electronic books.

Library of Congress Cataloging-in-Publication Data
(to follow)

British Library Cataloguing in Publication Data

A catalogue record for this book is available from the British Library

ISBN 10: 0-470-01295-1 (H/B)   ISBN 13: 978-0-470-01295-6 (H/B)
ISBN 10: 0-470-01296-X (P/B)   ISBN 13: 978-0-470-01296-3 (P/B)

Typeset in 10.5/12.5pt Times by Thomson Press (India) Limited, New Delhi, India.
Printed and bound in Great Britain by Antony Rowe Ltd, Chippenham, Wiltshire.
This book is printed on acid-free paper responsibly manufactured from sustainable forestry
in which at least two trees are planted for each one used for paper production.

Contents

4   Coupled Oscillations   79

5   Transverse Wave Motion   107

6   Longitudinal Waves   151

15   Non-linear Waves, Shocks and Solitons   505

Appendix 1: Normal Modes, Phase Space and Statistical Physics   533

Appendix 2: Kirchhoff's Integral Theorem   547

Appendix 3: Non-Linear Schrödinger Equation   551

Index   552

Introduction to First Edition

The opening session of the physics degree course at Imperial College includes an introduction to vibrations and waves where the stress is laid on the underlying unity of concepts which are studied separately and in more detail at later stages. The origin of this short textbook lies in that lecture course which the author has given for a number of years. Sections on Fourier transforms and non-linear oscillations have been added to extend the range of interest and application.

At the beginning no more than school-leaving mathematics is assumed and more advanced techniques are outlined as they arise. This involves explaining the use of exponential series, the notation of complex numbers and partial differentiation and putting trial solutions into differential equations. Only plane waves are considered and, with two exceptions, Cartesian coordinates are used throughout. Vector methods are avoided except for the scalar product and, on one occasion, the vector product.

Opinion canvassed amongst many undergraduates has argued for a 'working' as much as for a 'reading' book; the result is a concise text amplified by many problems over a wide range of content and sophistication. Hints for solution are freely given on the principle that an undergraduates gains more from being guided to a result of physical significance than from carrying out a limited arithmetical exercise.

The main theme of the book is that a medium through which energy is transmitted via wave propagation behaves essentially as a continuum of coupled oscillators. A simple oscillator is characterized by three parameters, two of which are capable of storing and exchanging energy, whilst the third is energy dissipating. This is equally true of any medium.

The product of the energy storing parameters determines the velocity of wave propagation through the medium and, in the absence of the third parameter, their ratio governs the impedance which the medium presents to the waves. The energy dissipating parameter introduces a loss term into the impedance; energy is absorbed from the wave system and it attenuates.

This viewpoint allows a discussion of simple harmonic, damped, forced and coupled oscillators which leads naturally to the behaviour of transverse waves on a string, longitudinal waves in a gas and a solid, voltage and current waves on a transmission line and electromagnetic waves in a dielectric and a conductor. All are amenable to this common treatment, and it is the wide validity of relatively few physical principles which this book seeks to demonstrate.

H. J. PAIN
May 1968

Introduction to Second Edition

The main theme of the book remains unchanged but an extra chapter on Wave Mechanics illustrates the application of classical principles to modern physics.

Any revision has been towards a simpler approach especially in the early chapters and additional problems. Reference to a problem in the course of a chapter indicates its relevance to the preceding text. Each chapter ends with a summary of its important results.

Constructive criticism of the first edition has come from many quarters, not least from successive generations of physics and engineering students who have used the book; a second edition which incorporates so much of this advice is the best acknowledgement of its value.

H. J. PAIN
June 1976

Introduction to Third Edition

Since this book was first published the physics of optical systems has been a major area of growth and this development is reflected in the present edition. Chapter 10 has been rewritten to form the basis of an introductory course in optics and there are further applications in Chapters 7 and 8.

The level of this book remains unchanged.

H. J. PAIN
January 1983

Introduction to Fourth Edition

Interest in non-linear dynamics has grown in recent years through the application of chaos theory to problems in engineering, economics, physiology, ecology, meteorology and astronomy as well as in physics, biology and fluid dynamics. The chapter on non-linear oscillations has been revised to include topics from several of these disciplines at a level appropriate to this book. This has required an introduction to the concept of phase space which combines with that of normal modes from earlier chapters to explain how energy is distributed in statistical physics. The book ends with an appendix on this subject.

H. J. PAIN
September 1992

Introduction to Fifth Edition

In this edition, three of the longer chapters of earlier versions have been split in two: Simple Harmonic Motion is now the first chapter and Damped Simple Harmonic Motion the second. Chapter 10 on waves in optical systems now becomes Chapters 11 and 12, Waves in Optical Systems, and Interference and Diffraction respectively through a reordering of topics. A final chapter on non-linear waves, shocks and solitons now follows that on non-linear oscillations and chaos.

New material includes matrix applications to coupled oscillations, optical systems and multilayer dielectric films. There are now sections on e.m. waves in the ionosphere and other plasmas, on the laser cavity and on optical wave guides. An extended treatment of solitons includes their role in optical transmission lines, in collisionless shocks in space, in non-periodic lattices and their connection with Schrödinger's equation.

H. J. PAIN
March 1998

Acknowledgement

The author is most grateful to Professor L. D. Roelofs of the Physics Department, Haverford College, Haverford, PA, USA. After using the last edition he provided an informed, extended and valuable critique that has led to many improvements in the text and questions of this book. Any faults remain the author's responsibility.

Introduction to Sixth Edition

This edition includes new material on electron waves in solids using the Kronig – Penney model to show how their allowed energies are limited to Brillouin zones. The role of phonons is also discussed. Convolutions are introduced and applied to optical problems via the Array Theorem in Young's experiment and the Optical Transfer Function. In the last two chapters the sections on Chaos and Solitons have been reduced but their essential contents remain.

I am grateful to my colleague Professor Robin Smith of Imperial College for his advice on the Optical Transfer Function. I would like to thank my wife for typing the manuscript of every edition except the first.

H. J. PAIN
January 2005, Oxford

Chapter Synopses

Chapter 1   Simple Harmonic Motion

Simple harmonic motion of mechanical and electrical oscillators (1) Vector representation of simple harmonic motion (6) Superpositions of two SHMs by vector addition (12) Superposition of two perpendicular SHMs (15) Polarization, Lissajous figures (17) Superposition of many SHMs (20) Complex number notation and use of exponential series (25) Summary of important results.

Chapter 2   Damped Simple Harmonic Motion

Damped motion of mechanical and electrical oscillators (37) Heavy damping (39) Critical damping (40) Damped simple harmonic oscillations (41) Amplitude decay   (43) Logarithmic decrement (44) Relaxation time (46) Energy decay (46) Q-value (46) Rate of energy decay equal to work rate of damping force (48) Summary of important results.

Chapter 3   The Forced Oscillator

The vector operator i (53) Electrical and mechanical impedance (56) Transient and steady state behaviour of a forced oscillator (58) Variation of displacement and velocity with frequency of driving force (60) Frequency dependence of phase angle between force and (a) displacement, (b) velocity (60) Vibration insulation (64) Power supplied to oscillator (68) Q-value as a measure of power absorption bandwidth (70) Q-value as amplification factor of low frequency response (71) Effect of transient term (74) Summary of important results.

Chapter 4   Coupled Oscillations

Spring coupled pendulums (79) Normal coordinates and normal modes of vibration (81) Matrices and eigenvalues (86) Inductance coupling of electrical oscillators (87) Coupling of many oscillators on a loaded string (90) Wave motion as the limit of coupled oscillations (95) Summary of important results.

Chapter 5   Transverse Wave Motion

Notation of partial differentiation (107) Particle and phase velocities (109) The wave equation (110) Transverse waves on a string (111) The string as a forced oscillator (115) Characteristic impedance of a string (117) Reflection and transmission of transverse waves at a boundary (117) Impedance matching (121) Insertion of quarter wave element (124) Standing waves on a string of fixed length (124) Normal modes and eigenfrequencies (125) Energy in a normal mode of oscillation (127) Wave groups (128) Group velocity (130) Dispersion (131) Wave group of many components (132) Bandwidth Theorem (134) Transverse waves in a periodic structure (crystal) (135) Doppler Effect (141) Summary of important results.

Chapter 6   Longitudinal Waves

Wave equation (151) Sound waves in gases (151) Energy distribution in sound waves (155) Intensity (157) Specific acoustic impedance (158) Longitudinal waves in a solid (159) Young's Modulus (159) Poisson's ratio (159) Longitudinal waves in a periodic structure (162) Reflection and transmission of sound waves at a boundary (163) Summary of important results.

Chapter 7   Waves on Transmission Lines

Ideal transmission line (173) Wave equation (174) Velocity of voltage and current waves (174) Characteristic impedance (175) Reflection at end of terminated line (177) Standing waves in short circuited line (178) Transmission line as a filter (179) Propagation constant (181) Real transmission line with energy losses (183) Attenuation coefficient (185) Diffusion equation (187) Diffusion coefficients (190) Attenuation (191) Wave equation plus diffusion effects (190) Summary of important results.

Chapter 8   Electromagnetic Waves

Permeability and permittivity of a medium (199) Maxwell's equations (202) Displacement current (202) Wave equations for electric and magnetic field vectors in a dielectric (204) Poynting vector (206) Impedance of a dielectric to e.m. waves (207) Energy density of e.m. waves (208) Electromagnetic waves in a conductor (208) Effect of conductivity adds diffusion equation to wave equation (209) Propagation and attenuation of e.m. waves in a conductor (210) Skin depth (211) Ratio of displacement current to conduction current as a criterion for dielectric or conducting behaviour (213) Relaxation time of a conductor (214) Impedance of a conductor to e.m. waves (215) Reflection and transmission of e.m. waves at a boundary (217) Normal incidence (217) Oblique incidence and Fresnel's equations (218) Reflection from a conductor (222) Connection between impedance and refractive index (219) E.m. waves in plasmas and the ionosphere (223) Summary of important results.

Chapter 9   Waves in More than One Dimension

Plane wave representation in 2 and 3 dimensions (239) Wave equation in 2- dimensions (240) Wave guide (242) Reflection of a 2-dimensional wave at rigid boundaries (242) Normal modes and method of separation of variables for 1, 2 and 3 dimensions (245) Normal modes in 2 dimensions on a rectangular membrane (247) Degeneracy (250) Normal modes in 3 dimensions (250) Number of normal modes per unit frequency interval per unit volume (251) Application to Planck's Radiation Law and Debye's Theory of Specific Heats (251) Reflection and transmission of an e.m. wave in 3 dimensions (254) Snell's Law (256) Total internal reflexion and evanescent waves (256) Summary of important results.

Chapter 10   Fourier Methods

Fourier series for a periodic function (267) Fourier series for any interval (271) Application to a plucked string (275) Energy in normal modes (275) Application to rectangular velocity pulse on a string (278) Bandwidth Theorem (281) Fourier integral of a single pulse (283) Fourier Transforms (285) Application to optical diffraction (287) Dirac function (292) Convolution (292) Convolution Theorem (297) Summary of important results.

Chapter 11   Waves in Optical Systems

Fermat's Principle (307) Laws of reflection and refraction (307) Wavefront propagation through a thin lens and a prism (310) Optical systems (313) Power of an optical surface (314) Magnification (316) Power of a thin lens (318) Principal planes of an optical system (320) Newton's equation (320) Optical Helmholtz equation (321) Deviation through a lens system (322) Location of principal planes (322) Matrix application to lens systems (325) Summary of important results.

Chapter 12   Interference and Diffraction

Interference (333) Division of amplitude (334) Fringes of constant inclination and thickness (335) Newton's Rings (337) Michelson's spectral interferometer (338) Fabry–Perot interferometer (341) Finesse (345) Resolving power (343) Free spectral range (345) Central spot scanning (346) Laser cavity (347) Multilayer dielectric films (350) Optical fibre wave guide (353) Division of wavefront (355) Two equal sources (355) Spatial coherence (360) Dipole radiation (362) Linear array of N equal sources (363) Fraunhofer diffraction (367) Slit (368) N slits (370) Missing orders (373) Transmission diffraction grating (373) Resolving power (374) Bandwidth theorem (376) Rectangular aperture (377) Circular aperture (379) Fraunhofer far field diffraction (383) Airy disc (385) Michelson Stellar Interferometer (386) Convolution Array Theorem (388) Optical Transfer Function (391) Fresnel diffraction (395) Straight edge (397) Cornu spiral (396) Slit (400) Circular aperture (401) Zone plate (402) Holography (403) Summary of important results.

Chapter 13   Wave Mechanics

Chapter 14   Non-linear Oscillations and Chaos

Chapter 15   Non-linear waves, Shocks and Solitons

Appendix 1   Normal Modes, Phase Space and Statistical Physics

Appendix 2   Kirchhoff's Integral Theorem (547)

Appendix 3   Non-linear Schrödinger Equation (551)

Index (553)

1

Simple Harmonic Motion

At first sight the eight physical systems in Figure 1.1 appear to have little in common.

1.1(a)   is a simple pendulum, a mass m swinging at the end of a light rigid rod of length l.

1.1(b)   is a flat disc supported by a rigid wire through its centre and oscillating through small angles in the plane of its circumference.

1.1(c)   is a mass fixed to a wall via a spring of stiffness s sliding to and fro in the x direction on a frictionless plane.

1.1(d)   is a mass m at the centre of a light string of length $2l$ fixed at both ends under a constant tension T. The mass vibrates in the plane of the paper.

1.1(e)   is a frictionless U-tube of constant cross-sectional area containing a length l of liquid, density ρ, oscillating about its equilibrium position of equal levels in each limb.

1.1(f)   is an open flask of volume V and a neck of length l and constant cross-sectional area A in which the air of density ρ vibrates as sound passes across the neck.

1.1(g)   is a hydrometer, a body of mass m floating in a liquid of density ρ with a neck of constant cross-sectional area cutting the liquid surface. When depressed slightly from its equilibrium position it performs small vertical oscillations.

1.1(h)   is an electrical circuit, an inductance L connected across a capacitance C carrying a charge q.

All of these systems are simple harmonic oscillators which, when slightly disturbed from their equilibrium or rest postion, will oscillate with simple harmonic motion. This is the most fundamental vibration of a single particle or one-dimensional system. A small displacement x from its equilibrium position sets up a restoring force which is proportional to x acting in a direction towards the equilibrium position.

Thus, this restoring force F may be written

$$F = -sx$$

where s, the constant of proportionality, is called the stiffness and the negative sign shows that the force is acting against the direction of increasing displacement and back towards

The Physics of Vibrations and Waves, 6th Edition   H. J. Pain
© 2005 John Wiley & Sons, Ltd

(a)

$$m\ddot{x} + mg\,\frac{x}{l} = 0$$

$$ml\ddot{\theta} + mg\,\theta = 0$$

$$\omega^2 = g/l$$

$$mg\sin\theta \approx mg\,\theta$$

$$\approx mg\,\frac{x}{l}$$

(b)

$$l\ddot{\theta} + c\,\theta = 0$$

$$\omega^2 = \frac{c}{l}$$

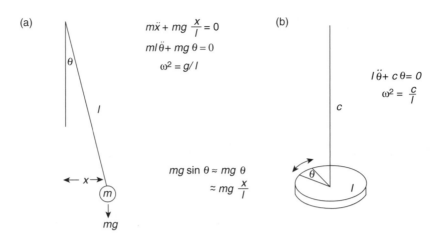

(c)

$$m\ddot{x} + sx = 0$$

$$\omega^2 = s/m$$

(d)

$$m\ddot{x} + 2T\,\frac{x}{l} = 0$$

$$\omega^2 = \frac{2T}{lm}$$

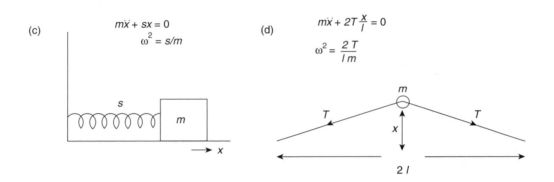

(e)

$$p\,l\ddot{x} + 2pg\,x = 0$$

$$\omega^2 = 2g/l$$

(f)

$$p\,Al\ddot{x} + \frac{\gamma\,pxA^2}{V} = 0$$

$$\omega^2 = \frac{\gamma\,pA}{l\,pV}$$

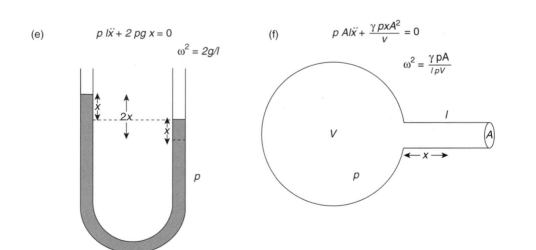

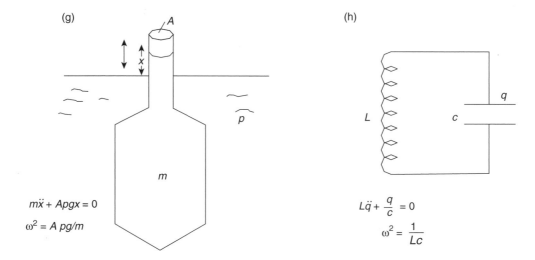

(g)

$m\ddot{x} + Apgx = 0$

$\omega^2 = A\,pg/m$

(h)

$L\ddot{q} + \dfrac{q}{c} = 0$

$\omega^2 = \dfrac{1}{Lc}$

Figure 1.1   Simple harmonic oscillators with their equations of motion and angular frequencies ω of oscillation. (a) A simple pendulum. (b) A torsional pendulum. (c) A mass on a frictionless plane connected by a spring to a wall. (d) A mass at the centre of a string under constant tension *T*. (e) A fixed length of non-viscous liquid in a U-tube of constant cross-section. (f) An acoustic Helmholtz resonator. (g) A hydrometer mass *m* in a liquid of density ρ. (h) An electrical *L C* resonant circuit

the equilibrium position. A constant value of the stiffness restricts the displacement *x* to small values (this is Hooke's Law of Elasticity). The stiffness *s* is obviously the restoring force per unit distance (or displacement) and has the dimensions

$$\frac{\text{force}}{\text{distance}} \equiv \frac{MLT^{-2}}{L}$$

The equation of motion of such a disturbed system is given by the dynamic balance between the forces acting on the system, which by Newton's Law is

$$\text{mass times acceleration} = \text{restoring force}$$

or

$$m\ddot{x} = -sx$$

where the acceleration

$$\ddot{x} = \frac{\mathrm{d}^2 x}{\mathrm{d}t^2}$$

This gives

$$m\ddot{x} + sx = 0$$

or

$$\ddot{x} + \frac{s}{m}x = 0$$

where the dimensions of

$$\frac{s}{m} \text{ are } \frac{MLT^{-2}}{ML} = T^{-2} = \nu^2$$

Here T is a time, or period of oscillation, the reciprocal of ν which is the frequency with which the system oscillates.

However, when we solve the equation of motion we shall find that the behaviour of x with time has a sinusoidal or cosinusoidal dependence, and it will prove more appropriate to consider, not ν, but the angular frequency $\omega = 2\pi\nu$ so that the period

$$T = \frac{1}{\nu} = 2\pi\sqrt{\frac{m}{s}}$$

where s/m is now written as ω^2. Thus the equation of simple harmonic motion

$$\ddot{x} + \frac{s}{m}x = 0$$

becomes

$$\boxed{\ddot{x} + \omega^2 x = 0} \tag{1.1}$$

(Problem 1.1)

Displacement in Simple Harmonic Motion

The behaviour of a simple harmonic oscillator is expressed in terms of its displacement x from equilibrium, its velocity $\dot{x}$, and its acceleration $\ddot{x}$ at any given time. If we try the solution

$$x = A \cos \omega t$$

where A is a constant with the same dimensions as x, we shall find that it satisfies the equation of motion

$$\ddot{x} + \omega^2 x = 0$$

for

$$\dot{x} = -A\omega \sin \omega t$$

and

$$\ddot{x} = -A\omega^2 \cos \omega t = -\omega^2 x$$

Another solution

$$x = B \sin \omega t$$

is equally valid, where B has the same dimensions as A, for then

$$\dot{x} = B\omega \cos \omega t$$

and

$$\ddot{x} = -B\omega^2 \sin \omega t = -\omega^2 x$$

The complete or general solution of equation (1.1) is given by the addition or superposition of both values for x so we have

$$x = A \cos \omega t + B \sin \omega t \qquad (1.2)$$

with

$$\ddot{x} = -\omega^2 (A \cos \omega t + B \sin \omega t) = -\omega^2 x$$

where A and B are determined by the values of x and $\dot{x}$ at a specified time. If we rewrite the constants as

$$A = a \sin \phi \quad \text{and} \quad B = a \cos \phi$$

where ϕ is a constant angle, then

$$A^2 + B^2 = a^2 (\sin^2 \phi + \cos^2 \phi) = a^2$$

so that

$$a = \sqrt{A^2 + B^2}$$

and

$$x = a \sin \phi \cos \omega t + a \cos \phi \sin \omega t$$
$$= a \sin (\omega t + \phi)$$

The maximum value of $\sin (\omega t + \phi)$ is unity so the constant a is the maximum value of x, known as the amplitude of displacement. The limiting values of $\sin (\omega t + \phi)$ are ± 1 so the system will oscillate between the values of $x = \pm a$ and we shall see that the magnitude of a is determined by the total energy of the oscillator.

The angle ϕ is called the 'phase constant' for the following reason. Simple harmonic motion is often introduced by reference to 'circular motion' because each possible value of the displacement x can be represented by the projection of a radius vector of constant length a on the diameter of the circle traced by the tip of the vector as it rotates in a positive

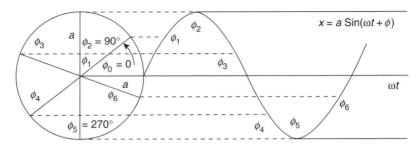

Figure 1.2 Sinusoidal displacement of simple harmonic oscillator with time, showing variation of starting point in cycle in terms of phase angle ϕ

anticlockwise direction with a constant angular velocity ω. Each rotation, as the radius vector sweeps through a phase angle of 2π rad, therefore corresponds to a complete vibration of the oscillator. In the solution

$$x = a \sin(\omega t + \phi)$$

the phase constant ϕ, measured in radians, defines the position in the cycle of oscillation at the time $t = 0$, so that the position in the cycle from which the oscillator started to move is

$$x = a \sin \phi$$

The solution

$$x = a \sin \omega t$$

defines the displacement only of that system which starts from the origin $x = 0$ at time $t = 0$ but the inclusion of ϕ in the solution

$$x = a \sin(\omega t + \phi)$$

where ϕ may take all values between zero and 2π allows the motion to be defined from any starting point in the cycle. This is illustrated in Figure 1.2 for various values of ϕ.

(Problems 1.2, 1.3, 1.4, 1.5)

Velocity and Acceleration in Simple Harmonic Motion

The values of the velocity and acceleration in simple harmonic motion for

$$x = a \sin(\omega t + \phi)$$

are given by

$$\frac{\mathrm{d}x}{\mathrm{d}t} = \dot{x} = a\omega \cos(\omega t + \phi)$$

and

$$\frac{\mathrm{d}^2 x}{\mathrm{d}t^2} = \ddot{x} = -a\omega^2 \sin(\omega t + \phi)$$

The maximum value of the velocity $a\omega$ is called the velocity *amplitude* and the *acceleration amplitude* is given by $a\omega^2$.

From Figure 1.2 we see that a positive phase angle of $\pi/2$ rad converts a sine into a cosine curve. Thus the velocity

$$\dot{x} = a\omega \cos(\omega t + \phi)$$

leads the displacement

$$x = a \sin(\omega t + \phi)$$

by a phase angle of $\pi/2$ rad and its maxima and minima are always a quarter of a cycle ahead of those of the displacement; the velocity is a maximum when the displacement is zero and is zero at maximum displacement. The acceleration is 'anti-phase' (π rad) with respect to the displacement, being maximum positive when the displacement is maximum negative and vice versa. These features are shown in Figure 1.3.

Often, the relative displacement or motion between two oscillators having the same frequency and amplitude may be considered in terms of their phase difference $\phi_1 - \phi_2$ which can have any value because one system may have started several cycles before the other and each complete cycle of vibration represents a change in the phase angle of $\phi = 2\pi$. When the motions of the two systems are diametrically opposed; that is, one has

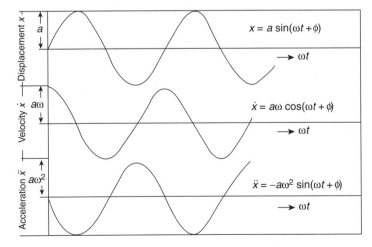

Figure 1.3 Variation with time of displacement, velocity and acceleration in simple harmonic motion. Displacement lags velocity by $\pi/2$ rad and is π rad out of phase with the acceleration. The initial phase constant ϕ is taken as zero

$x = +a$ whilst the other is at $x = -a$, the systems are 'anti-phase' and the total phase difference

$$\phi_1 - \phi_2 = n\pi \, \text{rad}$$

where n is an *odd* integer. Identical systems 'in phase' have

$$\phi_1 - \phi_2 = 2n\pi \, \text{rad}$$

where n is any integer. They have exactly equal values of displacement, velocity and acceleration at any instant.

(Problems 1.6, 1.7, 1.8, 1.9)

Non-linearity

If the stiffness s is constant, then the restoring force $F = -sx$, when plotted versus x, will produce a straight line and the system is said to be linear. The displacement of a linear simple harmonic motion system follows a sine or cosine behaviour. Non-linearity results when the stiffness s is not constant but varies with displacement x (see the beginning of Chapter 14).

Energy of a Simple Harmonic Oscillator

The fact that the velocity is zero at maximum displacement in simple harmonic motion and is a maximum at zero displacement illustrates the important concept of an exchange between kinetic and potential energy. In an ideal case the total energy remains constant but this is never realized in practice. If no energy is dissipated then all the potential energy becomes kinetic energy and vice versa, so that the values of (a) the total energy at any time, (b) the maximum potential energy and (c) the maximum kinetic energy will all be equal; that is

$$E_{\text{total}} = \text{KE} + \text{PE} = \text{KE}_{\text{max}} = \text{PE}_{\text{max}}$$

The solution $x = a \sin (\omega t + \phi)$ implies that the total energy remains constant because the amplitude of displacement $x = \pm a$ is regained every half cycle at the position of maximum potential energy; when energy is lost the amplitude gradually decays as we shall see later in Chapter 2. The potential energy is found by summing all the small elements of work $sx. \, dx$ (force sx times distance dx) *done by the system against the restoring force* over the range zero to x where $x = 0$ gives zero potential energy.

Thus the potential energy $=$

$$\int_0^x sx \cdot \mathrm{d}x = \tfrac{1}{2}sx^2$$

The kinetic energy is given by $\tfrac{1}{2}m\dot{x}^2$ so that the total energy

$$\boxed{E = \tfrac{1}{2}m\dot{x}^2 + \tfrac{1}{2}sx^2}$$

Since E is constant we have

$$\frac{\mathrm{d}E}{\mathrm{d}t} = (m\ddot{x} + sx)\dot{x} = 0$$

giving again the equation of motion

$$m\ddot{x} + sx = 0$$

The maximum potential energy occurs at $x = \pm a$ and is therefore

$$\mathrm{PE}_{\max} = \tfrac{1}{2}sa^2$$

The maximum kinetic energy is

$$\mathrm{KE}_{\max} = (\tfrac{1}{2}m\dot{x}^2)_{\max} = \tfrac{1}{2}ma^2\omega^2[\cos^2(\omega t + \phi)]_{\max}$$
$$= \tfrac{1}{2}ma^2\omega^2$$

when the cosine factor is unity.

But $m\omega^2 = s$ so the maximum values of the potential and kinetic energies are equal, showing that the energy exchange is complete.

The total energy at any instant of time or value of x is

$$E = \tfrac{1}{2}m\dot{x}^2 + \tfrac{1}{2}sx^2$$
$$= \tfrac{1}{2}ma^2\omega^2[\cos^2(\omega t + \phi) + \sin^2(\omega t + \phi)]$$
$$= \tfrac{1}{2}ma^2\omega^2$$
$$= \tfrac{1}{2}sa^2$$

as we should expect.

Figure 1.4 shows the distribution of energy versus displacement for simple harmonic motion. Note that the potential energy curve

$$\mathrm{PE} = \tfrac{1}{2}sx^2 = \tfrac{1}{2}ma^2\omega^2\sin^2(\omega t + \phi)$$

is parabolic with respect to x and is symmetric about $x = 0$, so that energy is stored in the oscillator both when x is positive and when it is negative, e.g. a spring stores energy whether compressed or extended, as does a gas in compression or rarefaction. The kinetic energy curve

$$\mathrm{KE} = \tfrac{1}{2}m\dot{x}^2 = \tfrac{1}{2}ma^2\omega^2\cos^2(\omega t + \phi)$$

is parabolic with respect to both x and $\dot{x}$. The inversion of one curve with respect to the other displays the $\pi/2$ phase difference between the displacement (related to the potential energy) and the velocity (related to the kinetic energy).

For any value of the displacement x the sum of the ordinates of both curves equals the total constant energy E.

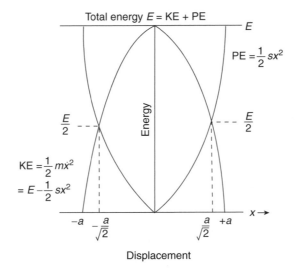

Figure 1.4 Parabolic representation of potential energy and kinetic energy of simple harmonic motion versus displacement. Inversion of one curve with respect to the other shows a 90° phase difference. At any displacement value the sum of the ordinates of the curves equals the total constant energy E

(Problems 1.10, 1.11, 1.12)

Simple Harmonic Oscillations in an Electrical System

So far we have discussed the simple harmonic motion of the mechanical and fluid systems of Figure 1.1, chiefly in terms of the inertial mass stretching the weightless spring of stiffness s. The stiffness s of a spring defines the difficulty of stretching; the reciprocal of the stiffness, the compliance C (where $s = 1/C$) defines the ease with which the spring is stretched and potential energy stored. This notation of compliance C is useful when discussing the simple harmonic oscillations of the electrical circuit of Figure 1.1(h) and Figure 1.5, where an inductance L is connected across the plates of a capacitance C. The force equation of the mechanical and fluid examples now becomes the voltage equation

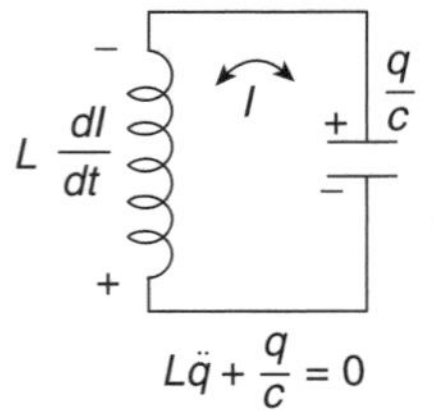

$$L\ddot{q} + \frac{q}{c} = 0$$

Figure 1.5 Electrical system which oscillates simple harmonically. The sum of the voltages around the circuit is given by Kirchhoff's law as $L\,dI/dt + q/C = 0$

(balance of voltages) of the electrical circuit, but the form and solution of the equations and the oscillatory behaviour of the systems are identical.

In the absence of resistance the energy of the electrical system remains constant and is exchanged between the *magnetic* field energy stored in the inductance and the *electric* field energy stored between the plates of the capacitance. At any instant, the voltage across the inductance is

$$V = -L\frac{\mathrm{d}I}{\mathrm{d}t} = -L\frac{\mathrm{d}^2 q}{\mathrm{d}t^2}$$

where I is the current flowing and q is the charge on the capacitor, the negative sign showing that the voltage opposes the increase of current. This equals the voltage q/C across the capacitance so that

$$L\ddot{q} + q/C = 0 \qquad \text{(Kirchhoff's Law)}$$

or

$$\ddot{q} + \omega^2 q = 0$$

where

$$\omega^2 = \frac{1}{LC}$$

The energy stored in the magnetic field or inductive part of the circuit throughout the cycle, as the current increases from 0 to I, is formed by integrating the power at any instant with respect to time; that is

$$E_{\mathrm{L}} = \int VI \cdot \mathrm{d}t$$

(where V is the magnitude of the voltage across the inductance).

So

$$E_{\mathrm{L}} = \int VI \, \mathrm{d}t = \int L\frac{\mathrm{d}I}{\mathrm{d}t} I \, \mathrm{d}t = \int_0^I LI \, \mathrm{d}I$$
$$= \tfrac{1}{2}LI^2 = \tfrac{1}{2}L\dot{q}^2$$

The potential energy stored mechanically by the spring is now stored electrostatically by the capacitance and equals

$$\tfrac{1}{2}CV^2 = \frac{q^2}{2C}$$

Comparison between the equations for the mechanical and electrical oscillators

$$\text{mechanical (force)} \rightarrow m\ddot{x} + sx = 0$$

$$\text{electrical (voltage)} \rightarrow L\ddot{q} + \frac{q}{C} = 0$$

$$\text{mechanical (energy)} \rightarrow \tfrac{1}{2}m\dot{x}^2 + \tfrac{1}{2}sx^2 = E$$

$$\text{electrical (energy)} \rightarrow \frac{1}{2}L\dot{q}^2 + \frac{1}{2}\frac{q^2}{C} = E$$

shows that magnetic field inertia (defined by the inductance L) controls the rate of change of current for a given voltage in a circuit in exactly the same way as the inertial mass controls the change of velocity for a given force. Magnetic inertial or inductive behaviour arises from the tendency of the magnetic flux threading a circuit to remain constant and reaction to any change in its value generates a voltage and hence a current which flows to oppose the change of flux. This is the physical basis of Fleming's right-hand rule.

Superposition of Two Simple Harmonic Vibrations in One Dimension

(1) Vibrations Having Equal Frequencies

In the following chapters we shall meet physical situations which involve the superposition of two or more simple harmonic vibrations on the same system.

We have already seen how the displacement in simple harmonic motion may be represented in magnitude and phase by a constant length vector rotating in the positive (anticlockwise) sense with a constant angular velocity ω. To find the resulting motion of a system which moves in the x direction under the simultaneous effect of two simple harmonic oscillations of equal angular frequencies but of different amplitudes and phases, we can represent each simple harmonic motion by its appropriate vector and carry out a vector addition.

If the displacement of the first motion is given by

$$x_1 = a_1 \cos(\omega t + \phi_1)$$

and that of the second by

$$x_2 = a_2 \cos(\omega t + \phi_2)$$

then Figure 1.6 shows that the resulting displacement amplitude R is given by

$$R^2 = (a_1 + a_2 \cos \delta)^2 + (a_2 \sin \delta)^2$$
$$= a_1^2 + a_2^2 + 2a_1 a_2 \cos \delta$$

where $\delta = \phi_2 - \phi_1$ is constant.

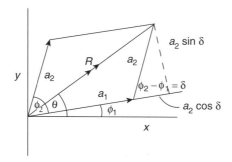

Figure 1.6 Addition of vectors, each representing simple harmonic motion along the *x* axis at angular frequency ω to give a resulting simple harmonic motion displacement $x = R\cos(\omega t + \theta)$ --- here shown for $t = 0$

The phase constant θ of R is given by

$$\tan\theta = \frac{a_1 \sin\phi_1 + a_2 \sin\phi_2}{a_1 \cos\phi_1 + a_2 \cos\phi_2}$$

so the resulting simple harmonic motion has a displacement

$$x = R\cos(\omega t + \theta)$$

an oscillation of the same frequency ω but having an amplitude R and a phase constant θ.

(Problem 1.13)

(2) Vibrations Having Different Frequencies

Suppose we now consider what happens when two vibrations of equal amplitudes but different frequencies are superposed. If we express them as

$$x_1 = a\sin\omega_1 t$$

and

$$x_2 = a\sin\omega_2 t$$

where

$$\omega_2 > \omega_1$$

then the resulting displacement is given by

$$x = x_1 + x_2 = a(\sin \omega_1 t + \sin \omega_2 t)$$

$$= 2a \sin \frac{(\omega_1 + \omega_2)t}{2} \cos \frac{(\omega_2 - \omega_1)t}{2}$$

This expression is illustrated in Figure 1.7. It represents a sinusoidal oscillation at the average frequency $(\omega_1 + \omega_2)/2$ having a displacement amplitude of $2a$ which modulates; that is, varies between $2a$ and zero under the influence of the cosine term of a much slower frequency equal to half the difference $(\omega_2 - \omega_1)/2$ between the original frequencies.

When ω_1 and ω_2 are almost equal the sine term has a frequency very close to both ω_1 and ω_2 whilst the cosine envelope modulates the amplitude $2a$ at a frequency $(\omega_2 - \omega_1)/2$ which is very slow.

Acoustically this growth and decay of the amplitude is registered as 'beats' of strong reinforcement when two sounds of almost equal frequency are heard. The frequency of the 'beats' is $(\omega_2 - \omega_1)$, the difference between the separate frequencies (not half the difference) because the maximum amplitude of $2a$ occurs twice in every period associated with the frequency $(\omega_2 - \omega_1)/2$. We shall meet this situation again when we consider the coupling of two oscillators in Chapter 4 and the wave group of two components in Chapter 5.

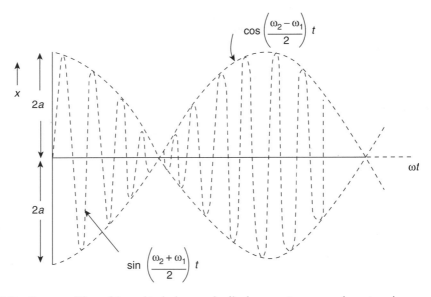

Figure 1.7 Superposition of two simple harmonic displacements $x_1 = a \sin \omega_1 t$ and $x_2 = a \sin \omega_2 t$ when $\omega_2 > \omega_1$. The slow $\cos [(\omega_2 - \omega_1)/2]t$ envelope modulates the $\sin [(\omega_2 + \omega_1)/2]t$ curve between the values $x = \pm 2a$

Superposition of Two Perpendicular Simple Harmonic Vibrations

(1) Vibrations Having Equal Frequencies

Suppose that a particle moves under the simultaneous influence of two simple harmonic vibrations of equal frequency, one along the x axis, the other along the perpendicular y axis. What is its subsequent motion?

This displacements may be written

$$x = a_1 \sin(\omega t + \phi_1)$$
$$y = a_2 \sin(\omega t + \phi_2)$$

and the path followed by the particle is formed by eliminating the time t from these equations to leave an expression involving only x and y and the constants ϕ_1 and ϕ_2.

Expanding the arguments of the sines we have

$$\frac{x}{a_1} = \sin \omega t \cos \phi_1 + \cos \omega t \sin \phi_1$$

and

$$\frac{y}{a_2} = \sin \omega t \cos \phi_2 + \cos \omega t \sin \phi_2$$

If we carry out the process

$$\left(\frac{x}{a_1} \sin \phi_2 - \frac{y}{a_2} \sin \phi_1\right)^2 + \left(\frac{y}{a_2} \cos \phi_1 - \frac{x}{a_1} \cos \phi_2\right)^2$$

this will yield

$$\frac{x^2}{a_1^2} + \frac{y^2}{a_2^2} - \frac{2xy}{a_1 a_2} \cos(\phi_2 - \phi_1) = \sin^2(\phi_2 - \phi_1) \tag{1.3}$$

which is the general equation for an ellipse.

In the most general case the axes of the ellipse are inclined to the x and y axes, but these become the principal axes when the phase difference

$$\phi_2 - \phi_1 = \frac{\pi}{2}$$

Equation (1.3) then takes the familiar form

$$\frac{x^2}{a_1^2} + \frac{y^2}{a_2^2} = 1$$

that is, an ellipse with semi-axes a_1 and a_2.

If $a_1 = a_2 = a$ this becomes the circle

$$x^2 + y^2 = a^2$$

When

$$\phi_2 - \phi_1 = 0, 2\pi, 4\pi, \text{ etc.}$$

the equation simplifies to

$$y = \frac{a_2}{a_1} x$$

which is a straight line through the origin of slope a_2/a_1.

Again for $\phi_2 - \phi_1 = \pi$, 3π, 5π, etc., we obtain

$$y = -\frac{a_2}{a_1} x$$

a straight line through the origin of equal but opposite slope.

The paths traced out by the particle for various values of $\delta = \phi_2 - \phi_1$ are shown in Figure 1.8 and are most easily demonstrated on a cathode ray oscilloscope.

When

$$\phi_2 - \phi_1 = 0, \ \pi, \ 2\pi, \text{ etc.}$$

and the ellipse degenerates into a straight line, the resulting vibration lies wholly in one plane and the oscillations are said to be *plane polarized*.

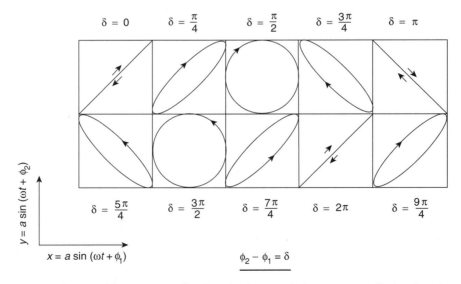

Figure 1.8 Paths traced by a system vibrating simultaneously in two perpendicular directions with simple harmonic motions of equal frequency. The phase angle δ is the angle by which the y motion leads the x motion

Convention defines the plane of polarization as that plane perpendicular to the plane containing the vibrations. Similarly the other values of

$$\phi_2 - \phi_1$$

yield *circular* or *elliptic* polarization where the tip of the vector resultant traces out the appropriate conic section.

(Problems 1.14, 1.15, 1.16)

*Polarization

Polarization is a fundamental topic in optics and arises from the superposition of two perpendicular simple harmonic optical vibrations. We shall see in Chapter 8 that when a light wave is plane polarized its electrical field oscillation lies within a single plane and traces a sinusoidal curve along the direction of wave motion. Substances such as quartz and calcite are capable of splitting light into two waves whose planes of polarization are perpendicular to each other. Except in a specified direction, known as the optic axis, these waves have different velocities. One wave, the ordinary or *O* wave, travels at the same velocity in all directions and its electric field vibrations are always perpendicular to the optic axis. The extraordinary or *E* wave has a velocity which is direction-dependent. Both ordinary and extraordinary light have their own refractive indices, and thus quartz and calcite are known as doubly refracting materials. When the ordinary light is faster, as in quartz, a crystal of the substance is defined as positive, but in calcite the extraordinary light is faster and its crystal is negative. The surfaces, spheres and ellipsoids, which are the loci of the values of the wave velocities in any direction are shown in Figure 1.9(a), and for a

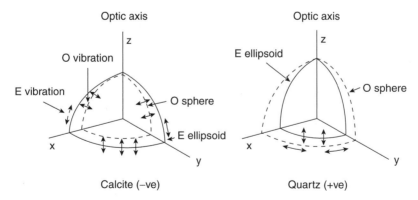

Figure 1.9a Ordinary (spherical) and extraordinary (elliposoidal) wave surfaces in doubly refracting calcite and quartz. In calcite the *E* wave is faster than the *O* wave, except along the optic axis. In quartz the *O* wave is faster. The *O* vibrations are always perpendicular to the optic axis, and the *O* and *E* vibrations are always tangential to their wave surfaces

*This section may be omitted at a first reading.

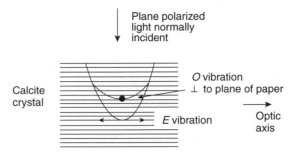

Figure 1.9b Plane polarized light normally incident on a calcite crystal face cut parallel to its optic axis. The advance of the *E* wave over the *O* wave is equivalent to a gain in phase

given direction the electric field vibrations of the separate waves are tangential to the surface of the sphere or ellipsoid as shown. Figure 1.9(b) shows plane polarized light normally incident on a calcite crystal cut parallel to its optic axis. Within the crystal the faster *E* wave has vibrations parallel to the optic axis, while the *O* wave vibrations are perpendicular to the plane of the paper. The velocity difference results in a phase gain of the *E* vibration over the *O* vibration which increases with the thickness of the crystal. Figure 1.9(c) shows plane polarized light normally incident on the crystal of Figure 1.9(b) with its vibration at an angle of 45° of the optic axis. The crystal splits the vibration into

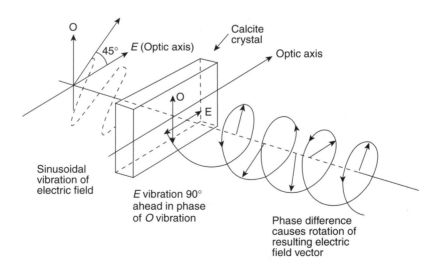

Figure 1.9c The crystal of Fig. 1.9c is thick enough to produce a phase gain of $\pi/2$ rad in the *E* wave over the *O* wave. Wave recombination on leaving the crystal produces circularly polarized light

equal E and O components, and for a given thickness the E wave emerges with a phase gain of 90° over the O component. Recombination of the two vibrations produces circularly polarized light, of which the electric field vector now traces a helix in the anticlockwise direction as shown.

(2) Vibrations Having Different Frequencies (Lissajous Figures)

When the frequencies of the two perpendicular simple harmonic vibrations are not equal the resulting motion becomes more complicated. The patterns which are traced are called Lissajous figures and examples of these are shown in Figure 1.10 where the axial frequencies bear the simple ratios shown and

$$\delta = \phi_2 - \phi_1 = 0 \text{ (on the left)}$$

$$= \frac{\pi}{2} \text{ (on the right)}$$

If the amplitudes of the vibrations are respectively a and b the resulting Lissajous figure will always be contained within the rectangle of sides $2a$ and $2b$. The sides of the rectangle will be tangential to the curve at a number of points and the ratio of the numbers of these tangential points along the x axis to those along the y axis is the inverse of the ratio of the corresponding frequencies (as indicated in Figure 1.10).

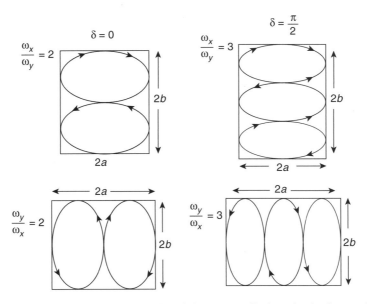

Figure 1.10 Simple Lissajous figures produced by perpendicular simple harmonic motions of different angular frequencies

Superposition of a Large Number n of Simple Harmonic Vibrations of Equal Amplitude a and Equal Successive Phase Difference δ

Figure 1.11 shows the addition of n vectors of equal length a, each representing a simple harmonic vibration with a constant phase difference δ from its neighbour. Two general physical situations are characterized by such a superposition. The first is met in Chapter 5 as a wave group problem where the phase difference δ arises from a small *frequency difference*, $\delta\omega$, between consecutive components. The second appears in Chapter 12 where the intensity of optical interference and diffraction patterns are considered. There, the superposed harmonic vibrations will have the same frequency but each component will have a constant phase difference from its neighbour because of the extra *distance* it has travelled.

The figure displays the mathematical expression

$$R\cos(\omega t + \alpha) = a\cos\omega t + a\cos(\omega t + \delta) + a\cos(\omega t + 2\delta)$$
$$+ \cdots + a\cos(\omega t + [n-1]\delta)$$

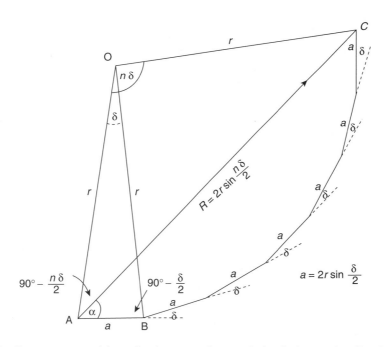

Figure 1.11   Vector superposition of a large number n of simple harmonic vibrations of equal amplitude a and equal successive phase difference δ. The amplitude of the resultant

$$R = 2r\sin\frac{n\delta}{2} = a\frac{\sin n\delta/2}{\sin \delta/2}$$

and its phase with respect to the first contribution is given by

$$\alpha = (n-1)\delta/2$$

where R is the magnitude of the resultant and α is its phase difference with respect to the first component $a \cos \omega t$.

Geometrically we see that each length

$$a = 2r \sin \frac{\delta}{2}$$

where r is the radius of the circle enclosing the (incomplete) polygon.

From the isosceles triangle OAC the magnitude of the resultant

$$R = 2r \sin \frac{n\delta}{2} = a \frac{\sin n\delta/2}{\sin \delta/2}$$

and its phase angle is seen to be

$$\alpha = \hat{OAB} - \hat{OAC}$$

In the isosceles triangle OAC

$$\hat{OAC} = 90° - \frac{n\delta}{2}$$

and in the isosceles triangle OAB

$$\hat{OAB} = 90° - \frac{\delta}{2}$$

so

$$\alpha = \left(90° - \frac{\delta}{2}\right) - \left(90° - \frac{n\delta}{2}\right) = (n-1)\frac{\delta}{2}$$

that is, half the phase difference between the first and the last contributions. Hence the resultant

$$R \cos(\omega t + \alpha) = a \frac{\sin n\delta/2}{\sin \delta/2} \cos \left[\omega t + (n-1)\frac{\delta}{2}\right]$$

We shall obtain the same result later in this chapter as an example on the use of exponential notation.

For the moment let us examine the behaviour of the magnitude of the resultant

$$R = a \frac{\sin n\delta/2}{\sin \delta/2}$$

which is not constant but depends on the value of δ. When n is very large δ is very small and the polygon becomes an arc of the circle centre O, of length $na = A$, with R as the chord. Then

$$\alpha = (n-1)\frac{\delta}{2} \approx \frac{n\delta}{2}$$

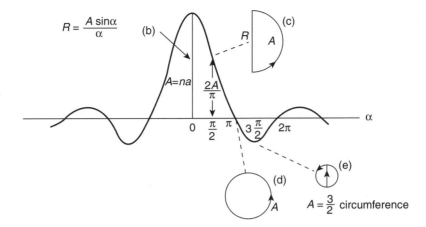

Figure 1.12 (a) Graph of $A \sin \alpha / \alpha$ versus α, showing the magnitude of the resultants for (b) $\alpha = 0$; (c) $\alpha = \pi/2$; (d) $\alpha = \pi$ and (e) $\alpha = 3\pi/2$

and

$$\sin\frac{\delta}{2} \rightarrow \frac{\delta}{2} \approx \frac{\alpha}{n}$$

Hence, in this limit,

$$R = a\frac{\sin n\delta/2}{\sin \delta/2} = a\frac{\sin\alpha}{\alpha/n} = na\frac{\sin \alpha}{\alpha} = \frac{A \sin \alpha}{\alpha}$$

The behaviour of $A \sin \alpha / \alpha$ versus α is shown in Figure 1.12. The pattern is symmetric about the value $\alpha = 0$ and is zero whenever $\sin \alpha = 0$ except at $\alpha \rightarrow 0$ that is, when $\sin \alpha / \alpha \rightarrow 1$. When $\alpha = 0$, $\delta = 0$ and the resultant of the n vectors is the straight line of length A, Figure 1.12(b). As δ increases A becomes the arc of a circle until at $\alpha = \pi/2$ the first and last contributions are out of phase $(2\alpha = \pi)$ and the arc A has become a semicircle of which the diameter is the resultant R Figure 1.12(c). A further increase in δ increases α and curls the constant length A into the circumference of a circle $(\alpha = \pi)$ with a zero resultant, Figure 1.12(d). At $\alpha = 3\pi/2$, Figure 1.12(e) the length A is now 3/2 times the circumference of a circle whose diameter is the amplitude of the first minimum.

*Superposition of *n* Equal SHM Vectors of Length *a* with Random Phase

When the phase difference between the successive vectors of the last section may take random values ϕ between zero and 2π (measured from the x axis) the vector superposition and resultant R may be represented by Figure 1.13.

*This section may be omitted at a first reading.

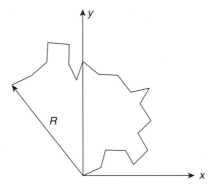

Figure 1.13 The resultant $R = \sqrt{n}a$ of n vectors, each of length a, having random phase. This result is important in optical incoherence and in energy loss from waves from random dissipation processes

The components of R on the x and y axes are given by

$$R_x = a\cos\phi_1 + a\cos\phi_2 + a\cos\phi_3 \ldots a\cos\phi_n$$

$$= a\sum_{i=1}^{n} \cos\phi_i$$

and

$$R_y = a\sum_{i=1}^{n} \sin\phi_i$$

where

$$R^2 = R_x^2 + R_y^2$$

Now

$$R_x^2 = a^2\left(\sum_{i=1}^{n} \cos\phi_i\right)^2 = a^2\left[\sum_{i=1}^{n} \cos^2\phi_i + \sum_{\substack{i=1 \\ i\neq j}}^{n} \cos\phi_i \sum_{j=1}^{n} \cos\phi_j\right]$$

In the typical term $2\cos\phi_i\cos\phi_j$ of the double summation, $\cos\phi_i$ and $\cos\phi_j$ have random values between ±1 and the averaged sum of sets of these products is effectively zero.

The summation

$$\sum_{i=1}^{n} \cos^2\phi_i = n\overline{\cos^2\phi}$$

that is, the number of terms n times the average value $\overline{\cos^2 \phi}$ which is the integrated value of $\cos^2 \phi$ over the interval zero to 2π divided by the total interval 2π, or

$$\overline{\cos^2 \phi} = \frac{1}{2\pi} \int_0^{2\pi} \cos^2 \phi \, d\phi = \frac{1}{2} = \overline{\sin^2 \phi}$$

So

$$R_x^2 = a^2 \sum_{i=1}^n \cos^2 \phi_i = na^2 \overline{\cos^2 \phi_i} = \frac{na^2}{2}$$

and

$$R_y^2 = a^2 \sum_{i=1}^n \sin^2 \phi_i = na^2 \overline{\sin^2 \phi_i} = \frac{na^2}{2}$$

giving

$$R^2 = R_x^2 + R_y^2 = na^2$$

or

$$R = \sqrt{n}a$$

Thus, the amplitude R of a system subjected to n equal simple harmonic motions of amplitude a with random phases in only $\sqrt{n}a$ whereas, if the motions were all in phase R would equal na.

Such a result illustrates a very important principle of random behaviour.

(Problem 1.17)

Applications

Incoherent Sources in Optics   The result above is directly applicable to the problem of coherence in optics. Light sources which are in phase are said to be coherent and this condition is essential for producing optical interference effects experimentally. If the amplitude of a light source is given by the quantity a its intensity is proportional to a^2, n coherent sources have a resulting amplitude na and a total intensity n^2a^2. Incoherent sources have random phases, n such sources each of amplitude a have a resulting amplitude $\sqrt{n}a$ and a total intensity of na^2.

Random Processes and Energy Absorption   From our present point of view the importance of random behaviour is the contribution it makes to energy loss or absorption from waves moving through a medium. We shall meet this in all the waves we discuss.

Random processes, for example collisions between particles, in Brownian motion, are of great significance in physics. Diffusion, viscosity or frictional resistance and thermal conductivity are all the result of random collision processes. These energy dissipating phenomena represent the transport of mass, momentum and energy, and change only in the direction of increasing disorder. They are known as 'thermodynamically irreversible' processes and are associated with the increase of entropy. Heat, for example, can flow only from a body at a higher temperature to one at a lower temperature. Using the earlier analysis where the length a is no longer a simple harmonic amplitude but is now the average distance a particle travels between random collisions (its mean free path), we see that after n such collisions (with, on average, equal time intervals between collisions) the particle will, on average, have travelled only a distance $\sqrt{n}a$ from its position at time $t = 0$, so that the distance travelled varies only with the square root of the time elapsed instead of being directly proportional to it. This is a feature of all random processes.

Not all the particles of the system will have travelled a distance $\sqrt{n}a$ but this distance is the most probable and represents a statistical average.

Random behaviour is described by the diffusion equation (see the last section of Chapter 7) and a constant coefficient called the diffusivity of the process will always arise. The dimensions of a diffusivity are always length^2/time and must be interpreted in terms of a characteristic distance of the process which varies only with the square root of time.

Some Useful Mathematics

The Exponential Series

By a 'natural process' of growth or decay we mean a process in which a quantity changes by a constant fraction of itself in a given interval of space or time. A 5% per annum compound interest represents a natural growth law; attenuation processes in physics usually describe natural decay.

The law is expressed differentially as

$$\frac{dN}{N} = \pm \alpha \, dx \quad \text{or} \quad \frac{dN}{N} = \pm \alpha \, dt$$

where N is the changing quantity, α is a constant and the positive and negative signs represent growth and decay respectively. The derivatives dN/dx or dN/dt are therefore proportional to the value of N at which the derivative is measured.

Integration yields $N = N_0 e^{\pm \alpha x}$ or $N = N_0 e^{\pm \alpha t}$ where N_0 is the value at x or $t = 0$ and e is the exponential or the base of natural logarithms. The exponential series is defined as

$$e^x = 1 + x + \frac{x^2}{2!} + \frac{x^3}{3!} + \cdots + \frac{x^n}{n!} + \cdots$$

and is shown graphically for positive and negative x in Figure 1.14. It is important to note that whatever the form of the index of the logarithmic base e, it is the power to which the

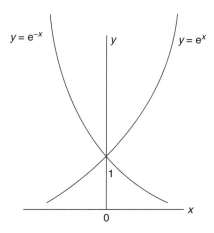

Figure 1.14 The behaviour of the exponential series $y = e^x$ and $y = e^{-x}$

base is raised, and is therefore always non-dimensional. Thus $e^{\alpha x}$ is non-dimensional and α must have the dimensions of x^{-1}. Writing

$$e^{\alpha x} = 1 + \alpha x + \frac{(\alpha x)^2}{2!} + \frac{(\alpha x)^3}{3!} + \cdots$$

it follows immediately that

$$\frac{d}{dx}(e^{\alpha x}) = \alpha + \frac{2\alpha^2}{2!}x + \frac{3\alpha^3}{3!}x^2 + \cdots$$

$$= \alpha \left[1 + \alpha x + \frac{(\alpha x)^2}{2!} + \frac{(\alpha x)^3}{3!} \right) + \cdots \right]$$

$$= \alpha e^{\alpha x}$$

Similarly

$$\frac{d^2}{dx^2}(e^{\alpha x}) = \alpha^2 e^{\alpha x}$$

In Chapter 2 we shall use $d(e^{\alpha t})/dt = \alpha e^{\alpha t}$ and $d^2(e^{\alpha t})/dt^2 = \alpha^2 e^{\alpha t}$ on a number of occasions.

By taking logarithms it is easily shown that $e^x e^y = e^{x+y}$ since $\log_e (e^x e^y) = \log_e e^x + \log_e e^y = x + y$.

The Notation $i = \sqrt{-1}$

The combination of the exponential series with the complex number notation $i = \sqrt{-1}$ is particularly convenient in physics. Here we shall show the mathematical convenience in expressing sine or cosine (oscillatory) behaviour in the form $e^{ix} = \cos x + i \sin x$.

In Chapter 3 we shall see the additional merit of i in its role of vector operator. The series representation of $\sin x$ is written

$$\sin x = x - \frac{x^3}{3!} + \frac{x^5}{5!} - \frac{x^7}{7!} \cdots$$

and that of $\cos x$ is

$$\cos x = 1 - \frac{x^2}{2!} + \frac{x^4}{4!} - \frac{x^6}{6!} \cdots$$

Since

$$i = \sqrt{-1}, i^2 = -1, i^3 = -i$$

etc. we have

$$e^{ix} = 1 + ix + \frac{(ix)^2}{2!} + \frac{(ix)^3}{3!} + \frac{(ix)^4}{4!} + \cdots$$
$$= 1 + ix - \frac{x^2}{2!} - \frac{ix^3}{3!} + \frac{x^4}{4!} + \cdots$$
$$= 1 - \frac{x^2}{2!} + \frac{x^4}{4!} + i\left(x - \frac{x^3}{3!} + \frac{x^5}{5!} + \cdots \right)$$
$$= \cos x + i \sin x$$

We also see that

$$\frac{d}{dx}(e^{ix}) = i e^{ix} = i \cos x - \sin x$$

Often we shall represent a sine or cosine oscillation by the form e^{ix} and recover the original form by taking that part of the solution preceded by i in the case of the sine, and the real part of the solution in the case of the cosine.

Examples

(1) In simple harmonic motion ($\ddot{x} + \omega^2 x = 0$) let us try the solution $x = a e^{i\omega t} e^{i\phi}$, where a is a constant length, and ϕ (and therefore $e^{i\phi}$) is a constant.

$$\frac{dx}{dt} = \dot{x} = i\omega a e^{i\omega t} e^{i\phi} = i\omega x$$
$$\frac{d^2 x}{dt^2} = \ddot{x} = i^2 \omega^2 a e^{i\omega t} e^{i\phi} = -\omega^2 x$$

Therefore

$$x = a e^{i\omega t} e^{i\phi} = a e^{i(\omega t + \phi)}$$
$$= a \cos(\omega t + \phi) + i a \sin(\omega t + \phi)$$

is a complete solution of $\ddot{x} + \omega^2 x = 0$.

On p. 6 we used the sine form of the solution; the cosine form is equally valid and merely involves an advance of $\pi/2$ in the phase ϕ.

(2)

$$e^{ix} + e^{-ix} = 2\left(1 - \frac{x^2}{2!} + \frac{x^4}{4!} - \cdots\right) = 2\cos x$$

$$e^{ix} - e^{-ix} = 2i\left(x - \frac{x^3}{3!} + \frac{x^5}{5!} - \cdots\right) = 2i\sin x$$

(3) On p. 21 we used a geometrical method to show that the resultant of the superposed harmonic vibrations

$$a\cos \omega t + a\cos(\omega t + \delta) + a\cos(\omega t + 2\delta) + \cdots + a\cos(\omega t + [n-1]\delta)$$

$$= a\frac{\sin n\delta/2}{\sin \delta/2}\cos\left\{\omega t + \left(\frac{n-1}{2}\right)\delta\right\}$$

We can derive the same result using the complex exponential notation and *taking the real part* of the series expressed as the geometrical progression

$$ae^{i\omega t} + ae^{i(\omega t+\delta)} + ae^{i(\omega t+2\delta)} + \cdots + ae^{i[\omega t+(n-1)\delta]}$$

$$= ae^{i\omega t}(1 + z + z^2 + \cdots + z^{(n-1)})$$

where $z = e^{i\delta}$.

Writing

$$S(z) = 1 + z + z^2 + \cdots + z^{n-1}$$

and

$$z[S(z)] = z + z^2 + \cdots + z^n$$

we have

$$S(z) = \frac{1 - z^n}{1 - z} = \frac{1 - e^{in\delta}}{1 - e^{i\delta}}$$

So

$$ae^{i\omega t}S(z) = ae^{i\omega t}\frac{1 - e^{in\delta}}{1 - e^{i\delta}}$$

$$= ae^{i\omega t}\frac{e^{in\delta/2}(e^{-in\delta/2} - e^{in\delta/2})}{e^{i\delta/2}(e^{-i\delta/2} - e^{i\delta/2})}$$

$$= ae^{i[\omega t+\left(\frac{n-1}{2}\right)\delta]}\frac{\sin n\delta/2}{\sin \delta/2}$$

with the real part

$$= a \cos \left[\omega t + \left(\frac{n-1}{2} \right) \delta \right] \frac{\sin n\delta/2}{\sin \delta/2}$$

which recovers the original cosine term from the complex exponential notation.

(Problem 1.18)

(4) Suppose we represent a harmonic oscillation by the complex exponential form

$$z = a \, e^{i\omega t}$$

where a is the amplitude. Replacing i by $-$i defines the *complex conjugate*

$$z^* = a \, e^{-i\omega t}$$

The use of this conjugate is discussed more fully in Chapter 3 but here we can note that the product of a complex quantity and its conjugate is always equal to the square of the amplitude for

$$zz^* = a^2 \, e^{i\omega t} \, e^{-i\omega t} = a^2 \, e^{(i-i)\omega t} = a^2 \, e^0$$
$$= a^2$$

(Problem 1.19)

Problem 1.1
The equation of motion

$$m\ddot{x} = -sx \quad \text{with} \quad \omega^2 = \frac{s}{m}$$

applies directly to the system in Figure 1.1(c).

If the pendulum bob of Figure 1.1(a) is displaced a small distance x show that the stiffness (restoring force per unit distance) is mg/l and that $\omega^2 = g/l$ where g is the acceleration due to gravity. Now use the small angular displacement θ instead of x and show that ω is the same.

In Figure 1.1(b) the angular oscillations are rotational so the mass is replaced by the moment of inertia I of the disc and the stiffness by the restoring couple of the wire which is C rad^{-1} of angular displacement. Show that $\omega^2 = C/I$.

In Figure 1.1(d) show that the stiffness is $2T/l$ and that $\omega^2 = 2T/lm$.

In Figure 1.1(e) show that the stiffness of the system in $2\rho Ag$, where A is the area of cross section and that $\omega^2 = 2g/l$ where g is the acceleration due to gravity.

In Figure 1.1(f) only the gas in the flask neck oscillates, behaving as a piston of mass ρAl. If the pressure changes are calculated from the equation of state use the adiabatic relation $pV^\gamma = \text{constant}$ and take logarithms to show that the pressure change in the flask is

$$\mathrm{d}p = -\gamma p \frac{\mathrm{d}V}{V} = -\gamma p \frac{Ax}{V},$$

where x is the gas displacement in the neck. Hence show that $\omega^2 = \gamma pA/l\rho V$. Note that γp is the stiffness of a gas (see Chapter 6).

In Figure 1.1(g), if the cross-sectional area of the neck is A and the hydrometer is a distance x above its normal floating level, the restoring force depends on the volume of liquid displaced (Archimedes' principle). Show that $\omega^2 = g\rho A/m$.

Check the dimensions of ω^2 for each case.

Problem 1.2
Show by the choice of appropriate values for A and B in equation (1.2) that equally valid solutions for x are

$$x = a\cos(\omega t + \phi)$$
$$x = a\sin(\omega t - \phi)$$
$$x = a\cos(\omega t - \phi)$$

and check that these solutions satisfy the equation

$$\ddot{x} + \omega^2 x = 0$$

Problem 1.3
The pendulum in Figure 1.1(a) swings with a displacement amplitude a. If its starting point from rest is

$$\text{(a)} \quad x = a$$
$$\text{(b)} \quad x = -a$$

find the different values of the phase constant ϕ for the solutions

$$x = a\sin(\omega t + \phi)$$
$$x = a\cos(\omega t + \phi)$$
$$x = a\sin(\omega t - \phi)$$
$$x = a\cos(\omega t - \phi)$$

For each of the different values of ϕ, find the values of ωt at which the pendulum swings through the positions

$$x = +a/\sqrt{2}$$
$$x = a/2$$

and

$$x = 0$$

for the first time after release from

$$x = \pm a$$

Problem 1.4

When the electron in a hydrogen atom bound to the nucleus moves a small distance from its equilibrium position, a restoring force per unit distance is given by

$$s = e^2/4\pi\epsilon_0 r^2$$

where $r = 0.05$ nm may be taken as the radius of the atom. Show that the electron can oscillate with a simple harmonic motion with

$$\omega_0 \approx 4.5 \times 10^{-16} \, \text{rad s}^{-1}$$

If the electron is forced to vibrate at this frequency, in which region of the electromagnetic spectrum would its radiation be found?

$$e = 1.6 \times 10^{-19}\,\text{C}, \text{ electron mass } m_e = 9.1 \times 10^{-31} \, \text{kg}$$
$$\epsilon_0 = 8.85 \times 10^{-12} \, \text{N}^{-1}\,\text{m}^{-2}\,\text{C}^2$$

Problem 1.5

Show that the values of ω^2 for the three simple harmonic oscillations (a), (b), (c) in the diagram are in the ratio $1 : 2 : 4$.

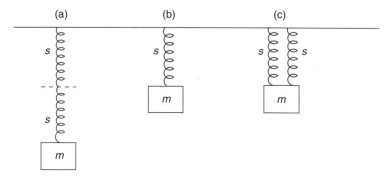

Problem 1.6

The displacement of a simple harmonic oscillator is given by

$$x = a \sin(\omega t + \phi)$$

If the oscillation started at time $t = 0$ from a position x_0 with a velocity $\dot{x} = v_0$ show that

$$\tan\phi = \omega x_0/v_0$$

and

$$a = (x_0^2 + v_0^2/\omega^2)^{1/2}$$

Problem 1.7

A particle oscillates with simple harmonic motion along the x axis with a displacement amplitude a and spends a time dt in moving from x to $x + dx$. Show that the probability of finding it between x and $x + dx$ is given by

$$\frac{dx}{\pi(a^2 - x^2)^{1/2}}$$

(in wave mechanics such a probability is not zero for $x > a$).

Problem. 1.8

Many identical simple harmonic oscillators are equally spaced along the x axis of a medium and a photograph shows that the locus of their displacements in the y direction is a sine curve. If the distance λ separates oscillators which differ in phase by 2π radians, what is the phase difference between two oscillators a distance x apart?

Problem 1.9

A mass stands on a platform which vibrates simple harmonically in a vertical direction at a frequency of 5 Hz. Show that the mass loses contact with the platform when the displacement exceeds 10^{-2}m.

Problem 1.10

A mass M is suspended at the end of a spring of length l and stiffness s. If the mass of the spring is m and the velocity of an element dy of its length is proportional to its distance y from the fixed end of the spring, show that the kinetic energy of this element is

$$\frac{1}{2}\left(\frac{m}{l}dy\right)\left(\frac{y}{l}v\right)^2$$

where v is the velocity of the suspended mass M. Hence, by integrating over the length of the spring, show that its total kinetic energy is $\frac{1}{6}mv^2$ and, from the total energy of the oscillating system, show that the frequency of oscillation is given by

$$\omega^2 = \frac{s}{M + m/3}$$

Problem 1.11

The general form for the energy of a simple harmonic oscillator is

$$E = \tfrac{1}{2}\text{mass (velocity)}^2 + \tfrac{1}{2}\text{stiffness (displacement)}^2$$

Set up the energy equations for the oscillators in Figure 1.1(a), (b), (c), (d), (e), (f) and (g), and use the expression

$$\frac{dE}{dt} = 0$$

to derive the equation of motion in each case.

Problem 1.12

The displacement of a simple harmonic oscillator is given by $x = a \sin \omega t$. If the values of the displacement x and the velocity $\dot{x}$ are plotted on perpendicular axes, eliminate t to show that the locus of the points $(x, \dot{x})$ is an ellipse. Show that this ellipse represents a path of constant energy.

Problem 1.13

In Chapter 12 the intensity of the pattern when light from two slits interferes (Young's experiment) will be seen to depend on the superposition of two simple harmonic oscillations of equal amplitude a and phase difference δ. Show that the intensity

$$I = R^2 \propto 4a^2 \cos^2 \delta/2$$

Between what values does the intensity vary?

Problem 1.14

Carry out the process indicated in the text to derive equation (1.3) on p. 15.

Problem 1.15

The co-ordinates of the displacement of a particle of mass m are given by

$$x = a \sin \omega t$$
$$y = b \cos \omega t$$

Eliminate t to show that the particle follows an elliptical path and show by adding its kinetic and potential energy at any position x, y that the ellipse is a path of constant energy equal to the sum of the separate energies of the simple harmonic vibrations.

Prove that the quantity $m(x\dot{y} - y\dot{x})$ is also constant. What does this quantity represent?

Problem 1.16

Two simple harmonic motions of the same frequency vibrate in directions perpendicular to each other along the x and y axes. A phase difference

$$\delta = \phi_2 - \phi_1$$

exists between them such that the principal axes of the resulting elliptical trace are inclined at an angle to the x and y axes. Show that the measurement of two separate values of x (or y) is sufficient to determine the phase difference.
(Hint: use equation (1.3) and measure $y(\max)$, and y for $(x = 0)$.)

Problem 1.17

Take a random group of $n > 7$ values of ϕ in the range $0 \leq \phi \leq \pi$ and form the product

$$\sum_{\substack{i=1 \\ i \neq j}}^{n} \cos \phi_i \sum_{j=1}^{n} \cos \phi_j$$

Show that the average value obtained for several such groups is negligible with respect to $n/2$.

Problem 1.18

Use the method of example (3) (p. 28) to show that

$$a \sin \omega t + a \sin (\omega t + \delta) + a \sin (\omega t + 2\delta) + \cdots + a \sin [\omega t + (n-1)\delta]$$

$$= a \sin \left[\omega t + \frac{(n-1)}{2} \delta \right] \frac{\sin n\delta/2}{\sin \delta/2}$$

Problem 1.19

If we represent the sum of the series

$$a \cos \omega t + a \cos (\omega t + \delta) + a \cos (\omega t + 2\delta) + \cdots + a \cos [\omega t + (n-1)\delta]$$

by the complex exponential form

$$z = a \, e^{i\omega t} (1 + e^{i\delta} + e^{i2\delta} + \cdots + e^{i(n-1)\delta})$$

show that

$$zz^* = a^2 \frac{\sin^2 n\delta/2}{\sin^2 \delta/2}$$

Summary of Important Results

Simple Harmonic Oscillator (mass m, stiffness s, amplitude a)
Equation of motion $\ddot{x} + \omega^2 x = 0$ where $\omega^2 = s/m$
Displacement $x = a \sin (\omega t + \phi)$
Energy $= \frac{1}{2} m \dot{x}^2 + \frac{1}{2} s x^2 = \frac{1}{2} m \omega^2 a^2 = \frac{1}{2} s a^2 = $ constant

Superposition (Amplitude and Phase) of two SHMs
One-dimensional

Equal ω, different amplitudes, phase difference δ, resultant R where $R^2 = a_1^2 + a_2^2 + 2a_1 a_2 \cos \delta$
Different ω, equal amplitude,

$$x = x_1 + x_2 = a(\sin \omega_1 t + \sin \omega_2 t)$$

$$= 2a \sin \frac{(\omega_1 + \omega_2)t}{2} \cos \frac{(\omega_2 - \omega_1)t}{2}$$

Two-dimensional: perpendicular axes
Equal ω, different amplitude—giving general conic section

$$\frac{x^2}{a_1^2} + \frac{y^2}{a_2^2} - \frac{2xy}{a_1 a_2} \cos (\phi_2 - \phi_1) = \sin^2 (\phi_2 - \phi_1)$$

(basis of optical polarization)

Superposition of n SHM Vectors (equal amplitude a, constant successive phase difference δ)

The resultant is $R \cos(\omega t + \alpha)$, where

$$R = a \frac{\sin n\delta/2}{\sin \delta/2}$$

and

$$\alpha = (n - 1)\delta/2$$

Important in optical diffraction and wave groups of many components.

2

Damped Simple Harmonic Motion

Initially we discussed the case of ideal simple harmonic motion where the total energy remained constant and the displacement followed a sine curve, apparently for an infinite time. In practice some energy is always dissipated by a resistive or viscous process; for example, the amplitude of a freely swinging pendulum will always decay with time as energy is lost. The presence of resistance to motion means that another force is active, which is taken as being proportional to the velocity. The frictional force acts in the direction opposite to that of the velocity (see Figure 2.1) and so Newton's Second law becomes

$$m\ddot{x} = -sx - r\dot{x}$$

where r is the constant of proportionality and has the dimensions of force per unit of velocity. The presence of such a term will always result in energy loss.

The problem now is to find the behaviour of the displacement x from the equation

$$m\ddot{x} + r\dot{x} + sx = 0 \tag{2.1}$$

where the coefficients m, r and s are constant.

When these coefficients are constant a solution of the form $x = C\,e^{\alpha t}$ can always be found. Obviously, since an exponential term is always nondimensional, C has the dimensions of x (a length, say) and α has the dimensions of inverse time, T^{-1}. We shall see that there are three possible forms of this solution, each describing a different behaviour of the displacement x with time. In two of these solutions C appears explicitly as a constant length, but in the third case it takes the form

$$C = A + Bt^{*}$$

---

* The number of constants allowed in the general solution of a differential equation is always equal to the order (that is, the highest differential coefficient) of the equation. The two values A and B are allowed because equation (2.1) is second order. The values of the constants are adjusted to satisfy the initial conditions.

The Physics of Vibrations and Waves, 6th Edition   H. J. Pain
© 2005 John Wiley & Sons, Ltd

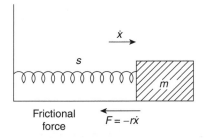

Figure 2.1 Simple harmonic motion system with a damping or frictional force $r\dot{x}$ acting against the direction of motion. The equation of motion is $m\ddot{x} + r\dot{x} + sx = 0$

where A is a length, B is a velocity and t is time, giving C the overall dimensions of a length, as we expect. From our point of view this case is not the most important.

Taking C as a constant length gives $\dot{x} = \alpha C e^{\alpha t}$ and $\ddot{x} = \alpha^2 C e^{\alpha t}$, so that equation (2.1) may be rewritten

$$C e^{\alpha t}(m\alpha^2 + r\alpha + s) = 0$$

so that either

$$x = C e^{\alpha t} = 0 \quad \text{(which is trivial)}$$

or

$$m\alpha^2 + r\alpha + s = 0$$

Solving the quadratic equation in α gives

$$\alpha = \frac{-r}{2m} \pm \sqrt{\frac{r^2}{4m^2} - \frac{s}{m}}$$

Note that $r/2m$ and $(s/m)^{1/2}$, and therefore, α, all have the dimensions of inverse time, T^{-1}, which we expect from the form of $e^{\alpha t}$.

The displacement can now be expressed as

$$x_1 = C_1 e^{-rt/2m+(r^2/4m^2-s/m)^{1/2}t}, \quad x_2 = C_2 e^{-rt/2m-(r^2/4m^2-s/m)^{1/2}t}$$

or the sum of both these terms

$$x = x_1 + x_2 = C_1 e^{-rt/2m+(r^2/4m^2-s/m)^{1/2}t} + C_2 e^{-rt/2m-(r^2/4m^2-s/m)^{1/2}t}$$

The bracket $(r^2/4m^2 - s/m)$ can be positive, zero or negative depending on the relative magnitude of the two terms inside it. Each of these conditions gives one of the three possible solutions referred to earlier and each solution describes a particular kind of behaviour. We shall discuss these solutions in order of *increasing* significance from our point of view; the third solution is the one we shall concentrate upon throughout the rest of this book.

The conditions are:

(1) *Bracket positive* $(r^2/4m^2 > s/m)$. Here the damping resistance term $r^2/4m^2$ dominates the stiffness term s/m, and heavy damping results in a *dead beat* system.
(2) *Bracket zero* $(r^2/4m^2 = s/m)$. The balance between the two terms results in a *critically damped* system.

Neither (1) nor (2) gives oscillatory behaviour.

(3) *Bracket negative* $(r^2/4m^2 < s/m)$. The system is lightly damped and gives oscillatory damped simple harmonic motion.

Case 1. Heavy Damping

Writing $r/2m = p$ and $(r^2/4m^2 - s/m)^{1/2} = q$, we can replace

$$x = C_1 e^{-rt/2m + (r^2/4m^2 - s/m)^{1/2} t} + C_2 e^{-rt/2m - (r^2/4m^2 - s/m)^{1/2} t}$$

by

$$x = e^{-pt}(C_1 e^{qt} + C_2 e^{-qt}),$$

where the C_1 and C_2 are arbitrary in value but have the same dimensions as C (note that two separate values of C are allowed because the differential equation (2.1) is second order).

If now $F = C_1 + C_2$ and $G = C_1 - C_2$, the displacement is given by

$$x = e^{-pt}\left[\frac{F}{2}(e^{qt} + e^{-qt}) + \frac{G}{2}(e^{qt} - e^{-qt})\right]$$

or

$$x = e^{-pt}(F \cosh qt + G \sinh qt)$$

This represents non-oscillatory behaviour, but the actual displacement will depend upon the initial (or boundary) conditions; that is, the value of x at time $t = 0$. If $x = 0$ at $t = 0$ then $F = 0$, and

$$x = G e^{-rt/2m} \sinh\left(\frac{r^2}{4m^2} - \frac{s}{m}\right)^{1/2} t$$

Figure 2.2 illustrates such behaviour when a heavily damped system is disturbed from equilibrium by a sudden impulse (that is, given a velocity at $t = 0$). It will return to zero

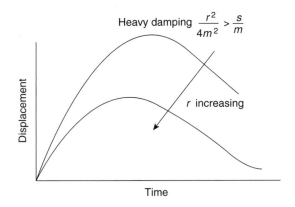

Figure 2.2 Non-oscillatory behaviour of damped simple harmonic system with heavy damping (where $r^2/4m^2 > s/m$) after the system has been given an impulse from a rest position $x = 0$

displacement quite slowly without oscillating about its equilibrium position. More advanced mathematics shows that the value of the velocity dx/dt vanishes only once so that there is only one value of maximum displacement.

(Problem 2.1)

Case 2. Critical Damping ($r^2/4m^2 = s/m$)

Using the notation of Case 1, we see that $q = 0$ and that $x = \mathrm{e}^{-pt}(C_1 + C_2)$. This is, in fact, the limiting case of the behaviour of Case I as q changes from positive to negative. In this case the quadratic equation in α has equal roots, which, in a differential equation solution, demands that C must be written $C = A + Bt$, where A is a constant length and B a given velocity which depends on the boundary conditions. It is easily verified that the value

$$x = (A + Bt)\mathrm{e}^{-rt/2m} = (A + Bt)\mathrm{e}^{-pt}$$

satisfies $m\ddot{x} + r\dot{x} + sx = 0$ when $r^2/4m^2 = s/m$.

(Problem 2.2)

Application to a Damped Mechanical Oscillator

Critical damping is of practical importance in mechanical oscillators which experience sudden impulses and are required to return to zero displacement in the minimum time. Suppose such a system has zero displacement at $t = 0$ and receives an impulse which gives it an initial velocity V.

Then $x = 0$ (so that $A = 0$) and $\dot{x} = V$ at $t = 0$. However,

$$\dot{x} = B[(-pt)\mathrm{e}^{-pt} + \mathrm{e}^{-pt}] = B \text{ at } t = 0$$

so that $B = V$ and the complete solution is

$$x = Vt\mathrm{e}^{-pt}$$

The maximum displacement x occurs when the system comes to rest before returning to zero displacement. At maximum displacement

$$\dot{x} = V\mathrm{e}^{-pt}(1 - pt) = 0$$

thus giving $(1 - pt) = 0$, i.e. $t = 1/p$.
At this time the displacement is therefore

$$x = Vt\mathrm{e}^{-pt} = \frac{V}{p}\mathrm{e}^{-1}$$

$$= 0.368\frac{V}{p} = 0.368\frac{2mV}{r}$$

The curve of displacement versus time is shown in Figure 2.3; the return to zero in a critically damped system is reached in *minimum* time.

Case 3. Damped Simple Harmonic Motion

When $r^2/4m^2 < s/m$ the damping is light, and this gives from the present point of view the most important kind of behaviour, *oscillatory damped simple harmonic motion.*

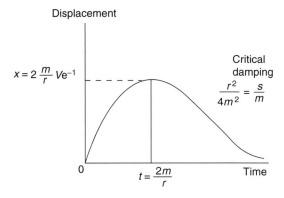

Figure 2.3 Limiting case of non-oscillatory behaviour of damped simple harmonic system where $r^2/4m^2 = s/m$ (critical damping)

The expression $(r^2/4m^2 - s/m)^{1/2}$ is an imaginary quantity, the square root of a negative number, which can be rewritten

$$\pm\left(\frac{r^2}{4m^2} - \frac{s}{m}\right)^{1/2} = \pm\sqrt{-1}\left(\frac{s}{m} - \frac{r^2}{4m^2}\right)^{1/2}$$

$$= \pm i\left(\frac{s}{m} - \frac{r^2}{4m^2}\right)^{1/2} \text{ (where } i = \sqrt{-1})$$

so the displacement

$$x = C_1 e^{-rt/2m} e^{+i(s/m - r^2/4m^2)^{1/2}t} + C_2 e^{-rt/2m} e^{-i(s/m - r^2/4m^2)^{1/2}t}$$

The bracket has the dimensions of inverse time; that is, of frequency, and can be written $(s/m - r^2/4m^2)^{1/2} = \omega'$, so that the second exponential becomes $e^{i\omega't} = \cos\omega't +$ $i\sin\omega't$. This shows that the behaviour of the displacement x is oscillatory with a new frequency $\omega' < \omega = (s/m)^{1/2}$, the frequency of ideal simple harmonic motion. To compare the behaviour of the damped oscillator with the ideal case we should like to express the solution in a form similar to $x = A\sin(\omega't + \phi)$ as in the ideal case, where ω has been replaced by ω'.

We can do this by writing

$$x = e^{-rt/2m}(C_1 e^{i\omega't} + C_2 e^{-i\omega't})$$

If we now choose

$$C_1 = \frac{A}{2i} e^{i\phi}$$

and

$$C_2 = -\frac{A}{2i} e^{-i\phi}$$

where A and ϕ (and thus $e^{i\phi}$) are constants which depend on the motion at $t = 0$, we find after substitution

$$x = A e^{-rt/2m}\frac{[e^{i(\omega't + \phi)} - e^{-i(\omega't + \phi)}]}{2i}$$

$$= A e^{-rt/2m}\sin(\omega't + \phi)$$

This procedure is equivalent to imposing the boundary condition $x = A\sin\phi$ at $t = 0$ upon the solution for x. The displacement therefore varies sinusoidally with time as in the case of simple harmonic motion, but now has a new frequency

$$\omega' = \left(\frac{s}{m} - \frac{r^2}{4m^2}\right)^{1/2}$$

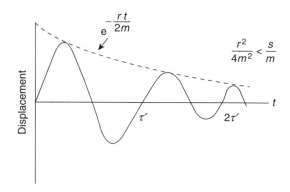

Figure 2.4 Damped oscillatory motion where $s/m > r^2/4m^2$. The amplitude decays with $e^{-rt/2m}$, and the reduced angular frequency is given by $\omega'^2 = s/m - r^2/4m^2$

and its amplitude A is modified by the exponential term $e^{-rt/2m}$, a term which decays with time.

If $x = 0$ at $t = 0$ then $\phi = 0$; Figure 2.4 shows the behaviour of x with time, its oscillations gradually decaying with the envelope of maximum amplitudes following the dotted curve $e^{-rt/2m}$. The constant A is obviously the value to which the amplitude would have risen at the first maximum if no damping were present.

The presence of the force term $r\dot{x}$ in the equation of motion therefore introduces a loss of energy which causes the amplitude of oscillation to decay with time as $e^{-rt/2m}$.

(Problem 2.3)

Methods of Describing the Damping of an Oscillator

Earlier in this chapter we saw that the energy of an oscillator is given by

$$E = \tfrac{1}{2}ma^2\omega^2 = \tfrac{1}{2}sa^2$$

that is, proportional to the square of its amplitude.

We have just seen that in the presence of a damping force $r\dot{x}$ the amplitude decays with time as

$$e^{-rt/2m}$$

so that the energy decay will be proportional to

$$\left(e^{-rt/2m}\right)^2$$

that is, $e^{-rt/m}$. The larger the value of the damping force r the more rapid the decay of the amplitude and energy. Thus we can use the exponential factor to express the rates at which the amplitude and energy are reduced.

Logarithmic Decrement

This measures the rate at which the *amplitude* dies away. Suppose in the expression

$$x = A\,e^{-rt/2m}\sin(\omega' t + \phi)$$

we choose

$$\phi = \pi/2$$

and we write

$$x = A_0\,e^{-rt/2m}\cos\omega' t$$

with $x = A_0$ at $t = 0$. Its behaviour will follow the curve in Figure 2.5.

If the period of oscillation is τ' where $\omega' = 2\pi/\tau'$, then one period later the amplitude is given by

$$A_1 = A_0\,e^{(-r/2m)\tau'}$$

so that

$$\frac{A_0}{A_1} = e^{r\tau'/2m} = e^{\delta}$$

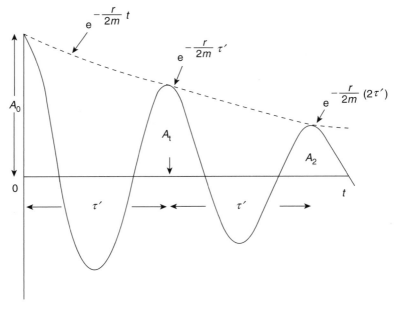

Figure 2.5 The logarithmic ratio of any two amplitudes one period apart is the logarithmic decrement, defined as $\delta = \log_e(A_n/A_{n+1}) = r\tau'/2m$

where

$$\delta = \frac{r}{2m} \tau' = \log_e \frac{A_0}{A_1}$$

is called the *logarithmic decrement*. (Note that this use of δ differs from that in Figure 1.11). The logarithmic decrement δ is the logarithm of the ratio of two amplitudes of oscillation which are separated by one period, the larger amplitude being the numerator since $e^{\delta} > 1$.

Similarly

$$\frac{A_0}{A_2} = e^{r(2\tau')/2m} = e^{2\delta}$$

and

$$\frac{A_0}{A_n} = e^{n\delta}$$

Experimentally, the value of δ is best found by comparing amplitudes of oscillations which are separated by n periods. The graph of

$$\log_e \frac{A_0}{A_n}$$

versus n for different values of n has a slope δ.

Relaxation Time or Modulus of Decay

Another way of expressing the damping effect is by means of the time taken for the amplitude to decay to

$$e^{-1} = 0.368$$

of its original value A_0. This time is called the *relaxation time* or *modulus of decay* and the amplitude

$$A_t = A_0 e^{-rt/2m} = A_0 e^{-1}$$

at a time $t = 2m/r$.

Measuring the natural decay in terms of the fraction e^{-1} of the original value is a very common procedure in physics. The time for a natural decay process to reach zero is, of course, theoretically infinite.

(Problem 2.4)

The Quality Factor or Q-value of a Damped Simple Harmonic Oscillator

This measures the rate at which the *energy* decays. Since the decay of the amplitude is represented by

$$A = A_0 e^{-rt/2m}$$

the decay of energy is proportional to

$$A^2 = A_0^2 e^{(-rt/2m)^2}$$

and may be written

$$E = E_0 e^{(-r/m)t}$$

where E_0 is the energy value at $t = 0$.

The time for the energy E to decay to $E_0 e^{-1}$ is given by $t = m/r$ s during which time the oscillator will have vibrated through $\omega' m/r$ rad.

We define the *quality factor*

$$Q = \frac{\omega' m}{r}$$

as the *number of radians through which the damped system oscillates as its energy decays to*

$$E = E_0 e^{-1}$$

If r is small, then Q is very large and

$$\frac{s}{m} \gg \frac{r^2}{4m^2}$$

so that

$$\omega' \approx \omega_0 = \left(\frac{s}{m}\right)^{1/2}$$

Thus, we write, to a very close approximation,

$$Q = \frac{\omega_0 m}{r}$$

which is a constant of the damped system.

Since r/m now equals ω_0/Q we can write

$$E = E_0 e^{(-r/m)t} = E_0 e^{-\omega_0 t/Q}$$

The fact that Q is a constant $(= \omega_0 m/r)$ implies that the ratio

$$\frac{\text{energy stored in system}}{\text{energy lost per cycle}}$$

is also a constant, for

$$\frac{Q}{2\pi} = \frac{\omega_0 m}{2\pi r} = \frac{\nu_0 m}{r}$$

is the number of *cycles* (or complete oscillations) through which the system moves in decaying to

$$E = E_0 \, \mathrm{e}^{-1}$$

and if

$$E = E_0 \, \mathrm{e}^{(-r/m)t}$$

the energy lost per cycle is

$$-\Delta E = \frac{\mathrm{d}E}{\mathrm{d}t}\,\Delta t = \frac{-r}{m}E\frac{1}{\nu'}$$

where $\Delta t = 1/\nu' = \tau'$, the period of oscillation.

Thus, the ratio

$$\frac{\text{energy stored in system}}{\text{energy lost per cycle}} = \frac{E}{-\Delta E} = \frac{\nu' m}{r} \approx \frac{\nu_0 m}{r}$$
$$= \frac{Q}{2\pi}$$

In the next chapter we shall meet the same quality factor Q in two other roles, the first as a measure of the power absorption bandwidth of a damped oscillator driven near its resonant frequency and again as the factor by which the displacement of the oscillator is amplified at resonance.

Example on the Q-value of a Damped Simple Harmonic Oscillator

An electron in an atom which is freely radiating power behaves as a damped simple harmonic oscillator.

If the radiated power is given by $P = q^2\omega^4 x_0^2/12\pi\varepsilon_0 c^3$ W at a wavelength of $0.6\,\mu\mathrm{m}$ (6000 Å), show that the Q-value of the atom is about 10^8 and that its free radiation lifetime is about 10^{-8}s (the time for its energy to decay to e^{-1} of its original value).

$$q = 1.6 \times 10^{-19}\mathrm{C}$$
$$1/4\pi\varepsilon_0 = 9 \times 10^9\,\mathrm{m\,F^{-1}}$$
$$m_e = 9 \times 10^{-31}\,\mathrm{kg}$$
$$c = 3 \times 10^8\,\mathrm{m\,s^{-1}}$$
$$x_0 = \text{maximum amplitude of oscillation}$$

The radiated power P is $-\nu\,\Delta E$, where $-\Delta E$ is the energy loss per cycle, and the energy of the oscillator is given by $E = \frac{1}{2}m_e\omega^2 x_0^2$.

Thus, $Q = 2\pi E / - \Delta E = \nu \pi m_e \omega^2 x_0^2 / P$, and inserting the values above with $\omega = 2\pi\nu = 2\pi c / \lambda$, where the wavelength λ is given, yields a Q value of $\sim 5 \times 10^7$.

The relation $Q = \omega t$ gives t, the radiation lifetime, a value of $\sim 10^{-8}$ s.

Energy Dissipation

We have seen that the presence of the resistive force reduces the amplitude of oscillation with time as energy is dissipated.

The total energy remains the sum of the kinetic and potential energies

$$E = \tfrac{1}{2}m\dot{x}^2 + \tfrac{1}{2}sx^2$$

Now, however, dE/dt is not zero but negative because energy is lost, so that

$$\frac{dE}{dt} = \frac{d}{dt}(\tfrac{1}{2}m\dot{x}^2 + \tfrac{1}{2}sx^2) = \dot{x}(m\ddot{x} + sx)$$
$$= \dot{x}(-r\dot{x}) \quad \text{for} \quad m\ddot{x} + r\dot{x} + sx = 0$$

i.e. $dE/dt = -r\dot{x}^2$, which is the rate of doing work against the frictional force (dimensions of force $\times$ velocity = force $\times$ distance/time).

(Problems 2.5, 2.6)

Damped SHM in an Electrical Circuit

The force equation in the mechanical oscillator is replaced by the voltage equation in the electrical circuit of inductance, resistance and capacitance (Figure 2.6).

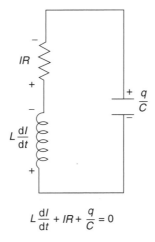

$$L\frac{dI}{dt} + IR + \frac{q}{C} = 0$$

Figure 2.6 Electrical circuit of inductance, capacitance and resistance capable of damped simple harmonic oscillations. The sum of the voltages around the circuit is given from Kirchhoff's law as $L\dfrac{dI}{dt} + RI + \dfrac{q}{C} = 0$

We have, therefore,

$$L\frac{\mathrm{d}I}{\mathrm{d}t} + RI + \frac{q}{C} = 0$$

or

$$L\ddot{q} + R\dot{q} + \frac{q}{C} = 0$$

and by comparison with the solutions for x in the mechanical case we know immediately that the charge

$$q = q_0\,\mathrm{e}^{-Rt/2L \pm (R^2/4L^2 - 1/LC)^{1/2}\,t}$$

which, for $1/LC > R^2/4L^2$, gives oscillatory behaviour at a frequency

$$\omega^2 = \frac{1}{LC} - \frac{R^2}{4L^2}$$

From the exponential decay term we see that R/L has the dimensions of inverse time T^{-1} or ω, so that ωL has the dimensions of R; that is, ωL is measured in ohms.

Similarly, since $\omega^2 = 1/LC$, $\omega L = 1/\omega C$, so that $1/\omega C$ is also measured in ohms. We shall use these results in the next chapter.

(Problems 2.7, 2.8, 2.9)

Problem 2.1

The heavily damped simple harmonic system of Figure 2.2 is displaced a distance F from its equilibrium position and released from rest. Show that in the expression for the displacement

$$x = \mathrm{e}^{-pt}(F\cosh qt + G\sinh qt)$$

where

$$p = \frac{r}{2m} \quad \text{and} \quad q = \left(\frac{r^2}{4m^2} - \frac{s}{m}\right)^{1/2}$$

that the ratio

$$\frac{G}{F} = \frac{r}{(r^2 - 4ms)^{1/2}}$$

Problem 2.2

Verify that the solution

$$x = (A + Bt)\mathrm{e}^{-rt/2m}$$

satisfies the equation

$$m\ddot{x} + r\dot{x} + sx = 0$$

when

$$r^2/4m^2 = s/m$$

Problem 2.3

The solution for damped simple harmonic motion is given by

$$x = e^{-rt/2m}(C_1 e^{i\omega't} + C_2 e^{-i\omega't})$$

If $x = A\cos\phi$ at $t = 0$, find the values of C_1 and C_2 to show that $\dot{x} \approx -\omega'A\sin\phi$ at $t = 0$ only if r/m is very small or $\phi \approx \pi/2$.

Problem 2.4

A capacitance C with a charge q_0 at $t = 0$ discharges through a resistance R. Use the voltage equation $q/C + IR = 0$ to show that the relaxation time of this process is RC s; that is,

$$q = q_0 e^{-t/RC}$$

(Note that t/RC is non-dimensional.)

Problem 2.5

The frequency of a damped simple harmonic oscillator is given by

$$\omega'^2 = \frac{s}{m} - \frac{r^2}{4m^2} = \omega_0^2 - \frac{r^2}{4m^2}$$

(a) If $\omega_0^2 - \omega'^2 = 10^{-6}\omega_0^2$ show that $Q = 500$ and that the logarithmic decrement $\delta = \pi/500$.
(b) If $\omega_0 = 10^6$ and $m = 10^{-10}$ Kg show that the stiffness of the system is $100\,\mathrm{N\,m^{-1}}$, and that the resistive constant r is $2 \times 10^{-7}\,\mathrm{N \cdot sm^{-1}}$.
(c) If the maximum displacement at $t = 0$ is 10^{-2} m, show that the energy of the system is 5×10^{-3} J and the decay to e^{-1} of this value takes 0.5 ms.
(d) Show that the energy loss in the first cycle is $2\pi \times 10^{-5}$ J.

Problem 2.6

Show that the fractional change in the resonant frequency $\omega_0(\omega_0^2 = s/m)$ of a damped simple harmonic mechanical oscillator is $\approx (8Q^2)^{-1}$ where Q is the quality factor.

Problem 2.7

Show that the quality factor of an electrical LCR series circuit is $Q = \omega_0 L/R$ where $\omega_0^2 = 1/LC$

Problem 2.8

A plasma consists of an ionized gas of ions and electrons of equal number densities ($n_i = n_e = n$) having charges of opposite sign $\pm e$, and masses m_i and m_e, respectively, where $m_i > m_e$. Relative displacement between the two species sets up a restoring

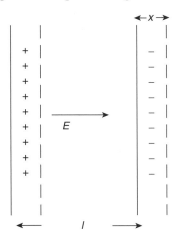

electric field which returns the electrons to equilibrium, the ions being considered stationary. In the diagram, a plasma slab of thickness l has all its electrons displaced a distance x to give a restoring electric field $E = nex/\varepsilon_0$, where ε_0 is constant. Show that the restoring force per unit area on the electrons is xn^2e^2l/ε_0 and that they oscillate simple harmonically with angular frequency $\omega_e^2 = ne^2/m_e\varepsilon_0$. This frequency is called the electron plasma frequency, and only those radio waves of frequency $\omega > \omega_e$ will propagate in such an ionized medium. Hence the reflection of such waves from the ionosphere.

Problem 2.9

A simple pendulum consists of a mass m at the end of a string of length l and performs small oscillations. The length is very slowly shortened whilst the pendulum oscillates many times at a constant amplitude $l\theta$ where θ is very small. Show that if the length is changed by $-\Delta l$ the work done is $-mg\,\Delta l$ (owing to the elevation of the position of equilibrium) together with an increase in the pendulum energy

$$\Delta E = \left(mg\frac{\overline{\theta^2}}{2} - ml\overline{\dot{\theta}^2} \right)\Delta l$$

where $\overline{\theta^2}$ is the average value of θ^2 during the shortening. If $\theta = \theta_0 \cos \omega t$, show that the energy of the pendulum at any instant may be written

$$E = \frac{ml^2\omega^2\theta_0^2}{2} = \frac{mgl\theta_0^2}{2}$$

and hence show that

$$\frac{\Delta E}{E} = -\frac{1}{2}\frac{\Delta l}{l} = \frac{\Delta \nu}{\nu}$$

that is, E/ν, the ratio of the energy of the pendulum to its frequency of oscillation remains constant during the slowly changing process. (This constant ratio under slowly varying conditions is important in quantum theory where the constant is written as a multiple of Planck's constant, h.)

Summary of Important Results

Damped Simple Harmonic Motion

Equation of motion $m\ddot{x} + r\dot{x} + sx = 0$
Oscillations when

$$\frac{s}{m} > \frac{r^2}{4m^2}$$

Displacement $x = A\,e^{-rt/2m}\cos(\omega't + \phi)$ where

$$\omega'^2 = \frac{s}{m} - \frac{r^2}{4m^2}$$

Amplitude Decay

Logarithmic decrement δ—the logarithm of the ratio of two successive amplitudes one period τ' apart

$$\delta = \log_e \frac{A_n}{A_{n+1}} = \frac{r\tau'}{2m}$$

Relaxation Time

Time for amplitude to decay to $A = A_0\,e^{-rt/2m} = A_0\,e^{-1}$; that is, $t = 2m/r$

Energy Decay

Quality factor Q is the number of radians during which energy decreases to $E = E_0\,e^{-1}$

$$Q = \frac{\omega_0 m}{r} = 2\pi\frac{\text{energy stored in system}}{\text{energy lost per cycle}}$$

$$E = E_0\,e^{-rt/m} = E_0\,e^{-1} \quad \text{when } Q = \omega_0 t$$

In damped SHM

$$\frac{dE}{dt} = (m\ddot{x} + sx)\dot{x} = -r\dot{x}^2 \quad \text{(work rate of resistive force)}$$

For equivalent expressions in electrical oscillators replace m by L, r by R and s by $1/C$. Force equations become voltage equations.

3

The Forced Oscillator

The Operation of i upon a Vector

We have already seen that a harmonic oscillation can be conveniently represented by the form $e^{i\omega t}$. In addition to its mathematical convenience i can also be used as a vector operator of physical significance. We say that when i precedes or operates on a vector the direction of that vector is turned through a positive angle (anticlockwise) of $\pi/2$, i.e. i acting as an operator advances the phase of a vector by 90°. The operator $-i$ rotates the vector clockwise by $\pi/2$ and retards its phase by 90°. The mathematics of i as an operator differs in no way from its use as $\sqrt{-1}$ and from now on it will play both roles.

The vector $\mathbf{r} = \mathbf{a} + i\mathbf{b}$ is shown in Figure 3.1, where the direction of $\mathbf{b}$ is perpendicular to that of $\mathbf{a}$ because it is preceded by i. The magnitude or modulus or $\mathbf{r}$ is written

$$r = |\mathbf{r}| = (a^2 + b^2)^{1/2}$$

and

$$r^2 = (a^2 + b^2) = (\mathbf{a} + i\mathbf{b})(\mathbf{a} - i\mathbf{b}) = \mathbf{r}\mathbf{r}^*,$$

where $(\mathbf{a} - i\mathbf{b}) = \mathbf{r}^*$ is defined as the complex conjugate of $(\mathbf{a} + i\mathbf{b})$; that is, the sign of i is changed.

The vector $\mathbf{r}^* = \mathbf{a} - i\mathbf{b}$ is also shown in Figure 3.1.

The vector $\mathbf{r}$ can be written as a product of its magnitude r (scalar quantity) and its phase or direction in the form (Figure 3.1)

$$\mathbf{r} = r\,e^{i\phi} = r(\cos\phi + i\sin\phi)$$
$$= \mathbf{a} + i\mathbf{b}$$

showing that $a = r\cos\phi$ and $b = r\sin\phi$.

The Physics of Vibrations and Waves, 6th Edition   H. J. Pain
© 2005 John Wiley & Sons, Ltd

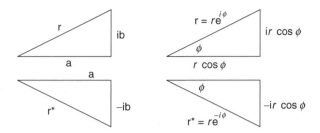

Figure 3.1 Vector representation using i operator and exponential index. Star superscript indicates complex conjugate where $-i$ replaces i

It follows that

$$\cos\phi = \frac{a}{r} = \frac{a}{(a^2 + b^2)^{1/2}}$$

and

$$\sin\phi = \frac{b}{r} = \frac{b}{(a^2 + b^2)^{1/2}}$$

giving $\tan\phi = b/a$.

Similarly

$$\mathbf{r}^* = r\,e^{-i\phi} = r(\cos\phi - i\sin\phi)$$

$$\cos\phi = \frac{a}{r}, \quad \sin\phi = \frac{-b}{r} \quad \text{and} \quad \tan\phi = \frac{-b}{a} \quad (\text{Figure 3.1})$$

The reader should confirm that the operator i rotates a vector by $\pi/2$ in the positive direction (as stated in the first paragraph of p. 53) by taking $\phi = \pi/2$ in the expression

$$\mathbf{r} = r\,e^{i\phi} = r(\cos\pi/2 + i\sin\pi/2)$$

Note that $\phi = -\pi/2$ in $\mathbf{r} = r\,e^{-i\pi/2}$ rotates the vector in the negative direction.

Vector form of Ohm's Law

Ohm's Law is first met as the scalar relation $V = IR$, where V is the voltage across the resistance R and I is the current through it. Its scalar form states that the voltage and current are always in phase. Both will follow a $\sin(\omega t + \phi)$ or a $\cos(\omega t + \phi)$ curve, and the value of ϕ will be the same for both voltage and current.

However, the presence of either or both of the other two electrical components, inductance L and capacitance C, will introduce a phase difference between voltage and

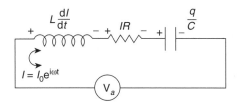

Figure 3.2a   An electrical forced oscillator. The voltage V_a is applied to the series *LCR* circuit giving $V_a = L\,\mathrm{d}I/\mathrm{d}t + IR + q/C$

current, and Ohm's Law takes the vector form

$$\mathbf{V} = \mathbf{IZ}_e,$$

where $\mathbf{Z}_e$, called the *impedance*, replaces the resistance, and is the vector sum of the effective resistances of R, L, and C in the circuit.

When an alternating voltage V_a of frequency ω is applied across a resistance, inductance and condenser in series as in Figure 3.2a, the balance of voltages is given by

$$V_a = IR + L\,\frac{\mathrm{d}I}{\mathrm{d}t} + q/C$$

and the current through the circuit is given by $I = I_0\,\mathrm{e}^{\mathrm{i}\omega t}$. The voltage across the inductance

$$V_L = L\,\frac{\mathrm{d}I}{\mathrm{d}t} = L\,\frac{\mathrm{d}}{\mathrm{d}t}\,I_0\,\mathrm{e}^{\mathrm{i}\omega t} = \mathrm{i}\omega L I_0\,\mathrm{e}^{\mathrm{i}\omega t} = \mathrm{i}\omega L I$$

But ωL, as we saw at the end of the last chapter, has the dimensions of ohms, being the value of the effective resistance presented by an inductance L to a current of frequency ω. The product $\omega L I$ with dimensions of ohms times current, i.e. volts, is preceded by i; this tells us that the phase of the voltage across the inductance is 90° ahead of that of the current through the circuit.

Similarly, the voltage across the condenser is

$$\frac{q}{C} = \frac{1}{C}\int I\,\mathrm{d}t = \frac{1}{C}\,I_0\int \mathrm{e}^{\mathrm{i}\omega t}\,\mathrm{d}t = \frac{1}{\mathrm{i}\omega C}\,I_0\,\mathrm{e}^{\mathrm{i}\omega t} = -\frac{\mathrm{i}I}{\omega C}$$

(since $1/\mathrm{i} = -\mathrm{i}$).

Again $1/\omega C$, measured in ohms, is the value of the effective resistance presented by the condenser to the current of frequency ω. Now, however, the voltage $I/\omega C$ across the condenser is preceded by $-\mathrm{i}$ and therefore lags the current by 90°. The voltage and current across the resistance are in phase and Figure 3.2b shows that the vector form of Ohm's Law may be written $\mathbf{V} = \mathbf{IZ}_e = I[R + \mathrm{i}(\omega L - 1/\omega C)]$, where the impedance $\mathbf{Z}_e = R + \mathrm{i}(\omega L - 1/\omega C)$. The quantities ωL and $1/\omega C$ are called *reactances* because they

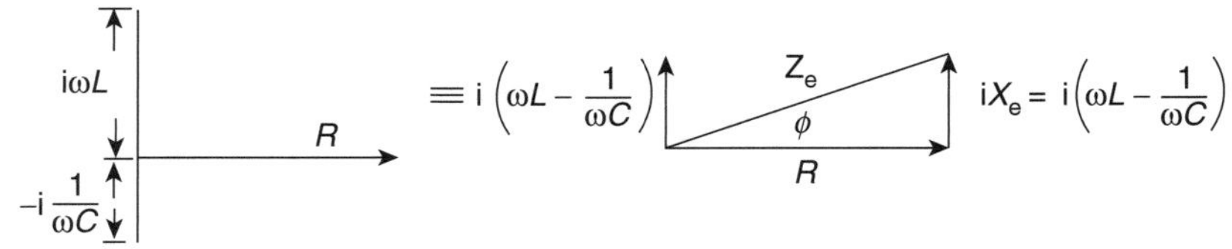

Figure 3.2b Vector addition of resistance and reactances to give the electrical impedance $\mathbf{Z}_e = R + i(\omega L - 1/\omega C)$

introduce a phase relationship as well as an effective resistance, and the bracket $(\omega L - 1/\omega C)$ is often written X_e, the reactive component of $\mathbf{Z}_e$.

The magnitude, in ohms, i.e. the value of the impedance, is

$$Z_e = \left[R^2 + \left(\omega L - \frac{1}{\omega C} \right)^2 \right]^{1/2}$$

and the vector $\mathbf{Z}_e$ may be represented by its magnitude and phase as

$$\mathbf{Z}_e = Z_e \, e^{i\phi} = Z_e(\cos\phi + i\sin\phi)$$

so that

$$\cos\phi = \frac{R}{Z_e}, \quad \sin\phi = \frac{X_e}{Z_e}$$

and

$$\tan\phi = X_e/R,$$

where ϕ is the phase difference between the total voltage across the circuit and the current through it.

The value of ϕ can be positive or negative depending on the relative value of ωL and $1/\omega C$: when $\omega L > 1/\omega C$, ϕ is positive, but the frequency dependence of the components show that ϕ can change both sign and size.

The magnitude of $\mathbf{Z}_e$ is also frequency dependent and has its minimum value $Z_e = R$ when $\omega L = 1/\omega C$.

In the vector form of Ohm's Law, $\mathbf{V} = \mathbf{I}\mathbf{Z}_e$. If $\mathbf{V} = V_0 \, e^{i\omega t}$ and $\mathbf{Z}_e = Z_e \, e^{i\phi}$, then we have

$$\mathbf{I} = \frac{V_0 \, e^{i\omega t}}{Z_e \, e^{i\phi}} = \frac{V_0}{Z_e} \, e^{i(\omega t - \phi)}$$

giving a current of amplitude V_0/Z_e which lags the voltage by a phase angle ϕ.

The Impedance of a Mechanical Circuit

Exactly similar arguments hold when we consider not an electrical oscillator but a mechanical circuit having mass, stiffness and resistance.

The mechanical impedance is defined as the force required to produce unit velocity in the oscillator, i.e. $\mathbf{Z}_m = \mathbf{F}/\mathbf{v}$ or $\mathbf{F} = \mathbf{v}\mathbf{Z}_m$.

Immediately, we can write the mechanical impedance as

$$\mathbf{Z}_m = r + i\left(\omega m - \frac{s}{\omega}\right) = r + iX_m$$

where

$$\mathbf{Z}_m = Z_m \, e^{i\phi}$$

and

$$\tan \phi = X_m/r$$

ϕ being the phase difference between the force and the velocity. The magnitude of $Z_m = [r^2 + (\omega m - s/\omega)^2]^{1/2}$.

Mass, like inductance, produces a positive reactance, and the stiffness behaves in exactly the same way as the capacitance.

Behaviour of a Forced Oscillator

We are now in a position to discuss the physical behaviour of a mechanical oscillator of mass m, stiffness s and resistance r being driven by an alternating force $F_0 \cos \omega t$, where F_0 is the amplitude of the force (Figure 3.3). The equivalent electrical oscillator would be an alternating voltage $V_0 \cos \omega t$ applied to the circuit of inductance L, capacitance C and resistance R in Figure 3.2a.

The mechanical equation of motion, i.e. the dynamic balance of forces, is given by

$$m\ddot{x} + r\dot{x} + sx = F_0 \cos \omega t$$

and the voltage equation in the electrical case is

$$L\ddot{q} + R\dot{q} + q/C = V_0 \cos \omega t$$

We shall analyse the behaviour of the mechanical system but the analysis fits the electrical oscillator equally well.

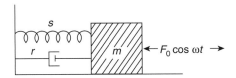

Figure 3.3 Mechanical forced oscillator with force $F_0 \cos \omega t$ applied to damped mechanical circuit of Figure 2.1

The complete solution for x in the equation of motion consists of two terms:

(1) a 'transient' term which dies away with time and is, in fact, the solution to the equation $m\ddot{x} + r\dot{x} + sx = 0$ discussed in Chapter 2. This contributes the term

$$x = C\,\mathrm{e}^{-rt/2m}\,\mathrm{e}^{\mathrm{i}(s/m - r^2/4m^2)^{1/2}t}$$

which decays with $\mathrm{e}^{-rt/2m}$. The second term

(2) is called the 'steady state' term, and describes the behaviour of the oscillator after the transient term has died away.

Both terms contribute to the solution initially, but for the moment we shall concentrate on the 'steady state' term which describes the ultimate behaviour of the oscillator.

To do this we shall rewrite the force equation in vector form and represent $\cos \omega t$ by $\mathrm{e}^{\mathrm{i}\omega t}$ as follows:

$$m\ddot{\mathbf{x}} + r\dot{\mathbf{x}} + s\mathbf{x} = F_0\,\mathrm{e}^{\mathrm{i}\omega t} \tag{3.1}$$

Solving for the vector $\mathbf{x}$ will give both its magnitude and phase with respect to the driving force $F_0\,\mathrm{e}^{\mathrm{i}\omega t}$. Initially, let us try the solution $\mathbf{x} = \mathbf{A}\,\mathrm{e}^{\mathrm{i}\omega t}$, where $\mathbf{A}$ may be complex, so that it may have components in and out of phase with the driving force.

The velocity

$$\dot{\mathbf{x}} = \mathrm{i}\omega\mathbf{A}\,\mathrm{e}^{\mathrm{i}\omega t} = \mathrm{i}\omega\mathbf{x}$$

so that

$$\ddot{\mathbf{x}} = \mathrm{i}^2\omega^2\mathbf{x} = -\omega^2\mathbf{x}$$

and equation (3.1) becomes

$$(-\mathbf{A}\omega^2 m + \mathrm{i}\omega\mathbf{A}r + \mathbf{A}s)\,\mathrm{e}^{\mathrm{i}\omega t} = F_0\,\mathrm{e}^{\mathrm{i}\omega t}$$

which is true for all t when

$$\mathbf{A} = \frac{F_0}{\mathrm{i}\omega r + (s - \omega^2 m)}$$

or, after multiplying numerator and denominator by $-\mathrm{i}$

$$\mathbf{A} = \frac{-\mathrm{i}F_0}{\omega[r + \mathrm{i}(\omega m - s/\omega)]} = \frac{-\mathrm{i}F_0}{\omega\mathbf{Z}_m}$$

Hence

$$\mathbf{x} = \mathbf{A}\,\mathrm{e}^{\mathrm{i}\omega t} = \frac{-\mathrm{i}F_0\,\mathrm{e}^{\mathrm{i}\omega t}}{\omega\mathbf{Z}_m} = \frac{-\mathrm{i}F_0\,\mathrm{e}^{\mathrm{i}\omega t}}{\omega Z_m\,\mathrm{e}^{\mathrm{i}\phi}}$$

$$= \frac{-\mathrm{i}F_0\,\mathrm{e}^{\mathrm{i}(\omega t - \phi)}}{\omega Z_m}$$

where

$$Z_m = [r^2 + (\omega m - s/\omega)^2]^{1/2}$$

This vector form of the *steady state* behaviour of x gives three pieces of information and completely defines the magnitude of the displacement x and its phase with respect to the driving force after the transient term dies away. It tells us

1. That the phase difference ϕ exists between x and the force because of the reactive part $(\omega m - s/\omega)$ of the mechanical impedance.

2. That an extra difference is introduced by the factor $-i$ and even if ϕ were zero the displacement x would lag the force $F_0 \cos \omega t$ by 90°.

3. That the maximum amplitude of the displacement x is $F_0/\omega Z_m$. We see that this is dimensionally correct because the velocity x/t has dimensions F_0/Z_m.

Having used $F_0 \, e^{i\omega t}$ to represent its real part $F_0 \cos \omega t$, we now take the real part of the solution

$$\mathbf{x} = \frac{-iF_0 \, e^{i(\omega t - \phi)}}{\omega Z_m}$$

to obtain the actual value of **x**. (If the force had been $F_0 \sin \omega t$, we would now take that part of **x** preceded by i.)

Now

$$\mathbf{x} = -\frac{iF_0}{\omega Z_m} e^{i(\omega t - \phi)}$$

$$= -\frac{iF_0}{\omega Z_m} [\cos(\omega t - \phi) + i \sin(\omega t - \phi)]$$

$$= -\frac{iF_0}{\omega Z_m} \cos(\omega t - \phi) + \frac{F_0}{\omega Z_m} \sin(\omega t - \phi)$$

The value of x resulting from $F_0 \cos \omega t$ is therefore

$$x = \frac{F_0}{\omega Z_m} \sin(\omega t - \phi)$$

[the value of x resulting from $F_0 \sin \omega t$ would be $-F_0 \cos(\omega t - \phi)/\omega Z_m$].

Note that both of these solutions satisfy the requirement that the total phase difference between displacement and force is ϕ plus the $-\pi/2$ term introduced by the $-i$ factor. When $\phi = 0$ the displacement $x = F_0 \sin \omega t/\omega Z_m$ *lags* the force $F_0 \cos \omega t$ by exactly 90°.

To find the velocity of the forced oscillation in the steady state we write

$$\mathbf{v} = \dot{\mathbf{x}} = (\mathrm{i}\omega)\frac{(-\mathrm{i}F_0)}{\omega Z_m}\,\mathrm{e}^{\mathrm{i}(\omega t - \phi)}$$

$$= \frac{F_0}{Z_m}\,\mathrm{e}^{\mathrm{i}(\omega t - \phi)}$$

We see immediately that

1. There is no preceding i factor so that the velocity **v** and the force differ in phase only by ϕ, and when $\phi = 0$ the velocity and force are in phase.

2. The amplitude of the velocity is F_0/Z_m, which we expect from the definition of mechanical impedance $\mathbf{Z}_m = \mathbf{F}/\mathbf{v}$.

Again we take the real part of the vector expression for the velocity, which will correspond to the real part of the force $F_0\,\mathrm{e}^{\mathrm{i}\omega t}$. This is

$$v = \frac{F_0}{Z_m}\cos{(\omega t - \phi)}$$

Thus, the *velocity is always exactly 90° ahead of the displacement in phase* and differs from the force only by a phase angle ϕ, where

$$\tan\phi = \frac{\omega m - s/\omega}{r} = \frac{X_m}{r}$$

so that a force $F_0\cos\omega t$ gives a displacement

$$x = \frac{F_0}{\omega Z_m}\sin{(\omega t - \phi)}$$

and a velocity

$$v = \frac{F_0}{Z_m}\cos{(\omega t - \phi)}$$

(Problems 3.1, 3.2, 3.3, 3.4)

Behaviour of Velocity v in Magnitude and Phase versus Driving Force Frequency ω

The velocity amplitude is

$$\frac{F_0}{Z_m} = \frac{F_0}{[r^2 + (\omega m - s/\omega)^2]^{1/2}}$$

so that the magnitude of the velocity will vary with the frequency ω because Z_m is frequency dependent.

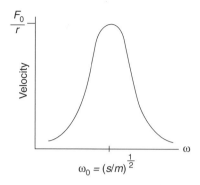

Figure 3.4   Velocity of forced oscillator versus driving frequency ω. Maximum velocity $v_{\text{max}} = F_0/r$ at $\omega_0^2 = s/m$

At low frequencies, the term $-s/\omega$ is the largest term in Z_m and the impedance is said to be *stiffness controlled*. At high frequencies ωm is the dominant term and the impedance is *mass controlled*. At a frequency ω_0 where $\omega_0 m = s/\omega_0$, the impedance has its minimum value $Z_m = r$ and is a real quantity with zero reactance.

The velocity F_0/Z_m then has its maximum value $v = F_0/r$, and ω_0 is said to be the frequency of *velocity resonance*. Note that $\tan\phi = 0$ at ω_0, the velocity and force being in phase.

The variation of the magnitude of the velocity with driving frequency, ω, is shown in Figure 3.4, the height and sharpness of the peak at resonance depending on r, which is the only effective term of Z_m at ω_0.

The expression

$$v = \frac{F_0}{Z_m}\cos\left(\omega t - \phi\right)$$

where

$$\tan\phi = \frac{\omega m - s/\omega}{r}$$

shows that for positive ϕ; that is, $\omega m > s/\omega$, the velocity v will lag the force because $-\phi$ appears in the argument of the cosine. When the driving force frequency ω is very high and $\omega \to \infty$, then $\phi \to 90°$ and the velocity lags the force by that amount.

When $\omega m < s/\omega$, ϕ is negative, the velocity is ahead of the force in phase, and at low driving frequencies as $\omega \to 0$ the term $s/\omega \to \infty$ and $\phi \to -90°$.

Thus, at low frequencies the velocity leads the force (ϕ negative) and at high frequencies the velocity lags the force (ϕ positive).

At the frequency ω_0, however, $\omega_0 m = s/\omega_0$ and $\phi = 0$, so that velocity and force are in phase. Figure 3.5 shows the variation of ϕ with ω for the velocity, the actual shape of the curves depending upon the value of r.

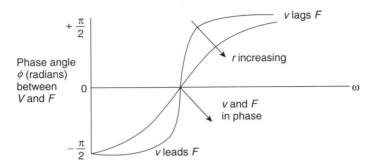

Figure 3.5 Variation of phase angle ϕ versus driving frequency, where ϕ is the phase angle between the velocity of the forced oscillator and the driving force. $\phi = 0$ at velocity resonance. Each curve represents a fixed resistance value

(Problem 3.5)

Behaviour of Displacement versus Driving Force Frequency ω

The phase of the displacement

$$x = \frac{F_0}{\omega Z_m} \sin(\omega t - \phi)$$

is at all times exactly 90° behind that of the velocity. Whilst the graph of ϕ versus ω remains the same, the total phase difference between the displacement and the force involves the extra 90° retardation introduced by the $-\mathrm{i}$ operator. Thus, at very low frequencies, where $\phi = -\pi/2$ rad and the velocity leads the force, the displacement and the force are in phase as we should expect. At high frequencies the displacement lags the force by π rad and is exactly out of phase, so that the curve showing the phase angle between the displacement and the force is equivalent to the ϕ versus ω curve, displaced by an amount equal to $\pi/2$ rad. This is shown in Figure 3.6.

The amplitude of the displacement $x = F_0/\omega Z_m$, and at low frequencies $Z_m = [r^2 + (\omega m - s/\omega)^2]^{1/2} \to s/\omega$, so that $x \approx F_0/(\omega s/\omega) = F_0/s$.

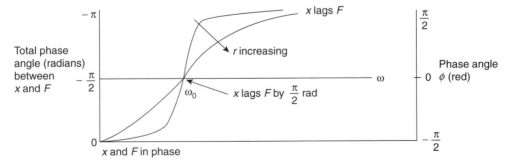

Figure 3.6 Variation of total phase angle between displacement and driving force versus driving frequency ω. The total phase angle is $-\phi - \pi/2$ rad

At high frequencies $Z_m \rightarrow \omega m$, so that $x \approx F_0/(\omega^2 m)$, which tends to zero as ω becomes very large. At very high frequencies, therefore, the displacement amplitude is almost zero because of the mass-controlled or inertial effect.

The velocity resonance occurs at $\omega_0^2 = s/m$, where the denominator Z_m of the velocity amplitude is a minimum, but the displacement resonance will occur, since $x = (F_0/\omega Z_m)\sin(\omega t - \phi)$, when the denominator ωZ_m is a minimum. This takes place when

$$\frac{\mathrm{d}}{\mathrm{d}\omega}(\omega Z_m) = \frac{\mathrm{d}}{\mathrm{d}\omega}\omega[r^2 + (\omega m - s/\omega)^2]^{1/2} = 0$$

i.e. when

$$2\omega r^2 + 4\omega m(\omega^2 m - s) = 0$$

or

$$2\omega[r^2 + 2m(\omega^2 m - s)] = 0$$

so that either

$$\omega = 0$$

or

$$\omega^2 = \frac{s}{m} - \frac{r^2}{2m^2} = \omega_0^2 - \frac{r^2}{2m^2}$$

Thus the *displacement resonance* occurs at a frequency slightly less than ω_0, the frequency of velocity resonance. For a small damping constant r or a large mass m these two resonances, for all practical purposes, occur at the frequency ω_0.

Denoting the displacement resonance frequency by

$$\omega_r = \left(\frac{s}{m} - \frac{r^2}{2m^2}\right)^{1/2}$$

we can write the maximum displacement as

$$x_{\max} = \frac{F_0}{\omega_r Z_m}$$

The value of $\omega_r Z_m$ at ω_r is easily shown to be equal to $\omega' r$ where

$$\omega'^2 = \frac{s}{m} - \frac{r^2}{4m^2} = \omega_0^2 - \frac{r^2}{4m^2}$$

The value of x at displacement resonance is therefore given by

$$x_{\max} = \frac{F_0}{\omega' r}$$

where

$$\omega' = \left(\omega_0^2 - \frac{r^2}{4m^2}\right)^{1/2}$$

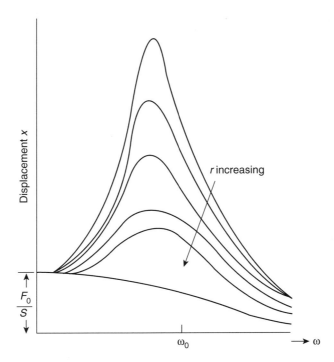

Figure 3.7   Variation of the displacement of a forced oscillator versus driving force frequency ω for various values of r

Since $x_{\max} = F_0/\omega'r$ at resonance, the amplitude at resonance is kept low by increasing r and the variation of x with ω for different values of r is shown in Figure 3.7. A negligible value of r produces a large amplification at resonance: this is the basis of high selectivity in a tuned radio circuit (see the section in this chapter on Q as an amplification factor). Keeping the resonance amplitude low is the principle of vibration insulation.

(Problems 3.6, 3.7)

Problem on Vibration Insulation

A typical vibration insulator is shown in Figure 3.8. A heavy base is supported on a vibrating floor by a spring system of stiffness s and viscous damper r. The insulator will generally operate at the mass controlled end of the frequency spectrum and the resonant frequency is designed to be lower than the range of frequencies likely to be met. Suppose the vertical vibration of the floor is given by $x = A \cos \omega t$ about its equilibrium position and y is the corresponding vertical displacement of the base about its rest position. The function of the insulator is to keep the ratio y/A to a minimum.

The equation of motion is given by

$$m\ddot{y} = -r(\dot{y} - \dot{x}) - s(y - x)$$

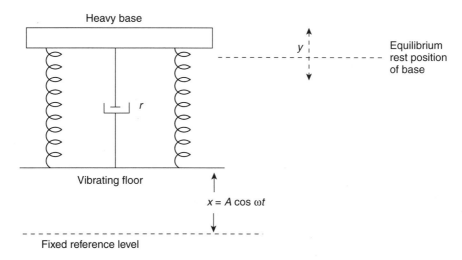

Figure 3.8 Vibration insulator. A heavy base supported by a spring and viscous damper system on a vibrating floor

which, if $y - x = X$, becomes

$$m\ddot{X} + r\dot{X} + sX = -m\ddot{x} = mA\omega^2 \cos \omega t$$
$$= F_0 \cos \omega t,$$

where

$$F_0 = mA\omega^2$$

Use the steady state solution of X to show that

$$y = \frac{F_0}{\omega Z_m} \sin(\omega t - \phi) + A \cos \omega t$$

and (noting that y is the superposition of two harmonic components with a constant phase difference) show that

$$\frac{y_{\max}}{A} = \frac{(r^2 + s^2/\omega^2)^{1/2}}{Z_m}$$

where

$$Z_m^2 = r^2 + (\omega m - s/\omega)^2$$

Note that

$$\frac{y_{\max}}{A} > 1 \quad \text{if} \quad \omega^2 < \frac{2s}{m}$$

so that s/m should be as low as possible to give protection against a given frequency ω.

(a) Show that

$$\frac{y_{\max}}{A} = 1 \quad \text{for} \quad \omega^2 = \frac{2s}{m}$$

(b) Show that

$$\frac{y_{\max}}{A} < 1 \quad \text{for} \quad \omega^2 > \frac{2s}{m}$$

(c) Show that if $\omega^2 = s/m$, then $y_{\max}/A > 1$ but that the damping term r is helpful in keeping the motion of the base to a reasonably low level.

(d) Show that if $\omega^2 > 2s/m$, then $y_{\max}/A < 1$ but damping is detrimental.

Significance of the Two Components of the Displacement Curve

Any single curve of Figure 3.7 is the superposition of the two component curves (a) and (b) in Figure 3.9, for the displacement x may be rewritten

$$x = \frac{F_0}{\omega Z_m} \sin(\omega t - \phi) = \frac{F_0}{\omega Z_m} (\sin \omega t \cos \phi - \cos \omega t \sin \phi)$$

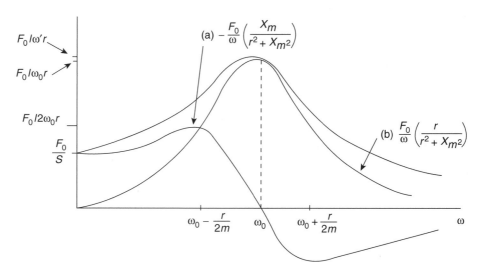

Figure 3.9 A typical curve of Figure 3.7 resolved into its 'anti-phase' component (curve (a)) and its '90° out of phase' component (curve (b)). Curve (b) represents the resistive fraction of the impedance and curve (a) the reactive fraction. Curve (b) corresponds to absorption and curve (a) to anomalous dispersion of an electromagnetic wave in a medium having an atomic or molecular resonant frequency equal to the frequency of the wave

or, since

$$\cos\phi = \frac{r}{Z_m} \quad \text{and} \quad \sin\phi = \frac{X_m}{Z_m}$$

as

$$x = \frac{F_0}{\omega Z_m}\frac{r}{Z_m}\sin\omega t - \frac{F_0}{\omega Z_m}\frac{X_m}{Z_m}\cos\omega t$$

The $\cos\omega t$ component (with a negative sign) is exactly anti-phase with respect to the driving force $F_0\cos\omega t$. Its amplitude, plotted as curve (a) may be expressed as

$$-\frac{F_0}{\omega}\frac{X_m}{Z_m^2} = \frac{F_0 m(\omega_0^2 - \omega^2)}{m^2(\omega_0^2 - \omega^2)^2 + \omega^2 r^2} \tag{3.2}$$

where $\omega_0^2 = s/m$ and ω_0 is the frequency of velocity resonance.

The $\sin\omega t$ component lags the driving force $F_0\cos\omega t$ by $90°$. Its amplitude plotted as curve (b) becomes

$$\frac{F_0}{\omega}\frac{r}{r^2 + X_m^2} = \frac{F_0\omega r}{m^2(\omega_0^2 - \omega^2)^2 + \omega^2 r^2}$$

We see immediately that at ω_0 curve (a) is zero and curve (b) is near its maximum but they combine to give a maximum at ω where

$$\omega^2 = \omega_0^2 - \frac{r^2}{2m^2}$$

the resonant frequency for amplitude displacement.

These curves are particularly familiar in the study of optical dispersion where the forced oscillator is an electron in an atom and the driving force is the oscillating field vector of an electromagnetic wave of frequency ω. When ω is the resonant frequency of the electron in the atom, the atom absorbs a large amount of energy from the electromagnetic wave and curve (b) is the shape of the characteristic absorption curve. Note that curve (b) represents the dissipating or absorbing fraction of the impedance

$$\frac{r}{(r^2 + X_m^2)^{1/2}}$$

and that part of the displacement which lags the driving force by $90°$. The velocity associated with this component will therefore be in phase with the driving force and it is this part of the velocity which appears in the energy loss term $r\dot{x}^2$ due to the resistance of the oscillator and which gives rise to absorption.

On the other hand, curve (a) represents the reactive or energy storing fraction of the impedance

$$\frac{X_m}{(r^2 + X_m^2)^{1/2}}$$

and the reactive components in a medium determine the velocity of the waves in the medium which in turn governs the refractive index n. In fact, curve (a) is a graph of the value of n^2 in a region of anomalous dispersion where the ω axis represents the value $n = 1$. These regions occur at every resonant frequency of the constituent atoms of the medium. We shall return to this topic later in the book.

(Problems 3.8, 3.9, 3.10)

Power Supplied to Oscillator by the Driving Force

In order to maintain the steady state oscillations of the system the driving force must replace the energy lost in each cycle because of the presence of the resistance. We shall now derive the most important result that:

'in the steady state the amplitude and phase of a driven oscillator adjust themselves so that the average power supplied by the driving force just equals that being dissipated by the frictional force'.

The *instantaneous power P* supplied is equal to the product of the *instantaneous driving force* and the *instantaneous velocity*; that is,

$$P = F_0 \cos \omega t \frac{F_0}{Z_m} \cos(\omega t - \phi)$$

$$= \frac{F_0^2}{Z_m} \cos \omega t \cos(\omega t - \phi)$$

The *average power*

$$P_{av} = \frac{\text{total work per oscillation}}{\text{oscillation period}}$$

$$\therefore P_{av} = \int_0^T \frac{P \, dt}{T} \quad \text{where } T = \text{oscillation period}$$

$$= \frac{F_0^2}{Z_m T} \int_0^T \cos \omega t \cos(\omega t - \phi) \, dt$$

$$= \frac{F_0^2}{Z_m T} \int_0^T [\cos^2 \omega t \cos \phi + \cos \omega t \sin \omega t \sin \phi] \, dt$$

$$= \frac{F_0^2}{2Z_m} \cos \phi$$

because

$$\int_0^T \cos \omega t \times \sin \omega t \, dt = 0$$

and

$$\frac{1}{T} \int_0^T \cos^2 \omega t \, dt = \frac{1}{2}$$

The power supplied by the driving force is not stored in the system, but dissipated as work expended in moving the system against the frictional force $r\dot{x}$.

The rate of working (instantaneous power) by the frictional force is

$$(r\dot{x})\dot{x} = r\dot{x}^2 = r \frac{F_0^2}{Z_m^2} \cos^2(\omega t - \phi)$$

and the average value of this over one period of oscillation

$$\frac{1}{2} \frac{rF_0^2}{Z_m^2} = \frac{1}{2} \frac{F_0^2}{Z_m} \cos \phi \quad \text{for} \quad \frac{r}{Z_m} = \cos \phi$$

This proves the initial statement that the power supplied equals the power dissipated.

In an electrical circuit the power is given by $VI \cos \phi$, where V and I are the instantaneous r.m.s. values of voltage and current and $\cos \phi$ is known as the *power factor*.

$$VI \cos \phi = \frac{V^2}{Z_e} \cos \phi = \frac{V_0^2}{2Z_e} \cos \phi$$

since

$$V = \frac{V_0}{\sqrt{2}}$$

(Problem 3.11)

Variation of P_{av} with ω. Absorption Resonance Curve

Returning to the mechanical case, we see that the average power supplied

$$P_{av} = (F_0^2/2Z_m) \cos \phi$$

is a maximum when $\cos \phi = 1$; that is, when $\phi = 0$ and $\omega m - s/\omega = 0$ or $\omega_0^2 = s/m$. The force and the velocity are then in phase and Z_m has its minimum value of r. Thus

$$P_{av}(\text{maximum}) = F_0^2/2r$$

A graph of P_{av} versus ω, the frequency of the driving force, is shown in Figure 3.10. Like the curve of displacement versus ω, this graph measures the response of the oscillator; the sharpness of its peak at resonance is also determined by the value of the damping constant r, which is the only term remaining in Z_m at the resonance frequency ω_0. The peak occurs at the frequency of velocity resonance when the power absorbed by the system from the driving force is a maximum; this curve is known as the absorption curve of the oscillator (it is similar to curve (b) of Figure 3.9).

The Q-Value in Terms of the Resonance Absorption Bandwidth

In the last chapter we discussed the quality factor of an oscillator system in terms of energy decay. We may derive the same parameter in terms of the curve of Figure 3.10, where the sharpness of the resonance is precisely defined by the ratio

$$Q = \frac{\omega_0}{\omega_2 - \omega_1},$$

where ω_2 and ω_1 are those frequencies at which the power supplied

$$P_{av} = \tfrac{1}{2} P_{av}(\text{maximum})$$

The frequency difference $\omega_2 - \omega_1$ is often called the bandwidth.

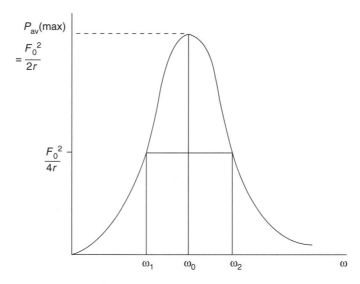

Figure 3.10 Graph of average power versus ω supplied to an oscillator by the driving force. Bandwidth $\omega_2 - \omega_1$ of resonance curve defines response in terms of the quality factor, $Q = \omega_0/(\omega_2 - \omega_1)$, where $\omega_0^2 = s/m$

Now

$$P_{av} = rF_0^2/2Z_m^2 = \tfrac{1}{2}P_{av} \text{ (maximum)} = \tfrac{1}{2}F_0^2/2r$$

when

$$Z_m^2 = 2r^2$$

that is, when

$$r^2 + X_m^2 = 2r^2 \quad \text{or} \quad X_m = \omega m - s/\omega = \pm r.$$

If $\omega_2 > \omega_1$, then

$$\omega_2 m - s/\omega_2 = +r$$

and

$$\omega_1 m - s/\omega_1 = -r$$

Eliminating s between these equations gives

$$\omega_2 - \omega_1 = r/m$$

so that

$$Q = \omega_0 m/r$$

Note that $\omega_1 = \omega_0 - r/2m$ and $\omega_2 = \omega_0 + r/2m$ are the two significant frequencies in Figure 3.9. The quality factor of an electrical circuit is given by

$$Q = \frac{\omega_0 L}{R},$$

where

$$\omega_0^2 = (LC)^{-1}$$

Note that for high values of Q, where the damping constant r is small, the frequency ω' used in the last chapter to define $Q = \omega' m/r$ moves very close to the frequency ω_0, and the two definitions of Q become equivalent to each other and to the third definition we meet in the next section.

The *Q*-Value as an Amplification Factor

We have seen that the value of the displacement at resonance is given by

$$A_{max} = \frac{F_0}{\omega' r} \quad \text{where} \quad \omega'^2 = \frac{s}{m} - \frac{r^2}{4m^2}$$

At low frequencies $(\omega \rightarrow 0)$ the displacement has a value $A_0 = F_0/s$, so that

$$\left(\frac{A_{\max}}{A_0}\right)^2 = \frac{F_0^2}{\omega'^2 r^2}\frac{s^2}{F_0^2} = \frac{m^2\omega_0^4}{r^2[\omega_0^2 - r^2/4m^2]}$$

$$= \frac{\omega_0^2 m^2}{r^2[1 - 1/4Q^2]^{1/2}} = \frac{Q^2}{[1 - 1/4Q^2]}$$

Hence:

$$\frac{A_{\max}}{A_0} = \frac{Q}{[1 - 1/4Q^2]^{1/2}} \approx Q\left[1 + \frac{1}{8Q^2}\right] \approx Q$$

for large Q.

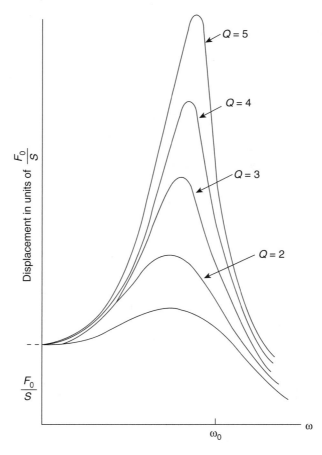

Figure 3.11   Curves of Figure 3.7 now given in terms of the quality factor Q of the system, where Q is amplification at resonance of low frequency response $x = F_0/s$

Thus, the displacement at low frequencies is amplified by a factor of Q at displacement resonance.

Figure 3.7 is now shown as Figure 3.11 where the Q-values have been attached to each curve. In tuning radio circuits, the Q-value is used as a measure of selectivity, where the sharpness of response allows a signal to be obtained free from interference from signals at nearby frequencies. In conventional radio circuits at frequencies of one megacycle,

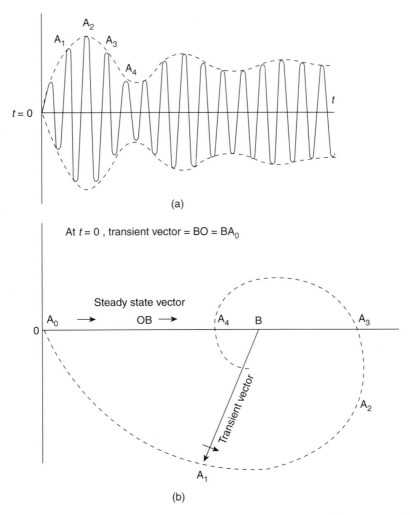

(a)

(b)

Figure 3.12 (a) The steady state oscillation (heavy curve) is modulated by the transient which decays exponentially with time. (b) In the vector diagram of (b) **OB** is the constant length steady state vector and **BA$_1$** is the transient vector. Each vector rotates anti-clockwise with its own angular velocity. At $t = 0$ the vectors **OB** and **BA$_0$** are equal and opposite on the horizontal axis and their vector sum is zero. At subsequent times the total amplitude is the length of OA$_1$ which changes as A traces a contracting spiral around B. The points A$_1$, A$_2$, A$_3$ and A$_4$ indicate how the amplitude is modified in (a)

Q-values are of the order of a few hundred; at higher radio frequencies resonant copper cavities have Q-values of about 30 000 and piezo-electric crystals can produce Q-values of 500 000. Optical absorption in crystals and nuclear magnetic resonances are often described in terms of Q-values. The Mössbauer effect in nuclear physics involves Q-values of 10^{10}.

The Effect of the Transient Term

Throughout this chapter we have considered only the steady state behaviour without accounting for the transient term mentioned on p. 58. This term makes an initial contribution to the total displacement but decays with time as $e^{-rt/2m}$. Its effect is best displayed by considering the vector sum of the transient and steady state components.

The steady state term may be represented by a vector of constant length rotating anticlockwise at the angular velocity ω of the driving force. The vector tip traces a circle. Upon this is superposed the transient term vector of diminishing length which rotates anti clockwise with angular velocity $\omega' = (s/m - r^2/4m^2)^{1/2}$. Its tip traces a contracting spiral.

The locus of the magnitude of the vector sum of these terms is the envelope of the varying amplitudes of the oscillator. This envelope modulates the steady state oscillations of frequency ω at a frequency which depends upon ω' and the relative phase between ωt and $\omega' t$.

Thus, in Figure 3.12(a) where the total oscillator displacement is zero at time $t = 0$ we have the steady state and transient vectors equal and opposite in Figure 3.12(b) but because $\omega \neq \omega'$ the relative phase between the vectors will change as the transient term decays. The vector tip of the transient term is shown as the dotted spiral and the total amplitude assumes the varying lengths OA_1, OA_2, OA_3, OA_4, etc.

(Problems 3.12, 3.13, 3.14, 3.15, 3.16, 3.17, 3.18)

Problem 3.1
Show, if $F_0 e^{i\omega t}$ represents $F_0 \sin \omega t$ in the vector form of the equation of motion for the forced oscillator that

$$x = -\frac{F_0}{\omega Z_m} \cos (\omega t - \phi)$$

and the velocity

$$v = \frac{F_0}{Z_m} \sin (\omega t - \phi)$$

Problem 3.2
The displacement of a forced oscillator is zero at time $t = 0$ and its rate of growth is governed by the rate of decay of the transient term. If this term decays to e^{-k} of its original value in a time t show that, for small damping, the average rate of growth of the oscillations is given by $x_0/t = F_0/2km\omega_0$ where x_0 is the maximum steady state displacement, F_0 is the force amplitude and $\omega_0^2 = s/m$.

Problem 3.3

The equation $m\ddot{x} + sx = F_0 \sin \omega t$ describes the motion of an undamped simple harmonic oscillator driven by a force of frequency ω. Show, by solving the equation in vector form, that the steady state solution is given by

$$x = \frac{F_0 \sin \omega t}{m(\omega_0^2 - \omega^2)} \quad \text{where} \quad \omega_0^2 = \frac{s}{m}$$

Sketch the behaviour of the amplitude of x versus ω and note that the change of sign as ω passes through ω_0 defines a phase change of π rad in the displacement. Now show that the general solution for the displacement is given by

$$x = \frac{F_0 \sin \omega t}{m(\omega_0^2 - \omega^2)} + A \cos \omega_0 t + B \sin \omega_0 t$$

where A and B are constant.

Problem 3.4

In problem 3.3, if $x = \dot{x} = 0$ at $t = 0$ show that

$$x = \frac{F_0}{m} \frac{1}{(\omega_0^2 - \omega^2)} \left(\sin \omega t - \frac{\omega}{\omega_0} \sin \omega_0 t \right)$$

and, by writing $\omega = \omega_0 + \Delta \omega$ where $\Delta \omega / \omega_0 \ll 1$ and $\Delta \omega t \ll 1$, show that near resonance,

$$x = \frac{F_0}{2m\omega_0^2} (\sin \omega_0 t - \omega_0 t \cos \omega_0 t)$$

Sketch this behaviour, noting that the second term increases with time, allowing the oscillations to grow (resonance between free and forced oscillations). Note that the condition $\Delta \omega t \ll 1$ focuses attention on the transient.

Problem 3.5

What is the general expression for the acceleration $\dot{v}$ of a simple damped mechanical oscillator driven by a force $F_0 \cos \omega t$? Derive an expression to give the frequency of maximum acceleration and show that if $r = \sqrt{sm}$, then the acceleration amplitude at the frequency of velocity resonance equals the limit of the acceleration amplitude at high frequencies.

Problem 3.6

Prove that the **exact** amplitude at the displacement resonance of a driven mechanical oscillator may be written $x = F_0 / \omega' r$ where F_0 is the driving force amplitude and

$$\omega'^2 = \frac{s}{m} - \frac{r^2}{4m^2}$$

Problem 3.7

In a forced mechanical oscillator show that the following are frequency independent (a) the displacement amplitude at low frequencies (b) the velocity amplitude at velocity resonance and (c) the acceleration amplitude at high frequencies, $(\omega \to \infty)$.

Problem 3.8
In Figure 3.9 show that for small r, the maximum value of curve (a) is $\approx F_0/2\omega_0 r$ at $\omega_1 = \omega_0 - r/2m$ and its minimum value is $\approx -F_0/2\omega_0 r$ at $\omega_2 = \omega_0 + r/2m$.

Problem 3.9
The equation $\ddot{x} + \omega_0^2 x = (-eE_0/m)\cos\omega t$ describes the motion of a bound undamped electric charge $-e$ of mass m under the influence of an alternating electric field $E = E_0 \cos\omega t$. For an electron number density n show that the induced polarizability per unit volume (the dynamic susceptibility) of a medium

$$\chi_e = -\frac{n\,ex}{\varepsilon_0 E} = \frac{n\,e^2}{\varepsilon_0 m(\omega_0^2 - \omega^2)}$$

(The permittivity of a medium is defined as $\varepsilon = \varepsilon_0(1 + \chi)$ where ε_0 is the permittivity of free space. The relative permittivity $\varepsilon_r = \varepsilon/\varepsilon_0$ is called the dielectric constant and is the square of the refractive index when E is the electric field of an electromagnetic wave.)

Problem 3.10
Repeat Problem 3.9 for the case of a damped oscillatory electron, by taking the displacement x as the component represented by curve (a) in Figure 3.9 to show that

$$\varepsilon_r = 1 + \chi = 1 + \frac{n\,e^2 m(\omega_0^2 - \omega^2)}{\varepsilon_0[m^2(\omega_0^2 - \omega^2)^2 + \omega^2 r^2]}$$

In fact, Figure 3.9(a) plots $\varepsilon_r = \varepsilon/\varepsilon_0$. Note that for

$$\omega \ll \omega_0, \quad \varepsilon_r \approx 1 + \frac{n\,e^2}{\varepsilon_0 m\omega_0^2}$$

and for

$$\omega \gg \omega_0, \quad \varepsilon_r \approx 1 - \frac{n\,e^2}{\varepsilon_0 m\omega^2}$$

Problem 3.11
Show that the energy dissipated per cycle by the frictional force $r\dot{x}$ at an angular frequency ω is given by $\pi r \omega x_{max}^2$.

Problem 3.12
Show that the bandwidth of the resonance absorption curve defines the phase angle range $\tan\phi = \pm 1$.

Problem 3.13
An alternating voltage, amplitude V_0 is applied across an *LCR* series circuit. Show that the voltage at current resonance across either the inductance or the condenser is QV_0.

Problem 3.14

Show that in a resonant *LCR* series circuit the maximum potential across the condenser occurs at a frequency $\omega = \omega_0(1 - 1/2Q_0^2)^{1/2}$ where $\omega_0^2 = (LC)^{-1}$ and $Q_0 = \omega_0 L/R$.

Problem 3.15

In Problem 3.14 show that the maximum potential across the inductance occurs at a frequency $\omega = \omega_0(1 - 1/2Q_0^2)^{-1/2}$.

Problem 3.16

Light of wavelength 0.6 μm (6000 Å) is emitted by an electron in an atom behaving as a lightly damped simple harmonic oscillator with a Q-value of 5×10^7. Show from the resonance bandwidth that the width of the spectral line from such an atom is 1.2×10^{-14} m.

Problem 3.17

If the Q-value of Problem 3.6 is high show that the width of the displacement resonance curve is approximately $\sqrt{3}r/m$ where the width is measured between those frequencies where $x = x_{\max}/2$.

Problem 3.18

Show that, in Problem 3.10, the mean rate of energy absorption per unit volume; that is, the power supplied is

$$P = \frac{n\,e^2 E_0^2}{2} \frac{\omega^2 r}{m^2(\omega_0^2 - \omega^2)^2 + \omega^2 r^2}$$

Summary of Important Results

Mechanical Impedance $\mathbf{Z}_m = \mathbf{F}/\mathbf{v}$ (force per unit velocity)

$$\mathbf{Z}_m = Z_m\,e^{i\phi} = r + i(\omega m - s/\omega)$$

where $Z_m^2 = r^2 + (\omega m - s/\omega)^2$

$$\sin\phi = \frac{\omega m - s/\omega}{Z_m}, \quad \cos\phi = \frac{r}{Z_m}, \quad \tan\phi = \frac{\omega m - s/\omega}{r}$$

ϕ is the phase angle between the force and velocity.

Forced Oscillator

Equation of motion $m\ddot{x} + r\dot{x} + sx = F_0\cos\omega t$
(Vector form) $m\ddot{\mathbf{x}} + r\dot{\mathbf{x}} + s\mathbf{x} = F_0\,e^{i\omega t}$
Use $\mathbf{x} = \mathbf{A}\,e^{i\omega t}$ to give steady state displacement

$$\mathbf{x} = -i\frac{F_0}{\omega Z_m}\,e^{i(\omega t - \phi)}$$

and velocity

$$\dot{\mathbf{x}} = \mathbf{v} = \frac{F_0}{Z_m} e^{i(\omega t - \phi)}$$

When $F_0 e^{i\omega t}$ represents $F_0 \cos \omega t$

$$x = \frac{F_0}{\omega Z_m} \sin(\omega t - \phi)$$

$$v = \frac{F_0}{Z_m} \cos(\omega t - \phi)$$

Maximum velocity $= \dfrac{F_0}{r}$ at **velocity** resonant frequency $\omega_0 = (s/m)^{1/2}$

Maximum displacement $= \dfrac{F_0}{\omega' r}$ where $\omega' = (s/m - r^2/4m^2)^{1/2}$ at **displacement** resonant frequency $\omega = (s/m - r^2/2m^2)^{1/2}$

Power Absorbed by Oscillator from Driving Force

Oscillator adjusts amplitude and phase so that power supplied equals power dissipated.

Power absorbed $= \frac{1}{2}(F_0^2/Z_m) \cos \phi$ ($\cos \phi$ is power factor)

Maximum power absorbed $= \dfrac{F_0^2}{2r}$ at ω_0

$\dfrac{\text{Maxmium power}}{2}$ absorbed $= \dfrac{F_0^2}{4r}$ at $\omega_1 = \omega_0 - \dfrac{r}{2m}$ and $\omega_2 = \omega_0 + \dfrac{r}{2m}$

Quality factor $Q = \dfrac{\omega_0 m}{r} = \dfrac{\omega_0}{\omega_2 - \omega_1}$

$$Q = \frac{\text{maximum displacement at displacement resonance}}{\text{displacement as } \omega \to 0}$$

$$= \frac{A(\max)}{F_0/s}$$

For equivalent expressions for electrical oscillators replace m by L, r by R, s by $1/C$ and F_0 by V_0 (voltage).

4

Coupled Oscillations

The preceding chapters have shown in some detail how a single vibrating system will behave. Oscillators, however, rarely exist in complete isolation; wave motion owes its existence to neighbouring vibrating systems which are able to transmit their energy to each other.

Such energy transfer takes place, in general, because two oscillators share a common component, capacitance or stiffness, inductance or mass, or resistance. Resistance coupling inevitably brings energy loss and a rapid decay in the vibration, but coupling by either of the other two parameters consumes no power, and continuous energy transfer over many oscillators is possible. This is the basis of wave motion.

We shall investigate first a mechanical example of stiffness coupling between two pendulums. Two atoms set in a crystal lattice experience a mutual coupling force and would be amenable to a similar treatment. Then we investigate an example of mass, or inductive, coupling, and finally we consider the coupled motion of an extended array of oscillators which leads us naturally into a discussion on wave motion.

Stiffness (or Capacitance) Coupled Oscillators

Figure 4.1 shows two identical pendulums, each having a mass m suspended on a light rigid rod of length l. The masses are connected by a light spring of stiffness s whose natural length equals the distance between the masses when neither is displaced from equilibrium. The small oscillations we discuss are restricted to the plane of the paper.

If x and y are the respective displacements of the masses, then the equations of motion are

$$m\ddot{x} = -mg\frac{x}{l} - s(x - y)$$

and

$$m\ddot{y} = -mg\frac{y}{l} + s(x - y)$$

The Physics of Vibrations and Waves, 6th Edition   H. J. Pain
© 2005 John Wiley & Sons, Ltd

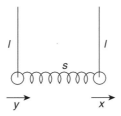

Figure 4.1 Two identical pendulums, each a light rigid rod of length l supporting a mass m and coupled by a weightless spring of stiffness s and of natural length equal to the separation of the masses at zero displacement

These represent the normal simple harmonic motion terms of each pendulum plus a coupling term $s(x - y)$ from the spring. We see that if $x > y$ the spring is extended beyond its normal length and will act against the acceleration of x but in favour of the acceleration of y.

Writing $\omega_0^2 = g/l$, where ω_0 is the natural vibration frequency of each pendulum, gives

$$\ddot{x} + \omega_0^2 x = -\frac{s}{m}(x - y) \tag{4.1}$$

$$\ddot{y} + \omega_0^2 y = -\frac{s}{m}(y - x) \tag{4.2}$$

Instead of solving these equations directly for x and y we are going to choose two new coordinates

$$X = x + y$$
$$Y = x - y$$

The importance of this approach will emerge as this chapter proceeds. Adding equations (4.1) and (4.2) gives

$$\ddot{x} + \ddot{y} + \omega_0^2(x + y) = 0$$

that is

$$\ddot{X} + \omega_0^2 X = 0$$

and subtracting (4.2) from (4.1) gives

$$\ddot{Y} + (\omega_0^2 + 2s/m)Y = 0$$

The motion of the coupled system is thus described in terms of the two coordinates X and Y, each of which has an equation of motion which is simple harmonic.

If $Y = 0$, $x = y$ at all times, so that the motion is completely described by the equation

$$\ddot{X} + \omega_0^2 X = 0$$

then the frequency of oscillation is the same as that of either pendulum in isolation and the stiffness of the coupling has no effect. This is because both pendulums are always swinging in phase (Figure 4.2a) and the light spring is always at its natural length.

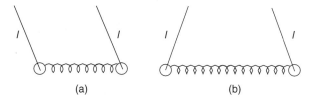

Figure 4.2 (a) The 'in phase' mode of vibration given by $\ddot{X} + \omega_0^2 X = 0$, where X is the normal coordinate $X = x + y$ and $\omega_0^2 = g/l$. (b) 'Out of phase' mode of vibration given by $\ddot{Y} + (\omega_0^2 + 2s/m)$ where Y is the normal coordinate $Y = x - y$

If $X = 0$, $x = -y$ at all times, so that the motion is completely described by

$$\ddot{Y} + (\omega_0^2 + 2s/m)Y = 0$$

The frequency of oscillation is greater because the pendulums are always out of phase (Figure 4.2b) so that the spring is either extended or compressed and the coupling is effective.

Normal Coordinates, Degrees of Freedom and Normal Modes of Vibration

The significance of choosing X and Y to describe the motion is that these parameters give a very simple illustration of normal coordinates.

- Normal coordinates are coordinates in which the equations of motion take the form of a set of linear differential equations with constant coefficients in which each equation contains *only one* dependent variable (our simple harmonic equations in X and Y).

- A vibration involving only one dependent variable X (or Y) is called a *normal mode of vibration* and has its own *normal frequency*. In such a *normal mode* all components of the system oscillate with the same *normal frequency*.

- The total energy of an undamped system may be expressed as a sum of the squares of the normal coordinates multiplied by constant coefficients and a sum of the squares of the first time derivatives of the coordinates multiplied by constant coefficients. The energy of a coupled system when the X and Y modes are both vibrating would then be expressed in terms of the squares of the velocities and displacements of X and Y.

- The importance of the normal modes of vibration is that they are entirely independent of each other. The energy associated with a normal mode is *never exchanged* with another mode; this is why we can add the energies of the separate modes to give the total energy. If only one mode vibrates the second mode of our system will always be at rest, acquiring no energy from the vibrating mode.

- Each independent way by which a system may acquire energy is called a *degree of freedom* to which is assigned its own particular normal coordinate. The number of such different ways in which the system can take up energy defines its number of degrees of freedom and its number of normal coordinates. Each harmonic oscillator has two degrees of freedom, it may take up both potential energy (normal coordinate X) and kinetic energy (normal coordinate $\dot{X}$). In our two normal modes the energies may be written

$$E_X = a\dot{X}^2 + bX^2 \tag{4.3a}$$

and

$$E_Y = c\dot{Y}^2 + dY^2 \tag{4.3b}$$

where a, b, c and d are constant.

Our system of two coupled pendulums has, then, four degrees of freedom and four normal coordinates.

Any configuration of our coupled system may be represented by the super-position of the two normal modes

$$X = x + y = X_0 \cos(\omega_1 t + \phi_1)$$

and

$$Y = x - y = Y_0 \cos(\omega_2 t + \phi_2)$$

where X_0 and Y_0 are the normal mode amplitudes, whilst $\omega_1^2 = g/l$ and $\omega_2^2 = (g/l + 2s/m)$ are the normal mode frequencies. To simplify the discussion let us choose

$$X_0 = Y_0 = 2a$$

and put

$$\phi_1 = \phi_2 = 0$$

The pendulum displacements are then given by

$$x = \tfrac{1}{2}(X + Y) = a\cos\omega_1 t + a\cos\omega_2 t$$

and

$$y = \tfrac{1}{2}(X - Y) = a\cos\omega_1 t - a\cos\omega_2 t$$

with velocities

$$\dot{x} = -a\omega_1 \sin\omega_1 t - a\omega_2 \sin\omega_2 t$$

and

$$\dot{y} = -a\omega_1 \sin\omega_1 t + a\omega_2 \sin\omega_2 t$$

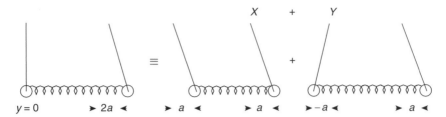

Figure 4.3 The displacement of one pendulum by an amount $2a$ is shown as the combination of the two normal coordinates $X + Y$

Now let us set the system in motion by displacing the right hand mass a distance $x = 2a$ and releasing both masses from rest so that $\dot{x} = \dot{y} = 0$ at time $t = 0$.

Figure 4.3 shows that our initial displacement $x = 2a$, $y = 0$ at $t = 0$ may be seen as a combination of the 'in phase' mode ($x = y = a$ so that $x + y = X_0 = 2a$) and of the 'out of phase' mode ($x = -y = a$ so that $Y_0 = 2a$). After release, the motion of the right hand pendulum is given by

$$x = a \cos \omega_1 t + a \cos \omega_2 t$$
$$= 2a \cos \frac{(\omega_2 - \omega_1)t}{2} \cos \frac{(\omega_1 + \omega_2)t}{2}$$

and that of the left hand pendulum is given by

$$y = a \cos \omega_1 t - a \cos \omega_2 t$$
$$= -2a \sin \frac{(\omega_1 - \omega_2)t}{2} \sin \frac{(\omega_1 + \omega_2)t}{2}$$
$$= 2a \sin \frac{(\omega_2 - \omega_1)t}{2} \sin \frac{(\omega_1 + \omega_2)t}{2}$$

If we plot the behaviour of the individual masses by showing how x and y change with time (Figure 4.4), we see that after drawing the first mass aside a distance $2a$ and releasing it x follows a consinusoidal behaviour at a frequency which is the average of the two normal mode frequencies, but its amplitude varies cosinusoidally with a low frequency which is half the difference between the normal mode frequencies. On the other hand, y, which started at zero, vibrates sinusoidally with the average frequency but its amplitude builds up to $2a$ and then decays sinusoidally at the low frequency of half the difference between the normal mode frequencies. In short, the y displacement mass acquires all the energy of the x displacement mass which is stationary when y is vibrating with amplitude $2a$, but the energy is then returned to the mass originally displaced. This *complete* energy exchange is only possible when the masses are identical and the ratio $(\omega_1 + \omega_2)/(\omega_2 - \omega_1)$ is an integer, otherwise neither will ever be quite stationary. The slow variation of amplitude at half the normal mode frequency difference is the phenomenon of 'beats' which occurs between two oscillations of nearly equal frequencies. We shall discuss this further in the section on wave groups in Chapter 5.

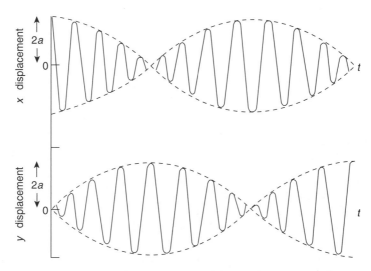

Figure 4.4 Behaviour with time of individual pendulums, showing complete energy exchange between the pendulums as *x* decreases from 2*a* to zero whilst *y* grows from zero to 2*a*

The important point to recognize, however, is that although the *individual* pendulums may exchange energy, there is *no* energy exchange between the normal modes. Figure 4.3 showed the initial configuration $x = 2a$, $y = 0$, decomposed into the X and Y modes. The higher frequency of the Y mode ensures that after a number of oscillations the Y mode will have gained half a vibration (a phase of π rad) on the X mode; this is shown in Figure 4.5. The combination of the X and Y modes then gives y the value of 2*a* and $x = 0$, and the process is repeated. When Y gains another half vibration then x equals 2*a* again. The pendulums may exchange energy; the normal modes do not.

To reinforce the importance of normal modes and their coordinates let us return to equations (4.3a) and (4.3b). If we modify our normal coordinates to read

$$X_q = \left(\frac{m}{2}\right)^{1/2}(x+y) \quad \text{and} \quad Y_q = \left(\frac{m}{2}\right)^{1/2}(x-y)$$

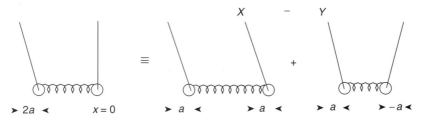

Figure 4.5 The faster vibration of the $\dot{Y}$ mode results in a phase gain of π rad over the X mode of vibration, to give $y = 2a$, which is shown here as a combination of the normal modes $X - Y$

then we find that the kinetic energy in those equations becomes

$$E_k = T = a\dot{X}^2 + c\dot{Y}^2 = \frac{1}{2}\dot{X}_q^2 + \frac{1}{2}\dot{Y}_q^2 \tag{4.4a}$$

and the potential energy

$$\begin{aligned}V = bX^2 + dY^2 &= \frac{1}{2}\left(\frac{g}{l}\right)X_q^2 + \frac{1}{2}\left(\frac{g}{l} + \frac{2s}{m}\right)Y_q^2 \\ &= \frac{1}{2}\omega_0^2 X_q^2 + \frac{1}{2}\omega_s^2 Y_q^2,\end{aligned} \tag{4.4b}$$

where $\omega_0^2 = g/l$ and $\omega_s^2 = g/l + 2s/m$.

Note that the coefficients of X_q^2 and Y_q^2 depend only on the mode frequencies and that the properties of individual parts of the system are no longer explicit.

The total energy of the system is the sum of the energies of each separate excited mode for there are no cross products $X_q Y_q$ in the energy expression of our example, i.e.,

$$E = T + V = \left(\frac{1}{2}\dot{X}_q^2 + \frac{1}{2}\omega_0^2 X_q^2\right) + \left(\frac{1}{2}\dot{Y}_q^2 + \frac{1}{2}\omega_s^2 Y_q^2\right)$$

Atoms in polyatomic molecules behave as the masses of our pendulums; the normal modes of two triatomic molecules CO_2 and H_2O are shown with their frequencies in Figure 4.6. Normal modes and their vibrations will occur frequently throughout this book.

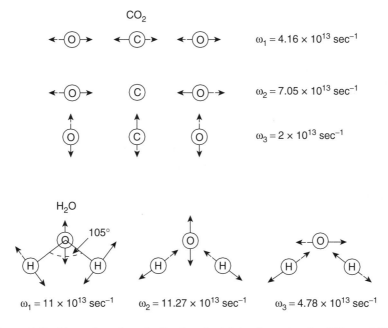

Figure 4.6   Normal modes of vibration for triatomic molecules CO_2 and H_2O

The General Method for Finding Normal Mode Frequencies, Matrices, Eigenvectors and Eigenvalues

We have just seen that when a coupled system oscillates in a *single* normal mode each component of the system will vibrate with frequency of that mode. This allows us to adopt a method which will always yield the values of the normal mode frequencies and the relative amplitudes of the individual oscillators at each frequency.

Suppose that our system of coupled pendulums in the last section oscillates in *only one* of its normal modes of frequency ω.

Then, in the equations of motion

$$m\ddot{x} + mg(x/l) + s(x - y) = 0$$

and

$$m\ddot{y} + mg(y/l) - s(x - y) = 0$$

If the pendulums start from test, we may assume the solutions

$$x = A\,\mathrm{e}^{i\omega t}$$

$$y = B\,\mathrm{e}^{i\omega t}$$

where A and B are the displacement amplitudes of x and y at the frequency ω. Using these solutions, the equations of motion become

$$\begin{aligned}
[-m\omega^2 A + (mg/l)A + s(A - B)]\,\mathrm{e}^{i\omega t} = 0 \\
[-m\omega^2 B + (mg/l)B - s(A - B)]\,\mathrm{e}^{i\omega t} = 0
\end{aligned} \tag{4.5}$$

The sum of these expressions gives

$$(A + B)(-m\omega^2 + mg/l) = 0$$

which is satisfied when $\omega^2 = g/l$, the first normal mode frequency. The difference between the expressions gives

$$(A - B)(-m\omega^2 + mg/l + 2s) = 0$$

which is satisfied when $\omega^2 = g/l + 2s/m$, the second normal mode frequency.

Inserting the value $\omega^2 = g/l$ in the pair of equations gives $A = B$ (the 'in phase' condition), whilst $\omega^2 = g/l + 2s/m$ gives $A = -B$ (the antiphase conditon).

These are the results we found in the previous section.

We may, however, by dividing through by $m\,\mathrm{e}^{i\omega t}$, rewrite equation (4.5) in matrix form as

$$\begin{bmatrix} \omega_0^2 + \omega_s^2 & -\omega_s^2 \\ -\omega_s^2 & \omega_0^2 + \omega_s^2 \end{bmatrix} \begin{bmatrix} A \\ B \end{bmatrix} = \omega^2 \begin{bmatrix} A \\ B \end{bmatrix} \tag{4.6}$$

where

$$\omega_0^2 = \frac{g}{l} \quad \text{and} \quad \omega_s^2 = \frac{s}{m}$$

This is called an *eigenvalue* equation. The value of ω^2 for which non-zero solutions exist are called the *eigenvalues* of the matrix. The column vector with components A and B is an *eigenvector* of the matrix.

Equation (4.6) may be written in the alternative form

$$\begin{bmatrix} (\omega_0^2 + \omega_s^2 - \omega^2) & -\omega_s^2 \\ -\omega_s^2 & (\omega_0^2 + \omega_s^2 - \omega^2) \end{bmatrix} \begin{bmatrix} A \\ B \end{bmatrix} = 0 \qquad (4.7)$$

and these equations have a non-zero solution if and only if the determinant of the matrix vanishes; that is, if

$$(\omega_0^2 + \omega_s^2 - \omega^2)^2 - \omega_s^4 = 0$$

or

$$(\omega_0^2 + \omega_s^2 - \omega^2) = \pm\omega_s^2$$

i.e.

$$\omega_1^2 = \omega_0^2 \quad \text{or} \quad \omega_2^2 = \omega_0^2 + 2\omega_s^2$$

as we expect.

The solution $\omega_1^2 = \omega_0^2$ in equation (4.6) yields $A = B$ as previously and $\omega_2^2 = \omega_0^2 + 2\omega_s^2$ yields $A = -B$.

Because the system started from rest we have been able to assume solutions of the simple form

$$x = A\,e^{i\omega t}$$
$$y = B\,e^{i\omega t}$$

When the pendulums have an initial velocity at $t = 0$, the boundary conditions require solutions of the form

$$x = Ae^{i(\omega t + \alpha_x)}$$
$$y = Be^{i(\omega t + \alpha_y)}$$

where each normal mode frequency ω has its own particular value of the phase constant α. The number of adjustable constants then allows the solutions to satisfy the arbitrary values of the initial displacements and velocities of both pendulums.

(Problems 4.1, 4.2, 4.3, 4.4, 4.5, 4.6, 4.7, 4.8, 4.9, 4.10, 4.11)

Mass or Inductance Coupling

In a later chapter we shall discuss the propagation of voltage and current waves along a transmission line which may be considered as a series of coupled electrical oscillators having identical values of inductance and of capacitance. For the moment we shall consider the energy transfer between two electrical circuits which are inductively coupled.

A mutual inductance (shared mass) exists between two electrical circuits when the magnetic flux from the current flowing on one circuit threads the second circuit. Any change of flux induces a voltage in both circuits.

A transformer depends upon mutual inductance for its operation. The power source is connected to the transformer primary coil of n_p turns, over which is wound in the same sense a secondary coil of n_s turns. If unit current flowing in a single turn of the primary coil produces a magnetic flux ϕ, then the flux threading each primary turn (assuming no flux leakage outside the coil) is $n_p\phi$ and the total flux threading all n_p turns of the primary is

$$L_p = n_p^2\phi$$

where L_p is the self inductance of the primary coil. If unit current in a single turn of the secondary coil produces a flux ϕ, then the flux threading each secondary turn is $n_s\phi$ and the total flux threading the secondary coil is

$$L_s = n_s^2\phi,$$

where L_s is the self inductance of the secondary coil.

If all the flux lines from unit current in the primary thread all the turns of the secondary, then the total flux lines threading the secondary defines the *mutual inductance*

$$M = n_s(n_p\phi) = \sqrt{L_pL_s}$$

In practice, because of flux leakage outside the coils, $M < \sqrt{L_pL_s}$ and the ratio

$$\frac{M}{\sqrt{L_pL_s}} = k, \text{the } coefficient \text{ of } coupling.$$

If the primary current I_p varies with $e^{i\omega t}$, a change of I_p gives an induced voltage $-L_p dI_p/\,dt = -i\omega LI_p$ in the primary and an induced voltage $-M\,dI_p/dt = -i\omega MI_p$ in the secondary.

If we consider now the two resistance-free circuits of Figure 4.7, where L_1 and L_2 are coupled by flux linkage and allowed to oscillate at some frequency ω (the voltage and current frequency of both circuits), then the voltage equations are

$$i\omega L_1 I_1 - i\frac{1}{\omega C_1} I_1 + i\omega MI_2 = 0 \qquad (4.8)$$

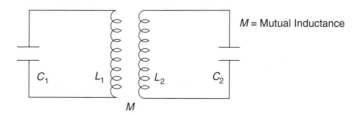

Figure 4.7   Inductively (mass) coupled *LC* circuits with mutual inductance *M*

and

$$i\omega L_2 I_2 - i\frac{1}{\omega C_2} I_2 + i\omega M I_1 = 0 \tag{4.9}$$

where M is the mutual inductance.

Multiplying (4.8) by ω/iL_1 gives

$$\omega^2 I_1 - \frac{I_1}{L_1 C_1} + \frac{M}{L_1}\omega^2 I_2 = 0$$

and multiplying (4.9) by ω/iL_2 gives

$$\omega^2 I_2 - \frac{I_2}{L_2 C_2} + \frac{M}{L_2}\omega^2 I_1 = 0,$$

where the natural frequencies of the circuit $\omega_1^2 = 1/L_1 C_1$ and $\omega_2^2 = 1/L_2 C_2$ give

$$(\omega_1^2 - \omega^2)I_1 = \frac{M}{L_1}\omega^2 I_2 \tag{4.10}$$

and

$$(\omega_2^2 - \omega^2)I_2 = \frac{M}{L_2}\omega^2 I_1 \tag{4.11}$$

The product of equations (4.10) and (4.11) gives

$$(\omega_1^2 - \omega^2)(\omega_2^2 - \omega^2) = \frac{M^2}{L_1 L_2}\omega^4 = k^2\omega^4, \tag{4.12}$$

where k is the coefficient of coupling.

Solving for ω gives the frequencies at which energy exchange between the circuits allows the circuits to resonate. If the circuits have equal natural frequencies $\omega_1 = \omega_2 = \omega_0$, say, then equation (4.12) becomes

$$(\omega_0^2 - \omega^2)^2 = k^2\omega^4$$

or

$$(\omega_0^2 - \omega^2) = \pm k\omega^2$$

that is

$$\omega = \pm\frac{\omega_0}{\sqrt{1 \pm k}}$$

The positive sign gives two frequencies

$$\omega' = \frac{\omega_0}{\sqrt{1+k}} \quad \text{and} \quad \omega'' = \frac{\omega_0}{\sqrt{1-k}}$$

at which, if we plot the current amplitude versus frequency, two maxima appear (Figure 4.8).

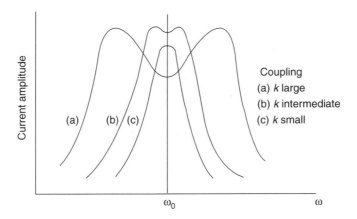

Figure 4.8 Variation of the current amplitude in each circuit near the resonant frequency. A small resistance prevents the amplitude at resonance from reaching infinite values but this has been ignored in the simple analysis. Flattening of the response curve maximum gives 'frequency band pass' coupling

In loose coupling k and M are small, and $\omega' \approx \omega'' \approx \omega_0$, so that both systems behave almost independently. In tight coupling the frequency difference $\omega'' - \omega'$ increases, the peak values of current are displaced and the dip between the peaks is more pronounced. In this simple analysis the effect of resistance has been ignored. In practice some resistance is always present to limit the amplitude maximum.

(Problems 4.12, 4.13, 4.14, 4.15, 4.16)

Coupled Oscillations of a Loaded String

As a final example involving a large number of coupled oscillators we shall consider a light string supporting n equal masses m spaced at equal distance a along its length. The string is fixed at both ends; it has a length $(n + 1)a$ and a constant tension T exists at all points and all times in the string.

Small simple harmonic oscillations of the masses are allowed in only one plane and the problem is to find the frequencies of the normal modes and the displacement of each mass in a particular normal mode.

This problem was first treated by Lagrange, its particular interest being the use it makes of normal modes and the light it throws upon the wave motion and vibration of a continuous string to which it approximates as the linear separation and the magnitude of the masses are progressively reduced.

Figure 4.9 shows the displacement y_r of the r th mass together with those of its two neighbours. The equation of motion of this mass may be written by considering the components of the tension directed towards the equilibrium position. The r th mass is pulled *downwards* towards the equilibrium position by a force $T \sin \theta_1$, due to the tension

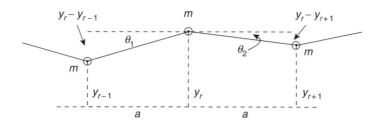

Figure 4.9 Displacements of three masses on a loaded string under tension T giving equation of motion $m\ddot{y}_r = T(y_{r+1} - 2y_r + y_{r-1})/a$

on its left and a force $T \sin \theta_2$ due to the tension on its right where

$$\sin \theta_1 = \frac{y_r - y_{r-1}}{a}$$

and

$$\sin \theta_2 = \frac{y_r - y_{r+1}}{a}$$

Hence the equation of motion is given by

$$m\frac{d^2 y_r}{dt^2} = -T (\sin \theta_1 + \sin \theta_2)$$
$$= -T \left(\frac{y_r - y_{r-1}}{a} + \frac{y_r - y_{r+1}}{a}\right)$$

so

$$\frac{d^2 y_r}{dt^2} = \ddot{y}_r = \frac{T}{ma}(y_{r-1} - 2y_r + y_{r+1}) \tag{4.13}$$

If, in a normal mode of oscillation of frequency ω, the time variation of y_r is simple harmonic about the equilibrium axis, we may write the displacement of the r th mass in this mode as

$$y_r = A_r e^{i\omega t}$$

where A_r is the maximum displacement. Similarly $y_{r+1} = A_{r+1} e^{i\omega t}$ and $y_{r-1} = A_{r-1} e^{i\omega t}$. Using these values of y in the equation of motion gives

$$-\omega^2 A_r e^{i\omega t} = \frac{T}{ma}(A_{r-1} - 2A_r + A_{r+1}) e^{i\omega t}$$

or

$$\boxed{-A_{r-1} + \left(2 - \frac{ma\omega^2}{T}\right) A_r - A_{r+1} = 0} \tag{4.14}$$

This is the fundamental equation.

The procedure now is to start with the first mass $r = 1$ and move along the string, writing out the set of similar equations as r assumes the values $r = 1, 2, 3, \ldots, n$ remembering that, because the ends are fixed

$$y_0 = A_0 = 0 \quad \text{and} \quad y_{n+1} = A_{n+1} = 0$$

Thus, when $r = 1$ the equation becomes

$$\left(2 - \frac{ma\omega^2}{T}\right) A_1 - A_2 = 0 \quad (A_0 = 0)$$

When $r = 2$ we have

$$-A_1 + \left(2 - \frac{ma\omega^2}{T}\right) A_2 - A_3 = 0$$

and when $r = n$ we have

$$-A_{n-1} + \left(2 - \frac{ma\omega^2}{T}\right) A_n = 0 \quad (A_{n+1} = 0)$$

Thus, we have a set of n equations which, when solved, will yield n different values of ω^2, each value of ω being the frequency of a normal mode, the number of normal modes being equal to the number of masses.

The formal solution of this set of n equations involves the theory of matrices. However, we may easily solve the simple cases for one or two masses on the string ($n = 1$ or 2) and, in additon, it is possible to show what the complete solution for n masses must be without using sophisticated mathematics.

First, when $n = 1$, one mass on a string of length $2a$, we need only the equation for $r = 1$ where the fixed ends of the string give $A_0 = A_2 = 0$.

Hence we have

$$\left(2 - \frac{ma\omega^2}{T}\right) A_1 = 0$$

giving

$$\omega^2 = \frac{2T}{ma}$$

a single allowed frequency of vibration (Figure 4.10a).

When $n = 2$, string length $3a$ (Figure 4.10b) we need the equations for both $r = 1$ and $r = 2$; that is

$$\left(2 - \frac{ma\omega^2}{T}\right) A_1 - A_2 = 0$$

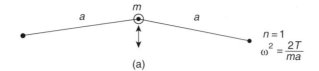

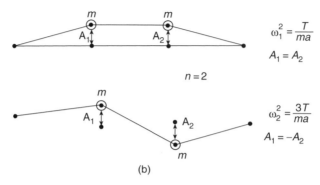

Figure 4.10 (a) Normal vibration of a single mass m on a string of length $2a$ at a frequency $\omega^2 = 2T/ma$. (b) Normal vibrations of two masses on a string of length $3a$ showing the loose coupled 'in phase' mode of frequency $\omega_1^2 = T/ma$ and the tighter coupled 'out of phase' mode of frequency $\omega_2^2 = 3T/ma$. The number of normal modes of vibration equals the number of masses

and

$$-A_1 + \left(2 - \frac{ma\omega^2}{T}\right)A_2 = 0 \quad (A_0 = A_3 = 0)$$

Eliminating A_1 or A_2 shows that these two equations may be solved (are consistent) when

$$\left(2 - \frac{ma\omega^2}{T}\right)^2 - 1 = 0$$

that is

$$\left(2 - \frac{ma\omega^2}{T} - 1\right)\left(2 - \frac{ma\omega^2}{T} + 1\right) = 0$$

Thus, there are two normal mode frequencies

$$\omega_1^2 = \frac{T}{ma} \quad \text{and} \quad \omega_2^2 = \frac{3T}{ma}$$

Using the values of ω_1 in the equations for $r = 1$ and $r = 2$ gives $A_1 = A_2$ the slow 'in phase' oscillation of Figure 4.10b, whereas ω_2 gives $A_1 = -A_2$ the faster 'anti-phase' oscillation resulting from the increased coupling.

To find the general solution for any value of n let us rewrite the equation

$$-A_{r-1} + \left(2 - \frac{ma\omega^2}{T}\right)A_r - A_{r+1} = 0$$

in the form

$$\frac{A_{r-1} + A_{r+1}}{A_r} = \frac{2\omega_0^2 - \omega^2}{\omega_0^2} \quad \text{where} \quad \omega_0^2 = \frac{T}{ma}$$

We see that for any particular *fixed* value of the normal mode frequency $\omega(\omega_j$ say) the right hand side of this equation is constant, independent of r, so the equation holds for all values of r. What values can we give to A_r which will satisfy this equation, meeting the boundary conditions $A_0 = A_{n+1} = 0$ at the end of the string?

Let us *assume* that we may express the amplitude of the rth mass at the frequency ω_j as

$$A_r = C\,e^{ir\theta}$$

where C is a constant and θ is some constant angle for a given value of ω_j. The left hand side of the equation then becomes

$$\frac{A_{r-1} + A_{r+1}}{A_r} = \frac{C(e^{i(r-1)\theta} + e^{i(r+1)\theta})}{C\,e^{ir\theta}} = (e^{-i\theta} + e^{i\theta})$$

$$= 2\cos\theta$$

which is constant and independent of r.

The value of θ_j (constant at ω_j) is easily found from the boundary conditions

$$A_0 = A_{n+1} = 0$$

which, using $\sin r\theta$ from $e^{ir\theta}$ gives

$$A_0 = C\sin r\theta = 0 \quad \text{(automatically at } r = 0)$$

and

$$A_{n+1} = C\sin(n+1)\theta = 0$$

when

$$(n+1)\,\theta_j = j\pi \quad \text{for} \quad j = 1, 2, \ldots, n$$

Hence

$$\theta_j = \frac{j\pi}{n+1}$$

and

$$A_r = C \sin r\theta_j = C \sin \frac{rj\pi}{n+1}$$

which is the amplitude of the rth mass at the fixed normal mode frequency ω_j.

To find the allowed values of ω_j we write

$$\frac{A_{r-1} + A_{r+1}}{A_r} = \frac{2\omega_0^2 - \omega_j^2}{\omega_0^2} = 2\cos\theta_j = 2\cos\frac{j\pi}{n+1}$$

giving

$$\omega_j^2 = 2\omega_0^2 \left[1 - \cos\frac{j\pi}{n+1} \right] \tag{4.15}$$

where j may take the values $j = 1, 2, \ldots, n$ and $\omega_0^2 = T/ma$.

Note that there is a maximum frequency of oscillation $\omega_j = 2\omega_0$. This is called the 'cut off' frequency and such an upper frequency limit is characteristic of all oscillating systems composed of similar elements (the masses) repeated periodically throughout the structure of the system. We shall meet this in the next chapter as a feature of wave propagation in crystals.

To summarize, we have found the normal modes of oscillation of n coupled masses on the string to have frequencies given by

$$\omega_j^2 = \frac{2T}{ma} \left[1 - \cos\frac{j\pi}{n+1} \right] \quad (j = 1, 2, 3 \ldots n)$$

At each frequency ω_j the r th mass has an amplitude

$$A_r = C \sin\frac{rj\pi}{n+1}$$

where C is a constant.

(Problems 4.17, 4.18, 4.19, 4.20, 4.21, 4.22)

The Wave Equation

Finally, in this chapter, we show how the coupled vibrations in the periodic structure of our loaded string become waves in a continuous medium.

We found the equation of motion of the r th mass to be

$$\frac{d^2 y_r}{dt^2} = \frac{T}{ma}(y_{r+1} - 2y_r + y_{r-1}) \qquad (4.13)$$

We know also that in a given normal mode all masses oscillate with the same mode frequency ω, so all y_r's have the same time dependence. However, as we see in Figure 4.10(b) where A_1 and A_2 are anti-phase, the transverse displacement y_r also depends upon the value of r; that is, the position of the r th mass on the string. In other words, y_r is a function of two independent variables, the time t and the location of r on the string.

If we use the separation $a \approx \delta x$ and let $\delta x \to 0$, the masses become closer and we can consider positions along the string in terms of a continuous variable x and any transverse displacement as $y(x, t)$, a function of both x and t.

The partial derivative notation $\partial y(x, t)/\partial t$ expresses the variation with time of $y(x, t)$ while x is kept constant.

The partial derivative $\partial y(x, t)/\partial x$ expresses the variation with x of $y(x, t)$ while the time t is kept constant. (Chapter 5 begins with an extended review of this process for students unfamiliar with this notation.)

In the same way, the second derivative $\partial^2 y(x, t)/\partial t^2$ continues to keep x constant and $\partial^2 y(x, t)/\partial x^2$ keeps t constant.

For example, if

$$y = e^{i(\omega t + kx)}$$

then

$$\frac{\partial y}{\partial t} = i\omega e^{i(\omega t + kx)} = i\omega y \quad \text{and} \quad \frac{\partial^2 y}{\partial t^2} = -\omega^2 y$$

while

$$\frac{\partial y}{\partial x} = ik e^{i(\omega t + kx)} = iky \quad \text{and} \quad \frac{\partial^2 y}{\partial x^2} = -k^2 y$$

If we now locate the transverse displacement y_r at a position $x = x_r$ along the string, then the left hand side of equation (4.13) becomes

$$\frac{\partial^2 y_r}{\partial t^2} \to \frac{\partial^2 y}{\partial t^2},$$

where y is evaluated at $x = x_r$ and now, as $a = \delta x \to 0$, we may write $x_r = x, x_{r+1} = x + \delta x$ and $x_{r-1} = x - \delta x$ with $y_r(t) \to y(x, t), y_{r+1}(t) \to y(x + \delta x, t)$ and $y_{r-1}(t) \to y(x - \delta x, t)$.

Using a Taylor series expansion to express $y(x \pm \delta x, t)$ in terms of partial derivates of y with respect to x we have

$$y(x \pm \delta x, t) = y(x) \pm \delta x \frac{\partial y}{\partial x} + \frac{1}{2}(\pm \delta x)^2 \frac{\partial^2 y}{\partial x^2}$$

and equation (4.13) becomes after substitution

$$\frac{\partial^2 y}{\partial t^2} = \frac{T}{m}\left(\frac{y_{r+1} - y_r}{a} - \frac{y_r - y_{r-1}}{a}\right)$$

$$= \frac{T}{m}\left(\frac{\delta x\frac{\partial y}{\partial x} + \frac{1}{2}(\delta x)^2\frac{\partial^2 y}{\partial x^2}}{\delta x} - \frac{\delta x\frac{\partial y}{\partial x} - \frac{1}{2}(\delta x)^2\frac{\partial^2 y}{\partial x^2}}{\delta x}\right)$$

so

$$\frac{\partial^2 y}{\partial t^2} = \frac{T}{m}\frac{(\delta x)^2}{\delta x}\frac{\partial^2 y}{\partial x^2} = \frac{T}{m}\delta x\frac{\partial^2 y}{\partial x^2}$$

If we now write $m = \rho\,\delta x$ where ρ is the linear density (mass per unit length) of the string, the masses must $\longrightarrow 0$ as $\delta x \longrightarrow 0$ to avoid infinite mass density. Thus, we have

$$\frac{\partial^2 y}{\partial t^2} = \frac{T}{\rho}\frac{\partial^2 y}{\partial x^2}$$

This is the *Wave Equation*.

T/ρ has the dimensions of the square of a velocity, the velocity with which the waves; that is, the phase of oscillation, is propagated. The solution for y at any particular point along the string is always that of a harmonic oscillation.

(Problem 4.23)

Problem 4.1
Show that the choice of new normal coordinates X_q and Y_q expresses equations (4.3a) and (4.3b) as equations (4.4a) and (4.4b).

Problem 4.2
Express the total energy of Problem 4.1 in terms of the pendulum displacements x and y as

$$E = (E_{kin} + E_{pot})_x + (E_{kin} + E_{pot})_y + (E_{pot})_{xy},$$

where the brackets give the energy of each pendulum expressed in its own coordinates and $(E_{pot})_{xy}$ is the coupling or interchange energy involving the product of these coordinates.

Problem 4.3
Figures 4.3 and 4.5 show how the pendulum configurations $x = 2a, y = 0$ and $x = 0, y = 2a$ result from the superposition of the normal modes X and Y. Using the same initial conditions

$(x = 2a, y = 0, \dot{x} = \dot{y} = 0)$ draw similar sketches to show how X and Y superpose to produce $x = -2a, y = 0$ and $x = 0, y = -2a$.

Problem 4.4
In the figure two masses m_1 and m_2 are coupled by a spring of stiffness s and natural length l. If x is the extension of the spring show that equations of motion along the x axis are

$$m_1 \ddot{x}_1 = sx$$

and

$$m_2 \ddot{x}_2 = -sx$$

and combine these to show that the system oscillates with a frequency

$$\omega^2 = \frac{s}{\mu},$$

where

$$\mu = \frac{m_1 m_2}{m_1 + m_2}$$

is called the reduced mass.

   The figure now represents a diatomic molecule as a harmonic oscillator with an effective mass equal to its reduced mass. If a sodium chloride molecule has a natural vibration frequency $= 1.14 \times 10^{13}$ Hz (in the infrared region of the electromagnetic spectrum) show that the interatomic force constant $s = 120\,\mathrm{N\,m^{-1}}$ (this simple model gives a higher value for s than more refined methods which account for other interactions within the salt crystal lattice)

Mass of Na atom $= 23$ a.m.u.
Mass of Cl atom $= 35$ a.m.u.
1 a.m.u. $= 1.67 \times 10^{-27}$ kg

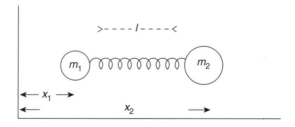

Problem 4.5
The equal masses in the figure oscillate in the vertical direction. Show that the frequencies of the normal modes of oscillation are given by

$$\omega^2 = (3 \pm \sqrt{5})\frac{s}{2m}$$

and that in the slower mode the ratio of the amplitude of the upper mass to that of the lower mass is $\frac{1}{2}(\sqrt{5} - 1)$ whilst in the faster mode this ratio is $-\frac{1}{2}(\sqrt{5} + 1)$.

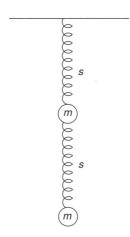

In the calculations it is not necessary to consider gravitational forces because they play no part in the forces responsible for the oscillation.

Problem 4.6

In the coupled pendulums of Figure 4.3 let us write the modulated frequency $\omega_m = (\omega_2 - \omega_1)/2$ and the average frequency $\omega_a = (\omega_2 + \omega_1)/2$ and assume that the spring is so weak that it stores a negligible amount of energy. Let the modulated amplitude

$$2a \cos \omega_m t \quad \text{or} \quad 2a \sin \omega_m t$$

be constant over one cycle at the average frequency ω_a to show that the energies of the masses may be written

$$E_x = 2ma^2\omega_a^2 \cos^2 \omega_m t$$

and

$$E_y = 2ma^2\omega_a^2 \sin^2 \omega_m t$$

Show that the total energy E remains constant and that the energy difference at any time is

$$E_x - E_y = E \cos(\omega_2 - \omega_1)t$$

Prove that

$$E_x = \frac{E}{2}[1 + \cos(\omega_2 - \omega_1)t]$$

and

$$E_y = \frac{E}{2}[1 - \cos(\omega_2 - \omega_1)t]$$

to show that the constant total energy is completely exchanged between the two pendulums at the beat frequency $(\omega_2 - \omega_1)$.

Problem 4.7

When the masses of the coupled pendulums of Figure 4.1 are no longer equal the equations of motion become

$$m_1\ddot{x} = -m_1(g/l)x - s(x - y)$$

and

$$m_2\ddot{y} = -m_2(g/l)y + s(x - y)$$

Show that we may choose the normal coordinates

$$X = \frac{m_1 x + m_2 y}{m_1 + m_2}$$

with a normal mode frequency $\omega_1^2 = g/l$ and $Y = x - y$ with a normal mode frequency $\omega_2^2 = g/l + s(1/m_1 + 1/m_2)$.

Note that X is the coordinate of the centre of mass of the system whilst the effective mass in the Y mode is the reduced mass μ of the system where $1/\mu = 1/m_1 + 1/m_2$.

Problem 4.8

Let the system of Problem 4.7 be set in motion with the initial conditions $x = A, y = 0, \dot{x} = \dot{y} = 0$ at $t = 0$. Show that the normal mode amplitudes are $X_0 = (m_1/M)A$ and $Y_0 = A$ to yield

$$x = \frac{A}{M}(m_1 \cos \omega_1 t + m_2 \cos \omega_2 t)$$

and

$$y = A\frac{m_1}{M}(\cos \omega_1 t - \cos \omega_2 t),$$

where $M = m_1 + m_2$.

Express these displacements as

$$x = 2A \cos \omega_m t \cos \omega_a t + \frac{2A}{M}(m_1 - m_2) \sin \omega_m t \sin \omega_a t$$

and

$$y = 2A\frac{m_1}{M} \sin \omega_m t \sin \omega_a t,$$

where $\omega_m = (\omega_2 - \omega_1)/2$ and $\omega_a = (\omega_1 + \omega_2)/2$.

Problem 4.9

Apply the weak coupling conditions of Problem 4.6 to the system of Problem 4.8 to show that the energies

$$E_x = \frac{E}{M^2}[m_1^2 + m_2^2 + 2m_1 m_2 \cos(\omega_2 - \omega_1)t]$$

and

$$E_y = E\left(\frac{2m_1 m_2}{M^2}\right)[1 - \cos(\omega_2 - \omega_1)t]$$

Note that E_x varies between a maximum of E (at $t = 0$) and a minimum of $[(m_1 - m_2)/M]^2 E$, whilst E_y oscillates between a minimum of zero at $t = 0$ and a maximum of $4(m_1 m_2/M^2)E$ at the beat frequency of $(\omega_2 - \omega_1)$.

Problem 4.10

In the figure below the right hand pendulum of the coupled system is driven by the horizontal force $F_0 \cos \omega t$ as shown. If a small damping constant r is included the equations of motion may be written

$$m\ddot{x} = -\frac{mg}{l}x - r\dot{x} - s(x - y) + F_0 \cos \omega t$$

and

$$m\ddot{y} = -\frac{mg}{l}y - r\dot{y} + s(x - y)$$

Show that the equations of motion for the normal coordinates $X = x + y$ and $Y = x - y$ are those for damped oscillators driven by a force $F_0 \cos \omega t$.

Solve these equations for X and Y and, by neglecting the effect of r, show that

$$x \approx \frac{F_0}{2m} \cos \omega t \left[\frac{1}{\omega_1^2 - \omega^2} + \frac{1}{\omega_2^2 - \omega^2} \right]$$

and

$$y \approx \frac{F_0}{2m} \cos \omega t \left[\frac{1}{\omega_1^2 - \omega^2} - \frac{1}{\omega_2^2 - \omega^2} \right]$$

where

$$\omega_1^2 = \frac{g}{l} \quad \text{and} \quad \omega_2^2 = \frac{g}{l} + \frac{2s}{m}$$

Show that

$$\frac{y}{x} \approx \frac{\omega_2^2 - \omega_1^2}{\omega_2^2 + \omega_1^2 - 2\omega^2}$$

and sketch the behaviour of the oscillator with frequency to show that outside the frequency range $\omega_2 - \omega_1$ the motion of y is attenuated.

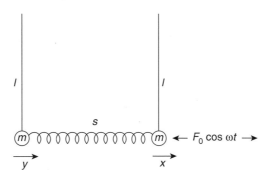

Problem 4.11

The diagram shows an oscillatory force $F_o \cos \omega t$ acting on a mass M which is part of a simple harmonic system of stiffness k and is connected to a mass m by a spring of stiffness s. If all oscillations are along the x axis show that the condition for M to remain stationary is $\omega^2 = s/m$. (This is a simple version of small mass loading in engineering to quench undesirable oscillations.)

$$F_0 \cos \omega t$$

Problem 4.12

The figure below shows two identical LC circuits coupled by a common capacitance C with the directions of current flow indicated by arrows. The voltage equations are

$$V_1 - V_2 = L\frac{dI_a}{dt}$$

and

$$V_2 - V_3 = L\frac{dI_b}{dt}$$

whilst the currents are given by

$$\frac{dq_1}{dt} = -I_a \quad \frac{dq_2}{dt} = I_a - I_b$$

and

$$\frac{dq_3}{dt} = I_b$$

Solve the voltage equations for the normal coordinates $(I_a + I_b)$ and $(I_a - I_b)$ to show that the normal modes of oscillation are given by

$$I_a = I_b \quad \text{at} \quad \omega_1^2 = \frac{1}{LC}$$

and

$$I_a = -I_b \quad \text{at} \quad \omega_2^2 = \frac{3}{LC}$$

Note that when $I_a = I_b$ the coupling capacitance may be removed and $q_1 = -q_2$. When $I_a = -I_b$, $q_2 = -2q_1 = -2q_3$.

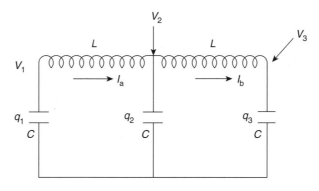

Problem 4.13

A generator of e.m.f. E is coupled to a load Z by means of an ideal transformer. From the diagram, Kirchhoff's Law gives

$$E = -e_1 = i\omega L_p I_1 - i\omega M I_2$$

and

$$I_2 Z_2 = e_2 = i\omega M I_1 - i\omega L_s I_2.$$

Show that E/I_1, the impedance of the whole system seen by the generator, is the sum of the primary impedance and a 'reflected impedance' from the secondary circuit of $\omega^2 M^2 / Z_s$ where $Z_s = Z_2 + i\omega L_s$.

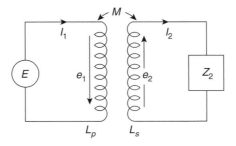

Problem 4.14

Show, for the perfect transformer of Problem 4.13, that the impedance seen by the generator consists of the primary impedance in parallel with an impedance $(n_p/n_s)^2 Z_2$, where n_p and n_s are the number of primary and secondary transformer coil turns respectively.

Problem 4.15

If the generator delivers maximum power when its load equals its own internal impedance show how an ideal transformer may be used as a device to match a load to a generator, e.g. a loudspeaker of a few ohms impedance to an amplifier output of $10^3 \, \Omega$ impedance.

Problem 4.16

The two circuits in the diagram are coupled by a variable mutual inductance M and Kirchhoff's Law gives

$$Z_1 I_1 + Z_M I_2 = E$$

and

$$Z_M I_1 + Z_2 I_2 = 0,$$

where

$$Z_M = +i\omega M$$

M is varied at a resonant frequency where the reactance $X_1 = X_2 = 0$ to give a maximum value of I_2. Show that the condition for this maximum is $\omega M = \sqrt{R_1 R_2}$ and that this defines a 'critical coefficient of coupling' $k = (Q_1 Q_2)^{-1/2}$, where the Q's are the quality factors of the circuits.

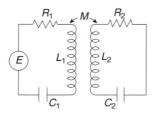

Problem 4.17

Consider the case when the number of masses on the loaded string of this chapter is $n = 3$. Use equation (4.15) to show that the normal mode frequencies are given by

$$\omega_1^2 = (2 - \sqrt{2})\omega_0^2; \qquad \omega_2^2 = 2\omega_0^2$$

and

$$\omega_3^2 = (2 + \sqrt{2})\omega_0^2$$

Repeat the problem using equation (4.14) (with $\omega_0^2 = T/ma$) in the matrix method of equation (4.7), where the eigenvector components are A_{r-1}, A_r and A_{r+1}.

Problem 4.18

Show that the relative displacements of the masses in the modes of Problem 4.17 are $1 : \sqrt{2} : 1$, $1 : 0 : -1$, and $1 : -\sqrt{2} : 1$. Show by sketching these relative displacements that tighter coupling increases the mode frequency.

Problem 4.19

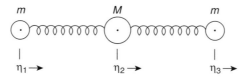

The figure represents a triatomic molecule with a heavy atom mass M bound to equal atoms of smaller mass m on either side. The binding is represented by springs of stiffness s and in equilibrium the atom centres are equally spaced along a straight line. Simple harmonic vibrations are considered only along this linear axis and are given by

$$\eta_J = \eta_J^0 e^{i\omega t}$$

where η_J is the displacement from equilibrium of the jth atom.

Set up the equation of motion for each atom and use the matrix method of equation (4.7) to show that the normal modes have frequencies

$$\omega_1^2 = 0, \omega_2^2 = \frac{s}{m} \quad \text{and} \quad \omega_3^2 = \frac{s(M + 2m)}{mM}$$

Describe the motion of the atoms in each normal mode.

Problem 4.20

Taking the maximum value of

$$\omega_j^2 = \frac{2T}{ma}\left(1 - \cos\frac{j\pi}{n+1}\right)$$

at $j = n$ as that produced by the strongest coupling, deduce the relative displacements of neighbouring masses and confirm your deduction by inserting your values in consecutive difference equations relating the displacements y_{r+1}, y_r and y_{r-1}. Why is your solution unlikely to satisfy the displacements of those masses near the ends of the string?

Problem 4.21

Expand the value of

$$\omega_j^2 = \frac{2T}{ma}\left(1 - \cos\frac{j\pi}{n+1}\right)$$

when $j \ll n$ in powers of $(j/n+1)$ to show that in the limit of very large values of n, a low frequency

$$\omega_J = \frac{j\pi}{l}\sqrt{\frac{T}{\rho}},$$

where $\rho = m/a$ and $l = (n+1)a$.

Problem 4.22

An electrical transmission line consists of equal inductances L and capacitances C arranged as shown. Using the equations

$$\frac{L\,dI_{r-1}}{dt} = V_{r-1} - V_r = \frac{q_{r-1} - q_r}{C}$$

and

$$I_{r-1} - I_r = \frac{dq_r}{dt},$$

show that an expression for I_r may be derived which is equivalent to that for y_r in the case of the mass-loaded string. (This acts as a low pass electric filter and has a cut-off frequency as in the case of the string. This cut-off frequency is a characteristic of wave propagation in periodic structures and electromagnetic wave guides.)

Problem 4.23

Show that

$$y = e^{i\omega t}\,e^{ikx}$$

satisfies the wave equation

$$\frac{\partial^2 y}{\partial t^2} = c^2\frac{\partial^2 y}{\partial x^2}, \quad \text{if} \quad \omega = ck$$

Summary of Important Results

In coupled systems each normal coordinate defines a degree of freedom, each degree of freedom defines a way in which a system may take up energy. The total energy of the system is the sum of the energies in its normal modes of oscillation because these remain separate and distinct, and energy is never exchanged between them.

A simple harmonic oscillator has two normal coordinates [velocity (or momentum) and displacement] and therefore two degrees of freedom, the first connected with kinetic energy, the second with potential energy.

n Equal Masses, Separation a, Coupled on a String under Constant Tension T

Equation of motion of the rth mass is

$$m\ddot{y}_r = (T/a)(y_{r-1} - 2y_r + y_{r+1})$$

which for $y_r = A_r\,e^{i\omega t}$ gives

$$-A_{r+1} + \left(\frac{2 - ma\omega^2}{T}\right)A_r - A_{r-1} = 0$$

There are n normal modes with frequencies ω_J given by

$$\omega_j^2 = \frac{2T}{ma}\left(1 - \cos\frac{j\pi}{n+1}\right)$$

In a normal mode of frequency ω_J the rth mass has an amplitude

$$A_r = C\sin\frac{rj\pi}{n+1}$$

where C is a constant.

Wave Equation

In the limit, as separation $a = \delta x \to 0$ equation of motion of the rth mass on a loaded string $m\ddot{y}_r = (T/a)(y_{r-1} - 2y_r + y_{r+1})$ becomes the wave equation

$$\frac{\partial^2 y}{\partial t^2} = \frac{T}{\rho}\frac{\partial^2 y}{\partial x^2} = c^2\frac{\partial^2 y}{\partial x^2}$$

where ρ is mass per unit length and c is the wave velocity.

5

Transverse Wave Motion

Partial Differentiation

From this chapter onwards we shall often need to use the notation of partial differentiation.

When we are dealing with a function of only one variable, $y = f(x)$ say, we write the differential coefficient

$$\frac{dy}{dx} = \lim_{\delta x \to 0} \frac{f(x + \delta x) - f(x)}{\delta x}$$

but if we consider a function of two or more variables, the value of this function will vary with a change in any or all of the variables. For instance, the value of the co-ordinate z on the surface of a sphere whose equation is $x^2 + y^2 + z^2 = a^2$, where a is the radius of the sphere, will depend on x and y so that z is a function of x and y written $z = z(x, y)$. The differential change of z which follows from a change of x and y may be written

$$dz = \left(\frac{\partial z}{\partial x} \right)_y dx + \left(\frac{\partial z}{\partial y} \right)_x dy$$

where $(\partial z / \partial x)_y$ means differentiating z with respect to x whilst y is kept constant, so that

$$\left(\frac{\partial z}{\partial x} \right)_y = \lim_{\delta x \to 0} \frac{z(x + \delta x, y) - z(x, y)}{\delta x}$$

The total change dz is found by adding the separate increments due to the change of each variable in turn whilst the others are kept constant. In Figure 5.1 we can see that keeping y constant isolates a plane which cuts the spherical surface in a curved line, and the incremental contribution to dz along this line is exactly as though z were a function of x only. Now by keeping x constant we turn the plane through $90°$ and repeat the process with y as a variable so that the total increment of dz is the sum of these two processes.

If only two independent variables are involved, the subscript showing which variable is kept constant is omitted without ambiguity.

The Physics of Vibrations and Waves, 6th Edition   H. J. Pain
© 2005 John Wiley & Sons, Ltd

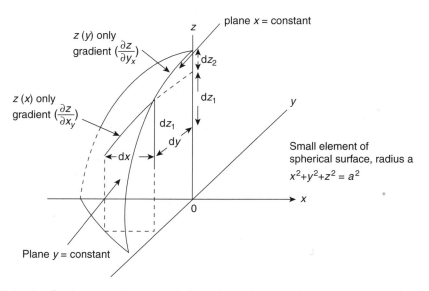

Figure 5.1 Small element of a Spherical Surface showing $dz = dz_1 + dz_2 = (\partial z/\partial x)_y \, dx + (\partial z/\partial y)_x \, dy$ where each gradient is calculated with one variable remaining constant

In wave motion our functions will be those of variables of distance and time, and we shall write $\partial/\partial x$ and $\partial^2/\partial x^2$ for the first or second derivatives with respect to x, whilst the time t remains constant. Again, $\partial/\partial t$ and $\partial^2/\partial t^2$ will denote first and second derivatives with respect to time, implying that x is kept constant.

Waves

One of the simplest ways to demonstrate wave motion is to take the loose end of a long rope which is fixed at the other end and to move the loose end quickly up and down. Crests and troughs of the waves move down the rope, and if the rope were infinitely long such waves would be called *progressive waves*–these are waves travelling in an unbounded medium free from possible reflection (Figure 5.2).

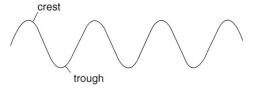

Progressive waves on infinitely long string

Figure 5.2 Progressive transverse waves moving along a string

If the medium is limited in extent; for example, if the rope were reduced to a violin string, fixed at both ends, the progressive waves travelling on the string would be reflected at both ends; the vibration of the string would then be the combination of such waves moving to and fro along the string and *standing waves* would be formed.

Waves on strings are *transverse waves* where the displacements or oscillations in the medium are transverse to the direction of wave propagation. When the oscillations are parallel to the direction of wave propagation the waves are *longitudinal*. Sound waves are longitudinal waves; a gas can sustain only longitudinal waves because transverse waves require a shear force to maintain them. Both transverse and longitudinal waves can travel in a solid.

In this book we are going to discuss *plane waves* only. When we see wave motion as a series of crests and troughs we are in fact observing the vibrational motion of the individual oscillators in the medium, and in particular all of those oscillators in a plane of the medium which, at the instant of observation, have the same phase in their vibrations.

If we take a plane perpendicular to the direction of wave propagation and all oscillators lying within that plane have a common phase, we shall observe with time how that plane of common phase progresses through the medium. Over such a plane, all parameters describing the wave motion remain constant. The crests and troughs are planes of maximum amplitude of oscillation which are π rad out of phase; a crest is a plane of maximum positive amplitude, while a trough is a plane of maximum negative amplitude. In formulating such wave motion in mathematical terms we shall have to relate the phase difference between any two planes to their physical separation in space. We have, in principle, already done this in our discussion on oscillators.

Spherical waves are waves in which the surfaces of common phase are spheres and the source of waves is a central point, e.g. an explosion; each spherical surface defines a set of oscillators over which the radiating disturbance has imposed a common phase in vibration. In practice, spherical waves become plane waves after travelling a very short distance. A small section of a spherical surface is a very close approximation to a plane.

Velocities in Wave Motion

At the outset we must be very clear about one point. The individual oscillators which make up the medium *do not* progress through the medium with the waves. Their motion is simple harmonic, limited to oscillations, transverse or longitudinal, about their equilibrium positions. It is their phase relationships we observe as waves, not their progressive motion through the medium.

There are three velocities in wave motion which are quite distinct although they are connected mathematically. They are

1. *The particle velocity*, which is the simple harmonic velocity of the oscillator about its equilibrium position.

2. *The wave or phase velocity*, the velocity with which planes of equal phase, crests or troughs, progress through the medium.

3. *The group velocity*. A number of waves of different frequencies, wavelengths and velocities may be superposed to form a group. Waves rarely occur as single

monochromatic components; a white light pulse consists of an infinitely fine spectrum of frequencies and the motion of such a pulse would be described by its group velocity. Such a group would, of course, 'disperse' with time because the wave velocity of each component would be different in all media except free space. Only in free space would it remain as white light. We shall discuss group velocity as a separate topic in a later section of this chapter. Its importance is that it is the velocity with which the energy in the wave group is transmitted. For a monochromatic wave the group velocity and the wave velocity are identical. Here we shall concentrate on particle and wave velocities.

The Wave Equation

This equation will dominate the rest of this text and we shall derive it, first of all, by considering the motion of transverse waves on a string.

We shall consider the vertical displacement y of a very short section of a uniform string. This section will perform vertical simple harmonic motions; it is our simple oscillator. The displacement y will, of course, vary with the time and also with x, the position along the string at which we choose to observe the oscillation.

The wave equation therefore will relate the displacement y of a single oscillator to distance x and time t. We shall consider oscillations only in the plane of the paper, so that our transverse waves on the string are plane polarized.

The mass of the uniform string per unit length or its linear density is ρ, and a constant tension T exists throughout the string although it is slightly extensible.

This requires us to consider such a short length and such small oscillations that we may linearize our equations. The effect of gravity is neglected.

Thus in Figure 5.3 the forces acting on the curved element of length ds are T at an angle θ to the axis at one end of the element, and T at an angle $\theta + \mathrm{d}\theta$ at the other end. The length of the curved element is

$$\mathrm{d}s = \left[1 + \left(\frac{\partial y}{\partial x}\right)^2\right]^{1/2} \mathrm{d}x$$

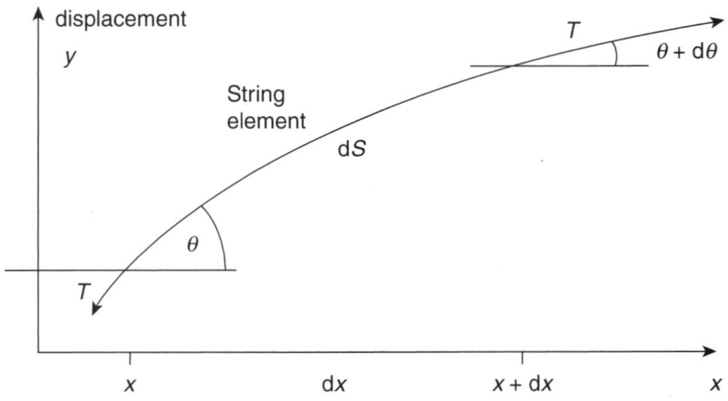

Figure 5.3 Displaced element of string of length d$s \approx$ dx with tension T acting at an angle θ at x and at $\theta + \mathrm{d}\theta$ at $x + \mathrm{d}x$

but within the limitations imposed $\partial y / \partial x$ is so small that we ignore its square and take $ds = dx$. The mass of the element of string is therefore $\rho ds = \rho dx$. Its equation of motion is found from Newton's Law, force equals mass times acceleration.

The perpendicular force on the element dx is $T \sin(\theta + d\theta) - T \sin \theta$ in the positive y direction, which equals the product of ρdx (mass) and $\partial^2 y / \partial t^2$ (acceleration).

Since θ is very small $\sin \theta \approx \tan \theta = \partial y / \partial x$, so that the force is given by

$$T \left[\left(\frac{\partial y}{\partial x} \right)_{x+dx} - \left(\frac{\partial y}{\partial x} \right)_{x} \right]$$

where the subscripts refer to the point at which the partial derivative is evaluated. The difference between the two terms in the bracket defines the differential coefficient of the partial derivative $\partial y / \partial x$ times the space interval dx, so that the force is

$$T \frac{\partial^2 y}{\partial x^2} \, dx$$

The equation of motion of the small element dx then becomes

$$T \frac{\partial^2 y}{\partial x^2} \, dx = \rho \, dx \, \frac{\partial^2 y}{\partial t^2}$$

or

$$\frac{\partial^2 y}{\partial x^2} = \frac{\rho}{T} \frac{\partial^2 y}{\partial t^2}$$

giving

$$\frac{\partial^2 y}{\partial x^2} = \frac{1}{c^2} \frac{\partial^2 y}{\partial t^2}$$

where T/ρ has the dimensions of a velocity squared, so c in the preceding equation is a velocity. THIS IS THE WAVE EQUATION.

It relates the acceleration of a simple harmonic oscillator in a medium to the second derivative of its displacement with respect to its position, x, in the medium. The position of the term c^2 in the equation is always shown by a rapid dimensional analysis.

So far we have not explicitly stated which velocity c represents. We shall see that it is the wave or phase velocity, the velocity with which planes of common phase are propagated. In the string the velocity arises as the ratio of the tension to the inertial density of the string. We shall see, whatever the waves, that the wave velocity can always be expressed as a function of the elasticity or potential energy storing mechanism in the medium and the inertia of the medium through which its kinetic or inductive energy is stored. For longitudinal waves in a solid the elasticity is measured by Young's modulus, in a gas by γP, where γ is the specific heat ratio and P is the gas pressure.

Solution of the Wave Equation

The solution of the wave equation

$$\frac{\partial^2 y}{\partial x^2} = \frac{1}{c^2}\frac{\partial^2 y}{\partial t^2}$$

will, of course, be a function of the variables x and t. We are going to show that any function of the form $y = f_1(ct - x)$ is a solution. Moreover, any function $y = f_2(ct + x)$ will be a solution so that, generally, their superposition $y = f_1(ct - x) + f_2(ct + x)$ is the complete solution.

If f_1' represents the differentiation of the function with respect to the bracket $(ct - x)$, then using the chain rule which also applies to *partial* differentiation

$$\frac{\partial y}{\partial x} = -f_1'(ct - x)$$

and

$$\frac{\partial^2 y}{\partial x^2} = f_1''(ct - x)$$

also

$$\frac{\partial y}{\partial t} = cf_1'(ct - x)$$

and

$$\frac{\partial^2 y}{\partial t^2} = c^2 f_1''(ct - x)$$

so that

$$\frac{\partial^2 y}{\partial x^2} = \frac{1}{c^2}\frac{\partial^2 y}{\partial t^2}$$

for $y = f_1(ct - x)$. When $y = f_2(ct + x)$ a similar result holds.

(Problems 5.1, 5.2)

If y is the simple harmonic displacement of an oscillator at position x and time t we would expect, from Chapter 1, to be able to express it in the form $y = a \sin (\omega t - \phi)$, and in fact all of the waves we discuss in this book will be described by sine or cosine functions.

The bracket $(ct - x)$ in the expression $y = f(ct - x)$ has the dimensions of a length and, for the function to be a sine or cosine, its argument must have the dimensions of radians so that $(ct - x)$ must be multiplied by a factor $2\pi/\lambda$, where λ is a length to be defined.

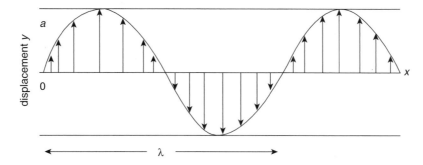

Figure 5.4 Locus of oscillator displacements in a continuous medium as a wave passes over them travelling in the positive x-direction. The wavelength λ is defined as the distance between any two oscillators having a phase difference of 2π rad

We can now write

$$y = a \sin(\omega t - \phi) = a \sin \frac{2\pi}{\lambda}(ct - x)$$

as a solution to the wave equation if $2\pi c/\lambda = \omega = 2\pi\nu$, where ν is the oscillation frequency and $\phi = 2\pi x/\lambda$.

This means that if a wave, moving to the right, passes over the oscillators in a medium and a photograph is taken at time $t = 0$, the locus of the oscillator displacements (Figure 5.4) will be given by the expression $y = a \sin(\omega t - \phi) = a \sin 2\pi(ct - x)/\lambda$. If we now observe the motion of the oscillator at the position $x = 0$ it will be given by $y = a \sin \omega t$.

Any oscillator to its right at some position x will be set in motion at some later time by the wave moving to the right; this motion will be given by

$$y = a \sin(\omega t - \phi) = a \sin \frac{2\pi}{\lambda}(ct - x)$$

having a phase lag of ϕ with respect to the oscillator at $x = 0$. This phase lag $\phi = 2\pi x/\lambda$, so that if $x = \lambda$ the phase lag is 2π rad that is, equivalent to exactly one complete vibration of an oscillator.

This defines λ as the *wavelength*, the separation in space between any two oscillators with a phase difference of 2π rad. The expression $2\pi c/\lambda = \omega = 2\pi\nu$ gives $c = \nu\lambda$, where c, the wave or phase velocity, is the product of the frequency and the wavelength. Thus, $\lambda/c = 1/\nu = \tau$, the period of oscillation, showing that the wave travels one wavelength in this time. An observer at any point would be passed by ν wavelengths per second, a distance per unit time equal to the velocity c of the wave.

If the wave is moving to the left the sign of ϕ is changed because the oscillation at x begins before that at $x = 0$. Thus, the bracket

$$(ct - x) \text{ denotes a wave moving to the right}$$

and

$(ct + x)$ gives a wave moving in the direction of negative x.

There are several equivalent expressions for $y = f(ct - x)$ which we list here as sine functions, although cosine functions are equally valid.

They are:

$$y = a \sin \frac{2\pi}{\lambda}(ct - x)$$

$$y = a \sin 2\pi \left(\nu t - \frac{x}{\lambda}\right)$$

$$y = a \sin \omega \left(t - \frac{x}{c}\right)$$

$$y = a \sin(\omega t - kx)$$

where $k = 2\pi/\lambda = \omega/c$ is called the *wave number*; also $y = a\,e^{i(\omega t - kx)}$, the exponential representation of both sine and cosine.

Each of the expressions above is a solution to the wave equation giving the displacement of an oscillator and its phase with respect to some reference oscillator. The changes of the displacements of the oscillators and the propagation of their phases are what we observe as wave motion.

The wave or phase velocity is, of course, $\partial x/\partial t$, the rate at which the disturbance moves across the oscillators; the oscillator or particle velocity is the simple harmonic velocity $\partial y/\partial t$.

Choosing any one of the expressions above for a right-going wave, e.g.

$$y = a \sin(\omega t - kx)$$

we have

$$\frac{\partial y}{\partial t} = \omega a \cos(\omega t - kx)$$

and

$$\frac{\partial y}{\partial x} = -ka \cos(\omega t - kx)$$

so that

$$\frac{\partial y}{\partial t} = -\frac{\omega}{k}\frac{\partial y}{\partial x} = -c\frac{\partial y}{\partial x} \left(= -\frac{\partial x}{\partial t}\frac{\partial y}{\partial x}\right)$$

The particle velocity $\partial y/\partial t$ is therefore given as the product of the wave velocity

$$c = \frac{\partial x}{\partial t}$$

and the gradient of the wave profile preceded by a negative sign for a right-going wave

$$y = f(ct - x)$$

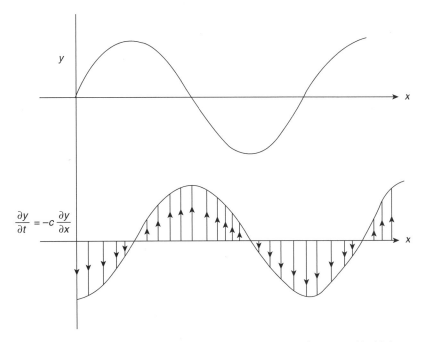

Figure 5.5 The magnitude and direction of the particle velocity $\partial y/\partial t = -c(\partial y/\partial x)$ at any point x is shown by an arrow in the right-going sine wave above

In Figure 5.5 the arrows show the direction of the particle velocity at various points of the right-going wave. It is evident that the particle velocity increases in the same direction as the transverse force in the wave and we shall see in the next section that this force is given by

$$-T\partial y/\partial x$$

where T is the tension in the string.

(Problem 5.3)

Characteristic Impedance of a String (the string as a forced oscillator)

Any medium through which waves propagate will present an impedance to those waves. If the medium is lossless, and possesses no resistive or dissipation mechanism, this impedance will be determined by the two energy storing parameters, inertia and elasticity, and it will be real. The presence of a loss mechanism will introduce a complex term into the impedance.

A string presents such an impedance to progressive waves and this is defined, because of the nature of the waves, as the transverse impedance

$$Z = \frac{\text{transverse force}}{\text{transverse velocity}} = \frac{F}{v}$$

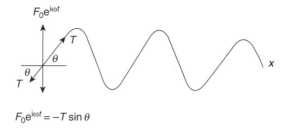

$$F_0 e^{i\omega t} = -T \sin\theta$$

Figure 5.6   The string as a forced oscillator with a vertical force $F_0\,e^{i\omega t}$ driving it at one end

The following analysis will emphasize the dual role of the string as a medium and as a forced oscillator.

In Figure 5.6 we consider progressive waves on the string which are generated at one end by an oscillating force, $F_0\,e^{i\omega t}$, which is restricted to the direction transverse to the string and operates only in the plane of the paper. The tension in the string has a constant value, T, and at the end of the string the balance of forces shows that the applied force is equal and opposite to $T \sin\theta$ at all time, so that

$$F_0\,e^{i\omega t} = -T \sin\theta \approx -T \tan\theta = -T \left(\frac{\partial y}{\partial x}\right)$$

where θ is small.

The displacement of the progressive waves may be represented exponentially by

$$\mathbf{y} = \mathbf{A}\,e^{i(\omega t - kx)}$$

where the amplitude $\mathbf{A}$ may be complex because of its phase relation with F. At the end of the string, where $x = 0$,

$$F_0\,e^{i\omega t} = -T\left(\frac{\partial y}{\partial x}\right)_{x=0} = ikT\mathbf{A}\,e^{i(\omega t - k\cdot 0)}$$

giving

$$\mathbf{A} = \frac{F_0}{ikT} = \frac{F_0}{i\omega}\left(\frac{c}{T}\right)$$

and

$$\mathbf{y} = \frac{F_0}{i\omega}\left(\frac{c}{T}\right)e^{i(\omega t - kx)}$$

(since $c = \omega/k$).

The transverse velocity

$$\mathbf{v} = \dot{\mathbf{y}} = F_0 \left(\frac{c}{T}\right) e^{i(\omega t - kx)}$$

where the velocity amplitude $v = F_0/Z$, gives a transverse impedance

$$Z = \frac{T}{c} = \rho c \ \text{(since } T = \rho c^2)$$

or *Characteristic Impedance* of the string.

Since the velocity c is determined by the inertia and the elasticity, the impedance is also governed by these properties.

(We can see that the amplitude of displacement $y = F_0/\omega Z$, with the phase relationship $-i$ with respect to the force, is in complete accord with our discussion in Chapter 3.)

Reflection and Transmission of Waves on a String at a Boundary

We have seen that a string presents a characteristic impedance ρc to waves travelling along it, and we ask how the waves will respond to a sudden change of impedance; that is, of the value ρc. We shall ask this question of all the waves we discuss, acoustic waves, voltage and current waves and electromagnetic waves, and we shall find a remarkably consistent pattern in their behaviour.

We suppose that a string consists of two sections smoothly joined at a point $x = 0$ with a constant tension T along the whole string. The two sections have different linear densities ρ_1 and ρ_2, and therefore different wave velocities $T/\rho_1 = c_1^2$ and $T/\rho_2 = c_2^2$. The specific impedances are $\rho_1 c_1$ and $\rho_2 c_2$, respectively.

An incident wave travelling along the string meets the discontinuity in impedance at the position $x = 0$ in Figure 5.7. At this position, $x = 0$, a part of the incident wave will be reflected and part of it will be transmitted into the region of impedance $\rho_2 c_2$.

We shall denote the impedance $\rho_1 c_1$ by Z_1 and the impedance $\rho_2 c_2$ by Z_2. We write the displacement of the incident wave as $y_i = A_1 e^{i(\omega t - kx)}$, a wave of real (not complex)

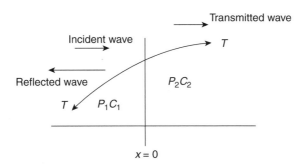

Figure 5.7 Waves on a string of impedance $\rho_1 c_1$ reflected and transmitted at the boundary $x = 0$ where the string changes to impedance $\rho_2 c_2$

amplitude A_1 travelling in the positive x-direction with velocity c_1. The displacement of the reflected wave is $y_r = B_1 e^{i(\omega t + k_1 x)}$, of amplitude B_1 and travelling in the negative x-direction with velocity c_1.

The transmitted wave displacement is given by $y_t = A_2 e^{i(\omega t - k_2 x)}$, of amplitude A_2 and travelling in the positive x-direction with velocity c_2.

We wish to find the reflection and transmission amplitude coefficients; that is, the relative values of B_1 and A_2 with respect to A_1. We find these via two boundary conditions which must be satisfied at the impedance discontinuity at $x = 0$.

The boundary conditions which apply at $x = 0$ are:

1. A geometrical condition that the displacement is the same immediately to the left and right of $x = 0$ for all time, so that there is no discontinuity of displacement.

2. A dynamical condition that there is a continuity of the transverse force $T(\partial y/\partial x)$ at $x = 0$, and therefore a continuous slope. This must hold, otherwise a finite difference in the force acts on an infinitesimally small mass of the string giving an infinite acceleration; this is not permitted.

Condition (1) at $x = 0$ gives

$$y_i + y_r = y_t$$

or

$$A_1 e^{i(\omega t - k_1 x)} + B_1 e^{i(\omega t + k_1 x)} = A_2 e^{i(\omega t - k_2 x)}$$

At $x = 0$ we may cancel the exponential terms giving

$$A_1 + B_1 = A_2 \tag{5.1}$$

Condition (2) gives

$$T \frac{\partial}{\partial x}(y_i + y_r) = T \frac{\partial}{\partial x} y_t$$

at $x = 0$ for all t, so that

$$-k_1 T A_1 + k_1 T B_1 = -k_2 T A_2$$

or

$$-\omega \frac{T}{c_1} A_1 + \omega \frac{T}{c_1} B_1 = -\omega \frac{T}{c_2} A_2$$

after cancelling exponentials at $x = 0$. But $T/c_1 = \rho_1 c_1 = Z_1$ and $T/c_2 = \rho_2 c_2 = Z_2$, so that

$$Z_1(A_1 - B_1) = Z_2 A_2 \tag{5.2}$$

Equations (5.1) and (5.2) give the

$$\text{Reflection coefficient of amplitude, } \frac{B_1}{A_1} = \frac{Z_1 - Z_2}{Z_1 + Z_2}$$

and the

$$\text{Transmission coefficient of amplitude, } \frac{A_2}{A_1} = \frac{2Z_1}{Z_1 + Z_2}$$

We see immediately that these coefficients are independent of ω and hold for waves of all frequencies; they are real and therefore free from phase changes other than that of π rad which will change the sign of a term. Moreover, these ratios depend entirely upon the ratios of the impedances. (See summary on p. 546). If $Z_2 = \infty$, this is equivalent to $x = 0$ being a fixed end to the string because no transmitted wave exists. This gives $B_1/A_1 = -1$, so that the incident wave is completely reflected (as we expect) with a phase change of π (phase reversal)—conditions we shall find to be necessary for standing waves to exist. A group of waves having many component frequencies will retain its shape upon reflection at $Z_2 = \infty$, but will suffer reversal (Figure 5.8). If $Z_2 = 0$, so that $x = 0$ is a free end of the string, then $B_1/A_1 = 1$ and $A_2/A_1 = 2$. This explains the 'flick' at the end of a whip or free ended string when a wave reaches it.

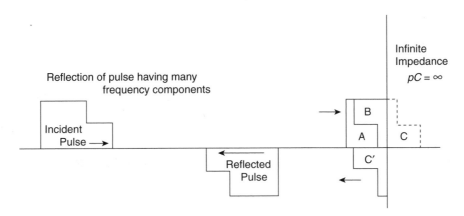

Figure 5.8 A pulse of arbitrary shape is reflected at an infinite impedance with a phase change of π rad, so that the reflected pulse is the inverted and reversed shape of the initial waveform. The pulse at reflection is divided in the figure into three sections A, B, and C. At the moment of observation section C has already been reflected and suffered inversion and reversal to become C'. The actual shape of the pulse observed at this instant is A being A + B − C' where B = C'. The displacement at the point of reflection must be zero.

(Problems 5.4, 5.5, 5.6)

Reflection and Transmission of Energy

Our interest in waves, however, is chiefly concerned with their function of transferring energy throughout a medium, and we shall now consider what happens to the energy in a wave when it meets a boundary between two media of different impedance values.

If we consider each unit length, mass ρ, of the string as a simple harmonic oscillator of maximum amplitude A, we know that its total energy will be $E = \frac{1}{2}\rho\omega^2 A^2$, where ω is the wave frequency.

The wave is travelling at a velocity c so that as each unit length of string takes up its oscillation with the passage of the wave the rate at which energy is being carried along the string is

$$(\text{energy} \times \text{velocity}) = \tfrac{1}{2}\rho\omega^2 A^2 c$$

Thus, the rate of energy arriving at the boundary $x = 0$ is the energy arriving with the incident wave; that is

$$\tfrac{1}{2}\rho_1 c_1 \omega^2 A_1^2 = \tfrac{1}{2}Z_1\omega^2 A_1^2$$

The rate at which energy leaves the boundary, via the reflected and transmitted waves, is

$$\tfrac{1}{2}\rho_1 c_1 \omega^2 B_1^2 + \tfrac{1}{2}\rho_2 c_2 \omega^2 A_2^2 = \tfrac{1}{2}Z_1\omega^2 B_1^2 + \tfrac{1}{2}Z_2\omega^2 A_2^2$$

which, from the ratio B_1/A_1 and A_2/A_1,

$$= \tfrac{1}{2}\omega^2 A_1^2 \frac{Z_1(Z_1 - Z_2)^2 + 4Z_1^2 Z_2}{(Z_1 + Z_2)^2} = \tfrac{1}{2}Z_1\omega^2 A_1^2$$

Thus, energy is conserved, and all energy arriving at the boundary in the incident wave leaves the boundary in the reflected and transmitted waves.

The Reflected and Transmitted Intensity Coefficients

These are given by

$$\frac{\text{Reflected Energy}}{\text{Incident Energy}} = \frac{Z_1 B_1^2}{Z_1 A_1^2} = \left(\frac{B_1}{A_1}\right)^2 = \left(\frac{Z_1 - Z_2}{Z_1 + Z_2}\right)^2$$

$$\frac{\text{Transmitted Energy}}{\text{Incident Energy}} = \frac{Z_2 A_2^2}{Z_1 A_1^2} = \frac{4Z_1 Z_2}{(Z_1 + Z_2)^2}$$

We see that if $Z_1 = Z_2$ no energy is reflected and the *impedances are said to be matched*.

(Problems 5.7, 5.8)

The Matching of Impedances

Impedance matching represents a very important practical problem in the transfer of energy. Long distance cables carrying energy must be accurately matched at all joints to avoid wastage from energy reflection. The power transfer from any generator is a maximum when the load matches the generator impedance. A loudspeaker is matched to the impedance of the power output of an amplifier by choosing the correct turns ratio on the coupling transformer. This last example, the insertion of a coupling element between two mismatched impedances, is of fundamental importance with applications in many branches of engineering physics and optics. We shall illustrate it using waves on a string, but the results will be valid for all wave systems.

We have seen that when a smooth joint exists between two strings of different impedances, energy will be reflected at the boundary. We are now going to see that the insertion of a particular length of another string between these two mismatched strings will allow us to eliminate energy reflection and match the impedances.

In Figure 5.9 we require to match the impedances $Z_1 = \rho_1 c_1$ and $Z_3 = \rho_3 c_3$ by the smooth insertion of a string of length l and impedance $Z_2 = \rho_2 c_2$. Our problem is to find the values of l and Z_2.

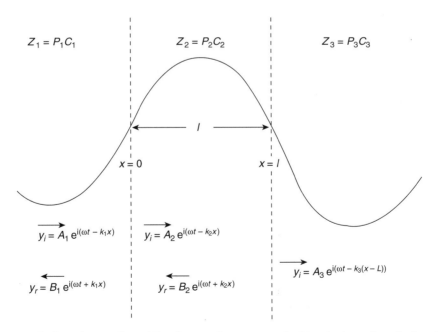

Figure 5.9 The impedances Z_1 and Z_3 of two strings are matched by the insertion of a length l of a string of impedance Z_2. The incident and reflected waves are shown for the boundaries $x = 0$ and $x = l$. The impedances are matched when $Z_2^2 = Z_1 Z_3$ and $l = \lambda/4$ in Z_2, results which are true for waves in all media

The incident, reflected and transmitted displacements at the junctions $x = 0$ and $x = l$ are shown in Figure 5.9 and we seek to make the ratio

$$\frac{\text{Transmitted energy}}{\text{Incident energy}} = \frac{Z_3 A_3^2}{Z_1 A_1^2}$$

equal to unity.

The boundary conditions are that y and $T(\partial y/\partial x)$ are continuous across the junctions $x = 0$ and $x = l$.

Between Z_1 and Z_2 the continuity of y gives

$$A_1 e^{i(\omega t - k_1 x)} + B_1 e^{i(\omega t + k_1 x)} = A_2 e^{i(\omega t - k_2 x)} + B_2 e^{i(\omega t + k_2 x)}$$

or

$$A_1 + B_1 = A_2 + B_2 \; (\text{at } x = 0) \tag{5.3}$$

Similarly the continuity of $T(\partial y/\partial x)$ at $x = 0$ gives

$$T(-ik_1 A_1 + ik_1 B_1) = T(-ik_2 A_2 + ik_2 B_2)$$

Dividing this equation by ω and remembering that $T(k/\omega) = T/c = \rho c = Z$ we have

$$Z_1(A_1 - B_1) = Z_2(A_2 - B_2) \tag{5.4}$$

Similarly at $x = l$, the continuity of y gives

$$A_2 e^{-ik_2 l} + B_2 e^{ik_2 l} = A_3 \tag{5.5}$$

and the continuity of $T(\partial y/\partial x)$ gives

$$Z_2(A_2 e^{-ik_2 l} - B_2 e^{ik_2 l}) = Z_3 A_3 \tag{5.6}$$

From the four boundary equations (5.3), (5.4), (5.5) and (5.6) we require the ratio A_3/A_1. We use equations (5.3) and (5.4) to eliminate B_1 and obtain A_1 in terms of A_2 and B_2. We then use equations (5.5) and (5.6) to obtain both A_2 and B_2 in terms of A_3. Equations (5.3) and (5.4) give

$$Z_1(A_1 - A_2 - B_2 + A_1) = Z_2(A_2 - B_2)$$

or

$$A_1 = \frac{A_2(r_{12} + 1) + B_2(r_{12} - 1)}{2r_{12}} \tag{5.7}$$

where

$$r_{12} = \frac{Z_1}{Z_2}$$

Equations (5.5) and (5.6) give

$$A_2 = \frac{r_{23}+1}{2r_{23}} A_3 e^{ik_2 l} \tag{5.8}$$

and

$$B_2 = \frac{r_{23}-1}{2r_{23}} A_3 e^{-ik_2 l}$$

where

$$r_{23} = \frac{Z_2}{Z_3}$$

Equations (5.7) and (5.8) give

$$\begin{aligned}
A_1 &= \frac{A_3}{4r_{12}r_{23}}[(r_{12}+1)(r_{23}+1)e^{ik_2 l} + (r_{12}-1)(r_{23}-1)e^{-ik_2 l}] \\
&= \frac{A_3}{4r_{13}}[(r_{13}+1)(e^{ik_2 l} + e^{-ik_2 l}) + (r_{12}+r_{23})(e^{ik_2 l} - e^{-ik_2 l})] \\
&= \frac{A_3}{2r_{13}}[(r_{13}+1)\cos k_2 l + i(r_{12}+r_{23})\sin k_2 l]
\end{aligned}$$

where

$$r_{12}r_{23} = \frac{Z_1}{Z_2}\frac{Z_2}{Z_3} = \frac{Z_1}{Z_3} = r_{13}$$

Hence

$$\left(\frac{A_3}{A_1}\right)^2 = \frac{4r_{13}^2}{(r_{13}+1)^2 \cos^2 k_2 l + (r_{12}+r_{23})^2 \sin^2 k_2 l}$$

or

$$\begin{aligned}
\frac{\text{transmitted energy}}{\text{incident energy}} &= \frac{Z_3 A_3^2}{Z_1 A_1^2} = \frac{1}{r_{13}}\frac{A_3^2}{A_1^2} \\
&= \frac{4r_{13}}{(r_{13}+1)^2 \cos^2 k_2 l + (r_{12}+r_{23})^2 \sin^2 k_2 l}
\end{aligned}$$

If we choose $l = \lambda_2/4$, $\cos k_2 l = 0$ and $\sin k_2 l = 1$ we have

$$\frac{Z_3 A_3^2}{Z_1 A_1^2} = \frac{4r_{13}}{(r_{12}+r_{23})^2} = 1$$

when

$$r_{12} = r_{23}$$

that is, when

$$\frac{Z_1}{Z_2} = \frac{Z_2}{Z_3} \quad \text{or} \quad Z_2 = \sqrt{Z_1 Z_3}$$

We see, therefore, that if the impedance of the coupling medium is the harmonic mean of the two impedances to be matched and the thickness of the coupling medium is

$$\frac{\lambda_2}{4} \quad \text{where} \quad \lambda_2 = \frac{2\pi}{k_2}$$

all the energy at frequency ω will be transmitted with zero reflection.

The thickness of the dielectric coating of optical lenses which eliminates reflections as light passes from air into glass is one quarter of a wavelength. The 'bloomed' appearance arises because exact matching occurs at only one frequency. Transmission lines are matched to loads by inserting quarter wavelength stubs of lines with the appropriate impedance.

(Problems 5.9, 5.10)

Standing Waves on a String of Fixed Length

We have already seen that a progressive wave is completely reflected at an infinite impedance with a π phase change in amplitude. A string of fixed length l with both ends rigidly clamped presents an infinite impedance at each end; we now investigate the behaviour of waves on such a string. Let us consider the simplest case of a monochromatic wave of one frequency ω with an amplitude a travelling in the positive x-direction and an amplitude b travelling in the negative x-direction. The displacement on the string at any point would then be given by

$$y = a\,e^{i(\omega t - kx)} + b\,e^{i(\omega t + kx)}$$

with the boundary condition that $y = 0$ at $x = 0$ and $x = l$ at all times.

The condition $y = 0$ at $x = 0$ gives $0 = (a + b)\,e^{i\omega t}$ for all t, so that $a = -b$. This expresses physically the fact that a wave in either direction meeting the infinite impedance at either end is completely reflected with a π phase change in amplitude. This is a general result for all wave shapes and frequencies.

Thus

$$y = a\,e^{i\omega t}(e^{-ikx} - e^{ikx}) = (-2i)a\,e^{i\omega t} \sin kx \tag{5.9}$$

an expression for y which satisfies the *standing wave time independent form* of the wave equation

$$\partial^2 y / \partial x^2 + k^2 y = 0$$

because $(1/c^2)(\partial^2 y / \partial t^2) = (-\omega^2/c^2)y = -k^2 y$. The condition that $y = 0$ at $x = l$ for all t requires

$$\sin kl = \sin \frac{\omega l}{c} = 0 \quad \text{or} \quad \frac{\omega l}{c} = n\pi$$

limiting the values of allowed frequencies to

$$\omega_n = \frac{n\pi c}{l}$$

or

$$\nu_n = \frac{nc}{2l} = \frac{c}{\lambda_n}$$

that is

$$l = \frac{n\lambda_n}{2}$$

giving

$$\sin \frac{\omega_n x}{c} = \sin \frac{n\pi x}{l}$$

These frequencies are the *normal frequencies or modes of vibration* we first met in Chapter 4. They are often called *eigenfrequencies*, particularly in wave mechanics.

Such allowed frequencies define the length of the string as an exact number of half wavelengths, and Figure 5.10 shows the string displacement for the first four *harmonics* ($n = 1, 2, 3, 4$). The value for $n = 1$ is called the *fundamental*.

As with the loaded string of Chapter 4, all normal modes may be present at the same time and the general displacement is the superposition of the displacements at each frequency. This is a more complicated problem which we discuss in Chapter 10 (Fourier Methods).

For the moment we see that for each single harmonic $n > 1$ there will be a number of positions along the string which are always at rest. These points occur where

$$\sin \frac{\omega_n x}{c} = \sin \frac{n\pi x}{l} = 0$$

or

$$\frac{n\pi x}{l} = r\pi \quad (r = 0, 1, 2, 3, \dots n)$$

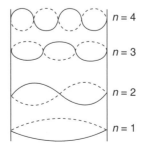

Figure 5.10 The first four harmonics, $n = 1, 2, 3, 4$ of the standing waves allowed between the two fixed ends of a string

The values $r=0$ and $r=n$ give $x=0$ and $x=l$, the ends of the string, but between the ends there are $n-1$ positions equally spaced along the string in the nth harmonic where the displacement is always zero. These positions are called *nodes* or *nodal points*, being the positions of zero motion in a system of *standing waves*. Standing waves arise when a single mode is excited and the incident and reflected waves are superposed. If the amplitudes of these progressive waves are equal and opposite (resulting from complete reflection), nodal points will exist. Often however, the reflection is not quite complete and the waves in the opposite direction do not cancel each other to give complete nodal points. In this case we speak of a *standing wave ratio* which we shall discuss in the next section but one.

Whenever nodal points exist, however, we know that the waves travelling in opposite directions are exactly equal in all respects so that the energy carried in one direction is exactly equal to that carried in the other. This means that the total energy flux; that is, the energy carried across unit area per second in a standing wave system, is zero.

Returning to equation (5.9), we see that the complete expression for the displacement of the nth harmonic is given by

$$y_n = 2a(-i)(\cos \omega_n t + i \sin \omega_n t) \sin \frac{\omega_n x}{c}$$

We can express this in the form

$$y_n = (A_n \cos \omega_n t + B_n \sin \omega_n t) \sin \frac{\omega_n x}{c} \tag{5.10}$$

where the amplitude of the nth mode is given by $(A_n^2 + B_n^2)^{1/2} = 2a$.

(Problem 5.11)

Energy of a Vibrating String

A vibrating string possesses both kinetic and potential energy. The kinetic energy of an element of length dx and linear density ρ is given by $\frac{1}{2}\rho \, dx \, \dot{y}^2$; the total kinetic energy is the integral of this along the length of the string.

Thus

$$E_{\text{kin}} = \frac{1}{2} \int_0^l \rho \dot{y}^2 \, dx$$

The potential energy is the work done by the tension T in extending an element dx to a new length ds when the string is vibrating.

Thus

$$E_{\text{pot}} = \int T(ds - dx) = \int T \left\{ \left[1 + \left(\frac{\partial y}{\partial x} \right)^2 \right]^{1/2} - 1 \right\} dx$$

$$= \frac{1}{2} T \int \left(\frac{\partial y}{\partial x} \right)^2 dx$$

if we neglect higher powers of $\partial y / \partial x$.

Now the change in the length of the element dx is $\frac{1}{2}(\partial y/\partial x)^2\,dx$, and if the string is elastic the change in tension is proportional to the change in length so that, provided $(\partial y/\partial x)$ in the wave is of the first order of small quantities, the change in tension is of the second order and T may be considered constant.

Energy in Each Normal Mode of a Vibrating String

The total displacement y in the string is the superposition of the displacements y_n of the individual harmonics and we can find the energy in each harmonic by replacing y_n for y in the results of the last section. Thus, the kinetic energy in the nth harmonic is

$$E_n(\text{kinetic}) = \frac{1}{2}\int_0^l \rho\dot{y}_n^2\,dx$$

and the potential energy is

$$E_n(\text{potential}) = \frac{1}{2}T\int_0^l \left(\frac{\partial y_n}{\partial x}\right)^2 dx$$

Since we have already shown for standing waves that

$$y_n = (A_n\cos\omega_n t + B_n\sin\omega_n t)\sin\frac{\omega_n x}{c}$$

then

$$\dot{y}_n = (-A_n\omega_n\sin\omega_n t + B_n\omega_n\cos\omega_n t)\sin\frac{\omega_n x}{c}$$

and

$$\frac{\partial y_n}{\partial x} = \frac{\omega_n}{c}(A_n\cos\omega_n t + B_n\sin\omega_n t)\cos\frac{\omega_n x}{c}$$

Thus

$$E_n(\text{kinetic}) = \frac{1}{2}\rho\omega_n^2[-A_n\sin\omega_n t + B_n\cos\omega_n t]^2\int_0^l \sin^2\frac{\omega_n x}{c}\,dx$$

and

$$E_n(\text{potential}) = \frac{1}{2}T\frac{\omega_n^2}{c^2}[A_n\cos\omega_n t + B_n\sin\omega_n t]^2\int_0^l \cos^2\frac{\omega_n x}{c}\,dx$$

Remembering that $T = \rho c^2$ we have

$$E_n(\text{kinetic} + \text{potential}) = \frac{1}{4}\rho l\omega_n^2(A_n^2 + B_n^2)$$
$$= \frac{1}{4}m\omega_n^2(A_n^2 + B_n^2)$$

where m is the mass of the string and $(A_n^2 + B_n^2)$ is the square of the maximum displacement (amplitude) of the mode. To find the exact value of the total energy E_n of the mode we would need to know the precise value of A_n and B_n and we shall evaluate these in Chapter 10 on Fourier Methods. The total energy of the vibrating string is, of course, the sum of all the E_n's of the normal modes.

(Problem 5.12)

Standing Wave Ratio

When a wave is completely reflected the superposition of the incident and reflected amplitudes will give nodal points (zero amplitude) where the incident and reflected amplitudes cancel each other, and points of maximum displacement equal to twice the incident amplitude where they reinforce.

If a progressive wave system is partially reflected from a boundary let the amplitude reflection coefficient B_1/A_1 of the earlier section be written as r, where $r < 1$.

The maximum amplitude at reinforcement is then $A_1 + B_1$; the minimum amplitude is given by $A_1 - B_1$. In this case the ratio of maximum to minimum amplitudes in the standing wave system is called the

$$\text{Standing Wave Ratio} = \frac{A_1 + B_1}{A_1 - B_1} = \frac{1 + r}{1 - r}$$

where $r = B_1/A_1$.

Measuring the values of the maximum and minimum amplitudes gives the value of the reflection coefficient for

$$r = B_1/A_1 = \frac{\text{SWR} - 1}{\text{SWR} + 1}$$

where SWR refers to the Standing Wave Ratio.

(Problem 5.13)

Wave Groups and Group Velocity

Our discussion so far has been limited to monochromatic waves—waves of a single frequency and wavelength. It is much more common for waves to occur as a mixture of a number or group of component frequencies; white light, for instance, is composed of a continuous visible wavelength spectrum extending from about 3000 Å in the blue to 7000 Å in the red. Examining the behaviour of such a group leads to the third kind of velocity mentioned at the beginning of this chapter; that is, the group velocity.

Superposition of Two Waves of Almost Equal Frequencies

We begin by considering a group which consists of two components of equal amplitude a but frequencies ω_1 and ω_2 which differ by a small amount.

Their separate displacements are given by

$$y_1 = a \cos (\omega_1 t - k_1 x)$$

and

$$y_2 = a \cos (\omega_2 t - k_2 x)$$

Superposition of amplitude and phase gives

$$y = y_1 + y_2 = 2a \cos \left[\frac{(\omega_1 - \omega_2)t}{2} - \frac{(k_1 - k_2)x}{2} \right] \cos \left[\frac{(\omega_1 + \omega_2)t}{2} - \frac{(k_1 + k_2)x}{2} \right]$$

a wave system with a frequency $(\omega_1 + \omega_2)/2$ which is very close to the frequency of either component but with a maximum amplitude of $2a$, modulated in space and time by a very slowly varying envelope of frequency $(\omega_1 - \omega_2)/2$ and wave number $(k_1 - k_2)/2$.

This system is shown in Figure 5.11 and shows, of course a behaviour similar to that of the equivalent coupled oscillators in Chapter 4. The velocity of the new wave is $(\omega_1 - \omega_2)/(k_1 - k_2)$ which, if the phase velocities $\omega_1/k_1 = \omega_2/k_2 = c$, gives

$$\frac{\omega_1 - \omega_2}{k_1 - k_2} = c \frac{(k_1 - k_2)}{k_1 - k_2} = c$$

so that the component frequencies and their superposition, or *group* will travel with the same velocity, the profile of their combination in Figure 5.11 remaining constant.

If the waves are sound waves the intensity is a maximum whenever the amplitude is a maximum of $2a$; this occurs twice for every period of the modulating frequency; that is, at a frequency $\nu_1 - \nu_2$.

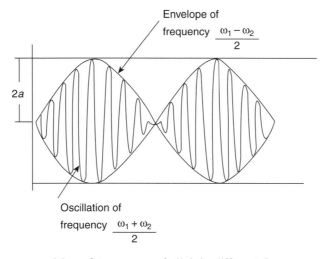

Figure 5.11 The superposition of two waves of slightly different frequency ω_1 and ω_2 forms a group. The faster oscillation occurs at the average frequency of the two components $(\omega_1 + \omega_2)/2$ and the slowly varying group envelope has a frequency $(\omega_1 - \omega_2)/2$, half the frequency difference between the components

The *beats* of maximum intensity fluctuations thus have a frequency equal to the difference $\nu_1 - \nu_2$ of the components. In the example here where the components have equal amplitudes a, superposition will produce an amplitude which varies between $2a$ and 0; this is called complete or 100% modulation.

More generally an amplitude modulated wave may be represented by

$$y = A \cos(\omega t - kx)$$

where the modulated amplitude

$$A = a + b \cos \omega' t$$

This gives

$$y = a \cos(\omega t - kx) + \frac{b}{2}\{[\cos(\omega + \omega')t - kx] + [\cos(\omega - \omega')t - kx]\}$$

so that here amplitude modulation has introduced two new frequencies $\omega \pm \omega'$, known as combination tones or sidebands. Amplitude modulation of a carrier frequency is a common form of radio transmission, but its generation of sidebands has led to the crowding of radio frequencies and interference between stations.

Wave Groups and Group Velocity

Suppose now that the two frequency components of the last section have different phase velocities so that $\omega_1/k_1 \neq \omega_2/k_2$. The velocity of the maximum amplitude of the group; that is, the *group velocity*

$$\frac{\omega_1 - \omega_2}{k_1 - k_2} = \frac{\Delta \omega}{\Delta k}$$

is now different from each of these velocities; the superposition of the two waves will no longer remain constant and the group profile will change with time.

A medium in which the phase velocity is frequency dependent (ω/k not constant) is known as a dispersive medium and a *dispersion relation* expresses the variation of ω as a function of k. If a group contains a number of components of frequencies which are nearly equal the original expression for the group velocity is written

$$\frac{\Delta \omega}{\Delta k} = \frac{d\omega}{dk}$$

The group velocity is that of the maximum amplitude of the group so that it is the velocity with which the energy in the group is transmitted. Since $\omega = kv$, where v is the phase velocity, the group velocity

$$v_g = \frac{d\omega}{dk} = \frac{d}{dk}(kv) = v + k\frac{dv}{dk}$$

$$= v - \lambda \frac{dv}{d\lambda}$$

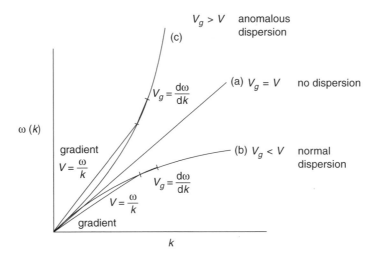

Figure 5.12 Curves illustrating dispersion relations: (a) a straight line representing a non-dispersive medium, $v = v_g$; (b) a normal dispersion relation where the gradient $v = \omega/k > v_g = d\omega/dk$; (c) an anomalous dispersion relation where $v < v_g$

where $k = 2\pi/\lambda$. Usually $dv/d\lambda$ is positive, so that $v_g < v$. This is called *normal dispersion*, but *anomalous dispersion* can arise when $dv/d\lambda$ is negative, so that $v_g > v$.

We shall see when we discuss electromagnetic waves that an electrical conductor is anomalously dispersive to these waves whilst a dielectric is normally dispersive except at the natural resonant frequencies of its atoms. In the chapter on forced oscillations we saw that the wave then acted as a driving force upon the atomic oscillators and that strong absorption of the wave energy was represented by the dissipation fraction of the oscillator impedance, whilst the anomalous dispersion curve followed the value of the reactive part of the impedance.

The three curves of Figure 5.12 represent

- A non-dispersive medium where ω/k is constant, so that $v_g = v$, for instance free space behaviour towards light waves.

- A normal dispersion relation $v_g < v$.

- An anomalous dispersion relation $v_g > v$.

Example. The electric vector of an electromagnetic wave propagates in a dielectric with a velocity $v = (\mu\varepsilon)^{-1/2}$ where μ is the permeability and ε is the permittivity. In free space the velocity is that of light, $c = (\mu_0\varepsilon_0)^{-1/2}$. The refractive index $n = c/v = \sqrt{\mu\varepsilon/\mu_0\varepsilon_0} = \sqrt{\mu_r\varepsilon_r}$ where $\mu_r = \mu/\mu_0$ and $\varepsilon_r = \varepsilon/\varepsilon_0$. For many substances μ_r is constant and ~ 1, but ε_r is frequency dependent, so that v depends on λ.

The group velocity

$$v_g = v - \lambda\, dv/d\lambda = v\left(1 + \frac{\lambda}{2\varepsilon_r}\frac{\partial\varepsilon_r}{\partial\lambda}\right)$$

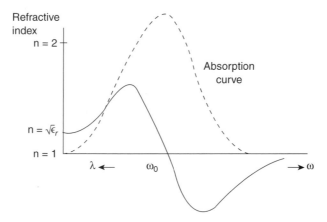

Figure 5.13 Anomalous dispersion showing the behaviour of the refractive index $n = \sqrt{\varepsilon_r}$ versus ω and λ, where ω_0 is a resonant frequency of the atoms of the medium. The absorption in such a region is also shown by the dotted line

so that $v_g > v$ (anomalous dispersion) when $\partial \varepsilon_r / \partial \lambda$ is $+ve$. Figure 5.13 shows the behaviour of the refractive index $n = \sqrt{\varepsilon_r}$ versus ω, the frequency, and λ, the wavelength, in the region of anomalous dispersion associated with a resonant frequency. The dotted curve shows the energy absorption (compare this with Figure 3.9).

(Problems 5.14, 5.15, 5.16, 5.17, 5.18, 5.19)

Wave Group of Many Components. The Bandwidth Theorem

We have so far considered wave groups having only two frequency components. We may easily extend this to the case of a group of many frequency components, each of amplitude a, lying within the narrow frequency range $\Delta \omega$.

We have already covered the essential physics of this problem on p. 20, where we found the sum of the series

$$R = \sum_{0}^{n-1} a \cos (\omega t + n\delta)$$

where δ was the constant phase difference between successive components. Here we are concerned with the constant phase difference $(\delta \omega)t$ which results from a constant frequency difference $\delta \omega$ between successive components. The spectrum or range of frequencies of this group is shown in Figure 5.14a and we wish to follow its behaviour with time.

We seek the amplitude which results from the superposition of the frequency components and write it

$$R = a \cos \omega_1 t + a \cos (\omega_1 + \delta \omega)t + a \cos (\omega_1 + 2\delta \omega)t + \cdots$$
$$+ a \cos [\omega_1 + (n-1)(\delta \omega)]t$$

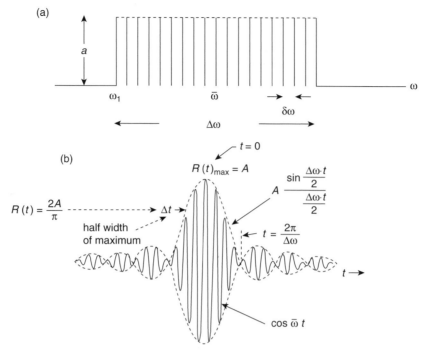

Figure 5.14 A rectangular wave band of width $\Delta\omega$ having n frequency components of amplitude a with a common frequency difference $\delta\omega$. (b) Representation of the frequency band on a time axis is a cosine curve at the average frequency $\bar{\omega}$, amplitude modulated by a $\sin\alpha/\alpha$ curve where $\alpha = \Delta\omega \cdot t/2$. After a time $t = 2\pi/\Delta\omega$ the superposition of the components gives a zero amplitude

The result is given on p. 21 by

$$R = a\frac{\sin\left[n(\delta\omega)t/2\right]}{\sin\left[(\delta\omega)t/2\right]}\cos\bar{\omega}t$$

where the average frequency in the group or band is

$$\bar{\omega} = \omega_1 + \tfrac{1}{2}(n-1)(\delta\omega)$$

Now $n(\delta\omega) = \Delta\omega$, the bandwidth, so the behaviour of the resultant R with time may be written

$$R(t) = a\frac{\sin\left(\Delta\omega \cdot t/2\right)}{\sin\left(\Delta\omega \cdot t/n2\right)}\cos\bar{\omega}t = na\frac{\sin\left(\Delta\omega \cdot t/2\right)}{\Delta\omega \cdot t/2}\cos\bar{\omega}t$$

when n is large,
or

$$R(t) = A\frac{\sin\alpha}{\alpha}\cos\bar{\omega}t$$

where $A = na$ and $\alpha = \Delta\omega \cdot t/2$ is half the phase difference between the first and last components at time t.

This expression gives us the time behaviour of the band and is displayed on a time axis in Figure 5.14b. We see that the amplitude $R(t)$ is given by the cosine curve of the average frequency $\bar{\omega}$ modified by the $A \sin\alpha/\alpha$ term.

At $t = 0$, $\sin\alpha/\alpha \to 1$ and all the components superpose with zero phase difference to give the maximum amplitude $R(t) = A = na$. After some time interval Δt when

$$\alpha = \frac{\Delta\omega\Delta t}{2} = \pi$$

the phases between the frequency components are such that the resulting amplitude $R(t)$ is zero.

The time Δt which is a measure of the width of the central pulse of Figure 5.14b is therefore given by

$$\frac{\Delta\omega\Delta t}{2} = \pi$$

or $\Delta\nu\,\Delta t = 1$ where $\Delta\omega = 2\pi\Delta\nu$.

The true width of the base of the central pulse is $2\Delta t$ but the interval Δt is taken as an arbitrary measure of time, centred about $t = 0$, during which the amplitude $R(t)$ remains significantly large $(> A/2)$. With this arbitrary definition the exact expression

$$\Delta\nu\,\Delta t = 1$$

becomes the approximation

$$\Delta\nu\,\Delta t \approx 1 \quad \text{or} \quad (\Delta\omega\,\Delta t \approx 2\pi)$$

and this approximation is known as the Bandwidth Theorem.

It states that the components of a band of width $\Delta\omega$ in the frequency range will superpose to produce a significant amplitude $R(t)$ only for a time Δt before the band decays from random phase differences. The greater the range $\Delta\omega$ the shorter the period Δt.

Alternatively, the theorem states that a single pulse of time duration Δt is the result of the superposition of frequency components over the range $\Delta\omega$; the shorter the period Δt of the pulse the wider the range $\Delta\omega$ of the frequencies required to represent it.

When $\Delta\omega$ is zero we have a single frequency, the monochromatic wave which is therefore required (in theory) to have an infinitely long time span.

We have chosen to express our wave group in the two parameters of frequency and time (having a product of zero dimensions), but we may just as easily work in the other pair of parameters wave number k and distance x.

Replacing ω by k and t by x would define the length of the wave group as Δx in terms of the range of component wavelengths $\Delta(1/\lambda)$.

The Bandwidth Theorem then becomes

$$\Delta x\,\Delta k \approx 2\pi$$

or

$$\Delta x \Delta(1/\lambda) \approx 1 \quad \text{i.e. } \Delta x \approx \lambda^2/\Delta\lambda$$

Note again that a monochromatic wave with $\Delta k = 0$ requires $\Delta x \to \infty$; that is, an infinitely long wavetrain.

In the wave group we have just considered the problem has been simplified by assuming all frequency components to have the same amplitude a. When this is not the case, the different values $a(\omega)$ are treated by Fourier methods as we shall see in Chapter 10.

We shall meet the ideas of this section several times in the course of this text, noting particularly that in modern physics the Bandwidth Theorem becomes Heisenberg's Uncertainty Principle.

(Problem 5.20)

Transverse Waves in a Periodic Structure

At the end of the chapter on coupled oscillations we discussed the normal transverse vibrations of n equal masses of separation a along a light string of length $(n+1)a$ under a tension T with both ends fixed. The equation of motion of the rth particle was found to be

$$m\ddot{y}_r = \frac{T}{a}(y_{r+1} + y_{r-1} - 2y_r)$$

and for n masses the frequencies of the normal modes of vibration were given by

$$\omega_j^2 = \frac{2T}{ma}\left(1 - \cos\frac{j\pi}{n+1}\right) \tag{4.15}$$

where $j = 1, 2, 3, \ldots, n$. When the separation a becomes infinitesimally small ($= \delta x$, say) the term in the equation of motion

$$\frac{1}{a}(y_{r+1} + y_{r-1} - 2y_r) \to \frac{1}{\delta x}(y_{r+1} + y_{r-1} - 2y_r)$$

$$= \frac{(y_{r+1} - y_r)}{\delta x} - \frac{(y_r - y_{r-1})}{\delta x} = \left(\frac{\partial y}{\partial x}\right)_{r+1/2} - \left(\frac{\partial y}{\partial x}\right)_{r-1/2} = \left(\frac{\partial^2 y}{\partial x^2}\right)_r dx$$

so that the equation of motion becomes

$$\frac{\partial^2 y}{\partial t^2} = \frac{T}{\rho}\frac{\partial^2 y}{\partial x^2},$$

the wave equation, where $\rho = m/\delta x$, the linear density and

$$y \propto e^{i(\omega t - kx)}$$

We are now going to consider the propagation of transverse waves along a linear array of atoms, mass m, in a crystal lattice where the tension T now represents the elastic force between the atoms (so that T/a is the stiffness) and a, the separation between the atoms, is about 1 Å or 10^{-10} m. When the clamped ends of the string are replaced by the ends of the crystal we can express the displacement of the rth particle due to the transverse waves as

$$y_r = A_r \, \mathrm{e}^{\mathrm{i}(\omega t - kx)} = A_r \, \mathrm{e}^{\mathrm{i}(\omega t - kra)},$$

since $x = ra$. The equation of motion then becomes

$$-\omega^2 m = \frac{T}{a}(\mathrm{e}^{\mathrm{i}ka} + \mathrm{e}^{-\mathrm{i}ka} - 2)$$

$$= \frac{T}{a}(\mathrm{e}^{\mathrm{i}ka/2} - \mathrm{e}^{-\mathrm{i}ka/2})^2 = -\frac{4T}{a} \sin^2 \frac{ka}{2}$$

giving the permitted frequencies

$$\omega^2 = \frac{4T}{ma} \sin^2 \frac{ka}{2} \tag{5.11}$$

This expression for ω^2 is equivalent to our earlier value at the end of Chapter 4:

$$\omega_j^2 = \frac{2T}{ma}\left(1 - \cos\frac{j\pi}{n+1}\right) = \frac{4T}{ma} \sin^2 \frac{j\pi}{2(n+1)} \tag{4.15}$$

if

$$\frac{ka}{2} = \frac{j\pi}{2(n+1)}$$

where $j = 1, 2, 3, \ldots, n$.

But $(n+1)a = l$, the length of the string or crystal, and we have seen that wavelengths λ are allowed where $p\lambda/2 = l = (n+1)a$.

Thus

$$\frac{ka}{2} = \frac{2\pi}{\lambda} \cdot \frac{a}{2} = \frac{\pi a}{\lambda} = \frac{ja\pi}{2(n+1)a} = \frac{j}{p} \cdot \frac{\pi a}{\lambda}$$

if $j = p$. When $j = p$, a unit change in j corresponds to a change from one allowed number of half wavelengths to the next so that the minimum wavelength is $\lambda = 2a$, giving a maximum frequency $\omega_m^2 = 4T/ma$. Thus, both expressions may be considered equivalent.

When $\lambda = 2a$, $\sin ka/2 = 1$ because $ka = \pi$, and neighbouring atoms are exactly π rad out of phase because

$$\frac{y_r}{y_{r+1}} \propto \mathrm{e}^{\mathrm{i}ka} = \mathrm{e}^{\mathrm{i}\pi} = -1$$

The highest frequency is thus associated with maximum coupling, as we expect.

If in equation (5.11) we plot $|\sin ka/2|$ against k (Figure 5.15) we find that when ka is increased beyond π the phase relationship is the same as for a negative value of

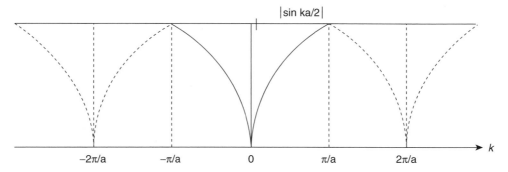

Figure 5.15 $|\sin \frac{ka}{2}|$ versus k from equation (5.11) shows the repetition of values beyond the region $\frac{-\pi}{a} \leq k \leq \frac{\pi}{a}$; this region defines a Brillouin zone

ka beyond $-\pi$. It is, therefore, sufficient to restrict the values of k to the region

$$\frac{-\pi}{a} \leq k \leq \frac{\pi}{a}$$

which is known as the first Brillouin zone. We shall use this concept in the section on electron waves in solids in Chapter 13.

For long wavelengths or low values of the wave number k, $\sin ka/2 \rightarrow ka/2$ so that

$$\omega^2 = \frac{4T}{ma}\frac{k^2 a^2}{4}$$

and the velocity of the wave is given by

$$c^2 = \frac{\omega^2}{k^2} = \frac{Ta}{m} = \frac{T}{\rho}$$

as before, where $\rho = m/a$.

In general the phase velocity is given by

$$v = \frac{\omega}{k} = c\left[\frac{\sin ka/2}{ka/2}\right] \tag{5.12}$$

a dispersion relation which is shown in Figure 5.16. Only at very short wavelengths does the atomic spacing of the crystal structure affect the wave propagation, and here the limiting or maximum value of the wave number $k_m = \pi/a \approx 10^{10}\,\mathrm{m}^{-1}$.

The elastic force constant T/a for a crystal is about 15 Nm^{-1}; a typical 'reduced' atomic mass is about 60×10^{-27} kg. These values give a maximum frequency

$$\omega^2 = \frac{4T}{ma} \approx \frac{60}{60 \times 10^{-27}} = 10^{27}\,\mathrm{rad}\,s^{-1}$$

that is, a frequency $\nu \approx 5 \times 10^{12}$ Hz.

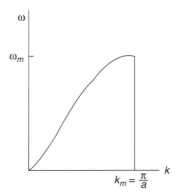

Figure 5.16 The dispersion relation $\omega(k)$ versus k for waves travelling along a linear one-dimensional array of atoms in a periodic structure

(Note that the value of T/a used here for the crystal is a factor of 8 lower than that found in Problem 4.4 for a single molecule. This is due to the interaction between neighbouring ions and the change in their equilibrium separation.)

This frequency is in the infrared region of the electromagnetic spectrum. We shall see in a later chapter that electromagnetic waves of frequency ω have a transverse electric field vector $E = E_0\,e^{i\omega t}$, where E_0 is the maximum amplitude, so that charged atoms or ions in a crystal lattice could respond as forced oscillators to radiation falling upon the crystal, which would absorb any radiation at the resonant frequency of its oscillating atoms.

Linear Array of Two Kinds of Atoms in an Ionic Crystal

We continue the discussion of this problem using a one dimensional line which contains two kinds of atoms with separation a as before, those atoms of mass M occupying the odd numbered positions, $2r - 1, 2r + 1$, etc. and those of mass m occupying the even numbered positions, $2r, 2r + 2$, etc. The equations of motion for each type are

$$m\ddot{y}_{2r} = \frac{T}{a}\left(y_{2r+1} + y_{2r-1} - 2y_{2r}\right)$$

and

$$M\ddot{y}_{2r+1} = \frac{T}{a}\left(y_{2r+2} + y_{2r} - 2y_{2r+1}\right)$$

with solutions

$$y_{2r} = A_m\,e^{i(\omega t - 2rka)}$$

$$y_{2r+1} = A_M\,e^{i(\omega t - (2r+1)ka)}$$

where A_m and A_M are the amplitudes of the respective masses.

The equations of motion thus become

$$-\omega^2 m A_m = \frac{TA_M}{a}\left(e^{-ika} + e^{ika}\right) - \frac{2TA_m}{a}$$

and

$$-\omega^2 M A_M = \frac{TA_m}{a}\left(e^{-ika} + e^{ika}\right) - \frac{2TA_M}{a}$$

equations which are consistent when

$$\omega^2 = \frac{T}{a}\left(\frac{1}{m} + \frac{1}{M}\right) \pm \frac{T}{a}\left[\left(\frac{1}{m} + \frac{1}{M}\right)^2 - \frac{4\sin^2 ka}{mM}\right]^{1/2} \tag{5.13}$$

Plotting the dispersion relation ω versus k for the positive sign and $m > M$ gives the upper curve of Figure 5.17 with

$$\omega^2 = \frac{2T}{a}\left(\frac{1}{m} + \frac{1}{M}\right) \quad \text{for} \quad k = 0$$

and

$$\omega^2 = \frac{2T}{aM} \quad \text{for} \quad k_m = \frac{\pi}{2a}\,(\text{minimum }\lambda = 4a)$$

The negative sign in equation (5.13) gives the lower curve of Figure 5.17 with

$$\omega^2 = \frac{2Tk^2 a^2}{a(M + m)} \quad \text{for very small } k$$

and

$$\omega^2 = \frac{2T}{am} \quad \text{for} \quad k = \frac{\pi}{2a}$$

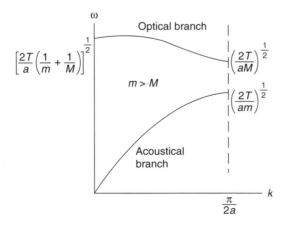

Figure 5.17 Dispersion relations for the two modes of transverse oscillation in a crystal structure

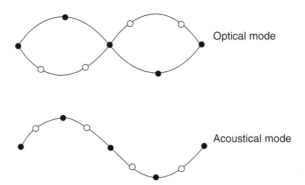

Figure 5.18 The displacements of the different atomic species in the two modes of transverse oscillations in a crystal structure (a) the optical mode, and (b) the acoustic mode

The upper curve is called the 'optical' branch and the lower curve is known as the 'acoustical' branch. The motions of the two types of atom for each branch are shown in Figure 5.18.

In the optical branch for long wavelengths and small k, $A_m/A_M = -M/m$, and the atoms vibrate against each other, so that the centre of mass of the unit cell in the crystal remains fixed. This motion can be generated by the action of an electromagnetic wave when alternate atoms are ions of opposite charge; hence the name 'optical branch'. In the acoustic branch, long wavelengths and small k give $A_m = A_M$, and the atoms and their centre of mass move together (as in longitudinal sound waves). We shall see in the next chapter that the atoms may also vibrate in a longitudinal wave.

The transverse waves we have just discussed are polarized in one plane; they may also vibrate in a plane perpendicular to the plane considered here. The vibrational energy of these two transverse waves, together with that of the longitudinal wave to be discussed in the next chapter, form the basis of the theory of the specific heats of solids, a topic to which we shall return in Chapter 9.

Absorption of Infrared Radiation by Ionic Crystals

Radiation of frequency 3×10^{12} Hz. gives an infrared wavelength of 100 μm (10^{-4} m) and a wave number $k = 2\pi/\lambda \approx 6.10^4 \, \text{m}^{-1}$. We found the cut-off frequency in the crystal lattice to give a wave number $k_m \approx 10^{10} \, \text{m}^{-1}$, so that the k value of infrared radiation is a negligible quantity relative to k_m and may be taken as zero. When the ions of opposite charge $\pm e$ move under the influence of the electric field vector $E = E_0 \, \text{e}^{i\omega t}$ of electromagnetic radiation, the equations of motion (with $k = 0$) become

$$-\omega^2 m A_m = \frac{2T}{a}(A_M - A_m) - eE_0$$

and

$$-\omega^2 M A_M = \frac{-2T}{a}(A_M - A_m) + eE_0$$

which may be solved to give

$$A_M = \frac{eE_0}{M(\omega_0^2 - \omega^2)} \quad \text{and} \quad A_m = \frac{-e}{m}\frac{E_0}{(\omega_0^2 - \omega^2)}$$

where

$$\omega_0^2 = \frac{2T}{a}\left(\frac{1}{m} + \frac{1}{M}\right)$$

the low k limit of the optical branch.

Thus, when $\omega = \omega_0$ infrared radiation is strongly absorbed by ionic crystals and the ion amplitudes A_M and A_m increase. Experimentally, sodium chloride is found to absorb strongly at $\lambda = 61\,\mu\text{m}$; potassium chloride has an absorption maximum at $\lambda = 71\,\mu\text{m}$.

(Problem 5.21)

Doppler Effect

In the absence of dispersion the velocity of waves sent out by a moving source is constant but the wavelength and frequency noted by a stationary observer are altered.

In Figure 5.19 a stationary source S emits a signal of frequency ν and wavelength λ for a period t so the distance to a stationary observer O is $\nu\lambda t$. If the source S' moves towards O at a velocity u during the period t then O registers a new frequency ν'.

We see that

$$\nu\lambda t = ut + \nu\lambda' t$$

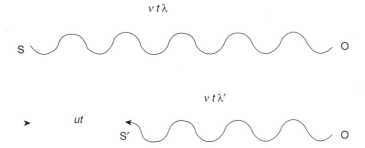

Figure 5.19 If waves from a stationary source S are received by a stationary observer O at frequency ν and wavelength λ the frequency is observed as ν' and the wavelength as λ' at O if the source S' moves during transmission. This is the Doppler effect

which, for

$$c = \nu\lambda = \nu'\lambda'$$

gives

$$\frac{c - u}{\nu} = \lambda' = \frac{c}{\nu'}$$

Hence

$$\nu' = \frac{\nu c}{c - u}$$

This observed change of frequency is called the *Doppler Effect*.

Suppose that the source S is now stationary but that an observer O' moves with a velocity v away from S. If we superimpose a velocity $-v$ on observer, source and waves, we bring the observer to rest; the source now has a velocity $-v$ and waves a velocity of $c - v$. Using these values in the expression for ν' gives a new observed frequency

$$\nu'' = \frac{\nu(c - v)}{c}$$

(Problems 5.22, 5.23, 5.24, 5.25, 5.26, 5.27, 5.28, 5.29, 5.30, 5.31)

Problem 5.1
Show that $y = f_2(ct + x)$ is a solution of the wave equation

$$\frac{\partial^2 y}{\partial x^2} = \frac{1}{c^2}\frac{\partial^2 y}{\partial t^2}$$

Problem 5.2
Show that the wave profile; that is,

$$y = f_1(ct - x)$$

remains unchanged with time when c is the wave velocity. To do this consider the expression for y at a time $t + \Delta t$ where $\Delta t = \Delta x/c$.

Repeat the problem for $y = f_2(ct + x)$.

Problem 5.3
Show that

$$\frac{\partial y}{\partial t} = +c\frac{\partial y}{\partial x}$$

for a left-going wave drawing a diagram to show the particle velocities as in Figure 5.5 (note that c is a magnitude and does not change sign).

Problem 5.4
A triangular shaped pulse of length l is reflected at the fixed end of the string on which it travels ($Z_2 = \infty$). Sketch the shape of the pulse (see Figure 5.8) after a length (a) $l/4$ (b) $l/2$ (c) $3l/4$ and (d) l of the pulse has been reflected.

Problem 5.5

A point mass M is concentrated at a point on a string of characteristic impedance ρc. A transverse wave of frequency ω moves in the positive x direction and is partially reflected and transmitted at the mass. The boundary conditions are that the string displacements just to the left and right of the mass are equal $(y_i + y_r = y_t)$ and that the difference between the transverse forces just to the left and right of the mass equal the mass times its acceleration. If A_1, B_1 and A_2 are respectively the incident, reflected and transmitted wave amplitudes show that

$$\frac{B_1}{A_1} = \frac{-iq}{1+iq} \quad \text{and} \quad \frac{A_2}{A_1} = \frac{1}{1+iq}$$

where $q = \omega M / 2\rho c$ and $i^2 = -1$.

Problem 5.6

In problem 5.5, writing $q = \tan \theta$, show that A_2 lags A_1 by θ and that B_1 lags A_1 by $(\pi/2 + \theta)$ for $0 < \theta < \pi/2$.

Show also that the reflected and transmitted energy coefficients are represented by $\sin^2 \theta$ and $\cos^2 \theta$, respectively.

Problem 5.7

If the wave on the string in Figure 5.6 propagates with a displacement

$$y = a \sin (\omega t - kx)$$

Show that the average rate of working by the force (average value of transverse force times transverse velocity) equals the rate of energy transfer along the string.

Problem 5.8

A transverse harmonic force of peak value 0.3 N and frequency 5 Hz initiates waves of amplitude 0.1 m at one end of a very long string of linear density 0.01 kg/m. Show that the rate of energy transfer along the string is $3\pi/20$ W and that the wave velocity is $30/\pi\,\mathrm{m\,s}^{-1}$.

Problem 5.9

In the figure, media of impedances Z_1 and Z_3 are separated by a medium of intermediate impedance Z_2 and thickness $\lambda/4$ measured in this medium. A normally incident wave in the first medium has unit amplitude and the reflection and transmission coefficients for multiple reflections are shown. Show that the total reflected amplitude in medium 1 which is

$$R + tTR'(1 + rR' + r^2R'^2 \ldots)$$

is zero at $R = R'$ and show that this defines the condition

$$Z_2^2 = Z_1Z_3$$

(Note that for zero total reflection in medium 1, the first reflection R is cancelled by the sum of all subsequent reflections.)

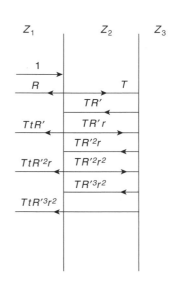

Problem 5.10

The relation between the impedance Z and the refractive index n of a dielectric is given by $Z = 1/n$. Light travelling in free space enters a glass lens which has a refractive index of 1.5 for a free space wavelength of 5.5×10^{-7} m. Show that reflections at this wavelength are avoided by a coating of refractive index 1.22 and thickness 1.12×10^{-7} m.

Problem 5.11

Prove that the displacement y_n of the standing wave expression in equation (5.10) satisfies the time independent form of the wave equation

$$\frac{\partial^2 y}{\partial x^2} + k^2 y = 0.$$

Problem 5.12

The total energy E_n of a normal mode may be found by an alternative method. Each section dx of the string is a simple harmonic oscillator with total energy equal to the maximum kinetic energy of oscillation

$$k.e._{\max} = \tfrac{1}{2}\rho\, dx (\dot{y}_n^2)_{\max} = \tfrac{1}{2}\rho\, dx \omega_n^2 (y_n^2)_{\max}$$

Now the value of $(y_n^2)_{\max}$ at a point x on the string is given by

$$(y_n^2)_{\max} = (A_n^2 + B_n^2)\sin^2\frac{\omega_n x}{c}$$

Show that the sum of the energies of the oscillators along the string; that is, the integral

$$\tfrac{1}{2}\rho\omega_n^2 \int_0^l (y_n^2)_{\max}\, dx$$

gives the expected result.

Problem 5.13

The displacement of a wave on a string which is fixed at both ends is given by

$$y(x,t) = A\cos(\omega t - kx) + rA\cos(\omega t + kx)$$

where r is the coefficient of amplitude reflection. Show that this may be expressed as the superposition of standing waves

$$y(x,t) = A(1+r)\cos\omega t \cos kx + A(1-r)\sin\omega t \sin kx.$$

Problem 5.14

A wave group consists of two wavelengths λ and $\lambda + \Delta\lambda$ where $\Delta\lambda/\lambda$ is very small.

Show that the number of wavelengths λ contained between two successive zeros of the modulating envelope is $\approx \lambda/\Delta\lambda$.

Problem 5.15

The phase velocity v of transverse waves in a crystal of atomic separation a is given by

$$v = c\left(\frac{\sin(ka/2)}{(ka/2)}\right)$$

where k is the wave number and c is constant. Show that the value of the group velocity is

$$c \cos \frac{ka}{2}$$

What is the limiting value of the group velocity for long wavelengths?

Problem 5.16

The dielectric constant of a gas at a wavelength λ is given by

$$\varepsilon_r = \frac{c^2}{v^2} = A + \frac{B}{\lambda^2} - D\lambda^2$$

where A, B and D are constants, c is the velocity of light in free space and v is its phase velocity. If the group velocity is V_g show that

$$V_g \varepsilon_r = v(A - 2D\lambda^2)$$

Problem 5.17

Problem 3.10 shows that the relative permittivity of an ionized gas is given by

$$\varepsilon_r = \frac{c^2}{v^2} = 1 - \left(\frac{\omega_e}{\omega}\right)^2$$

where v is the phase velocity, c is the velocity of light and ω_e is the constant value of the electron plasma frequency. Show that this yields the dispersion relation $\omega^2 = \omega_e^2 + c^2 k^2$, and that as $\omega \to \omega_e$ the phase velocity exceeds that of light, c, but that the group velocity (the velocity of energy transmission) is always less than c.

Problem 5.18

The electron plasma frequency of Problem 5.17 is given by

$$\omega_e^2 = \frac{n_e e^2}{m_e \varepsilon_0}.$$

Show that for an electron number density $n_e \sim 10^{20} (10^{-5}$ of an atmosphere), electromagnetic waves must have wavelengths $\lambda < 3 \times 10^{-3}$ m (in the microwave region) to propagate. These are typical wavelengths for probing thermonuclear plasmas at high temperatures.

$$\varepsilon_0 = 8.8 \times 10^{-12} \, \mathrm{F\,m^{-1}}$$
$$m_e = 9.1 \times 10^{-31} \, \mathrm{kg}$$
$$e = 1.6 \times 10^{-19} \, \mathrm{C}$$

Problem 5.19

In relativistic wave mechanics the dispersion relation for an electron of velocity $v = \hbar k/m$ is given by $\omega^2/c^2 = k^2 + m^2 c^2/\hbar^2$, where c is the velocity of light, m is the electron mass (considered constant at a given velocity) $\hbar = h/2\pi$ and h is Planck's constant. Show that the product of the group and particle velocities is c^2.

Problem 5.20

The figure shows a pulse of length Δt given by $y = A \cos \omega_0 t$.

Show that the frequency representation

$$y(\omega) = a \cos \omega_1 t + a \cos (\omega_1 + \delta\omega)t \cdots + a \cos [\omega_1 + (n-1)(\delta\omega)]t$$

is centred on the average frequency ω_0 and that the range of frequencies making significant contributions to the pulse satisfy the criterion

$$\Delta\omega \, \Delta t \approx 2\pi$$

Repeat this process for a pulse of length Δx with $y = A \cos k_0 x$ to show that in k space the pulse is centred at k_0 with the significant range of wave numbers Δk satisfying the criterion $\Delta x \, \Delta k \approx 2\pi$.

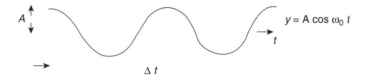

Problem 5.21

The elastic force constant for an ionic crystal is $\sim 15 \, \text{N m}^{-1}$. Show that the experimental values for the frequencies of infrared absorption quoted at the end of this chapter for NaCl and KCl are in reasonable agreement with calculated values.

$$1 \, \text{a.m.u.} = 1.66 \times 10^{-27} \, \text{kg}$$
$$\text{Na mass} = 23 \, \text{a.m.u.}$$
$$\text{K mass} = 39 \, \text{a.m.u.}$$
$$\text{Cl mass} = 35 \, \text{a.m.u.}$$

Problem 5.22

Show that, in the Doppler effect, the change of frequency noted by a stationary observer O as a moving source S' passes him is given by

$$\Delta\nu = \frac{2\nu c u}{(c^2 - u^2)}$$

where $c = \nu\lambda$, the signal velocity and u is the velocity of S'.

Problem 5.23

Suppose, in the Doppler effect, that a source S' and an observer O' move in the same direction with velocities u and v, respectively. Bring the observer to rest by superimposing a velocity $-v$ on the system to show that O' now registers a frequency

$$\nu''' = \frac{\nu(c - v)}{(c - u)}$$

Problem 5.24

Light from a star of wavelength $6 \times 10^{-7} \, \text{m}$ is found to be shifted $10^{-11} \, \text{m}$ towards the red when compared with the same wavelength from a laboratory source. If the velocity of light is $3 \times 10^8 \, \text{m s}^{-1}$ show that the earth and the star are separating at a velocity of $5 \, \text{Km s}^{-1}$.

Problem 5.25

An aircraft flying on a level course transmits a signal of 3×10^9 Hz which is reflected from a distant point ahead on the flight path and received by the aircraft with a frequency difference of 15 kHz. What is the aircraft speed?

Problem 5.26

Light from hot sodium atoms is centred about a wavelength of 6×10^{-7} m but spreads 2×10^{-12} m on either side of this wavelength due to the Doppler effect as radiating atoms move towards and away from the observer. Calculate the thermal velocity of the atoms to show that the gas temperature is ~ 900 K.

Problem 5.27

Show that in the Doppler effect when the source and observer are not moving in the same direction that the frequencies

$$\nu' = \frac{\nu c}{c - u'}, \quad \nu'' = \frac{\nu(c - v)}{c}$$

and

$$\nu''' = \nu\left(\frac{c - v}{c - u}\right)$$

are valid if u and v are not the actual velocities but the components of these velocities along the direction in which the waves reach the observer.

Problem 5.28

In extending the Doppler principle consider the accompanying figure where O is a stationary observer at the origin of the coordinate system $O(x, t)$ and O' is an observer situated at the origin of the system $O'(x', t')$ which moves with a constant velocity v in the x direction relative to the system O. When O and O' are coincident at $t = t' = 0$ a light source sends waves in the x direction with constant velocity c. These waves obey the relation

$$0 \equiv x^2 - c^2 t^2 (\text{seen by } O) \equiv x'^2 - c^2 t'^2 (\text{seen by } O'). \tag{1}$$

Since there is only one relative velocity v, the transformation

$$x' = k(x - vt) \tag{2}$$

and

$$x = k'(x' + vt') \tag{3}$$

must also hold. Use (2) and (3) to eliminate x' and t' from (1) and show that this identity is satisfied only by $k = k' = 1/(1 - \beta^2)^{1/2}$, where $\beta = v/c$. (Hint—in the identity of equation (1) equate coefficients of the variables to zero.).

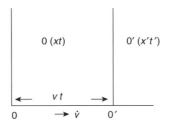

This is the Lorentz transformation in the theory of relativity giving

$$x' = \frac{(x - vt)}{(1 - \beta^2)^{1/2}}, \quad x = \frac{x' + vt'}{(1 - \beta^2)^{1/2}}$$

$$t' = \frac{(t - (v/c^2)x)}{(1 - \beta^2)^{1/2}}, \quad t = \frac{(t' + (v/c^2)x')}{(1 - \beta^2)^{1/2}}$$

Problem 5.29
Show that the interval $\Delta t = t_2 - t_1$ seen by O in Problem 5.28 is seen as $\Delta t' = k\Delta t$ by O' and that the length $l = x_2 - x_1$ seen by O is seen by O' as $l' = l/k$.

Problem 5.30
Show that two simultaneous events at x_2 and $x_1(t_2 = t_1)$ seen by O in the previous problems are not simultaneous when seen by O' (that is, $t'_1 \neq t'_2$).

Problem 5.31
Show that the order of events seen by $O(t_2 > t_1)$ of the previous problems will not be reversed when seen by O' (that is, $t'_2 > t'_1$) as long as the velocity of light c is the greatest velocity attainable.

Summary of Important Rules

Wave Equation $\dfrac{\partial^2 y}{\partial x^2} = \dfrac{1}{c^2}\dfrac{\partial^2 y}{\partial t^2}$

Wave (phase) velocity $= c = \dfrac{\omega}{k} = \dfrac{\partial x}{\partial t}$

k = wave number $= \dfrac{2\pi}{\lambda}$

where the wavelength λ defines separation between two oscillations with phase difference of 2π rad.

Particle velocity $\dfrac{\partial y}{\partial t} = -c\dfrac{\partial y}{\partial x}$

Displacement $y = a\,e^{i(\omega t - kx)}$,
where a is wave amplitude.

===

Characteristic Impedance of a String

$$Z = \frac{\text{transverse force}}{\text{transverse velocity}} = -T\frac{\partial y}{\partial x}\Big/\frac{\partial y}{\partial t} = \rho c$$

Reflection and Transmission Coefficients

$$\frac{\text{Reflected Amplitude}}{\text{Incident Amplitude}} = \frac{Z_1 - Z_2}{Z_1 + Z_2}$$

$$\frac{\text{Transmitted Amplitude}}{\text{Incident Amplitude}} = \frac{2Z_1}{Z_1 + Z_2}$$

$$\frac{\text{Reflected Energy}}{\text{Incident Energy}} = \left(\frac{Z_1 - Z_2}{Z_1 + Z_2}\right)^2$$

$$\frac{\text{Transmitted Energy}}{\text{Incident Energy}} = \frac{4Z_1 Z_2}{(Z_1 + Z_2)^2}$$

Impedance Matching

Impedances Z_1 and Z_3 are matched by insertion of impedance Z_2 where $Z_2^2 = Z_1 Z_3$
Thickness of Z_2 is $\lambda/4$ measured in Z_2.

Standing Waves. Normal Modes. Harmonics

Solution of wave equation separates time and space dependence to satisfy time independent wave equation

$$\frac{\partial^2 y}{\partial x^2} + k^2 y = 0 \quad (\text{cancel } e^{i\omega t})$$

Standing waves on string of length l have wavelength λ_n where

$$n\frac{\lambda_n}{2} = l$$

Displacement of nth harmonic is

$$y_n = (A_n \cos \omega_n t + B_n \sin \omega_n t) \sin \frac{\omega_n x}{c}$$

Energy of nth harmonic (string mass m)

$$E_n = KE_n + PE_n = \frac{1}{4}m\omega_n^2(A_n^2 + B_n^2)$$

Group Velocity

In a dispersive medium the wave velocity v varies with frequency ω (wave number k). The energy of a group of such waves travels with the group velocity

$$v_g = \frac{d\omega}{dk} = v + \frac{k \, dv}{dk} = v - \lambda \frac{dv}{d\lambda}$$

Rectangular Wave Group of n Frequency Components Amplitude a, Width $\Delta\omega$, represented in time by

$$R(t) = a \cdot \frac{\sin(\Delta\omega \cdot t/2)}{\sin(\Delta\omega \cdot t/n \cdot 2)} \cos \bar{\omega} t$$

where $\bar{\omega}$ is average frequency. $R(t)$ is zero when

$$\frac{\Delta\omega \cdot t}{2} = \pi$$

i.e. *Bandwidth Theorem* gives

$$\Delta\omega \cdot \Delta t = 2\pi$$

or

$$\Delta x \Delta k = 2\pi$$

A pulse of duration Δt requires a frequency band width $\Delta\omega$ to define it in frequency space and vice versa.

Doppler Effect

Signal of frequency ν and velocity c transmitted by a stationary source S and received by a stationary observer O becomes

$$\nu' = \frac{\nu c}{c - u}$$

when source is no longer stationary but moves towards O with a velocity u.

6

Longitudinal Waves

In deriving the wave equation

$$\frac{\partial^2 y}{\partial x^2} = \frac{1}{c^2}\frac{\partial^2 y}{\partial t^2}$$

in Chapter 5, we used the example of a transverse wave and continued to discuss waves of this type on a vibrating string. In this chapter we consider longitudinal waves, waves in which the particle or oscillator motion is in the same direction as the wave propagation. Longitudinal waves propagate as sound waves in all phases of matter, plasmas, gases, liquids and solids, but we shall concentrate on gases and solids. In the case of gases, limitations of thermodynamic interest are imposed; in solids the propagation will depend on the dimensions of the medium. Neither a gas nor a liquid can sustain the transverse shear necessary for transverse waves, but a solid can maintain both longitudinal and transverse oscillations.

Sound Waves in Gases

Let us consider a fixed mass of gas, which at a pressure P_0 occupies a volume V_0 with a density ρ_0. These values define the equilibrium state of the gas which is disturbed, or deformed, by the compressions and rarefactions of the sound waves. Under the influence of the sound waves

the pressure P_0 becomes $P = P_0 + p$

the volume V_0 becomes $V = V_0 + v$

and

the density ρ_0 becomes $\rho = \rho_0 + \rho_d$.

The excess pressure p_m is the maximum pressure amplitude of the sound wave and p is an alternating component superimposed on the equilibrium gas pressure P_0.

The Physics of Vibrations and Waves, 6th Edition   H. J. Pain
© 2005 John Wiley & Sons, Ltd

The fractional change in volume is called the *dilatation*, written $v/V_0 = \delta$, and the fractional change of density is called the *condensation*, written $\rho_d/\rho_0 = s$. The values of δ and s are $\approx 10^{-3}$ for ordinary sound waves, and a value of $p_m = 2 \times 10^{-5} \, \text{N m}^{-2}$ (about 10^{-10} of an atmosphere) gives a sound wave which is still audible at 1000 Hz. Thus, the changes in the medium due to sound waves are of an extremely small order and define limitations within which the wave equation is appropriate.

The fixed mass of gas is equal to

$$\rho_0 V_0 = \rho V = \rho_0 V_0 (1 + \delta)(1 + s)$$

so that $(1 + \delta)(1 + s) = 1$, giving $s = -\delta$ to a very close approximation. The elastic property of the gas, a measure of its compressibility, is defined in terms of its *bulk modulus*

$$B = -\frac{\mathrm{d}P}{\mathrm{d}V/V} = -V\frac{\mathrm{d}P}{\mathrm{d}V}$$

the difference in pressure for a fractional change in volume, a volume increase with fall in pressure giving the negative sign. The value of B depends on whether the changes in the gas arising from the wave motion are adiabatic or isothermal. They must be thermodynamically reversible in order to avoid the energy loss mechanisms of diffusion, viscosity and thermal conductivity. The complete absence of these random, entropy generating processes defines an adiabatic process, a thermodynamic cycle with a 100% efficiency in the sense that none of the energy in the wave, potential or kinetic, is lost. In a sound wave such thermodynamic concepts restrict the excess pressure amplitude; too great an amplitude raises the local temperature in the gas at the amplitude peaks and thermal conductivity removes energy from the wave system. Local particle velocity gradients will also develop, leading to diffusion and viscosity.

Using a constant value of the adiabatic bulk modulus limits sound waves to small oscillations since the total pressure $P = P_0 + p$ is taken as constant; larger amplitudes lead to non-linear effects and shock waves, which we shall discuss separately in Chapter 15.

All adiabatic changes in the gas obey the relation $PV^\gamma = \text{constant}$, where γ is the ratio of the specific heats at constant pressure and volume, respectively.

Differentiation gives

$$V^\gamma \, \mathrm{d}P + \gamma P V^{\gamma-1} \, \mathrm{d}V = 0$$

or

$$-V\frac{\mathrm{d}P}{\mathrm{d}V} = \gamma P = B_a \text{ (where the subscript } a \text{ denotes adiabatic)}$$

so that the elastic property of the gas is γP, considered to be constant. Since $P = P_0 + p$, then $\mathrm{d}P = p$, the excess pressure, giving

$$B_a = -\frac{p}{v/V_0} \quad \text{or} \quad p = -B_a\delta = B_a s$$

In a sound wave the particle displacements and velocities are along the *x*-axis and we choose the co-ordinate η to define the displacement where $\eta(x, t)$.

In obtaining the wave equation we consider the motion of an element of the gas of thickness Δx and unit cross section. Under the influence of the sound wave the behaviour

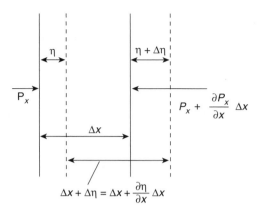

Figure 6.1 Thin element of gas of unit cross-section and thickness Δx displaced an amount η and expanded by an amount $(\delta\eta/\partial x)\Delta x$ under the influence of a pressure differene $-(\partial P_x/\partial x)\Delta x$

of this element is shown in Figure 6.1. The particles in the layer x are displaced a distance η and those at $x + \Delta x$ are displaced a distance $\eta + \Delta\eta$, so that the increase in the thickness Δx of the element of unit cross section (which therefore measures the increase in volume) is

$$\Delta\eta = \frac{\partial\eta}{\partial x}\Delta x$$

and

$$\delta = \frac{v}{V_0} = \left(\frac{\partial\eta}{\partial x}\right)\Delta x/\Delta x = \frac{\partial\eta}{\partial x} = -s$$

where $\partial\eta/\delta x$ is called the *strain*.

The medium is deformed because the pressures along the x-axis on either side of the thin element are not in balance (Figure 6.1). The net force acting on the element is given by

$$P_x - P_{x+\Delta x} = \left[P_x - \left(P_x + \frac{\partial P_x}{\partial x}\Delta x\right)\right]$$

$$= -\frac{\partial P_x}{\partial x}\Delta x = -\frac{\partial}{\partial x}(P_0 + p)\Delta x = -\frac{\partial p}{\partial x}\Delta x$$

The mass of the element is $\rho_0\Delta x$ and its acceleration is given, to a close approxmation, by $\partial^2\eta/\partial t^2$.

From Newton's Law we have

$$-\frac{\partial p}{\partial x}\Delta x = \rho_0\Delta x\frac{\partial^2\eta}{\partial t^2}$$

where

$$p = -B_a \delta = -B_a \frac{\partial \eta}{\partial x}$$

so that

$$-\frac{\partial p}{\partial x} = B_a \frac{\partial^2 \eta}{\partial x^2}, \quad \text{giving} \quad B_a \frac{\partial^2 \eta}{\partial x^2} = \rho_0 \frac{\partial^2 \eta}{\partial t^2}$$

But $B_a / \rho_0 = \gamma P / \rho_0$ is the ratio of the elasticity to the inertia or density of the gas, and this ratio has the dimensions

$$\frac{\text{force}}{\text{area}} \cdot \frac{\text{volume}}{\text{mass}} = (\text{velocity})^2, \quad \text{so} \quad \frac{\gamma P}{\rho_0} = c^2$$

where c is the sound wave velocity.

Thus

$$\frac{\partial^2 \eta}{\partial x^2} = \frac{1}{c^2} \frac{\partial^2 \eta}{\partial t^2}$$

is the wave equation. Writing η_m as the maximum amplitude of displacement we have the following expressions for a wave in the *positive x-direction*:

$$\eta = \eta_m \, e^{i(\omega t - kx)} \quad \dot{\eta} = \frac{\partial \eta}{\partial t} = i\omega\eta$$

$$\delta = \frac{\partial \eta}{\partial x} = -ik\eta = -s \quad (\text{so } s = ik\eta)$$

$$p = B_a s = iB_a k\eta$$

The phase relationships between these parameters (Figure 6.2a) show that when the wave is in the positive x-direction, the excess pressure p, the fractional density increase s and the particle velocity $\dot{\eta}$ are all $\pi/2$ rad in phase ahead of the displacement η, whilst the volume change (π rad out of phase with the density change) is $\pi/2$ rad behind the displacement. These relationships no longer hold when the wave direction is reversed (Figure 6.2b); for a *wave in the negative x-direction*

$$\eta = \eta_m \, e^{i(\omega t + kx)} \quad \dot{\eta} = \frac{\partial \eta}{\partial t} = i\omega\eta$$

$$\delta = \frac{\partial \eta}{\partial x} = -ik\eta = -s \quad (\text{so } s = ik\eta)$$

$$p = B_a s = -iB_a k\eta$$

In both waves the particle displacement η is measured in the positive x-direction and the thin element Δx of the gas oscillates about the value $\eta = 0$, which defines its central position. For a wave in the positive x-direction the value $\eta = 0$, with $\dot{\eta}$ a maximum in the positive x-direction, gives a *maximum positive excess pressure* (*compression*) with a maximum condensation s_m (maximum density) and a minimum volume. For a wave in the

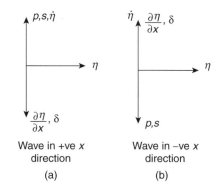

Figure 6.2 Phase relationships between the particle displacement η, particle velocity $\dot{\eta}$, excess pressure p and condensation $s = -\delta$ (the dilatation) for waves travelling in the positive and negative x directions. The displacement η is taken in the positive x direction for both waves

negative x-direction, the same value $\eta = 0$, with $\dot{\eta}$ a maximum in the positive x-direction, gives a *maximum negative excess pressure (rarefaction)*, a maximum volume and a minimum density. To produce a compression in a wave moving in the negative x-direction the particle velocity $\dot{\eta}$ must be a maximum in the negative x-direction at $\eta = 0$. This distinction is significant when we are defining the impedance of the medium to the waves. A change of sign is involved with a change of direction—a convention we shall also have to follow when discussing the waves of Chapters 7 and 8.

Energy Distribution in Sound Waves

The kinetic energy in the sound wave is found by considering the motion of the individual gas elements of thickness Δx.

Each element will have a kinetic energy per unit cross section

$$\Delta E_{\text{kin}} = \tfrac{1}{2} \rho_0 \, \Delta x \, \dot{\eta}^2$$

where $\dot{\eta}$ will depend upon the position x of the element. The average value of the kinetic energy density is found by taking the value of $\dot{\eta}^2$ averaged over a region of n wavelengths.

Now

$$\dot{\eta} = \dot{\eta}_m \sin \frac{2\pi}{\lambda} (ct - x)$$

so that

$$\overline{\dot{\eta}^2} = \frac{\dot{\eta}_m^2 \int_0^{n\lambda} \sin^2 2\pi (ct - x)/\lambda \, \Delta x}{n\lambda} = \tfrac{1}{2} \dot{\eta}_m^2$$

so that the *average kinetic energy density* in the medium is

$$\overline{\Delta E}_{\text{kin}} = \tfrac{1}{4} \rho_0 \dot{\eta}_m^2 = \tfrac{1}{4} \rho_0 \omega^2 \eta_m^2$$

(a simple harmonic oscillator of maximum amplitude a has an average kinetic energy over one cycle of $\frac{1}{4}m\omega^2 a^2$).

The potential energy density is found by considering the work $P\,dV$ done on the fixed mass of gas of volume V_0 during the adiabatic changes in the sound wave. This work is expressed for the complete cycle as

$$\Delta E_{\text{pot}} = -\int P dV = -\frac{-1}{2\pi}\int_0^{2\pi} pv\,\mathrm{d}(\omega t) = \frac{p_m v_m}{2} : \left[\frac{p}{p_m} = \frac{-v}{v_m} = \sin(\omega t - kx)\right]$$

The negative sign shows that the potential energy change is positive in both a compression (p positive, dV negative) and a rarefaction (p negative, dV positive) Figure 6.3.

The condensation

$$s = \frac{-\int \mathrm{d}v}{V_0} = \frac{-v}{V_0} = -\delta$$

we write

$$\frac{s}{s_m} = \frac{-\delta}{\delta_m} = \sin(\omega t - kx) \text{ and } -v = V_0 s$$

which, with

$$p = B_a s$$

gives

$$\Delta E_{\text{pot}} = \frac{-1}{2\pi}\int_0^{2\pi} pv\,\mathrm{d}(\omega t) = \frac{B_a V_0}{2\pi}\int_0^{2\pi} s^2\mathrm{d}(\omega t)$$

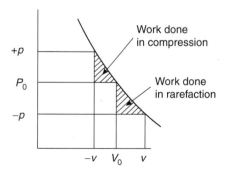

Figure 6.3 Shaded triangles show that potential energy $\frac{pv}{2} = \frac{p_m v_m}{4}$ gained by gas in compression equals that gained in rarefaction when both p and v change sign

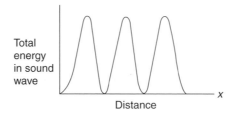

Figure 6.4 Energy distribution in space for a sound wave in a gas. Both potential and kinetic energies are at a maximum when the particle velocity $\dot{\eta}$ is a maximum and zero at $\dot{\eta} = 0$

where $s = -\delta$ and the thickness Δx of the element of unit cross section represents its volume V_0.

Now

$$\eta = \eta_m \, e^{i(\omega t \pm kx)}$$

so that

$$\delta = \frac{\partial \eta}{\partial x} = \pm \frac{1}{c} \frac{\partial \eta}{\partial t}, \quad \text{where} \quad c = \frac{\omega}{k}$$

Thus

$$\Delta E_{\text{pot}} = \frac{1}{2} \frac{B_a}{c^2} \dot{\eta}^2 \Delta x = \frac{1}{2} \rho_0 \dot{\eta}^2 \Delta x$$

and its average value over $n\lambda$ gives the potential energy density

$$\overline{\Delta E}_{\text{pot}} = \tfrac{1}{4} \rho_0 \dot{\eta}_m^2$$

We see that the average values of the kinetic and potential energy density in the sound wave are equal, but more important, since the value of each for the element Δx is $\frac{1}{2} \rho_0 \dot{\eta}^2 \Delta x$, we observe that the element possesses maximum (or minimum) potential and kinetic energy at the same time. A compression or rarefaction produces a maximum in the energy of the element since the value $\dot{\eta}$ governs the energy content. Thus, the energy in the wave is distributed in the wave system with distance as shown in Figure 6.4. Note that this distribution is non-uniform with distance unlike that for a transverse wave.

Intensity of Sound Waves

This is a measure of the energy flux, the rate at which energy crosses unit area, so that it is the product of the energy density (kinetic plus potential) and the wave velocity c. Normal sound waves range in intensity between 10^{-12} and $1 \, \text{W m}^{-2}$, extremely low levels which testify to the sensitivity of the ear. The roar of a large football crowd greeting a goal will just about heat a cup of coffee.

The intensity may be written

$$I = \tfrac{1}{2}\rho_0 c\dot{\eta}_m^2 = \tfrac{1}{2}\rho_0 c\omega^2\eta_m^2 = \rho_0 c\dot{\eta}_{rms}^2 = p_{rms}^2/\rho_0 c = p_{rms}\dot{\eta}_{rms}$$

A commonly used standard of sound intensity is given by

$$I_0 = 10^{-2}\,\mathrm{W\,m^{-2}}$$

which is about the level of the average conversational tone between two people standing next to each other. Shouting at this range raises the intensity by a factor of 100 and in the range $100\,I_0$ to $1000\,I_0$ ($10\,\mathrm{W\,m^{-2}}$) the sound is painful.

Whenever the sound intensity increases by a factor of 10 it is said to have increased by 1 B so the dynamic range of the ear is about 12 B. An intensity increase by a factor of

$$10^{0.1} = 1 \cdot 26$$

increases the intensity by 1 dB, a change of loudness which is just detected by a person with good hearing. dB is a decibel.

We see that the product $\rho_0 c$ appears in most of the expressions for the intensity; its significance becomes apparent when we define the impedance of the medium to the waves as the

$$\textit{Specific Acoustic Impedance} = \frac{\text{excess pressure}}{\text{particle velocity}} = \frac{p}{\dot{\eta}}$$

(the ratio of a force per unit area to a velocity).

Now, for a wave in the positive x-direction.

$$p = B_a s = iB_a k\eta \quad \text{and} \quad \dot{\eta} = i\omega\eta$$

so that,

$$\frac{p}{\dot{\eta}} = \frac{B_a k}{\omega} = \frac{B_a}{c} = \rho_o c$$

Thus, the acoustic impedance presented by the medium to these waves, as in the case of the transverse waves on the string, is given by the product of the density and the wave velocity and is governed by the elasticity and inertia of the medium. For a wave in the negative x-direction, the specific acoustic impedance

$$\frac{p}{\dot{\eta}} = -\frac{iB_a k\eta}{i\omega\eta} = -\rho_0 c$$

with a change of sign because of the changed phase relationship.

The units of $\rho_0 c$ are normally stated as $\mathrm{kg\,m^{-2}\,s^{-1}}$ in books on practical acoustics; in these units air has a specific acoustic impedance value of 400, water a value of 1.45×10^6 and steel a value of 3.9×10^7. These values will become more significant when we use them later in examples on the reflection and transmission of sound waves.

Although the specific acoustic impedance $\rho_0 c$ is a real quantity for plane sound waves, it has an added reactive component ik/r for spherical waves, where r is the distance travelled by the wavefront. This component tends to zero with increasing r as the spherical wave becomes effectively plane.

(Problems 6.1, 6.2, 6.3, 6.4, 6.5, 6.6, 6.7, 6.8)

Longitudinal Waves in a Solid

The velocity of longitudinal waves in a solid depends upon the dimensions of the specimen in which the waves are travelling. If the solid is a thin bar of finite cross section the analysis for longitudinal waves in a gas is equally valid, except that the bulk modulus B_a is replaced by Young's modulus Y, the ratio of the longitudinal stress in the bar to its longitudinal strain.

The wave equation is then

$$\frac{\partial^2 \eta}{\partial x^2} = \frac{1}{c^2} \frac{\partial^2 \eta}{\partial t^2}, \quad \text{with} \quad c^2 = \frac{Y}{\rho}$$

A longitudinal wave in a medium compresses the medium and distorts it laterally. Because a solid can develop a shear force in any direction, such a lateral distortion is accompanied by a transverse shear. The effect of this upon the wave motion in solids of finite cross section is quite complicated and has been ignored in the very thin specimen above. In bulk solids, however, the longitudinal and transverse modes may be considered separately.

We have seen that the longitudinal compression produces a strain $\partial \eta / \partial x$; the accompanying lateral distortion produces a strain $\partial \beta / \partial y$ (of opposite sign to $\partial \eta / \partial x$ and perpendicular to the x-direction).

Here β is the displacement in the y-direction and is a function of both x and y. The ratio of these strains

$$-\frac{\partial \beta}{\partial y} \bigg/ \frac{\partial \eta}{\partial x} = \sigma$$

is known as Poisson's ratio and is expressed in terms of Lamé's elastic constants λ and μ for a solid as

$$\sigma = \frac{\lambda}{2(\lambda + \mu)} \quad \text{where} \quad \lambda = \frac{\sigma Y}{(1 + \sigma)(1 - 2\sigma)}$$

These constants are always positive, so that $\sigma < \frac{1}{2}$, and is commonly $\approx \frac{1}{3}$. In terms of these constants Young's modulus becomes

$$Y = (\lambda + 2\mu - 2\lambda\sigma)$$

The constant μ is the transverse coefficient of rigidity; that is, the ratio of the transverse stress to the transverse strain. It plays the role of the elasticity in the propagation of pure

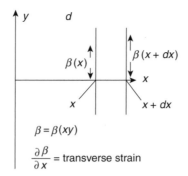

Figure 6.5 Shear in a bulk solid producing a transverse wave. The transverse shear strain is $\partial\beta/\partial x$ and the transverse shear stress is $\mu\,\partial\beta/\partial x$, where μ is the shear modulus of rigidity

transverse waves in a bulk solid which Young's modulus plays for longitudinal waves in a thin specimen. Figure 6.5 illustrates the shear in a transverse plane wave, where the transverse strain is defined by $\partial\beta/\partial x$. The transverse stress at x is therefore $T_x = \mu\,\partial\beta/\partial x$. The equation of transverse motion of the thin element dx is then given by

$$T_{x+dx} - T_{dx} = \rho\,dx\ddot{y}$$

where ρ is the density, or

$$\frac{\partial}{\partial x}\left(\mu\frac{\partial\beta}{\partial x}\right) = \rho\ddot{y}$$

but $\ddot{y} = \partial^2\beta/\partial t^2$, hence

$$\frac{\partial^2\beta}{\partial x^2} = \frac{\rho}{\mu}\frac{\partial^2\beta}{\partial t^2}$$

the wave equation with a velocity given by $c^2 = \mu/\rho$.

The effect of the transverse rigidity μ is to stiffen the solid and increase the elastic constant governing the propagation of longitudinal waves. In a bulk solid the velocity of these waves is no longer given by $c^2 = Y/\rho$, but becomes

$$c^2 = \frac{\lambda + 2\mu}{\rho}$$

Since Young's modulus $Y = \lambda + 2\mu - 2\lambda\sigma$, the elasticity is increased by the amount $2\lambda\sigma \approx \lambda$, so that longitudinal waves in a bulk solid have a higher velocity than the same waves along a thin specimen.

In an isotropic solid, where the velocity of propagation is the same in all directions, the concept of a bulk modulus, used in the discussion on waves in gases, holds equally well. Expressed in terms of Lamé's elastic constants the bulk modulus for a solid is written

$$B = \lambda + \tfrac{2}{3}\mu = Y[3(1 - 2\sigma)]^{-1}$$

the longitudinal wave velocity for a bulk solid becomes

$$c_L = \left(\frac{B + (4/3)\mu}{\rho}\right)^{1/2}$$

whilst the transverse velocity remains as

$$c_T = \left(\frac{\mu}{\rho}\right)^{1/2}$$

Application to Earthquakes

The values of these velocities are well known for seismic waves generated by earthquakes. Near the surface of the earth the longitudinal waves have a velocity of 8 km s^{-1} and the transverse waves travel at 4.45 km s^{-1}. The velocity of the longitudinal waves increases with depth until, at a depth of about 1800 miles, no waves are transmitted because of a discontinuity and severe mismatch of impedances associated with the fluid core.

At the surface of the earth the transverse wave velocity is affected by the fact that stress components directed through the surface are zero there and these waves, known as Rayleigh Waves, travel with a velocity given by

$$c = f(\sigma)\left(\frac{\mu}{\rho}\right)^{1/2}$$

where

$$f(\sigma) = 0.9194 \quad \text{when} \quad \sigma = 0 \cdot 25$$

and

$$f(\sigma) = 0.9553 \quad \text{when} \quad \sigma = 0 \cdot 5$$

The energy of the Rayleigh Waves is confined to two dimensions; their amplitude is often much higher than that of the three dimensional longitudinal waves and therefore they are potentially more damaging.

In an earthquake the arrival of the fast longitudinal waves is followed by the Rayleigh Waves and then by a complicated pattern of reflected waves including those affected by the stratification of the earth's structure, known as Love Waves.

(Problem 6.9)

Longitudinal Waves in a Periodic Structure

Lamé's elastic constants, λ and μ, which are used to define such macroscopic quantities as Young's modulus and the bulk modulus, are themselves determined by forces which operate over interatomic distances. The discussion on transverse waves in a periodic structure has already shown that in a one-dimensional array representing a crystal lattice a stiffness $s = T/a$ dyn cm^{-1} can exist between two atoms separated by a distance a.

When the waves along such a lattice are longitudinal the atomic displacements from equilibrium are represented by η (Figure 6.6). An increase in the separation between two atoms from a to $a + \eta$ gives a strain $\varepsilon = \eta/a$, and a stress normal to the face area a^2 of a unit cell in a crystal equal to $s\eta/a^2 = s\varepsilon/a$, a force per unit area.

Now Young's modulus is the ratio of this longitudinal stress to the longitudinal strain, so that $Y = s\varepsilon/\varepsilon a$ or $s = Ya$. The longitudinal vibration frequency of the atoms of mass m connected by stiffness constants s is given, very approximately by

$$\nu = \frac{\omega}{2\pi} = \frac{1}{2\pi}\sqrt{\frac{s}{m}} \approx \frac{1}{2\pi a}\sqrt{\frac{Y}{\rho}} \approx \frac{c_0}{2\pi a}$$

where $m = \rho a^3$ and c_0 is the velocity of sound in a solid. The value of $c_0 \approx 5 \times 10^3$ m s^{-1}, and $a \approx 2 \times 10^{-10}$ m, so that $\nu \approx 3 \times 10^{12}$ Hz, which is almost the same value as the frequency of the transverse wave in the infrared region of the electromagnetic spectrum. The highest ultrasonic frequency generated so far is about a factor of 10 lower than $\nu = c_0/2\pi a$. At frequencies $\approx 5 \times 10^{12}$ to 10^{13} Hz many interesting experimental results must be expected. A more precise mathematical treatment yields the same equation of motion for the r th particle as in the transverse wave; namely

$$m\ddot{\eta}_r = s(\eta_{r+1} + \eta_{r-1} - 2\eta_r)$$

where $s = T/a$ and

$$\eta_r = \eta_{\max} e^{i(\omega t - kra)}$$

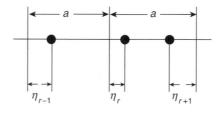

Figure 6.6 Displacement of atoms in a linear array due to a longitudinal wave in a crysal structure

The results are precisely the same as in the case of transverse waves and the shape of the dispersion curve is also similar. The maximum value of the cut-off frequency ω_m is, however, higher for the longitudinal than for the transverse waves. This is because the longitudinal elastic constant Y is greater than the transverse constant μ; that is, the force required for a given displacement in the longitudinal direction is greater than that for the same displacement in the transverse direction.

Reflection and Transmission of Sound Waves at Boundaries

When a sound wave meets a boundary separating two media of different acoustic impedances two boundary conditions must be met in considering the reflection and transmission of the wave. They are that

(i) the particle velocity $\dot{\eta}$

and

(ii) the acoustic excess pressure p

are both continuous across the boundary. Physically this ensures that the two media are in complete contact everywhere across the boundary.

Figure 6.7 shows that we are considering a plane sound wave travelling in a medium of specific acoustic impedance $Z_1 = \rho_1 c_1$ and meeting, at normal incidence, an infinite plane boundary separating the first medium from another of specific acoustic impedance $Z_2 = \rho_2 c_2$. If the subscripts i, r and t denote incident, reflected and transmitted respectively, then the boundary conditions give

$$\eta_i + \dot{\eta}_r = \dot{\eta}_t \tag{6.1}$$

and

$$p_i + p_r = p_t \tag{6.2}$$

For the incident wave $p_i = \rho_1 c_1 \dot{\eta}_i$ and for the reflected wave $p_r = -\rho_1 c_1 \dot{\eta}_r$, so equation (6.2) becomes

$$\rho_1 c_1 \dot{\eta}_i - \rho_1 c_1 \dot{\eta}_r = \rho_2 c_2 \dot{\eta}_t$$

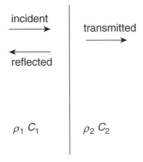

Figure 6.7 Incident, reflected and transmitted sound waves at a plane boundary between media of specific acoustic impedances $\rho_1 c_1$ and $\rho_2 c_2$

or

$$Z_1\dot{\eta}_i - Z_1\dot{\eta}_r = Z_2\dot{\eta}_t \tag{6.3}$$

Eliminating $\dot{\eta}_t$ from (6.1) and (6.3) gives

$$\frac{\dot{\eta}_r}{\dot{\eta}_i} = \frac{\omega\eta_r}{\omega\eta_i} = \frac{\eta_r}{\eta_i} = \frac{Z_1 - Z_2}{Z_1 + Z_2}$$

Eliminatiing $\dot{\eta}_r$ from (6.1) and (6.3) gives

$$\frac{\dot{\eta}_t}{\dot{\eta}_i} = \frac{\eta_t}{\eta_i} = \frac{2Z_1}{Z_1 + Z_2}$$

Now

$$\frac{p_r}{p_i} = -\frac{Z_1\dot{\eta}_r}{Z_1\dot{\eta}_i} = \frac{Z_2 - Z_1}{Z_1 + Z_2} = -\frac{\dot{\eta}_r}{\dot{\eta}_i}$$

and

$$\frac{p_t}{p_i} = \frac{Z_2\dot{\eta}_t}{Z_1\dot{\eta}_i} = \frac{2Z_2}{Z_1 + Z_2}$$

We see that if $Z_1 > Z_2$ the incident and reflected particle velocities are in phase, whilst the incident and reflected acoustic pressures are out of phase. The superposition of incident and reflected velocities which are in phase leads to a cancellation of pressure (a pressure node in a standing wave system). If $Z_1 < Z_2$ the pressures are in phase and the velocities are out of phase.

The transmitted particle velocity and acoustic pressure are always in phase with their incident counterparts.

At a rigid wall, where Z_2 is infinite, the velocity $\dot{\eta} = 0 = \dot{\eta}_i + \dot{\eta}_r$, which leads to a doubling of pressure at the boundary. (See Summary on p. 546.)

Reflection and Transmission of Sound Intensity

The intensity coefficients of reflection and transmission are given by

$$\frac{I_r}{I_i} = \frac{Z_1(\dot{\eta}_r^2)_{\text{rms}}}{Z_1(\dot{\eta}_i^2)_{\text{rms}}} = \left(\frac{Z_1 - Z_2}{Z_1 + Z_2}\right)^2$$

and

$$\frac{I_t}{I_i} = \frac{Z_2(\dot{\eta}_t^2)_{\text{rms}}}{Z_1(\dot{\eta}_i^2)_{\text{rms}}} = \frac{Z_2}{Z_1}\left(\frac{2Z_1}{Z_1 + Z_2}\right)^2 = \frac{4Z_1Z_2}{(Z_1 + Z_2)^2}$$

The conservation of energy gives

$$\frac{I_r}{I_i} + \frac{I_t}{I_i} = 1 \quad \text{or} \quad I_i = I_t + I_r$$

The great disparity between the specific acoustic impedance of air on the one hand and water or steel on the other leads to an extreme mismatch of impedances when the transmission of acoustic energy between these media is attempted.

There is an almost total reflection of sound wave energy at an air-water interface, independent of the side from which the wave approaches the boundary. Only 14% of acoustic energy can be transmitted at a steel-water interface, a limitation which has severe implications for underwater transmission and detection devices which rely on acoustics.

(Problems 6.10, 6.11, 6.12, 6.13, 6.14, 6.15, 6.16, 6.17)

Problem 6.1

Show that in a gas at temperature T the average thermal velocity of a molecule is approximatley equal to the velocity of sound.

Problem 6.2

The velocity of sound in air of density 1.29 kg m^{-3} may be taken as 330 m s^{-1}. Show that the acoustic pressure for the painful sound of 10 W m$^{-2} \approx 6.5 \times 10^{-4}$ of an atmosphere.

Problem 6.3

Show that the displacement amplitude of an air molecule at a painful sound level of 10 W m^{-2} at 500 Hz $\approx 6.9 \times 10^{-5}$ m.

Problem 6.4

Barely audible sound in air has an intensity of $10^{-10} I_0$. Show that the displacement amplitude of an air molecule for sound at this level at 500 Hz is $\approx 10^{-10}$ m; that is, about the size of the molecular diameter.

Problem 6.5

Hi-fi equipment is played very loudly at an intensity of $100 I_0$ in a small room of cross section 3 m $\times$ 3 m. Show that this audio output is about 10 W.

Problem 6.6

Two sound waves, one in water and one in air, have the same intensity. Show that the ratio of their pressure amplitudes (p water/p air) is about 60. When the pressure amplitudes are equal show that the intensity ratio is $\approx 3 \times 10^{-2}$.

Problem 6.7

A spring of mass m, stiffness s and length L is stretched to a length $L + l$. When longitudinal waves propagate along the spring the equation of motion of a length dx may be written

$$\rho \, \mathrm{d}x \frac{\partial^2 \eta}{\partial t^2} = \frac{\partial F}{\partial x} \, \mathrm{d}x$$

where ρ is the mass per unit length of the spring, η is the longitudinal displacement and F is the restoring force. Derive the wave equation to show that the wave velocity v is given by

$$v^2 = s(L + l)/\rho$$

Problem 6.8

In Problem 1.10 we showed that a mass M suspended by a spring of stiffness s and mass m oscillated simple harmonically at a frequency given by

$$\omega^2 = \frac{s}{M + m/3}$$

We may consider the same problem in terms of standing waves along the vertical spring with displacement

$$\eta = (A \cos kx + B \sin kx) \sin \omega t$$

where $k = \omega/v$ is the wave number. The boundary conditions are that $\eta = 0$ at $x = 0$ (the top of the spring) and

$$M\frac{\partial^2 \eta}{\partial t^2} = -sL\frac{\partial \eta}{\partial x} \quad \text{at } x = L$$

(the bottom of the spring). Show that these lead to the expression

$$kL \tan kL = \frac{m}{M}$$

and expand $\tan kL$ in powers of kL to show that, in the second order approximation

$$\omega^2 = \frac{s}{M + m/3}$$

The value of v is given in Problem 6.7.

Problem 6.9

A solid has a Poissons ratio $\sigma = 0.25$. Show that the ratio of the longitudinal wave velocity to the transverse wave velocity is $\sqrt{3}$. Use the values of these velocities given in the text to derive an appropriate value of σ for the earth.

Problem 6.10

Show that when sound waves are normally incident on a plane steel water interface 86% of the energy is reflected. If the waves are travelling in water and are normally incident on a plane water-ice interface show that 82.3% of the energy is transmitted.

$$(\rho c \text{ values in kg m}^{-2}\text{s}^{-1})$$

$$\text{water} = 1.43 \times 10^6$$
$$\text{ice} = 3.49 \times 10^6$$
$$\text{steel} = 3.9 \times 10^7$$

Problem 6.11

Use the boundary conditions for standing acoustic waves in a tube to confirm the following:

	Particle displacement		Pressure	
	closed end	open end	closed end	open end
Phase change on reflection	180°	0	0	180°
	node	antinode	antinode	node

Problem 6.12

Standing acoustic waves are formed in a tube of length l with (a) both ends open and (b) one end open and the other closed. If the particle displacement

$$\eta = (A \cos kx + B \sin kx) \sin \omega t$$

and the boundary conditions are as shown in the diagrams, show that for

$$\text{(a)} \quad \eta = A \cos kx \sin \omega t \quad \text{with} \quad \lambda = 2l/n$$

and for

$$\text{(b)} \quad \eta = A \cos kx \sin \omega t \quad \text{with} \quad \lambda = 4l/(2n+1)$$

Sketch the first three harmonics for each case.

Problem 6.13

On p. 121 we discussed the problem of matching two strings of impedances Z_1 and Z_3 by the insertion of a quarter wave element of impedance

$$Z_2 = (Z_1 Z_3)^{1/2}$$

Repeat this problem for the acoustic case where the expressions for the string displacements

$$y_i, y_r, y_t$$

now represent the appropriate acoustic pressures p_i, p_r and p_t.

Show that the boundary condition for pressure continuity at $x = 0$ is

$$A_1 + B_1 = A_2 + B_2$$

and that for continuity of particle velocity is

$$Z_2(A_1 - B_1) = Z_1(A_2 - B_2)$$

Similarly, at $x = l$, show that the boundary conditions are

$$A_2 e^{-ik_2 l} + B_2 e^{ik_2 l} = A_3$$

and

$$Z_3(A_2 e^{-ik_2 l} - B_2 e^{ik_2 l}) = Z_2 A_3$$

Hence prove that the coefficient of sound transmission

$$\frac{Z_1}{Z_3} \frac{A_3^2}{A_1^2} = 1$$

when

$$Z_2^2 = Z_1 Z_3 \quad \text{and} \quad l = \frac{\lambda_2}{4}$$

(Note that the expressions for both boundary conditions and transmission coefficient differ from those in the case of the string.)

Problem 6.14
For sound waves of high amplitude the adiabatic bulk modulus may no longer be considered as a constant. Use the adiabatic condition that

$$\frac{P}{P_0} = \left[\frac{V_0}{V_0(1 + \delta)} \right]^{\gamma}$$

in deriving the wave equation to show that each part of the high amplitude wave has its own sound velocity $c_0(1 + s)^{(\gamma+1)/2}$, where $c_0^2 = \gamma P_0/\rho_0$, δ is the dilatation, s the condensation and γ the ratio of the specific heats at constant pressure and volume.

Problem 6.15
Some longitudinal waves in a plasma exhibit a combination of electrical and acoustical phenomena. They obey a dispersion relation at temperature T of $\omega^2 = \omega_e^2 + 3aTk^2$, where ω_e is the constant electron plasma frequency (see Problem 5.18) and the Boltzmann constant is written as a to avoid confusion with the wave number k. Show that the product of the phase and group velocities is related to the average thermal energy of an electron (found from $pV = RT$).

Problem 6.16
It is possible to obtain the wave equation for tidal waves (long waves in shallow water) by the method used in deriving the acoustic wave equation. In the figure a constant mass of fluid in an element of unit width, height h and length Δx moves a distance η and assumes

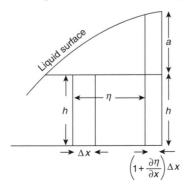

a new height $h + \alpha$ and length $(1 + \partial\eta\partial x)\Delta x$, but retains unit width. Show that, to a first approximation,

$$\alpha = -h\frac{\partial\eta}{\partial x}$$

Neglecting surface tension, the force on the element face of height $h + \alpha$ arises from the product of the height and the mean hydrostatic pressure. Show, if $\rho g h \ll P_0$ (i.e. $h \ll 10$ m) and $\alpha \ll h$, that the net force on the liquid element is given by

$$-\frac{\partial F}{\partial x}\Delta x = -\rho g h \frac{\partial\alpha}{\partial x}\Delta x$$

Continue the derivation using the acoustic case as a model to show that these waves are non-dispersive with a phase velocity given by $v^2 = gh$.

Problem 6.17

Waves near the surface of a non-viscous incompressible liquid of density ρ have a phase velocity given by

$$v^2(k) = \left[\frac{g}{k} + \frac{Tk}{\rho}\right]\tanh kh$$

where g is the acceleration due to gravity, T is the surface tension, k is the wave number and h is the liquid depth. When $h \ll \lambda$ the liquid is shallow; when $h \gg \lambda$ the liquid is deep.

(a) Show that, when gravity and surface tension are equally important and $h \gg \lambda$, the wave velocity is a minimum at $v^4 = 4gT/\rho$, and show that this occurs for a 'critical' wavelength $\lambda_c = 2\pi(T/\rho g)^{1/2}$.
(b) The condition $\lambda \gg \lambda_c$ defines a *gravity* wave, and surface tension is negligible. Show that gravity waves in a shallow liquid are non-dispersive with a velocity $v = \sqrt{gh}$ (see Problem 6.16).
(c) Show that gravity waves in a deep liquid have a phase velocity $v = \sqrt{g/k}$ and a group velocity of half this value.
(d) The condition $\lambda < \lambda_c$ defines a ripple (dominated by surface tension). Show that short ripples in a deep liquid have a phase velocity $v = \sqrt{Tk/\rho}$ and a group velocity of $\frac{3}{2}v$. (Note the anomalous dispersion).

Summary of Important Results

Wave Velocity

$$c^2 = \frac{\text{Bulk Modulus}}{\rho} = \frac{\gamma P}{\rho}$$

Specific Acoustic Impedance

$$Z = \frac{\text{acoustic pressure}}{\text{particle velocity}}$$

$$Z = \rho c \text{ (for right-going wave)}$$

$$= -\rho c \text{ (for left-going wave because pressure}$$
$$\text{and particle velocity become anti-phase)}$$

$$Intensity = \tfrac{1}{2}\rho c \dot{\eta}_m^2 = \frac{p_{\text{rms}}^2}{\rho c} = p_{\text{rms}} \dot{\eta}_{\text{rms}}$$

Reflection and Transmission Coefficients

$$\frac{\text{Reflected Amplitude}}{\text{Incident Amplitude}} \left\{ \begin{array}{l} \text{displacement} \\ \text{and velocity} \end{array} \right\} = \frac{Z_1 - Z_2}{Z_1 + Z_2} = -\frac{\text{Reflected pressure}}{\text{Incident pressure}}$$

$$\frac{\text{Transmitted Amplitude}}{\text{Incident Amplitude}} \left\{ \begin{array}{l} \text{displacement} \\ \text{and velocity} \end{array} \right\} = \frac{2Z_1}{Z_1 + Z_2} = \frac{Z_1}{Z_2} \times \frac{\text{Transmitted pressure}}{\text{Incident pressure}}$$

$$\frac{\text{Reflected Intensity}}{\text{Incident Intensity}} \text{(energy)} = \left(\frac{Z_1 - Z_2}{Z_1 + Z_2} \right)^2$$

$$\frac{\text{Transmitted Intensity}}{\text{Incident Intensity}} \text{(energy)} = \frac{4Z_1 Z_2}{(Z_1 + Z_2)^2}$$

7

Waves on Transmission Lines

In the wave motion discussed so far four major points have emerged. They are

1. Individual particles in the medium oscillate about their equilibrium positions with simple harmonic motion but do not propagate through the medium.

2. Crests and troughs and all planes of equal phase are transmitted through the medium to give the wave motion.

3. The wave or phase velocity is governed by the product of the inertia of the medium and its capacity to store potential energy; that is, its elasticity.

4. The impedance of the medium to this wave motion is governed by the ratio of the inertia to the elasticity (see table on p. 546).

In this chapter we wish to investigate the wave propagation of voltages and currents and we shall see that the same physical features are predominant. Voltage and current waves are usually sent along a geometrical configuration of wires and cables known as transmission lines. The physical scale or order of magnitude of these lines can vary from that of an oscilloscope cable on a laboratory bench to the electric power distribution lines supported on pylons over hundreds of miles or the submarine telecommunication cables lying on an ocean bed.

Any transmission line can be simply represented by a pair of parallel wires into one end of which power is fed by an a.c. generator. Figure 7.1a shows such a line at the instant when the generator terminal A is positive with respect to terminal B, with current flowing out of the terminal A and into terminal B as the generator is doing work. A half cycle later the position is reversed and B is the positive terminal, the net result being that along each of the two wires there will be a distribution of charge as shown, reversing in sign at each half cycle due to the oscillatory simple harmonic motion of the charge carriers (Figure 7.1b). These carriers move a distance equal to a fraction of a wavelength on either side of their equilibrium positions. As the charge moves current flows, having a maximum value where the product of charge density and velocity is greatest.

The Physics of Vibrations and Waves, 6th Edition   H. J. Pain
© 2005 John Wiley & Sons, Ltd

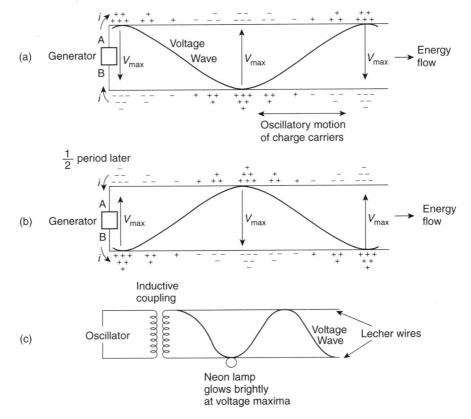

Figure 7.1 Power fed continuously by a generator into an infinitely long transmission line. Charge distribution and voltage waves for (a) generator terminal positive at A and (b) a half period later, generator terminal positive at B. Laboratory demonstration (c) of voltage maxima along a Lecher wire system. The neon lamp glows when held near a position of V_{max}

The existence along the cable of maximum and minimum current values varying simple harmonically in space and time describes a current wave along the cable. Associated with these currents there are voltage waves (Figure 7.1a), and if the voltage and current at the generator are always in phase then power is continuously fed into the transmission line and the waves will always be carrying energy away from the generator. In a laboratory the voltage and current waves may be shown on a Lecher Wire sysem (Figure 7.1c).

In deriving the wave equation for both voltage and current to obtain the velocity of wave propagation we shall concentrate our attention on a short element of the line having a length very much less than that of the waves. Over this element we may consider the variables to change linearly to the first order and we can use differentials.

The currents which flow will generate magnetic flux lines which thread the region between the cables, giving rise to a self inductance L_0 per unit length measured in henries per metre. Between the lines, which form a condenser, there is an electrical capacitance C_0

per unit length measured in farads per metre. In the absence of any resistance in the line these two parameters completely describe the line, which is known as *ideal* or *lossless*.

Ideal or Lossless Transmission Line

Figure 7.2 represents a short element of zero resistance of an ideal transmission line length $dx \ll \lambda$ (the voltage or current wavelength). The self inductance of the element is $L_0 \, dx$ and its capacitance if $C_0 \, dx$ F.

If the rate of change of voltage per unit length at constant time is $\partial V / \partial x$, then the voltage difference between the ends of the element dx is $\partial V / \partial x \, dx$, which equals the voltage drop from the self inductance $-(L_0 \, dx) \partial I / \partial t$.

Thus

$$\frac{\partial V}{\partial x} \, dx = -(L_0 \, dx) \frac{\partial I}{\partial t}$$

or

$$\frac{\partial V}{\partial x} = -L_0 \frac{\partial I}{\partial t} \tag{7.1}$$

If the rate of change of current per unit length at constant time is $\partial I / \partial x$ there is a loss of current along the length dx of $-\partial I / \partial x \, dx$ because some current has charged the capacitance $C_0 \, dx$ of the line to a voltage V.

If the amount of charge is $q = (C_0 \, dx) V$,

$$dI = \frac{dq}{dt} = \frac{\partial}{\partial t} (C_0 \, dx) V$$

so that

$$\frac{-\partial I}{\partial x} \, dx = \frac{\partial}{\partial t} (C_0 \, dx) V$$

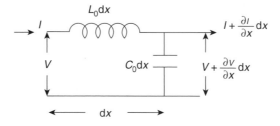

Figure 7.2 Representation of element of an ideal transmission line of inductance L_0 H per unit length and capacitance C_0 F per unit length. The element length $\ll \lambda$, the voltage and current wavelength

or

$$-\frac{\partial I}{\partial x} = C_0 \frac{\partial V}{\partial t} \tag{7.2}$$

Since $\partial^2/\partial x \partial t = \partial^2/\partial t\, \partial x$ it follows, by taking $\partial/\partial x$ of equation (7.1) and $\partial/\partial t$ of equation (7.2) that

$$\frac{\partial^2 V}{\partial x^2} = L_0 C_0 \frac{\partial^2 V}{\partial t^2} \tag{7.3}$$

a pure wave equation for the voltage with a velocity of propagation given by $v^2 = 1/L_0 C_0$.
  Similarly $\partial/\partial t$ of (7.1) and $\partial/\partial x$ of (7.2) gives

$$\frac{\partial^2 I}{\partial x^2} = L_0 C_0 \frac{\partial^2 I}{\partial t^2} \tag{7.4}$$

showing that the current waves propagate with the same velocity $v^2 = 1/L_0 C_0$. We must remember here, in checking dimensions, that L_0 and C_0 are defined per unit length.
  So far then, the oscillatory motion of the charge carriers (our particles in a medium) has led to the propagation of voltage and current waves with a velocity governed by the product of the magnetic inertia or inductance of the medium and its capactiy to store potential energy.

Coaxial Cables

Many transmission lines are made in the form of coaxial cables, e.g. a cylinder of dielectric material such as polythene having one conductor along its axis and the other surrounding its outer surface. This configuration has an inductance per unit length of

$$L_0 = \frac{\mu}{2\pi} \log_e \frac{r_2}{r_1} \text{ H}$$

where r_1 and r_2 are the radii of the inner and outer conductors respectively and μ is the magnetic permeability of the dielectric (henries per metre). Its capacitance per unit length

$$C_0 = \frac{2\pi\varepsilon}{\log_e r_2/r_1} \text{ F}$$

where ε is the permittivity of the dielectric (farads per metre) so that $v^2 = 1/L_0 C_0 = 1/\mu\varepsilon$.
  The velocity of the voltage and current waves along such a cable is wholly determined by the properties of the dielectric medium. We shall see in the next chapter on electromagnetic waves that μ and ε represent the inertial and elastic properties of any medium in which such waves are propagating; the velocity of these waves will be given by $v^2 = 1/\mu\varepsilon$. In free space these parameters have the values

$$\mu_0 = 4\pi \times 10^{-7}\,\text{H m}^{-1}$$

$$\varepsilon_0 = (36\pi \times 10^9)^{-1}\,\text{F m}^{-1}$$

and v^2 becomes $c^2 = (\mu_0\varepsilon_0)^{-1}$ where c is the velocity of light, equal to 3×10^8 m s^{-1}.

As we shall see in the next section the ratio of the voltage to the current in the waves travelling along the cable is

$$\frac{V}{I} = Z_0 = \sqrt{\frac{L_0}{C_0}}$$

where Z_0 defines the impedance seen by the waves moving down an infinitely long cable. It is called the Characteristic Impedance.

We write $\varepsilon = \varepsilon_r \varepsilon_0$ where ε_r is the relative permittivity (dielectric constant) of a material and $\mu = \mu_r \mu_0$, where μ_r is the relative permeability. Polythene, which commonly fills the space between r_1 and r_2, has $\varepsilon_r \approx 10$ and $\mu_r \approx 1$.

Hence

$$Z_0 = \sqrt{\frac{L_0}{C_0}} = \frac{1}{2\pi} \sqrt{\frac{\mu}{\varepsilon}} \log_e \frac{r_2}{r_1} = \frac{1}{2\pi} \frac{1}{\sqrt{\varepsilon_r}} \log_e \frac{r_2}{r_1} \sqrt{\frac{\mu_0}{\varepsilon_0}}$$

where

$$\sqrt{\frac{\mu_0}{\varepsilon_0}} = 376.6 \ \Omega$$

Typically, the ratio r_2/r_1 varies between 2 and 10^2 and for a laboratory cable using polythene $Z_0 \approx 50-75 \ \Omega$ with a signal speed $\approx c/3$ where c is the speed of light.

Coaxial cables can be made to a very high degree of precision and the time for an electrical signal to travel a given length can be accurately calculated because the velocity is known.

Such a cable can be used as a 'delay line' in order to separate the arrival of signals at a given point by very small intervals of time.

Characteristic Impedance of a Transmission Line

The solutions to equations (7.3) and (7.4) are, of course,

$$V_+ = V_{0+} \sin \frac{2\pi}{\lambda} (vt - x)$$

and

$$I_+ = I_{0+} \sin \frac{2\pi}{\lambda} (vt - x)$$

where V_0 and I_0 are the maximum values and where the subscript + refers to a wave moving in the positive x-direction. Equation (7.1), $\partial V/\partial x = -L_0 \, \partial I/\partial t$, therefore gives $-V'_+ = -vL_0 I'_+$, where the superscript refers to differentiation with respect to the bracket $(vt - x)$.

Integration of this equation gives

$$V_+ = vL_0 I_+$$

where the constant of integration has no significance because we are considering only oscillatory values of voltage and current whilst the constant will change merely the d.c. level.

The ratio

$$\frac{V_+}{I_+} = vL_0 = \sqrt{\frac{L_0}{C_0}} \, \Omega$$

and the value of $\sqrt{L_0/C_0}$, written as Z_0, is a constant for a transmission line of given properties and is called the *characteristic impedance*. Note that it is a pure resistance (no dimensions of length are involved) and it is the impedance seen by the wave system propagating along an infinitely long line, just as an acoustic wave experiences a specific acoustic impedance ρc. The physical correspondence between ρc and $L_0 v = \sqrt{L_0/C_0} = Z_0$ is immediately evident.

The value of Z_0 for the coaxial cable considered earlier can be shown to be

$$Z_0 = \frac{1}{2\pi} \sqrt{\frac{\mu}{\varepsilon}} \log_e \frac{r_2}{r_1}$$

Electromagnetic waves in free space experience an impedance $Z_0 = \sqrt{\mu_0/\varepsilon_0} = 376.6 \, \Omega$.

So far we have considered waves travelling only in the x-direction. Waves which travel in the negative x-direction will be represented (from solving the wave equation) by

$$V_- = V_{0-} \sin \frac{2\pi}{\lambda} (vt + x)$$

and

$$I_- = I_{0-} \sin \frac{2\pi}{\lambda} (vt + x)$$

where the negative subscript denotes the negative x-direction of propagation.

Equation (7.1) then yields the results that

$$\frac{V_-}{I_-} = -vL_0 = -Z_0$$

so that, in common with the specific acoustic impedance, a negative sign is introduced into the ratio when the waves are travelling in the negative x-direction.

When waves are travelling in both directions along the transmission line the total voltage and current at any point will be given by

$$V = V_+ + V_-$$

and

$$I = I_+ + I_-$$

When a transmission line has waves only in the positive direction the voltage and current waves are always in phase, energy is propagated and power is being fed into the line by the generator at all times. This situation is destroyed when waves travel in both directions;

waves in the negative *x*-direction are produced by reflection at a boundary when a line is terminated or mismatched; we shall now consider such reflections.

(Problems 7.1, 7.2)

Reflections from the End of a Transmission Line

Suppose that a transmission line of characteristic impedance Z_0 has a finite length and that the end opposite that of the generator is terminated by a load of impedance Z_L as shown in Figure 7.3.

A wave travelling to the right (V_+, I_+) may be reflected to produce a wave (V_-, I_-)

The boundary conditions at Z_L must be $V_+ + V_- = V_L$, where V_L is the voltage across the load and $I_+ + I_- = I_L$. In addition $V_+/I_+ = Z_0$, $V_-/I_- = -Z_0$ and $V_L/I_L = Z_L$. It is easily shown that these equations yield

$$\frac{V_-}{V_+} = \frac{Z_L - Z_0}{Z_L + Z_0}$$

(the voltage amplitude reflection coefficient),

$$\frac{I_-}{I_+} = \frac{Z_0 - Z_L}{Z_L + Z_0}$$

(the current amplitude reflection coefficient),

$$\frac{V_L}{V_+} = \frac{2Z_L}{Z_L + Z_0}$$

and

$$\frac{I_L}{I_+} = \frac{2Z_0}{Z_L + Z_0}$$

in complete correspondence with the reflection and transmission coefficients we have met so far. (See Summary on p. 546.)

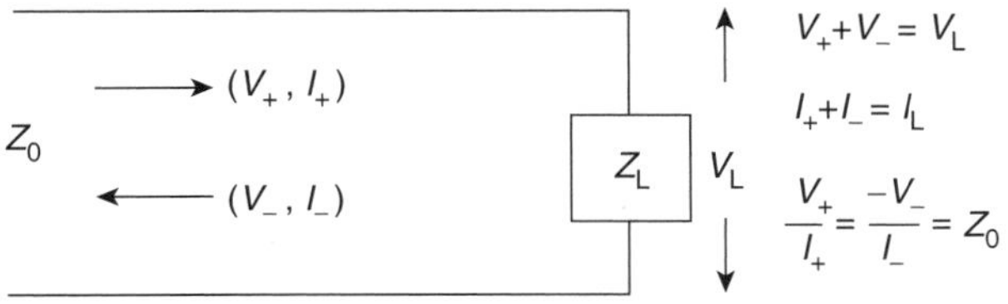

Figure 7.3 Transmission line terminated by impedance Z_L to produce reflected waves unless $Z_L = Z_0$, the characteristic impedance

We see that if the line is terminated by a load $Z_L = Z_0$, its characteristic impedance, the line is matched, all the energy propagating down the line is absorbed and there is no reflected wave. When $Z_L = Z_0$, therefore, the wave in the positive direction continues to behave as though the transmission line were infinitely long.

Short Circuited Transmission Line ($Z_L = 0$)

If the ends of the transmission line are short circuited (Figure 7.4), $Z_L = 0$, and we have

$$V_L = V_+ + V_- = 0$$

so that $V_+ = -V_-$, and there is total reflection with a phase change of π, But this is the condition, as we saw in an earlier chapter, for the existence of standing waves; we shall see that such waves exist on the transmission line.

At any position x on the line we may express the two voltage waves by

$$V_+ = Z_0 I_+ = V_{0+}\, e^{i(\omega t - kx)}$$

and

$$V_- = -Z_0 I_- = V_{0-}\, e^{i(\omega t + kx)}$$

where, with total reflection and π phase change, $V_{0+} = -V_{0-}$. The total voltage at x is

$$V_x = (V_+ + V_-) = V_{0+}(e^{-ikx} - e^{ikx})\, e^{i\omega t} = (-i)2V_{0+}\sin kx\, e^{i\omega t}$$

and the total current at x is

$$I_x = (I_+ + I_-) = \frac{V_{0+}}{Z_0}(e^{-ikx} + e^{ikx})\, e^{i\omega t} = \frac{2V_{0+}}{Z_0}\cos kx\, e^{i\omega t}$$

We see then that at any point x along the line the voltage V_x varies as $\sin kx$ and the current I_x varies as $\cos kx$, so that voltage and current are $90°$ out of phase in space. In addition the $-i$ factor in the voltage expression shows that the voltage lags the current $90°$ in time, so that if we take the voltage to vary with $\cos \omega t$ from the $e^{i\omega t}$ term, then the current

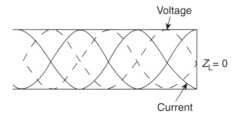

Figure 7.4 Short circuited transmission line of length $(2n+1)\lambda/4$ produces a standing wave with a current maximum and zero voltage at end of line

will vary with $-\sin \omega t$. If we take the time variation of voltage to be as $\sin \omega t$ the current will change with $\cos \omega t$.

Voltage and current at all points are 90° out of phase in space and time, and the power factor $\cos \phi = \cos 90° = 0$, so that no power is consumed. A standing wave system exists with equal energy propagated in each direction and the total energy propagation equal to zero. Nodes of voltage and current are spaced along the transmission line as shown in Figure 7.4, with I always a maximum where $V = 0$ and vice versa.

If the current I varies with $\cos \omega t$ it will be at a maximum when $V = 0$; when V is a maximum the current is zero. The energy of the system is therefore completely exchanged each quarter cycle between the magnetic inertial energy $\frac{1}{2}L_0 I^2$ and the electric potential energy $\frac{1}{2}C_0 V^2$.

(Problems 7.3, 7.4, 7.5, 7.6, 7.7, 7.8, 7.9, 7.10, 7.11)

The Transmission Line as a Filter

The transmission line is a continuous network of impedances in series and parallel combination. The unit section is shown in Figure 7.5(a) and the continuous network in Figure 7.5(b).

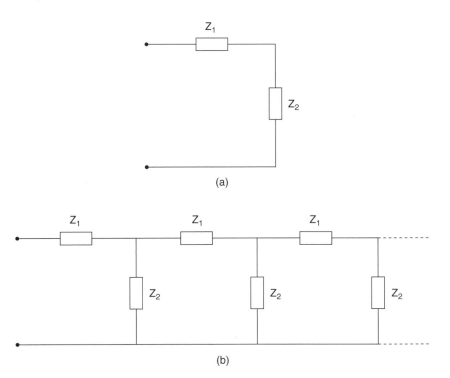

Figure 7.5 (a) The elementary unit of a transmission line. (b) A transmission line formed by a series of such units

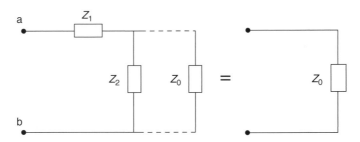

Figure 7.6 A infinite series of elemenetary units presents a characteristic impedance Z_0 to a wave travelling down the transmission line. Adding an extra unit at the input terminal leaves Z_0 unchanged

If we add an infinite series of such sections a wave travelling down the line will meet its characteristic impedance Z_0. Figure 7.6 shows that, adding an extra section to the beginning of the line does not change Z_0. The impedance in Figure 7.6 is

$$Z = Z_1 + \left(\frac{1}{Z_2} + \frac{1}{Z_0} \right)^{-1}$$

or

$$Z = Z_1 + \frac{Z_2 Z_0}{Z_2 + Z_0} = Z_0$$

so the characteristic impedance is

$$Z_0 = \frac{Z_1}{2} + \sqrt{\frac{Z_1^2}{4} + Z_1 Z_2}$$

Note that $Z_1/2$ is half the value of the first impedance in the line so if we measure the impedance from a point half way along this impedance we have

$$Z_0 = \left(\frac{Z_1^2}{4} + Z_1 Z_2 \right)^{1/2}$$

We shall, however, use the larger value of Z_0 in what follows.

In Figure 7.7 we now consider the currents and voltages at the far end of the transmission line. Any V_n since it is across Z_0 is given by $V_n = I_n Z_0$

Moreover

$$V_n - V_{n+1} = I_n Z_1 = V_n \frac{Z_1}{Z_0}$$

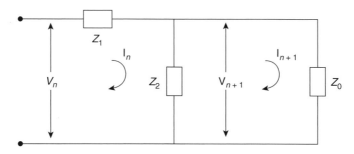

Figure 7.7 The propagation constant $\alpha = V_{n+1}/V_n = Z_0 - 1/Z_0$ for all sections of the transmission line

So

$$\frac{V_{n+1}}{V_n} = 1 - \frac{Z_1}{Z_0} = \frac{Z_0 - Z_1}{Z_0}$$

a result which is the same for all sections of the line.

We define a propagation factor

$$\alpha = \frac{V_{n+1}}{V_n} = \frac{Z_0 - Z_1}{Z_0}$$

which, with

$$Z_0 = \frac{Z_1}{2} + \left(\frac{Z_1^2}{4} + Z_1 Z_2\right)^{1/2}$$

gives

$$\alpha = \frac{\left(\sqrt{Z_0} - \dfrac{Z_1}{2}\right)}{\left(\sqrt{Z_0} + \dfrac{Z_1}{2}\right)}$$

$$= 1 + \frac{Z_1}{2Z_2} - \left[\left(1 + \frac{Z_1}{2Z_2}\right)^2 - 1\right]^{1/2}$$

In all practical cases Z_1/Z_2 is real since

1. there is either negligible resistance so that Z_1 and Z_2 are imaginary or

2. the impedances are purely resistive.

So, given (1) or (2) we see that if

(a) $\left(1 + \dfrac{Z_1}{2Z_2}\right)^2 = \left[1 + \dfrac{Z_1}{Z_2}\left(1 + \dfrac{Z_1}{4Z_2}\right)\right] \geq 1$ then α is real, and

(b) $\left(1 + \dfrac{Z_1}{2Z_2}\right)^2 < 1$ then α is complex.

For α real we have $Z_1/4Z_2 \geq 0$ or ≤ -1.

If $Z_1/4Z_2 \geq 0$, then $0 < \alpha < 1$, the currents in successive sections decrease progressively and since α is real and positive there is no phase change from one section to another.

If $Z_1/4Z_2 \leq -1$, then $\alpha \leq 0$, and there is again a progressive decrease in current amplitudes along the network but here α is negative and there is a π phase change for each successive section.

When α is complex we have

$$-1 < \frac{Z_1}{4Z_2} < 0$$

and

$$\alpha = 1 + \frac{Z_1}{2Z_2} - i\left[1 - \left(1 + \frac{Z_1}{2Z_2}\right)^2\right]^{1/2}$$

Note that $|\alpha| = 1$ so we can write

$$\alpha = \cos\beta - i\sin\beta = e^{-i\beta}$$

where

$$\cos\beta = 1 + \frac{Z_1}{2Z_2}$$

The current amplitude remains constant along the transmission line but the phase is retarded by β with each section. If Z_1 and Z_2 are purely resistive α is fixed and the attenuation is constant for all voltage inputs.

If Z_1 is an inductance with Z_2 a capacitance (or vice versa) the division between α real and α complex occurs at certain frequencies governed by their relative magnitudes.

If $Z_1 = i\omega L$ and $Z_2 = 1/i\omega C$ for an input voltage $V = V_0 e^{i\omega t}$ then $|\alpha| = 1$ when $0 \leq \omega^2 LC \leq 4$.

So the line behaves as a low pass filter with a cut-off frequency $\omega_c = 2/\sqrt{LC}$ Above this frequency there is a progressive decrease in amplitude with a phase change of π in each section, Figure 7.8a.

If the positions of Z_1 and Z_2 are now interchanged so that $Z_1 = 1/i\omega C$ is now a capacitance and Z_2 is now an inductance with $Z_2 = i\omega L$ the transmisson line becomes a

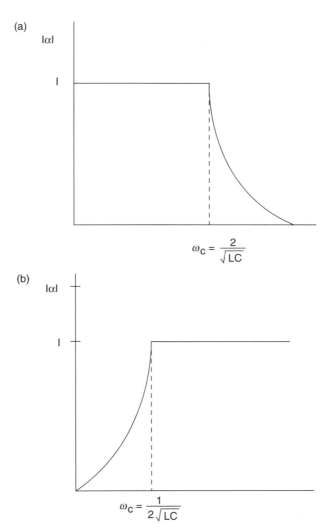

Figure 7.8 (a) When $Z_1 = i\omega L$ and $Z_2 = (i\omega L)^{-1}$ the transmission line acts as a low-pass filter. (b) Reversing the positions of Z_1 and Z_2 changes the transmission line into a high-pass filter

high pass filter with zero attenuation for $0 \leq 1/\omega^2 LC \leq 4$ that is for all frequencies above $\omega_C = (1/2\sqrt{LC})$ Figure 7.8b.

(Problem 7.12)

Effect of Resistance in a Transmission Line

The discussion so far has concentrated on a transmission line having only inductance and capacitance, i.e. wattless components which consume no power. In practice, of course, no

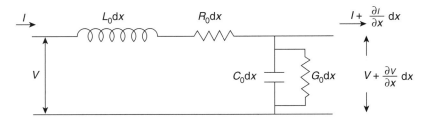

Figure 7.9 Real transmission line element includes a series resistance $R_0 \, \Omega$ per unit length and a shunt conductance G_0 S per unit length

such line exists: there is always some resistance in the wires which will be responsible for energy losses. We shall take this resistance into account by supposing that the transmission line has a series resistance $R_0 \Omega$ per unit length and a short circuiting or shunting resistance between the wires, which we express as a shunt conductance (inverse of resistance) written as G_0, where G_0 has the dimensions of siemens per metre. Our model of the short element of length dx of the transmission line now appears in Figure 7.9, with a resistance $R_0 \, dx$ in series with $L_0 \, dx$ and the conductance $G_0 \, dx$ shunting the capacitance $C_0 \, dx$. Current will now leak across the transmission line because the dielectric is not perfect. We have seen that the time-dependence of the voltage and current variations along a transmission line may be written

$$V = V_0 \, \mathrm{e}^{\mathrm{i}\omega t} \quad \text{and} \quad I = I_0 \, \mathrm{e}^{\mathrm{i}\omega t}$$

so that

$$L_0 \frac{\partial I}{\partial t} = \mathrm{i}\omega L_0 I \quad \text{and} \quad C_0 \frac{\partial V}{\partial t} = \mathrm{i}\omega C_0 V$$

The voltage and current changes across the line element length dx are now given by

$$\frac{\partial V}{\partial x} = -L_0 \frac{\partial I}{\partial t} - R_0 I = -(R_0 + \mathrm{i}\omega L_0)I \tag{7.1a}$$

$$\frac{\partial I}{\partial x} = -C_0 \frac{\partial V}{\partial t} - G_0 V = -(G_0 + \mathrm{i}\omega C_0)V \tag{7.2a}$$

since $(G_0 \, dx)V$ is the current shunted across the condenser. Inserting $\partial/\partial x$ of equation (7.1a) into equation (7.2a) gives

$$\frac{\partial^2 V}{\partial x^2} = -(R_0 + \mathrm{i}\omega L_0)\frac{\partial I}{\partial x} = (R_0 + \mathrm{i}\omega L_0)(G_0 + \mathrm{i}\omega C_0)V = \gamma^2 V$$

where $\gamma^2 = (R_0 + \mathrm{i}\omega L_0)(G_0 + \mathrm{i}\omega C_0)$, so that γ is a complex quantity which may be written

$$\gamma = \alpha + \mathrm{i}k$$

Inserting $\partial/\partial x$ of equation (7.2a) into equation (7.1a) gives

$$\frac{\partial^2 I}{\partial x^2} = -(G_0 + i\omega C_0)\frac{\partial V}{\partial x} = (R_0 + i\omega L_0)(G_0 + i\omega C_0)I = \gamma^2 I$$

an equation similar to that for V.

The equation

$$\frac{\partial^2 V}{\partial x^2} - \gamma^2 V = 0 \tag{7.5}$$

has solutions for the x-dependence of V of the form

$$V = A\,e^{-\gamma x} \quad \text{or} \quad V = B\,e^{+\gamma x}$$

where A and B are constants.

We know already that the time-dependence of V is of the form $e^{i\omega t}$, so that the complete solution for V may be written

$$V = (A\,e^{-\gamma x} + B\,e^{\gamma x})\,e^{i\omega t}$$

or, since $\gamma = \alpha + ik$,

$$V = (A\,e^{-\alpha x}\,e^{-ikx} + B\,e^{\alpha x}\,e^{+ikx})\,e^{i\omega t}$$
$$= A\,e^{-\alpha x}\,e^{i(\omega t - kx)} + B\,e^{\alpha x}\,e^{i(\omega t + kx)}$$

The behaviour of V is shown in Figure 7.10—a wave travelling to the right with an amplitude decaying exponentially with distance because of the term $e^{-\alpha x}$ and a wave travelling to the left with an amplitude decaying exponentially with distance because of the term $e^{\alpha x}$.

In the expression $\gamma = \alpha + ik$, γ is called the propagation constant, α is called the attenuation or absorption coefficient and k is the wave number.

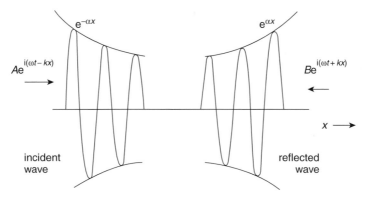

Figure 7.10 Voltage and current waves in both directions along a transmission line with resistance. The effect of the dissipation term is shown by the exponentially decaying wave in each direction

The behaviour of the current wave I is exactly similar and since power is the product VI, the power loss with distance varies as $(e^{-\alpha x})^2$; that is, as $e^{-2\alpha x}$.

We would expect this behaviour from our discussion of damped simple harmonic oscillations. When the transmission line properties are purely inductive (inertial) and capacitative (elastic), a pure wave equation with a sine or cosine solution will follow. The introduction of a resistive or loss element produces an exponential decay with distance along the transmission line in exactly the same way as an oscillator is damped with time.

Such a loss mechanism, resistive, viscous, frictional or diffusive, will always result in energy loss from the propagating wave. These are all examples of random collision processes which operate in only one direction in the sense that they are thermodynamically irreversible. At the end of this chapter we shall discuss their effects in more detail.

Characteristic Impedance of a Transmission Line with Resistance

In a lossless line we saw that the ratio $V_+/I_+ = Z_0 = \sqrt{L_0/C_0} = Z_0\,\Omega$, a purely resistive term. In what way does the introduction of the resistance into the line affect the characteristic impedance?

The solution to the equation $\partial^2 I/\partial x^2 = \gamma^2 I$ may be written (for the x-dependence of I) as

$$I = (A' e^{-\gamma x} + B' e^{\gamma x})$$

so that equation (7.2a)

$$\frac{\partial I}{\partial x} = -(G_0 + i\omega C_0)V$$

gives

$$-\gamma(A' e^{-\gamma x} - B' e^{\gamma x}) = -(G_0 + i\omega C_0)V$$

or

$$\frac{\sqrt{(R_0 + i\omega L_0)(G_0 + i\omega C_0)}}{G_0 + i\omega C_0}(A' e^{-\gamma x} - B' e^{\gamma x}) = V = V_+ + V_-$$

But, except for the $e^{i\omega t}$ term,

$$A' e^{-\gamma x} = I_+$$

the current wave in the positive x-direction, so that

$$\sqrt{\frac{R_0 + i\omega L_0}{G_0 + i\omega C_0}}I_+ = V_+$$

or

$$\frac{V_+}{I_+} = \sqrt{\frac{R_0 + i\omega L_0}{G_0 + i\omega C_0}} = Z_0'$$

for a transmission line with resistance. Similarly $B' e^{\gamma x} = I_-$ and

$$\frac{V_-}{I_-} = -\sqrt{\frac{R_0 + i\omega L_0}{G_0 + i\omega C_0}} = -Z_0'$$

The presence of the resistance term in the complex characteristic impedance means that power will be lost through Joule dissipation and that energy will be absorbed from the wave system.

We shall discuss this aspect in some detail in the next chapter on electromagnetic waves, but for the moment we shall examine absorption from a different (although equivalent) viewpoint.

(Problems 7.13, 7.14)

The Diffusion Equation and Energy Absorption in Waves

On p. 23 of Chapter 1 we discussed quite briefly the effect of random processes. We shall now look at this in more detail. The wave equation

$$\frac{\partial^2 \phi}{\partial x^2} = \frac{1}{c^2} \frac{\partial^2 \phi}{\partial t^2}$$

is only one of a family of equations which have a double differential with respect to space on the left hand side.

In three dimensions the left hand side would be of the form

$$\frac{\partial^2 \phi}{\partial x^2} + \frac{\partial^2 \phi}{\partial y^2} + \frac{\partial^2 \phi}{\partial z^2}$$

which, in vector language, is called the divergence of the gradient or div grad and is written $\nabla^2 \phi$.

Five members of this family of equations may be written (in one dimension) as

1. Laplace's Equation

$$\frac{\partial^2 \phi}{\partial x^2} = 0 \quad (\text{for } \phi(x) \text{ only})$$

2. Poisson's Equation

$$\frac{\partial^2 \phi}{\partial x^2} = \text{constant} \quad (\text{for } \phi(x) \text{ only})$$

3. Helmholtz Equation

$$\frac{\partial^2 \phi}{\partial x^2} = \text{constant} \times \phi$$

4.   Diffusion Equation

$$\frac{\partial^2 \phi}{\partial x^2} = +\text{ve constant} \times \frac{\partial \phi}{\partial t}$$

5.   Wave Equation

$$\frac{\partial^2 \phi}{\partial x^2} = +\text{ve constant} \times \frac{\partial^2 \phi}{\partial t^2}$$

Laplace's and Poisson's equations occur very often in electrostatic field theory and are used to find the values of the electric field and potential at any point. We have already met the Helmholtz equation in this chapter as equation (7.5), where the constant was positive (written γ^2) and we have seen its behaviour when the constant is negative, for it is then equivalent to the equation for standing waves (p. 124). The constant in the wave equation is of course $1/c^2$ where c is the wave velocity. Where the wave equation has an 'acceleration' or $\partial^2 \phi / \partial t^2$ term on the right hand side, the diffusion equation has a 'velocity' or $\partial \phi / \partial t$ term.

All these equations, however, have the same term $\partial^2 \phi / \partial x^2$ on the left hand side, and we must ask: 'What is its physical significance?'

We know that the values of the scalar ϕ will depend upon the point in space at which it is measured. Suppose we choose some point at which ϕ has the value ϕ_0 and surround this point by a small cube of side l, over the volume of which ϕ may take other values. If the average value of ϕ over the small cube is written $\bar{\phi}$, then the difference between the average $\bar{\phi}$ and the value at the centre of the cube ϕ_0 is given by

$$\bar{\phi} - \phi_0 = \text{constant} \times \left(\frac{\partial^2 \phi}{\partial x^2} + \frac{\partial^2 \phi}{\partial y^2} + \frac{\partial^2 \phi}{\partial z^2} \right)_0$$

This statement is proved in the appendix at the end of this chapter and is readily understood by those familiar with triple integration. The left hand side of any of these equations therefore measures the value

$$\bar{\phi} - \phi_0$$

In Laplace's equation the difference is zero, so that ϕ has a constant value over the volume considered. Poisson's equation tells us that the difference is constant and Helmholtz equation states that the value of ϕ at any point in the volume is proportional to this difference. The first two equations are 'steady state', i.e. they do not vary with time.

The Helmholtz equation states that if the constant is positive the behaviour of ϕ with space grows or decays exponentially, e.g. γ^2 is positive in equation (7.5), but if the constant is negative, ϕ will vary sinusoidally or cosinusoidally with space as the displacement varies with time in simple harmonic motion and the equation becomes the time independent wave equation for standing waves. This equation says nothing about the time behaviour of ϕ, which will depend only upon the function ϕ itself.

Both the diffusion and wave equations are time-derivative dependent. The diffusion equation states that the 'velocity' or change of ϕ with time at a point in the volume is proportional to the difference $\bar{\phi} - \phi_0$, whereas the wave equation states that the 'acceleration' $\partial^2 \phi / \partial t^2$ depends on this difference.

The wave equation recalls the simple harmonic oscillator, where the difference from the centre $(\bar{x} = 0)$ was a measure of the force or acceleration term; both the oscillator and the wave equation have time varying sine and cosine solutions with maximum velocity $\partial\phi/\partial t$ at the zero displacement from equilibrium; that is, where the difference $\bar{\phi} - \phi_0 = 0$.

The diffusion equation, however, describes a different kind of behaviour. It describes a non-equilibrium situation which is moving towards equilibrium at a rate governed by its distance from equilibrium, so that it reaches equilibrium in a time which is theoretically infinite. Readers will have already met this situation in Newton's Law of Cooling, where a hot body at temperatue T_0 stands in a room of lower temperature $\bar{T}$. The rate at which the body cools, i.e. the value of $\partial T/\partial t$, depends on $\bar{T} - T_0$; a cooling graph of this experiment is given in Figure 7.11. The greatest rate of cooling occurs when the temperature difference is greatest and the process slows down as the system approaches equilibrium. Here, of course, $\bar{T} - T_0$ and $\partial T/\partial t$ are both negative.

All non-equilibrium processes of this kind are unidirectional in the sense that they are thermodynamically irreversible. They involve the transport of mass in diffusion, the transport of momentum in friction or viscosity and the transport of energy in conductivity. All such processes involve the loss of useful energy and the generation of entropy.

They are all processes which are governed by random collisions, and we found in the first chapter, where we added vectors of constant length and random phase, that the average distance travelled by particles involved in these processes was proportional, not to the time, but to the square root of the time.

Rewriting the diffusion equation as

$$\frac{\partial^2 \phi}{\partial x^2} = \frac{1}{d} \frac{\partial \phi}{\partial t}$$

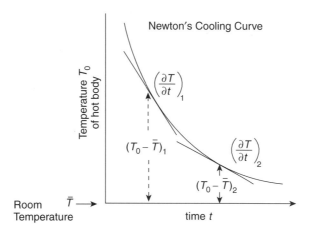

Figure 7.11 Newton's cooling curve shows that the rate of cooling of a hot body $\partial T/\partial t$ depends on the temperature difference between the body and its surrounding, this difference being directly measured by $\partial^2 T/\partial x^2$

we see that the dimensions of the constant d, called the diffusivity, are given by

$$\frac{\phi}{\text{length}^2} = \frac{1}{d}\frac{\phi}{\text{time}}$$

so that d has the dimensions of $\text{length}^2/\text{time}$. The interpretation of this as the square of a characteristic length varying with the square root of time has already been made in Chapter 1.

In a viscous process d is given by η/ρ, where η is the coefficient of viscosity and ρ is the density. In thermal conductivity $d = K/\rho C_p$, where K is the coefficient of thermal conductivity, ρ is the density and C_p is the specific heat at constant pressure.

A magnetic field which is non-uniformly distributed in a conductor has a diffusivity $d = (\mu\sigma)^{-1}$, where μ is the permeability and σ is the conductivity.

Brownian motion is one of the best known examples of random collision processes. The distance x travelled in time t by a particle suffering multiple random collisions is given by Einstein's diffusivity relation

$$d = \frac{\bar{x}^2}{t} = \frac{2RT}{6\pi\eta N}$$

The gas law, $pV = RT$, gives RT as the energy of a mole of such particles at temperature T; a mole contains N particles, where N is Avogadro's number and $RT/N = kT$, the average energy of the individual particles, where k is Boltzmann's constant.

The process is governed, therefore, by the ratio of the energy of the particles to the coefficient of viscosity, which measures the frictional force. The higher the temperature, the greater is the energy, the less the effect of the frictional force and the greater the average distance travelled.

Wave Equation with Diffusion Effects

In natural systems we can rarely find pure waves which propagate free from the energy-loss mechanisms we have been discussing, but if these losses are not too serious we can describe the total propagation in space and time by a combination of the wave and diffusion equations.

If we try to solve the combined equation

$$\frac{\partial^2 \phi}{\partial x^2} = \frac{1}{c^2}\frac{\partial^2 \phi}{\partial t^2} + \frac{1}{d}\frac{\partial \phi}{\partial t}$$

we shall not obtain a pure sine or cosine solution.

Let us try the solution

$$\phi = \phi_m e^{i(\omega t - \gamma x)}$$

where ϕ_m is the maximum amplitude. This gives

$$i^2\gamma^2 = i^2\frac{\omega^2}{c^2} + i\frac{\omega}{d}$$

or

$$\gamma^2 = \frac{\omega^2}{c^2} - i\frac{\omega}{d}$$

giving a complex value for γ. But $\omega^2/c^2 = k^2$, where k is the wave number, and if we put $\gamma = k - i\alpha$ we obtain

$$\gamma^2 = k^2 - 2ik\alpha - \alpha^2 \approx k^2 - i\,2\,k\alpha \quad if \ \alpha \ll k$$

The solution for ϕ then becomes

$$\phi = \phi_m\, e^{i(\omega t - \gamma x)} = \phi_m\, e^{-\alpha x}\, e^{i(\omega t - kx)}$$

i.e. a sine or cosine oscillation of maximum amplitude ϕ_m which decays exponentially with distance. The physical significance of the condition $\alpha \ll k = 2\pi/\lambda$ is that many wavelengths λ are contained in the distance $1/\alpha$ before the amplitude decays to $\phi_m e^{-1}$ at $x = 1/\alpha$. Diffusion mechanisms will cause attenuation or energy loss from the wave; the energy in a wave is proportional to the square of its amplitude and therefore decays as $e^{-2\alpha x}$.

(Problems 7.15, 7.16, 7.17)

Appendix

Physical interpretation of

$$\frac{\partial^2\phi}{\partial x^2} + \frac{\partial^2\phi}{\partial y^2} + \frac{\partial^2\phi}{\partial z^2} \equiv \nabla^2\phi$$

At a certain point O of the scalar field, $\phi = \phi_0$. Constructing a cube around the point O having sides of length l gives for the average value over the cube volume

$$\bar{\phi}l^3 = \iiint_{-l/2}^{+l/2} \phi\, \mathrm{d}x\, \mathrm{d}y\, \mathrm{d}z$$

Expanding ϕ about the point O by a Taylor series gives

$$\begin{aligned}
\phi = \phi_0 &+ \left(\frac{\partial\phi}{\partial x}\right)_0 x + \left(\frac{\partial\phi}{\partial y}\right)_0 y + \left(\frac{\partial\phi}{\partial z}\right)_0 z \\
&+ \frac{1}{2}\left[\left(\frac{\partial^2\phi}{\partial x^2}\right)_0 x^2 + \left(\frac{\partial^2\phi}{\partial y^2}\right)_0 y^2 + \left(\frac{\partial^2\phi}{\partial z^2}\right)_0 z^2\right] \\
&+ \left(\frac{\partial^2\phi}{\partial x\partial y}\right)_0 xy + \left(\frac{\partial^2\phi}{\partial y\partial z}\right)_0 yz + \left(\frac{\partial^2\phi}{\partial z\partial x}\right)_0 zx + \cdots
\end{aligned}$$

Integrating from $-l/2$ to $+l/2$ removes all the functions of the form

$$\left(\frac{\partial\phi}{\partial x}\right)_0 x \quad \text{and} \quad \left(\frac{\partial^2\phi}{\partial x\partial y}\right)_0 xy$$

whose integrals are zero, leaving, since

$$\iiint_{-l/2}^{+l/2} x^2\, dx\, dy\, dz = \frac{l^5}{12}$$

$$\bar{\phi}l^3 = \phi_0 l^3 + \frac{l^5}{24}\left(\frac{\partial^2\phi}{\partial x^2} + \frac{\partial^2\phi}{\partial y^2} + \frac{\partial^2\phi}{\partial z^2}\right)_0$$

i.e.

$$\bar{\phi} - \phi_0 = \frac{l^2}{24}(\nabla^2\phi)_0$$

where l is a constant.

Problem 7.1

The figure shows the mesh representation of a transmission line of inductance L_0 per unit length and capacitance C_0 per unit length. Use equations of the form

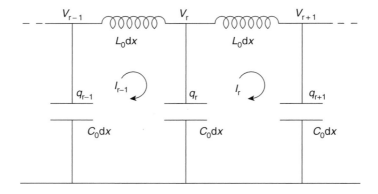

$$I_{r-1} - I_r = \frac{d}{dt}q_r = C_0\, dx\, \frac{d}{dt}V_r$$

and

$$L_0\, dx\, \frac{d}{dt}I_r = V_r - V_{r+1}$$

together with the method of the final section of Chapter 4 to show that the voltage and current wave equations are

$$\frac{\partial^2 V}{\partial x^2} = L_0 C_0 \frac{\partial^2 V}{\partial t^2}$$

and

$$\frac{\partial^2 I}{\partial x^2} = L_0 C_0 \frac{\partial^2 I}{\partial t^2}$$

Problem 7.2
Show that the characteristic impedance for a pair of Lecher wires of radius r and separation d in a medium of permeability μ and permittivity ε is given by

$$Z_0 = \frac{1}{\pi}\sqrt{\frac{\mu}{\varepsilon}}\log_e\frac{d}{r}$$

Problem 7.3
In a short-circuited lossless transmission line integrate the magnetic (inductive) energy $\frac{1}{2}L_0 I^2$ and the electric (potential) energy $\frac{1}{2}C_0 V^2$ over the last quarter wavelength (0 to $-\lambda/4$) to show that they are equal.

Problem 7.4
Show, in Problem 7.3, that the sum of the instantaneous values of the two energies over the last quarter wavelength is equal to the maximum value of either.

Problem 7.5
Show that the impedance of a real transmission line seen from a position x on the line is given by

$$Z_x = Z_0 \frac{A e^{-\gamma x} - B e^{+\gamma x}}{A e^{-\gamma x} + B e^{+\gamma x}}$$

where γ is the propagation constant and A and B are the current amplitudes at $x = 0$ of the waves travelling in the positive and negative x-directions respectively. If the line has a length l and is terminated by a load Z_L, show that

$$Z_L = Z_0 \frac{A e^{-\gamma l} - B e^{\gamma l}}{A e^{-\gamma l} + B e^{\gamma l}}$$

Problem 7.6
Show that the input impedance of the line of Problem 7.5; that is, the impedance of the line at $x = 0$, is given by

$$Z_i = Z_0 \left(\frac{Z_0 \sinh \gamma l + Z_L \cosh \gamma l}{Z_0 \cosh \gamma l + Z_L \sinh \gamma l}\right)$$

$$(Note : 2\cosh\gamma l = e^{\gamma l} + e^{-\gamma l}$$
$$2\sinh\gamma l = e^{\gamma l} - e^{-\gamma l})$$

Problem 7.7
If the transmission line of Problem 7.6 is short-circuited, show that its input impedance is given by

$$Z_{sc} = Z_0 \tanh \gamma l$$

and when it is open-circuited the input impedance is

$$Z_{0c} = Z_0 \coth \gamma l$$

By taking the product of these quantities, suggest a method for measuring the characteristic impedance of the line.

Problem 7.8

Show that the input impedance of a short-circuited loss-free line of lenght l is given by

$$Z_i = i\sqrt{\frac{L_0}{C_0}} \tan \frac{2\pi l}{\lambda}$$

and by sketching the variation of the ratio $Z_i/\sqrt{L_0/C_0}$ with l, show that for l just greater than $(2n+1)\lambda/4$, Z_i is capacitive, and for l just greater than $n\lambda/2$ it is inductive. (This provides a positive or negative reactance to match another line.)

Problem 7.9

Show that a line of characteristic impedance Z_0 may be matched to a load Z_L by a loss-free quarter wavelength line of characteristic impedance Z_m if $Z_m^2 = Z_0 Z_L$.

   (Hint—calculate the input impedance at the $Z_0 Z_m$ junction.)

Problem 7.10

Show that a short-circuited quarter wavelength loss-free line has an infinite impedance and that if it is bridged across another transmission line it will not affect the fundamental wavelength but will short-circuit any undesirable second harmonic.

Problem 7.11

Show that a loss-free line of characteristic impedance Z_0 and length $n\lambda/2$ may be used to couple two high frequency circuits without affecting other impedances.

Problem 7.12

A transmission line has $Z_1 = i\omega L$ and $Z_2 = (i\omega C)^{-1}$. If, for a range of frequencies ω, the phase shift per section β is very small show that $\beta = k$ the wave number and that the phase velocity is independent of the frequency.

Problem 7.13

In a transmission line with losses where $R_0/\omega L_0$ and $G_0/\omega C_0$ are both small quantities expand the expression for the propagation constant

$$\gamma = [(R_0 + i\omega L_0)(G_0 + i\omega C_0)]^{1/2}$$

to show that the attenuation constant

$$\alpha = \frac{R_0}{2}\sqrt{\frac{C_0}{L_0}} + \frac{G_0}{2}\sqrt{\frac{L_0}{C_0}}$$

and the wave number

$$k = \omega\sqrt{L_0 C_0} = \frac{\omega}{v}$$

Show that for $G_0 = 0$ the Q value of such a line is given by $k/2\alpha$.

Problem 7.14

Expand the expression for the characteristic impedance of the transmission line of Problem 7.13 in terms of the characteristic impedance of a lossless line to show that if

$$\frac{R_0}{L_0} = \frac{G_0}{C_0}$$

the impedance remains real because the phase effects introduced by the series and shunt losses are equal but opposite.

Problem 7.15

The wave description of an electron of total energy E in a potential well of depth V over the region $0 < x < l$ is given by Schrödinger's time independent wave equation

$$\frac{\partial^2 \psi}{\partial x^2} + \frac{8\pi^2 m}{h^2}(E - V)\psi = 0$$

where m is the electron mass and h is Planck's constant. (Note that $V = 0$ within the well.)

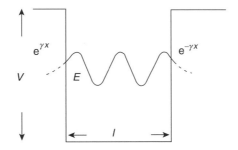

Show that for $E > V$ (inside the potential well) the solution for ψ is a standing wave solution but for $E < V$ (outside the region $0 < x < l$) the x dependence of ψ is $e^{\pm \gamma x}$, where

$$\gamma = \frac{2\pi}{h}\sqrt{2m(V - E)}$$

Problem 7.16

A localized magnetic field H in an electrically conducting medium of permeability μ and conductivity σ will diffuse through the medium in the x-direction at a rate given by

$$\frac{\partial H}{\partial t} = \frac{1}{\mu\sigma}\frac{\partial^2 H}{\partial x^2}$$

Show that the time of decay of the field is given approximately by $L^2\mu\sigma$, where L is the extent of the medium, and show that for a copper sphere of radius 1 m this time is less than 100s.

$$\mu \,(\text{copper}) = 1 \cdot 26 \times 10^{-6} \ \text{H m}^{-1}$$
$$\sigma \,(\text{copper}) = 5 \cdot 8 \times 10^{7} \ \text{S m}^{-1}$$

(If the earth's core were molten iron its field would freely decay in approximately 15×10^3 years. In the sun the local field would take 10^{10} years to decay. When σ is very high the local field will change only by being carried away by the movement of the medium—such a field is said to be 'frozen' into the medium—the field lines are stretched and exert a restoring force against the motion.)

Problem 7.17

A point x_0 at the centre of a large slab of material of thermal coductivity k, specific heat C and density ρ has an infinitely high temperature T at a time t_0. If the heat diffuses through the medium at a rate given by

$$\frac{\partial T}{\partial t} = \frac{k}{\rho C} \frac{\partial^2 T}{\partial x^2} = d \frac{\partial^2 T}{\partial x^2}$$

show that the heat flow along the x-aixs is given by

$$f(\alpha, t) = \frac{r}{\sqrt{\pi}} e^{-(r\alpha)^2},$$

where

$$\alpha = (x - x_0) \quad \text{and} \quad r = \frac{1}{2\sqrt{dt}}$$

by inserting this solution in the differential equation. The solution is a Guassian function; its behaviour with x and t in this problem is shown in Fig. 10.12. At (x_0, t_0) the function is the Dirac delta function. The Guassian curves decay in height and widen with time as the heat spreads through the medium, the total heat, i.e. the area under the Gaussian curve, remaining constant.

Summary of Important Results

Lossless Transmission Line

Inductance per unit length $= L_0$ or μ

Capacitance per unit length $= C_0$ or ε

Wave Equation

$$\frac{\partial^2 V}{\partial x^2} = \frac{1}{v^2} \frac{\partial^2 V}{\partial t^2} \text{ (voltage)}$$

$$\frac{\partial^2 I}{\partial x^2} = \frac{1}{v^2} \frac{\partial^2 I}{\partial t^2} \text{ (current)}$$

Phase Velocity

$$v^2 = \frac{1}{L_0 C_0} \quad \text{or} \quad \frac{1}{\mu\varepsilon}$$

Characteristic Impedance

$$Z_0 = \frac{V}{I} = \sqrt{\frac{L_0}{C_0}} \quad \text{or} \quad \sqrt{\frac{\mu}{\varepsilon}} \quad \text{(for right-going wave)}$$

$$(-Z_0 \text{ for left-going wave})$$

Transmission Line with Losses

Resistane R_0 per unit length
Shunt conductance G_0 per unit length
Wave equation takes form

$$e^{i\omega t}\left(\frac{\partial^2 V}{\partial x^2} - \gamma^2 V\right) = 0 \quad \text{(same for } I\text{)}$$

where $\gamma = \alpha + ik$ is the propagation constant

$$\alpha = \text{attenuation coefficient}$$
$$k = \text{wave number}$$

giving

$$V = A\,e^{-\alpha x}\,e^{i(\omega t - kx)} + B\,e^{\alpha x}\,e^{i(\omega t + kx)}$$

Characteristic Impedance

$$Z_0' = \frac{V}{I} = \sqrt{\frac{R_0 + i\omega L_0}{G_0 + i\omega C_0}} \quad \text{(right-going wave)}$$
$$(-Z_0' \text{ for left-going wave})$$

Wave Attenuation

Energy absorption in a medium described by diffusion equation

$$\frac{\partial^2 \phi}{\partial x^2} = \frac{1}{d}\frac{\partial \phi}{\partial t}$$

Add to wave equation to account for attenuation giving

$$\frac{\partial^2 \phi}{\partial x^2} = \frac{1}{c^2}\frac{\partial^2 \phi}{\partial t^2} + \frac{1}{d}\frac{\partial \phi}{\partial t}$$

with exponentially decaying solution

$$\phi = \phi_m\,e^{-\alpha x}\,e^{i(\omega t - kx)}$$

8

Electromagnetic Waves

Earlier chapters have shown that the velocity of waves through a medium is determined by the inertia and the elasticity of the medium. These two properties are capable of storing wave energy in the medium, and in the absence of energy dissipation they also determine the impedance presented by the medium to the waves. In addition, when there is no loss mechanism a pure wave equation with a sine or cosine solution will always be obtained, but this equation will be modified by any resistive or loss term to give an oscillatory solution which decays with time or distance.

These physical processes describe exactly the propagation of electromagnetic waves through a medium. The magnetic inertia of the medium, as in the case of the transmission line, is provided by the inductive property of the medium, i.e. the permeability μ, which has the units of henries per metre. The elasticity or capacitive property of the medium is provided by the permittivity ε, with units of farads per metre. The storage of magnetic energy arises through the permeability μ; the potential or electric field energy is stored through the permittivity ε.

If the material is defined as a dielectric, only μ and ε are effective and a pure wave equation for both the magnetic field vector H and the electric field vector E will result. If the medium is a conductor, having conductivity σ (the inverse of resistivity) with dimensions of siemens per metre or (ohms m)$^{-1}$, in addition to μ and ε, then some of the wave energy will be dissipated and absorption will take place.

In this chapter we will consider first the propagation of electromagnetic waves in a medium characterized by μ and ε only, and then treat the general case of a medium having μ, ε and σ properties.

Maxwell's Equations

Electromagnetic waves arise whenever an electric charge changes its velocity. Electrons moving from a higher to a lower energy level in an atom will radiate a wave of a particular frequency and wavelength. A very hot ionized gas consisting of charged particles will radiate waves over a continuous spectrum as the paths of individual particles are curved in

The Physics of Vibrations and Waves, 6th Edition   H. J. Pain
© 2005 John Wiley & Sons, Ltd

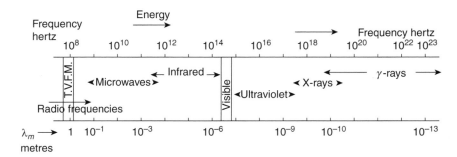

Figure 8.1 Wavelengths and frequencies in the electromagnetic spectrum

mutual collisions. This radiation is called 'Bremsstrahlung'. The radiation of electro-magnetic waves from an aerial is due to the oscillatory motion of charges in an alternating current flowing in the aerial.

Figure 8.1 shows the frequency spectrum of electromagnetic waves. All of these waves exhibit the same physical characteristics.

It is quite remarkable that the whole of electromagnetic theory can be described by the four vector relations in Maxwell's equations. In examining these relations in detail we shall see that two are steady state; that is, independent of time, and that two are time-varying.

The two time-varying equations are mathematically sufficient to produce separate wave equations for the electric and magnetic field vectors, E and H, but the steady state equations help to identify the wave nature as transverse.

The first time-varying equation relates the *time* variation of the magnetic induction, $\mu H = B$, with the *space* variation of E; that is

$$\frac{\partial}{\partial t}(\mu H) \text{ is connected with} \frac{\partial E}{\partial z}(\text{say})$$

This is nothing but a form of Lenz's or Faraday's Law, as we shall see.

The second time-varying equation states that the *time* variation of εE defines the *space* variation of H, that is

$$\frac{\partial}{\partial t}(\varepsilon E) \text{ is connected with} \frac{\partial H}{\partial z}(\text{say})$$

Again we shall see that this is really a statement of Ampere's Law.

These equations show that the variations of E in time and space affect those of H and vice versa. E and H cannot be considered as isolated quantities but are interdependent.

The product εE has dimensions

$$\frac{\text{farads}}{\text{metre}} \times \frac{\text{volts}}{\text{metre}} = \frac{\text{charge}}{\text{area}}$$

This charge per unit area is called the displacement charge $D = \varepsilon E$.

Physically it appears in a dielectric when an applied electric field polarizes the constituent atoms or molecules and charge moves across any plane in the dielectric which

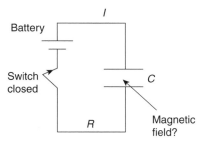

Figure 8.2 In this circuit, when the switch is closed the conduction current charges the condenser. Throughout charging the quantity $\varepsilon\mathbf{E}$ in the volume of the condenser is changing and the displacement current per unit area $\partial/\partial t\,(\varepsilon\mathbf{E})$ is associated with the magnetic field present between the condenser plates

is normal to the applied field direction. If the applied field is varying or alternating with time we see that the dimensions of

$$\frac{\partial\mathbf{D}}{\partial t} = \frac{\partial}{\partial t}(\varepsilon\mathbf{E}) = \frac{\text{charge}}{\text{time} \times \text{area}}$$

current per unit area. This current is called the displacement current. It is comparatively simple to visualize this current in a dielectric where physical charges may move—it is not easy to associate a displacement current with free space in the absence of a material but it may always be expressed as $I_d = \varepsilon(\partial\phi_E/\partial t)$, where ϕ_E is the electric field flux through a surface.

Consider what happens in the electric circuit of Figure 8.2 when the switch is closed and the battery begins to charge the condenser C to a potential V. A current I obeying Ohm's Law ($V = IR$) will flow through the connecting leads as long as the condenser is charging and a compass needle or other magnetic field detector placed near the leads will show the presence of the magnetic field associated with that current. But suppose a magnetic field detector (shielded from all outside effects) is placed in the region between the condenser plates where no ohmic or conduction current is flowing. Would it detect a magnetic field? The answer is yes; all the magnetic field effects from a current exist in this region as long as the condenser is charging, that is, as long as the potential difference and the electric field between the condenser plates are changing.

It was Maxwell's major contribution to electromagnetic theory to assert that the existence of a time-changing electric field in free space gave rise to a displacement current. The same result follows from considering the conservation of charge. The flow of charge into any small volume in space must equal that flowing out. If the volume includes the top plate of the condenser the ohmic current through the leads produces the flow into the volume, while the displacement current represents the flow out.

In future, therefore, two different kinds of current will have to be considered:

1. The familar conduction current obeying Ohm's Law ($V = IR$) and
2. The displacement current of density $\partial\mathbf{D}/\partial t$.

In a medium of permeability μ and permittivity ε, but where the conductivity $\sigma = 0$, *the displacement current will be the only current flowing*. In this case a pure wave equation for E and H will follow and there will be no energy loss or attenuation.

When $\sigma \neq 0$ a resistive element allows the conduction current to flow, energy loss will follow, a diffusion term is added to the wave equation and the wave amplitude will attenuate exponentially with distance. We shall see that the relative magnitude of these two currents is frequency-dependent and that their ratio governs whether the medium behaves as a conductor or as a dielectric.

Electromagnetic Waves in a Medium having Finite Permeability μ and Permittivity ε but with Conductivity $\sigma = 0$

We shall consider a system of plane waves and choose the plane xy as that region over which the wave properties are constant. These properties will not vary with respect to x and y and all derivatives $\partial/\partial x$ and $\partial/\partial y$ will be zero.

The first time-varying equation of Maxwell is written in vector notation as

$$\text{curl } \mathbf{E} = \nabla \times \mathbf{E} = -\frac{\partial \mathbf{B}}{\partial t} = -\mu \frac{\partial \mathbf{H}}{\partial t}$$

This represents three component equations:

$$\left.\begin{array}{l} -\mu\dfrac{\partial}{\partial t}H_x = \dfrac{\partial}{\partial y}E_z - \dfrac{\partial}{\partial z}E_y \\[2mm] -\mu\dfrac{\partial}{\partial t}H_y = \dfrac{\partial}{\partial z}E_x - \dfrac{\partial}{\partial x}E_z \\[2mm] -\mu\dfrac{\partial}{\partial t}H_z = \dfrac{\partial}{\partial x}E_y - \dfrac{\partial}{\partial y}E_x \end{array}\right\} \tag{8.1}$$

where the subscripts represent the component directions. E_x, E_y and E_z are, respectively, the magnitudes of $\mathbf{E}_x\mathbf{E}_y$ and $\mathbf{E}_z$. Similarly, H_x, H_y and H_z are the magnitudes of $\mathbf{H}_x\mathbf{H}_y$ and $\mathbf{H}_z$. The dimensions of these equations may be written

$$-\frac{\mu H}{\text{time}} = \frac{E}{\text{length}}$$

and multiplying each side by (length)^2 gives

$$-\frac{\mu H}{\text{time}} \times \text{area} = E \times \text{length}$$

i.e.

$$\frac{\text{total magnetic flux}}{\text{time}} = \text{volts}$$

This is dimensionally of the form of Lenz's or Faraday's Law.

The second time-varying equation of Maxwell is written in vector notation as

$$\text{curl } \mathbf{H} = \nabla \times \mathbf{H} = \frac{\partial \mathbf{D}}{\partial t} = \varepsilon \frac{\partial \mathbf{E}}{\partial t}$$

This represents three component equations:

$$\left.\begin{aligned}
\varepsilon \frac{\partial}{\partial t} E_x &= \frac{\partial}{\partial y} H_z - \frac{\partial}{\partial z} H_y \\
\varepsilon \frac{\partial}{\partial t} E_y &= \frac{\partial}{\partial z} H_x - \frac{\partial}{\partial x} H_z \\
\varepsilon \frac{\partial}{\partial t} E_z &= \frac{\partial}{\partial x} H_y - \frac{\partial}{\partial y} H_x
\end{aligned}\right\} \tag{8.2}$$

The dimensions of these equations may be written

$$\frac{\text{current } I}{\text{area}} = \frac{H}{\text{length}}$$

and multiplying both sides by a length gives

$$\frac{\text{current}}{\text{length}} = \frac{I}{\text{length}} = H$$

which is dimensionally of the form of Ampere's Law (i.e. the circular magnetic field at radius r due to the current I flowing in a straight wire is given by $H = I/2\pi r$). Maxwell's first steady state equation may be written

$$\text{div } \mathbf{D} = \nabla \cdot \mathbf{D} = \varepsilon \left(\frac{\partial E_x}{\partial x} + \frac{\partial E_y}{\partial y} + \frac{\partial E_z}{\partial z} \right) = \rho \tag{8.3}$$

where ε is constant and ρ is the charge density. This states that over a small volume element $\mathrm{d}x \, \mathrm{d}y \, \mathrm{d}z$ of charge density ρ the change of displacement depends upon the value of ρ.

When $\rho = 0$ the equation becomes

$$\varepsilon \left(\frac{\partial E_x}{\partial x} + \frac{\partial E_y}{\partial y} + \frac{\partial E_z}{\partial z} \right) = 0 \tag{8.3a}$$

so that if the displacement $D = \varepsilon E$ is graphically represented by flux lines which must begin and end on electric charges, the number of flux lines entering the volume element $\mathrm{d}x \, \mathrm{d}y \, \mathrm{d}z$ must equal the number leaving it.

The second steady state equation is written

$$\text{div } \mathbf{B} = \nabla \cdot \mathbf{B} = \mu \left(\frac{\partial H_x}{\partial x} + \frac{\partial H_y}{\partial y} + \frac{\partial H_z}{\partial z} \right) = 0 \tag{8.4}$$

Again this states that an equal number of magnetic induction lines enter and leave the volume $dx \, dy \, dz$. This is a physical consequence of the non-existence of isolated magnetic poles, i.e. a single north pole or south pole.

Whereas the charge density ρ in equation (8.3) can be positive, i.e. a source of flux lines (or displacement), or negative, i.e. a sink of flux lines (or displacement), no separate source or sink of magnetic induction can exist in isolation, every source being matched by a sink of equal strength.

The Wave Equation for Electromagnetic Waves

Since, with these plane waves, all derivatives with respect to x and y are zero. equations (8.1) and (8.4) give

$$-\mu \frac{\partial H_z}{\partial t} = 0 \quad \text{and} \quad \frac{\partial H_z}{\partial z} = 0$$

therefore, H_z is constant in space and time and because we are considering only the oscillatory nature of H a constant H_z can have no effect on the wave motion. We can therefore put $H_z = 0$. A similar consideration of equations (8.2) and (8.3a) leads to the result that $E_z = 0$.

The absence of variation in H_z and E_z means that the oscillations or variations in H and E occur in directions perpendicular to the z-direction. We shall see that this leads to the conclusion that electromagnetic waves are transverse waves.

In addition to having plane waves we shall simplify our picture by considering only *plane-polarized* waves.

We can choose the electric field vibration to be in either the x or y direction. Let us consider E_x only, with $E_y = 0$. In this case equations (8.1) give

$$-\mu \frac{\partial H_y}{\partial t} = \frac{\partial E_x}{\partial z} \tag{8.1a}$$

and equations (8.2) give

$$\varepsilon \frac{\partial E_x}{\partial t} = -\frac{\partial H_y}{\partial z} \tag{8.2a}$$

Using the fact that

$$\frac{\partial^2}{\partial z \partial t} = \frac{\partial^2}{\partial t \partial z}$$

it follows by taking $\partial/\partial t$ of equation (8.1a) and $\partial/\partial z$ of equation (8.2a) that

$$\frac{\partial^2}{\partial z^2} H_y = \mu \varepsilon \frac{\partial^2}{\partial t^2} H_y \quad \text{(the wave equation for } H_y)$$

Similarly, by taking $\partial/\partial t$ of (8.2a) and $\partial/\partial z$ of (8.1a), we obtain

$$\frac{\partial^2}{\partial z^2}E_x = \mu\varepsilon\frac{\partial^2}{\partial t^2}E_x \quad \text{(the wave equation for } E_x)$$

Thus, the vectors E_x and H_y both obey the same wave equation, propagating in the z-direction with the same velocity $v^2 = 1/\mu\varepsilon$. In free space the velocity is that of light, that is, $c^2 = 1/\mu_0\varepsilon_0$, where μ_0 is the permeability of free space and ε_0 is the permittivity of free space.

The solutions to these wave equations may be written, for plane waves, as

$$E_x = E_0 \sin\frac{2\pi}{\lambda}(vt - z)$$

$$H_y = H_0 \sin\frac{2\pi}{\lambda}(vt - z)$$

where E_0 and H_0 are the maximum amplitude values of E and H. Note that the sine (or cosine) solutions means that no attenuation occurs: only displacement currents are involved and there are no conductive or ohmic currents.

We can represent the electromagnetic wave (E_x, H_y) travelling in the z-direction in Figure 8.3, and recall that because E_z and H_z are constant (or zero) the electromagnetic wave is a transverse wave.

The direction of propagation of the waves will always be in the $\mathbf{E}\times\mathbf{H}$ direction; in this case, $\mathbf{E}\times\mathbf{H}$ has magnitude, E_xH_y and is in the z-direction.

This product has the dimensions

$$\frac{\text{voltage} \times \text{current}}{\text{length} \times \text{length}} = \frac{\text{electrical power}}{\text{area}}$$

measured in units of watts per square metre.

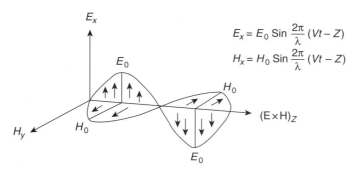

$$E_x = E_0 \sin\frac{2\pi}{\lambda}(Vt - Z)$$

$$H_x = H_0 \sin\frac{2\pi}{\lambda}(Vt - Z)$$

Figure 8.3 In a plane-polarized electromagnetic wave the electric field vector E_x and magnetic field vector H_y are perpendicular to each other and vary sinusoidally. In a non-conducting medium they are in phase. The vector product, $\mathbf{E}\times\mathbf{H}$, gives the direction of energy flow

The vector product, $E \times H$ gives the direction of energy flow. The energy flow per second across unit area is given by the Poynting vector:

$$\frac{1}{2}E \times H^*$$

(Problem 8.1)

Illustration of Poynting Vector

We can illustrate the flow of electromagnetic energy in terms of the Poynting vector by considering the simple circuit of Figure 8.4, where the parallel plate condenser of area A and separation d, containing a dielectric of permittivity ε, is being charged to a voltage V.

Throughout the charging process current flows, and the electric and magnetic field vectors show that the Poynting vector is always directed into the volume Ad occupied by the dielectric.

The capacitance C of the condenser is $\varepsilon A/d$ and the total energy of the condenser at potential V is $\frac{1}{2}CV^2$ joules, which is stored as electrostatic energy. But $V = Ed$, where E is the final value of the electric field, so that the total energy

$$\frac{1}{2}CV^2 = \frac{1}{2}\left(\frac{\varepsilon A}{d}\right)E^2d^2 = \frac{1}{2}(\varepsilon E^2)Ad$$

where Ad is the volume of the condenser.

The electrostatic energy per unit volume stored in the condenser is therefore $\frac{1}{2}\varepsilon E^2$ and results from the flow of electromagnetic energy during charging.

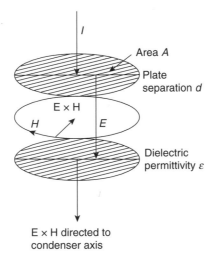

Figure 8.4 During charging the vector **E**×**H** is directed into the condenser volume. At the end of the charging the energy is totally electrostatic and equals the product of the condenser volume, Ad, and the electrostatic energy per unit volume, $\frac{1}{2}\varepsilon E^2$

Impedance of a Dielectric to Electromagnetic Waves

If we put the solutions

$$E_x = E_0 \sin \frac{2\pi}{\lambda}(vt - z)$$

and

$$H_y = H_0 \sin \frac{2\pi}{\lambda}(vt - z)$$

in equation (8.1a) where

$$-\mu \frac{\partial H_y}{\partial t} = \frac{\partial E_x}{\partial z}$$

then

$$-\mu v H_y = -E_x, \quad \text{and since} \quad v^2 = \frac{1}{\mu \varepsilon}$$

$$\sqrt{\mu} H_y = \sqrt{\varepsilon} E_x$$

that is

$$\frac{E_x}{H_y} = \sqrt{\frac{\mu}{\varepsilon}} = \frac{E_0}{H_0}$$

which has the dimensions of ohms.

The value $\sqrt{\mu/\varepsilon}$ therefore represents the *characteristic impedance* of the medium to electromagnetic waves (compare this with the equivalent result $V/I = \sqrt{L_0/C_0} = Z_0$ for the transmission line of the previous chapter).

In free space

$$\frac{E_x}{H_y} = \sqrt{\frac{\mu_0}{\varepsilon_0}} = 376.7\,\Omega$$

so that free space presents an impedance of $376.7\,\Omega$ to electromagnetic waves travelling through it.

It follows from

$$\frac{E_x}{H_y} = \sqrt{\frac{\mu}{\varepsilon}} \quad \text{that} \quad \frac{E_x^2}{H_y^2} = \frac{\mu}{\varepsilon}$$

and therefore

$$\varepsilon E_x^2 = \mu H_y^2$$

Both of these quantities have the dimensions of energy per unit volume, for instance εE_x^2 has dimensions

$$\frac{\text{farads}}{\text{metre}} \times \frac{\text{volts}^2}{\text{metres}^2} = \frac{\text{joules}}{\text{metres}^3}$$

as we saw in the illustration of the Poynting vector. Thus, for a dielectric the electrostatic energy $\frac{1}{2}\varepsilon E_x^2$ per unit volume in an electromagnetic wave equals the magnetic energy per unit volume $\frac{1}{2}\mu H_y^2$ and the total energy is the sum $\frac{1}{2}\varepsilon E_x^2 + \frac{1}{2}\mu H_y^2$.

This gives the instantaneous value of the energy per unit volume and we know that, in the wave,

$$E_x = E_0 \sin{(2\pi/\lambda)}(vt - z)$$

and

$$H_y = H_0 \sin{(2\pi/\lambda)}(vt - z)$$

so that the *time average value* of the energy per unit volume is

$$\frac{1}{2}\varepsilon \bar{E}_x^2 + \frac{1}{2}\mu \bar{H}_y^2 = \frac{1}{4}\varepsilon E_0^2 + \frac{1}{4}\mu H_0^2$$
$$= \frac{1}{2}\varepsilon E_0^2 \, \text{J m}^{-3}$$

Now the amount of energy in an electromagnetic wave which crosses unit area in unit time is called the intensity, I, of the wave and is evidently $(\frac{1}{2}\varepsilon E_0^2)v$ where v is the velocity of the wave.

This gives the time averaged value of the Poynting vector and, for an electromagnetic wave in free space we have

$$I = \frac{1}{2}c\varepsilon_0 E_0^2 = \frac{1}{2}c\mu_0 H_0^2 \, \text{W m}^{-2}$$

(Problems 8.2, 8.3, 8.4, 8.5, 8.6, 8.7, 8.8, 8.9, 8.10, 8.11)

Electromagnetic Waves in a Medium of Properties μ, ε and σ (where $\sigma \neq 0$)

From a physical point of view the electric vector in electromagnetic waves plays a much more significant role than the magnetic vector, e.g. most optical effects are associated with the electric vector. We shall therefore concentrate our discussion on the electric field behaviour.

In a medium of conductivity $\sigma = 0$ we have obtained the wave equation

$$\frac{\partial^2 E_x}{\partial z^2} = \mu\varepsilon \frac{\partial^2 E_x}{\partial t^2}$$

where the right hand term, rewritten

$$\mu \frac{\partial}{\partial t}\left[\frac{\partial}{\partial t}(\varepsilon E_x)\right]$$

shows that we are considering a term

$$\mu \frac{\partial}{\partial t}\left[\frac{\text{displacement current}}{\text{area}}\right]$$

When $\sigma \neq 0$ we must also consider the conduction currents which flow. These currents are given by Ohm's Law as $I = V/R$, and we define the current density; that is, the current per unit area, as

$$J = \frac{I}{\text{Area}} = \frac{1}{R \times \text{Length}} \times \frac{V}{\text{Length}} = \sigma E$$

where σ is the conductivity $1/(R \times \text{Length})$ and E is the electric field. $J = \sigma E$ is another form of Ohm's Law.

With both displacement and conduction currents flowing, Maxwell's second time-varying equation reads, in vector form,

$$\nabla \times \mathbf{H} = \frac{\partial}{\partial t}\mathbf{D} + \mathbf{J} \tag{8.5}$$

each term on the right hand side having dimensions of current per unit area. The presence of the conduction current modifies the wave equation by adding a second term of the same form to its righthand side, namely

$$\mu\frac{\partial}{\partial t}\left(\frac{\text{current}}{\text{area}}\right) \text{ which is } \mu\frac{\partial}{\partial t}(\mathbf{J}) = \mu\frac{\partial}{\partial t}(\sigma\mathbf{E})$$

The final equation is therefore given by

$$\boxed{\frac{\partial^2}{\partial z^2}E_x = \mu\varepsilon\frac{\partial^2}{\partial t^2}E_x + \mu\sigma\frac{\partial}{\partial t}E_x} \tag{8.6}$$

and this equation may be derived formally by writing the component equation of (8.5) as

$$\varepsilon\frac{\partial E_x}{\partial t} + \sigma E_x = -\frac{\partial H_y}{\partial z} \tag{8.5a}$$

together with

$$-\mu\frac{\partial H_y}{\partial t} = \frac{\partial E_x}{\partial z} \tag{8.1a}$$

and taking $\partial/\partial t$ of (8.5a) and $\partial/\partial z$ of (8.1a). We see immediately that the presence of the resistive or dissipation term, which allows conduction currents to flow, will add a diffusion term of the type discussed in the last chapter to the pure wave equation. The product $(\mu\sigma)^{-1}$ is called the magnetic diffusivity, and has the dimensions $L^2 T^{-1}$, as we expect of all diffusion coefficients.

We are now going to look for the behaviour of E_x in this new equation, with the assumption that its time-variation is simple harmonic, so that $E_x = E_0\,e^{i\omega t}$. Using this value in equation (8.6) gives

$$\frac{\partial^2 E_x}{\partial z^2} - (i\omega\mu\sigma - \omega^2\mu\varepsilon)E_x = 0$$

which is in the form of equation (7.5), written

$$\frac{\partial^2 E_x}{\partial z^2} - \gamma^2 E_x = 0$$

where $\gamma^2 = i\omega\mu\sigma - \omega^2\mu\varepsilon$.

We saw in Chapter 7 that this produced a solution with the term $e^{-\gamma z}$ or $e^{+\gamma z}$, but we concentrate on the E_x oscillation in the positive z-direction by writing

$$E_x = E_0 e^{i\omega t} e^{-\gamma z}$$

In order to assign a suitable value to γ we must go back to equation (8.6) and consider the relative magnitudes of the two right hand side terms. If the medium is a dielectric, only displacement currents will flow. When the medium is a conductor, the ohmic currents of the second term on the right hand side will be dominant. The ratio of the magnitudes of the conduction current density to the displacement current density is the ratio of the two right hand side terms. This ratio is

$$\frac{\mathbf{J}}{\partial\mathbf{D}/\partial t} = \frac{\sigma E_x}{\partial/\partial t(\varepsilon E_x)} = \frac{\sigma E_x}{\partial/\partial t(\varepsilon E_0 e^{i\omega t})} = \frac{\sigma E_x}{i\omega\varepsilon E_x} = \frac{\sigma}{i\omega\varepsilon}$$

We see immediately from the presence of i that the phase of the displacement current is 90° ahead of that of the ohmic or conduction current. It is also 90° ahead of the electric field E_x so the displacement current dissipates no power.

For a conductor, where $\mathbf{J} \gg \partial\mathbf{D}/\partial t$, we have $\sigma \gg \omega\varepsilon$, and $\gamma^2 = i\sigma(\omega\mu) - \omega\varepsilon(\omega\mu)$ becomes

$$\gamma^2 \approx i\sigma\omega\mu$$

to a high order of accuracy.
Now

$$\sqrt{i} = \frac{1+i}{\sqrt{2}}$$

so that

$$\gamma = (1+i)\left(\frac{\omega\mu\sigma}{2}\right)^{1/2}$$

and

$$E_x = E_0 e^{i\omega t} e^{-\gamma z}$$
$$= E_0 e^{-(\omega\mu\sigma/2)^{1/2}z} e^{i[\omega t - (\omega\mu\sigma/2)^{1/2}z]}$$

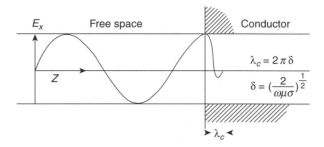

Figure 8.5 Electromagnetic waves in a dielectric strike the plane surface of a conductor, and the electric field vector E_0 is damped to a value $E_0\,e^{-1}$ in a distance of $(2/\omega\mu\sigma)^{1/2}$, the 'skin depth'. This explains the electrical shielding properties of a conductor. λ_c is the wavelength in the conductor

a progressive wave in the positive z-direction with an amplitude decaying with the factor $e^{-(\omega\mu\sigma/2)^{1/2}z}$.

Note that the product $\omega\mu\sigma$ has dimensions L^{-2}.

(Problem 8.12)

Skin Depth

After travelling a distance

$$\delta = \left(\frac{2}{\omega\mu\sigma}\right)^{1/2}$$

in the conductor the electric field vector has decayed to a value $E_x = E_0\,e^{-1}$; this distance is called the *skin depth* (Figure 8.5).

For copper, with $\mu \approx \mu_0$ and $\sigma = 5.8 \times 10^7$ S m^{-1} at a frequency of 60 Hz, $\delta \approx 9$ mm; at 1 MHz, $\delta \approx 6.6 \times 10^{-5}$ m and at 30 000 MHz (radar wavelength of 1 cm), $\delta \approx 3.8 \times 10^{-7}$ m.

Thus, high frequency electromagnetic waves propagate only a very small distance in a conductor. The electric field is confined to a very small region at the surface; significant currents will flow only at the surface and the resistance of the conductor therefore increases with frequency. We see also why a conductor can act to 'shield' a region from electromagnetic waves.

Electromagnetic Wave Velocity in a Conductor and Anomalous Dispersion

The phase velocity of the wave v is given by

$$v = \frac{\omega}{k} = \frac{\omega}{(\omega\mu\sigma/2)^{1/2}} = \omega\delta = \left(\frac{2\omega}{\mu\sigma}\right)^{1/2} = \nu\lambda_c$$

When δ is small, v is small, and the refractive index c/v of a conductor can be very large. We shall see later that this can explain the high optical reflectivities of good conductors. The velocity $v = \omega\delta = 2\pi\nu\delta$, so that λ_c in the conductor is $2\pi\delta$ and can be very small. Since v is a function of the frequency an electrical conductor is a dispersive medium to electromagnetic waves. Moreover, as the table below shows us, $\partial v/\partial\lambda$ is negative, so that the conductor is anomalously dispersive and the group velocity is greater than the wave velocity. Since $c^2/v^2 = \mu\varepsilon/\mu_0\varepsilon_0 = \mu_r\varepsilon_r$, where the subscript r defines non-dimensional relative values; that is, $\mu/\mu_0 = \mu_r$, $\varepsilon/\varepsilon_0 = \varepsilon_r$, then for $\mu_r \approx 1$

$$\varepsilon_r v^2 = c^2$$

and

$$\frac{\partial}{\partial\lambda}\varepsilon_r = -\frac{2}{v}\varepsilon_r\frac{\partial v}{\partial\lambda}$$

which confirms our statement in the chapter on group velocity that for $\partial\varepsilon_r/\partial\lambda$ positive a medium is anomalously dispersive. We see too that $c^2/v^2 = \varepsilon_r = n^2$, where n is the refractive index, so that the curve in Figure 3.9 showing the reactive behaviour of the oscillator impedance at displacement resonance is also showing the behaviour of n. This relative value of the permittivity is, of course, familiarly known as the dielectric constant when the frequency is low. This identity is lost at higher frequencies because the permittivity is frequency-dependent.

Note that $\lambda_c = 2\pi\delta$ is very small, and that when an electromagnetic wave strikes a conducting surface the electric field vector will drop to about 1% of its surface value in a distance equal to $\frac{3}{4}\lambda_c = 4.6\,\delta$. Effectively, therefore, the electromagnetic wave travels less than one wavelength into the conductor.

Frequency	$\lambda_{\text{free space}}$	δ (m)	$v_{\text{conductor}} = \omega\delta$ (m/s)	Refractive index ($c/v_{\text{conductor}}$)
60	5000 km	9×10^{-3}	3.2	9.5×10^7
10^6	300 m	6.6×10^{-5}	4.1×10^2	7.3×10^5
3×10^{10}	10^{-2} m	3.9×10^{-7}	7.1×10^4	4.2×10^3

(Problems 8.13, 8.14, 8.15)

When is a Medium a Conductor or a Dielectric?

We have already seen that in any medium having $\mu\varepsilon$ and σ properties the magnitude of the ratio of the conduction current density to the displacement current density

$$\frac{J}{\partial D/\partial t} = \frac{\sigma}{\omega\varepsilon}$$

a non-dimensional quantity.

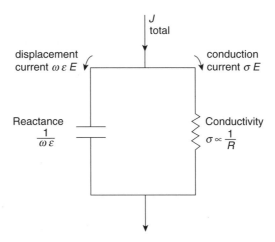

Figure 8.6   A simple circuit showing the response of a conducting medium to an electromagnetic wave. The total current density J is divided by the parallel circuit in the ratio $\sigma/\omega\varepsilon$ (the ratio of the conduction current density to the displacement current density). A large conductance σ (small resistance) gives a large conduction current while a small capacitative reactance $1/\omega\varepsilon$ allows a large displacement current to flow. For a conductor $\sigma/\omega\varepsilon \geq 100$; for a dielectric $\omega\varepsilon/\sigma \geq 100$. Note the frequency dependence of this ratio. At $\omega \approx 10^{20}$ rad/s copper is a dielectric to X-rays

We may therefore represent the medium by the simple circuit in Figure 8.6 where the total current is divided between the two branches, a capacitative branch of reactance $1/\omega\varepsilon$ (ohms-metres) and a resistive branch of conductance σ (siemens/metre). If σ is large the resistivity is small, and most of the current flows through the σ branch and is conductive. If the capacitative reactance $1/\omega\varepsilon$ is so small that it takes most of the current, this current is the displacement current and the medium behaves as a dielectric.

Quite arbitrarily we say that if

$$\frac{J}{\partial D/\partial t} = \frac{\sigma}{\omega\varepsilon} > 100$$

then conduction currents dominate and the medium is a conductor. If

$$\frac{\partial D/\partial t}{J} = \frac{\omega\varepsilon}{\sigma} > 100$$

then displacement currents dominate and the material behaves as a dielectric. Between these values exist a range of quasi-conductors; some of the semi-conductors fall into this category.

The ratio $\sigma/\omega\varepsilon$ is, however, frequency dependent, and a conductor at one frequency may be a dielectric at another.

For copper, which has $\sigma = 5.8 \times 10^7$ S m^{-1} and $\varepsilon \approx \varepsilon_0 = 9 \times 10^{-12}$ F m^{-1},

$$\frac{\sigma}{\omega\varepsilon} \approx \frac{10^{18}}{\text{frequency}}$$

so up to a frequency of 10^{16} Hz (the frequency of ultraviolet light) $\sigma/\omega\varepsilon > 100$, and copper is a conductor. At a frequency of 10^{20} Hz, however (the frequency of X-rays), $\omega\varepsilon/\sigma > 100$, and copper behaves as a dielectric. This explains why X-rays travel distances equivalent to many wavelengths in copper.

Typically, an insulator has $\sigma \approx 10^{-15}$ S m^{-1} and $\varepsilon \approx 10^{-11}$F m^{-1}, which gives

$$\frac{\omega\varepsilon}{\sigma} \approx 10^{4}\omega$$

so the conduction current is negligible at all frequencies.

Why will an Electromagnetic Wave not Propagate into a Conductor?

To answer this question we need only consider the simple circuit where a condenser C discharges through a resistance R. The voltage equation gives

$$\frac{q}{C} + IR = 0$$

and since $I = dq/dt$, we have

$$\frac{dq}{dt} = -\frac{q}{RC} \quad \text{or} \quad q = q_0\,e^{-t/RC}$$

where q_0 is the initial charge.

We see that an electric field will exist between the plates of the condenser only for a time $t \sim RC$ and will disappear when the charge has had time to distribute itself uniformly throughout the circuit. An electric field can only exist in the presence of a non-uniform charge distribution.

If we take a slab of any medium and place a charge of density q at a point within the slab, the medium will behave as an RC circuit and the equation

$$q = q_0\,e^{-t/RC}$$

becomes

$$q = q_0\,e^{-\sigma/\omega\varepsilon} \rightarrow q_0\,e^{-\sigma t/\varepsilon} \left(\begin{array}{c} \varepsilon \equiv C \\ \hline \sigma \equiv 1/R \end{array} \right)$$

The charge will distribute itself uniformly in a time $t \sim \varepsilon/\sigma$, and the electric field will be maintained for that time only. The time ε/σ is called the *relaxation time* of the medium (RC time of the electrical circuit) and it is a measure of the maximum time for which an electric field can be maintained before the charge distribution becomes uniform.

Any electric field of a frequency ν, where $1/\nu = t > \varepsilon/\sigma$, will not be maintained; only a high frequency field where $1/\nu = t < \varepsilon/\sigma$ will establish itself.

Impedance of a Conducting Medium to Electromagnetic Waves

The impedance of a lossless medium is a real quantity. For the transmission line of Chapter 7 the characteristic impedance

$$Z_0 = \frac{V_+}{I_+} = \sqrt{\frac{L_0}{C_0}}\,\Omega\,;$$

for an electromagnetic wave in a dielectric

$$Z = \frac{E_x}{H_y} = \sqrt{\frac{\mu}{\varepsilon}}\,\Omega$$

with E_x and H_y in phase.

We saw in the case of the transmission line that when the loss mechanisms of a series resistance R_0 and a shunt conductance G_0 were introduced the impedance became the complex quantity

$$\mathbf{Z} = \sqrt{\frac{R_0 + i\omega L_0}{G_0 + i\omega C_0}}$$

We now ask what will be the impedance of a conducting medium of properties μ, ε and σ to electromagnetic waves? If the ratio of E_x to H_y is a complex quantity, it implies that a phase difference exists between the two field vectors.

We have already seen that in a conductor

$$E_x = E_0\,\mathrm{e}^{i\omega t}\,\mathrm{e}^{-\gamma z}$$

where $\gamma = (1 + i)\,(\omega\mu\sigma/2)^{1/2}$, and we shall now write $H_y = H_0\,\mathrm{e}^{i(\omega t - \phi)}\,\mathrm{e}^{-\gamma z}$, suggesting that H_y lags E_x by a phase angle ϕ. This gives the impedance of the conductor as

$$\mathbf{Z}_c = \frac{E_x}{H_y} = \frac{E_0}{H_0}\,\mathrm{e}^{i\phi}$$

Equation (8.1a) gives

$$\frac{\partial E_x}{\partial z} = -\mu\frac{\partial H_y}{\partial t}$$

so that

$$-\gamma E_x = -i\omega\mu H_y$$

and

$$\begin{aligned}
\mathbf{Z}_c &= \frac{E_x}{H_y} = \frac{i\omega\mu}{\gamma} = \frac{i(\omega\mu)}{(1 + i)(\omega\mu\sigma/2)^{1/2}} = \frac{i(1 - i)}{(1 + i)(1 - i)}\left(\frac{2\omega\mu}{\sigma}\right)^{1/2} \\
&= \frac{(1 + i)}{2}\left(\frac{2\omega\mu}{\sigma}\right)^{1/2} = \frac{1 + i}{\sqrt{2}}\left(\frac{\omega\mu}{\sigma}\right)^{1/2} \\
&= \left(\frac{\omega\mu}{\sigma}\right)^{1/2}\left(\frac{1}{\sqrt{2}} + i\frac{1}{\sqrt{2}}\right) = \left(\frac{\omega\mu}{\sigma}\right)^{1/2}\mathrm{e}^{i\phi}
\end{aligned}$$

a vector of magnitude $(\omega\mu/\sigma)^{1/2}$ and phase angle $\phi = 45°$. Thus the magnitude

$$Z_c = \frac{E_0}{H_0} = \left(\frac{\omega\mu}{\sigma}\right)^{1/2}$$

and H_y lags E_x by 45°.

We can also express $\mathbf{Z}_c$ by

$$\mathbf{Z}_c = R + \mathrm{i}X = \left(\frac{\omega\mu}{2\sigma}\right)^{1/2} + \mathrm{i}\left(\frac{\omega\mu}{2\sigma}\right)^{1/2}$$

and also write it

$$\mathbf{Z}_c = \frac{1 + \mathrm{i}}{\sqrt{2}}\left(\frac{\omega\mu}{\sigma}\right)^{1/2}$$

$$= \sqrt{\frac{\mu_0}{\varepsilon_0}\frac{\varepsilon_0}{\varepsilon}\frac{\mu}{\mu_0}\frac{\omega\varepsilon}{\sigma}}\,\mathrm{e}^{\mathrm{i}\phi}$$

of magnitude

$$|Z_c| = 376.6\,\Omega\,\sqrt{\frac{\mu_r}{\varepsilon_r}}\sqrt{\frac{\omega\varepsilon}{\sigma}}$$

At a wavelength $\lambda = 10^{-1}$ m, i.e. at a frequency $\nu = 3000$ MHz, the value of $\omega\varepsilon/\sigma$ for copper is 2.9×10^{-9} and $\mu_r \approx \varepsilon_r \approx 1$. This gives a magnitude $Z_c = 0.02\,\Omega$ at this frequency; for $\sigma = \infty$, $Z_c = 0$, and the electric field vector E_x vanishes, so we can say that when Z_c is small or zero the conductor behaves as a short circuit to the electric field. This sets up large conduction currents and the magnetic energy is increased.

In a dielectric, the impedance

$$Z = \frac{E_x}{H_y} = \sqrt{\frac{\mu}{\varepsilon}}$$

led to the equivalence of the electric and magnetic field energy densities; that is, $\frac{1}{2}\mu H_y^2 = \frac{1}{2}\varepsilon E_x^2$. In a conductor, the magnitude of the impedance

$$Z_c = \left|\frac{E_x}{H_y}\right| = \left(\frac{\omega\mu}{\sigma}\right)^{1/2}$$

so that the ratio of the magnetic to the electric field energy density in the wave is

$$\frac{\frac{1}{2}\mu H_y^2}{\frac{1}{2}\varepsilon E_x^2} = \frac{\mu}{\varepsilon}\frac{\sigma}{\omega\mu} = \frac{\sigma}{\omega\varepsilon}$$

We already know that this ratio is very large for a conductor for it is the ratio of conduction to displacement currents, so that in a conductor the magnetic field energy dominates the electric field energy and increases as the electric field energy decreases.

Reflection and Transmission of Electromagnetic Waves at a Boundary

Normal Incidence

An infinite plane boundary separates two media of impedances Z_1 and Z_2 (real or complex) in Figure 8.7.

The electromagnetic wave normal to the boundary has the components shown where subscripts i, r and t denote incident, reflected and transmitted, respectively. Note that the vector direction $(\mathbf{E}_r \times \mathbf{H}_r)$ must be opposite to that of $(\mathbf{E}_i \times \mathbf{H}_i)$ to satisfy the energy flow condition of the Poynting vector.

The boundary conditions, from electromagnetic theory, are that the components of the field vectors $\mathbf{E}$ and $\mathbf{H}$ tangential or parallel to the boundary are continuous across the boundary.

Thus

$$E_i + E_r = E_t$$

and

$$H_i + H_r = H_t$$

where

$$\frac{E_i}{H_i} = Z_1, \quad \frac{E_r}{H_r} = -Z_1 \quad \text{and} \quad \frac{E_t}{H_t} = Z_2$$

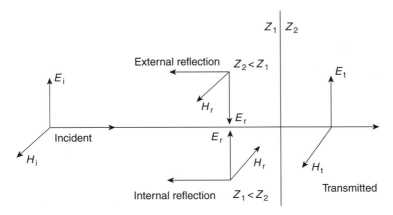

Figure 8.7 Reflection and transmission of an electromagnetic wave incident normally on a plane between media of impedances Z_1 and Z_2. The Poynting vector of the reflected wave $(\mathbf{E} \times \mathbf{H})_r$ shows that either **E** or **H** may be reversed in phase, depending on the relative magnitudes of Z_1 and Z_2

From these relations it is easy to show that the amplitude reflection coefficient

$$R = \frac{E_r}{E_i} = \frac{Z_2 - Z_1}{Z_2 + Z_1}$$

and the amplitude transmission coefficient

$$T = \frac{E_t}{E_i} = \frac{2Z_2}{Z_2 + Z_1}$$

in agreement with the reflection and transmission coefficients we have found for the acoustic pressure p (Chapter 6) and voltage V (Chapter 7). If the wave is travelling in air and strikes a perfect conductor of $Z_2 = 0$ at normal incidence then

$$\frac{E_r}{E_i} = \frac{Z_2 - Z_1}{Z_2 + Z_1} = -1$$

giving complete reflection and

$$\frac{E_t}{E_i} = \frac{2Z_2}{Z_2 + Z_1} = 0$$

Thus, good conductors are very good reflectors of electromagnetic waves, e.g. lightwaves are well reflected from metal surfaces. (See Summary on p. 550.)

Oblique Incidence and Fresnel's Equations for Dielectrics

When the incident wave is oblique and not normal to the infinite boundary of Figure 8.7 we may still use the boundary conditions of the preceding section for these apply to the *tangential* components of **E** and **H** at the boundary and remain valid.

In Figure 8.8(a) **H** is perpendicular to the plane of the paper with tangential components H_i, H_r and H_t but the tangential components of **E** become

$$E_i \cos\theta, \ E_r \cos\theta \quad and \quad E_t \cos\phi, \text{ respectively.}$$

In Figure 8.8(b) **E** is perpendicular to the plane of the paper with tangential components E_i, E_r and E_t but the tangential components of **H** become $H_i \cos\theta$, $H_r \cos\theta$ and $H_t \cos\phi$.

Using these components in the expressions for the reflextion and transmission coefficients we have, for Figure 8.8(a)

$$\frac{E_r \cos\theta}{E_i \cos\theta} = \frac{E_t \cos\phi/H_t - E_i \cos\theta/H_i}{E_t \cos\phi/H_t + E_i \cos\theta/H_i}$$

so

$$R_{\parallel} = \frac{E_r}{E_i} = \frac{Z_2 \cos\phi - Z_1 \cos\theta}{Z_2 \cos\phi + Z_1 \cos\theta}$$

where $R_{\parallel}$ is the reflection coefficient amplitude when **E** lies in the plane of incidence.

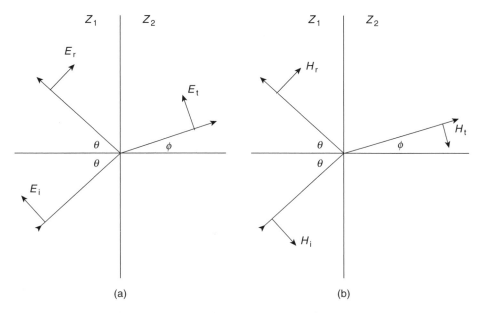

Figure 8.8 Incident, reflected and transmitted components of a plane polarized electromagnetic wave at oblique incidence to the plane boundary separating media of impedances Z_1 and Z_2. The electric vector lies in the plane of incidence in (a) and is perpendicular to the plane of incidence in (b)

For the transmission coefficient in Figure 8.8(a)

$$\frac{E_t \cos \phi}{E_i \cos \theta} = \frac{2E_t \cos \phi / H_t}{E_i \cos \theta / H_i + E_t \cos \phi / H_t}$$

so

$$T_{\parallel} = \frac{E_t}{E_i} = \frac{2Z_2 \cos \theta}{Z_1 \cos \theta + Z_2 \cos \phi}$$

A similar procedure for Figure 8.8(b) where **E** is perpendicular to the plane of incidence yields

$$R_{\perp} = \frac{Z_2 \cos \theta - Z_1 \cos \phi}{Z_2 \cos \theta + Z_1 \cos \phi}$$

and

$$T_{\perp} = \frac{2Z_2 \cos \theta}{Z_2 \cos \theta + Z_1 \cos \phi}$$

Now the relation between the refractive index n of the dielectric and its impedance Z is given by

$$n = \frac{c}{v} = \sqrt{\frac{\mu \varepsilon}{\mu_0 \varepsilon_0}} = \sqrt{\varepsilon_r} = \frac{Z(\text{free space})}{Z(\text{dielectric})}$$

where

$$\frac{\mu}{\mu_0} = \mu_r \approx 1.$$

Hence we have

$$\frac{Z_1}{Z_2} = \frac{n_2}{n_1} = \frac{\sin \theta}{\sin \phi}$$

from Snell's Law and we may write the reflection and transmission amplitude coefficients as

$$R_\parallel = \frac{\tan (\phi - \theta)}{\tan (\phi + \theta)}, \qquad T_\parallel = \frac{4 \sin \phi \cos \theta}{\sin 2\phi + \sin 2\theta}$$

$$R_\perp = \frac{\sin (\phi - \theta)}{\sin (\phi + \theta)}, \qquad T_\perp = \frac{2 \sin \phi \cos \theta}{\sin (\phi + \theta)}$$

In this form the expressions for the coefficients are known as Fresnel's Equations. They are plotted in Figure 8.9 for $n_2/n_1 = 1.5$ and they contain several significant features.

When θ is very small and incidence approaches the normal we have $\theta \to 0$ and $\phi \to 0$ so that

$$\sin (\phi - \theta) \sim \tan (\phi - \theta) \sim (\phi - \theta)$$

and

$$R_\parallel \sim R_\perp \sim \frac{(\phi - \theta)}{(\phi + \theta)} \sim \frac{\dfrac{1}{n_2} - \dfrac{1}{n_1}}{\dfrac{1}{n_2} + \dfrac{1}{n_1}} = \frac{n_1 - n_2}{n_1 + n_2}$$

Thus, the reflected intensity

$$R_{\theta \to 0}^2 = \frac{I_r}{I_i} = \left(\frac{n_1 - n_2}{n_1 + n_2} \right)^2$$

$$\sim 0.4 \quad \text{at an air-glass interface.}$$

We note also that when $\tan (\theta + \phi) = \infty$ and $\theta + \phi = 90°$ then $R_\parallel = 0$.

In this case only $R_\perp$ is finite and the reflected light is completely plane polarized with the electric vector perpendicular to the plane of incidence. This condition defines the value of the Brewster or polarizing angle θ_B for, when θ and ϕ are complementary $\cos \theta_B = \sin \phi$ so

$$n_1 \sin \theta_B = n_2 \sin \phi = n_2 \cos \theta_B$$

and

$$\tan \theta_B = n_2/n_1$$

which, for air to glass defines $\theta_B = 56°$.

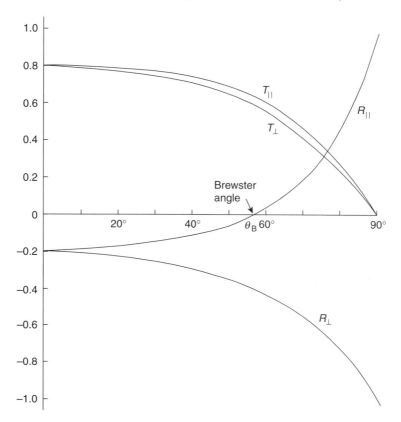

Figure 8.9 Amplitude coefficient R and T of reflection and transmission for $n_2/n_1 = 1.5$. $R_{\parallel}$ and $T_{\parallel}$ refer to the case when the electric field vector E lies in the plane of incidence. $R_{\perp}$ and $T_{\perp}$ apply when E is perpendicular to the plane of incidence. The Brewster angle θ_B defines $\theta + \phi = 90°$ when $R_{\parallel} = 0$ and the reflected light is polarized with the E vector perpendicular to the plane of incidence. $R_{\parallel}$ changes sign (phase) at θ_B. When $\theta < \theta_B$, tan $(\phi - \theta)$ is negative for $n_2/n_1 = 1.5$. When $\theta + \phi \geq 90°$, tan $(\phi + \theta)$ is also negative

A typical modern laboratory use of the Brewster angle is the production of linearly polarized light from a He-Ne laser. If the window at the end of the laser tube is tilted so that the angle of incidence for the emerging light is θ_B and $R_{\parallel} = 0$, then the light with its electric vector parallel to the plane of incidence is totally transmitted while some of the light with transverse polarization ($R_{\perp}$) is reflected back into the laser off-axis. If the light makes multiple transits along the length of the tube before it emerges the transmitted beam is strongly polarized in one plane.

More general but less precise uses involve the partial polarization of light reflected from wet road and other shiny surfaces where refractive indices are in the range $n = 1.3 - 1.6$. Polarized windscreens and spectacles are effective in reducing the glare from such reflections.

Reflection from a Conductor (Normal Incidence)

For Z_2 a conductor and Z_1 free space, the refractive index

$$n = \frac{Z_1}{Z_2} = \frac{\beta}{\alpha + i\alpha}$$

is complex, where

$$\beta = \sqrt{\frac{\mu_0}{\varepsilon_0}}$$

and

$$\alpha = \left(\frac{\omega\mu}{2\sigma}\right)^{1/2}$$

A complex refractive index must always be interpreted in terms of absorption because a complex impedance is determined by a complex propagation constant, e.g. here $Z_2 = i\omega\mu/\gamma$, so that

$$n = \frac{Z_1}{Z_2} = \sqrt{\frac{\mu_0}{\varepsilon_0}}\frac{1}{i\omega\mu}(1+i)\left(\frac{\omega\mu\sigma}{2}\right)^{1/2} = (1-i)\left(\frac{\sigma}{2\omega\varepsilon_0}\right)^{1/2}$$

where

$$\frac{(\mu\mu_0)^{1/2}}{\mu} \approx 1$$

The ratio E_r/E_i is therefore complex (there is a phase difference between the incident and reflected vectors) with a value

$$\frac{E_r}{E_i} = \frac{Z_2 - Z_1}{Z_2 + Z_1} = \frac{\alpha + i\alpha - \beta}{\alpha + i\alpha + \beta} = \frac{1 - \beta/\alpha + i}{1 + \beta/\alpha + i}$$

where $\beta/\alpha \gg 1$.

Since E_r/E_i is complex, the value of the reflected intensity $I_r = (E_r/E_i)^2$ is found by taking the ratio the squares of the moduli of the numerator and the denominator, so that

$$I_r = \frac{|E_r|^2}{|E_i|^2} = \frac{|Z_2 - Z_1|^2}{|Z_2 + Z_1|^2} = \frac{(1 - \beta/\alpha)^2 + 1}{(1 + \beta/\alpha)^2 + 1}$$

$$= 1 - \frac{4\beta/\alpha}{2 + 2\beta/\alpha + (\beta/\alpha)^2} \rightarrow 1 - \frac{4\alpha}{\beta} \quad \text{(for } \beta/\alpha \gg 1\text{)}$$

so that

$$I_r = 1 - 4\left(\frac{\omega\mu}{2\sigma}\right)^{1/2}\left(\frac{\varepsilon_0}{\mu_0}\right)^{1/2} \approx 1 - 2\sqrt{\frac{2\omega\varepsilon_0}{\sigma}}$$

For copper $\sigma = 6 \times 10^7 (\text{ohm m}^{-1})$ and $(2\omega\varepsilon_0/\sigma)^{1/2} \approx 0.01$ at infra-red frequencies. The emission from an electric heater at 10^3K has a peak at $\lambda \approx 2.5 \times 10^{-6}\text{m}$. A metal reflector behind the heater filament reflects $\approx 97\%$ of these infra-red rays with 3% entering the metal to be lost as Joule heating between the metal surface and the skin depth. (see Problem 8.20)

(Problems 8.16, 8.17, 8.18, 8.19, 8.20, 8.21, 8.22, 8.23, 8.24)

Electromagnetic Waves in a Plasma

We saw in Problem 1.4 that when an electron in an atom or, quantum mechanically the charge centre of an electron cloud, moves a small distance from its equilibrium position, the charge separation creates an electric field which acts as a linear restoring force and the resulting motion is simple harmonic with an angular frequency ω_0. For a hydrogen atom

$$\omega_0 \approx 4.5 \times 10^{16} \text{ rad s}^{-1}$$

When a steady electric field is applied to a dielectric, the resulting charge separation between an electron and the rest of its atom induces a polarization field of magnitude

$$P = \frac{n_e e x}{\varepsilon_0}$$

where P defines the dipole moment per unit volume. Here, n_e is the electron number density, x is the displacement from equilibrium and ε_0 is the permittivity of free space.

The value of P per unit electric field is called the susceptibility

$$\chi = \frac{n_e e x}{\varepsilon_0 E}$$

and the permittivity of the dielectric is given by

$$\varepsilon = \varepsilon_0(1 + \chi)$$

The relative permittivity or dielectric constant

$$\varepsilon_r = \frac{\varepsilon}{\varepsilon_0} = (1 + \chi) = \left(1 + \frac{n_e e x}{\varepsilon_0 E}\right) \tag{8.7}$$

A steady electric field E defines a static susceptibility. An alternating electric field E defines a dynamic susceptibility in which case the relative permittivity.

$$\varepsilon_r = n^2$$

where n is the refractive index of the medium.

There may be resistive or damping effects to the electric field within the medium and it is here that our discussion of the forced damped oscillator on p. 66 becomes significant (see Figure 3.9).

If the electric field is that of an electromagnetic wave of angular frequency ω we have $E = E_0 e^{i\omega t}$ and the value of x in equatin (8.7) is that given by equation (3.2) on p. 67 representing curve (a) in Figure 3.9 where F_0 is now the force Ee acting on each electron.

Equation (8.7) now becomes

$$\varepsilon_r = 1 + \chi = 1 + \frac{n_e e^2 m_e (\omega_0^2 - \omega^2)}{\varepsilon_0 [m_e^2 (\omega_0^2 - \omega^2)^2 + \omega^2 r^2]}$$

where m_e is the electron mass, ω_0 is its harmonic frequency within the atom, ω is the electromagnetic wave frequency and r is the damping constant.

This is the solution given to problem 3.10.

Note that for

$$\omega \ll \omega_0$$

$$\varepsilon_r \approx 1 + \frac{n_e e^2}{\varepsilon_0 m_e \omega_0^2} \tag{8.8}$$

and for

$$\omega \gg \omega_0$$

$$\varepsilon_r \approx 1 - \frac{n_e e^2}{\varepsilon_0 m_e \omega^2} \tag{8.9}$$

The factor $n_e e^2 / \varepsilon_0 m_e$ in the second term of ε_r has a particular significance if the material is not a solid but an ionized gas called a plasma. Such a gas consists of ions and electrons of equal number densities $n_i = n_e$ with charges of opposite signs $\pm e$ and masses m_i and m_e where $m_i \gg m_e$. Relative displacements between ions and electrons set up a restoring electric field which returns the electrons to equilibrium. The relatively heavy ions are considered as stationary. The result in Figure 8.10 shows a sheet of negative charge $-n_e ex$

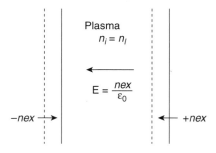

Figure 8.10 In an ionized gas with equal number densities of ions and electrons ($n_i = n_e$) and $m_i \gg m_e$, relative displacements between ions and electrons form thin sheaths of charge $\pm nex$, which generate an electric field $E = nex/\varepsilon_0$ acting on each electron. The motion of each electron is simple harmonic with an electron plasma frequency ω_p where $\omega_p^2 = n_e e^2 / \varepsilon_0 m_e$ rad s^{-1}

per unit area on one side of the plasma slab with the stationary ions producing a sheet of positive charge $+n_e ex$ on the other side (where $n_i = n_e$).

This charge separation generates an electric field E in the plasma of magnitude

$$E = \frac{n_e ex}{\varepsilon_0}$$

which produces an electric force $-n_e e^2 x/\varepsilon_0$ acting on each electron in the direction of its equilibrium position.

The equation of motion of each electron is therefore

$$m_e \ddot{x} + \frac{n_e e^2 x}{\varepsilon_0} = 0$$

and the electron motion is simple harmonic with an angular frequency ω_p where

$$\omega_p^2 = \frac{n_e e^2}{\varepsilon_0 m_e}$$

The angular frequency ω_p is called the electron plasma frequency and plays a significant role in the propagation of electromagnetic waves in the plasma.

In the expression for the refractive index

$$\varepsilon_r = n^2 \approx 1 + \frac{\omega_p^2}{\omega_0^2} \tag{8.8}$$

n is real for all values of ω and waves of that frequency will propagate. However, when

$$\varepsilon_r = n^2 \approx 1 - \frac{\omega_p^2}{\omega^2} \tag{8.9}$$

waves will propagate only when $\omega > \omega_p$

When $\omega_p^2/\omega^2 > 1$

$$n^2 = \frac{c^2}{v^2} = \frac{c^2 k^2}{\omega^2} = 1 - \frac{\omega_p^2}{\omega^2}$$

is negative and the wave number k is considered to be complex with

$$k = k' - i\alpha.$$

In this case, electromagnetic waves incident on the plasma will be attenuated within the plasma, or if α is large enough, will be reflected at the plasma surface.

The electric field of the wave $E = E_0 e^{i(\omega t - kz)}$ becomes $E = E_0 e^{-\alpha z} e^{i(\omega t - k z)}$ and is reduced to $E_0 e^{-1}$ when $z = 1/\alpha = \delta$ the penetration depth. When $\alpha \gg k'$, the penetration is extremely small and

$$k^2 \to -\alpha^2 = \left(1 - \frac{\omega_{\mathrm{p}}}{\omega^2}\right)\frac{\omega^2}{c^2}$$

so that

$$\alpha^2 = \frac{\omega_{\mathrm{p}}^2}{c^2}\left(1 - \frac{\omega^2}{\omega_{\mathrm{p}}^2}\right)$$

and

$$\delta = \frac{1}{\alpha} = \frac{c}{\omega_{\mathrm{p}}}\left(1 - \frac{\omega^2}{\omega_{\mathrm{p}}^2}\right)^{-1/2}$$

When

$$\omega \ll \omega_{\mathrm{p}}, \quad \delta \approx c/\omega_{\mathrm{p}}$$

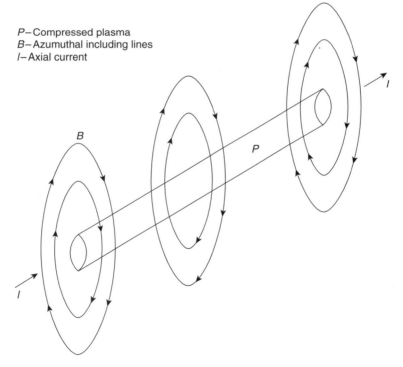

P–Compressed plasma
B–Azumuthal including lines
I–Axial current

Figure 8.11 The pinch effect. A plasma is formed when a large electrical current I is discharged along the axis of a cylindrical tube of gas. The azimuthal magnetic field lines compress the plasma and when the conductivity of the plasma is very high the penetration of the field lines into the plasma is very small

On a laboratory scale number densities of the order $n_e \approx 10^{-6}-10^{-10}$ m^{-3} are produced with electron plasma frequencies in the range $\omega_p \approx 6 \times 10^{10}-6 \times 10^{12}$ rad s^{-1}, several orders below that of visible light.

For these values of ω_p, electromagnetic waves have a penetration depth

$$\delta \approx \frac{c}{\omega_p} \approx 5 \times 10^{-3}-5 \times 10^{-5} \text{ m}$$

The analysis above provides an experimental method of measuring the electron number density of a plasma using electromagnetic waves as a probe. The angular frequency of the transmitted wave is varied until propagation no longer occurs and a reflected wave is detected.

The rejection of magnetic fields by a plasma is exploited in laboratory experiments on controlled thermonuclear fusion. In these a strong magnetic induction **B** is used as the confining mechanism to keep the plasma from the walls of its containing vessel. The magnetic energy per unit volume is given by $B^2/2\mu$ and this has the dimensions of a pressure which opposes and often exceeds that of the hot ionized gas.

The well-known 'pinch effect', Figure 8.11, results when a large current is discharged along the axis of gas contained in a cylindrical tube. The current ionizes the gas and its azimuthal field compresses the plasma in the radial direction towards the axis, increasing its temperature even further. Typical magnitudes in such an experiment are $T \sim 10^8$ K and $n_e \sim 10^{21}$ m^{-3}. This corresponds to a pressure of ~ 14 atmospheres which requires a discharge current $\sim 10^3 R$ A where R m is the radius of the cylinder.

Electromagnetic Waves in the Ionosphere

The simple expression

$$n^2 = 1 - \frac{\omega_p^2}{\omega^2} \tag{8.9}$$

for the index of refraction of a plasma is modified by the presence of an external static magnetic field. This situation exists in the ionosphere which consists of bands of low density ionized gas approximately 300 km above the earth and located within the earth's dipole field of magnetic induction $\mathbf{B}_0$.

A charged particle of velocity **v** in such a field experiences an electric field $\mathbf{E} = \mathbf{v} \times \mathbf{B}_0$ and when **v** is in the plane perpendicular to $\mathbf{B}_0$ it rotates around the field line with an angular frequency $\omega = eB_0/m$, where e is the particle charge and m is its mass. This is most easily seen by considering the force mv^2/r in a circular orbit balancing the electric force $\varepsilon \mathbf{E} = e \cdot \mathbf{v} \times \mathbf{B}_0$.

From $mv^2/r = evB_0$
we have

$$\frac{v}{r} = \frac{eB_0}{m} = 2\pi\left(\frac{v}{2\pi r}\right) = 2\pi f = \omega_{\mathbf{B}}$$

where f is the frequency of precession or the number of orbits per second made by the particle.

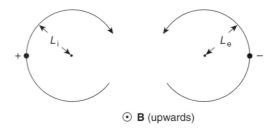

⊙ **B** (upwards)

Figure 8.12 Charged particles of velocity **v** perpendicular to a magnetic field line **B** are bound to the field line and orbit around it due to the Lorentz force $e(\mathbf{v} \times \mathbf{B})$. The radius L of the orbit, the Larmor radius, is given by $L = mv/eB$ and the orbital Larmor frequency is $\omega_{\mathbf{B}} = eB/m$ rad s^{-1}

Figure 8.12 shows the direction of motion for positive and negative charges around a magnetic field line which points upwards out of the paper.

We consider the simplest case of electromagnetic wave propagation along the direction $\mathbf{B}_0$ and assume that

- The amplitude of electron motion is small.

- The value of n_e is low enough to neglect collisional damping.

- The magnetic induction $\mathbf{B}_0 \gg$ the magnetic induction of the electromagnetic wave.

If we consider the electric field to be that of a circularly polarized transverse electromagnetic wave, then we may write $\mathbf{E} = E(\mathbf{r}_1 + i\mathbf{r}_2)$, where $\mathbf{r}_1$ and $\mathbf{r}_2$ are orthogonal (mutually perpendicular) unit vectors and $\mathbf{B}_0$ is along the $\mathbf{r}_3$ direction.

The equation of motion for an electron of velocity **v** is given by

$$m\frac{d\mathbf{v}}{dt} = \mathbf{E}\, e^{i\omega t} + e\mathbf{v} \times \mathbf{B}_0$$

If we take the steady state electron velocity to be of the form

$$\mathbf{v} = v(\mathbf{r}_1 + i\mathbf{r}_2)\, e^{i\omega t}$$

we find that

$$v = \frac{-ie}{m(\omega \pm \omega_{\mathbf{B}})} E$$

satisfies the equation of motion

This means that the electron precessing around $\mathbf{B}_0$ with an angular frequency $\omega_{\mathbf{B}}$ is driven by a rotating electric field of effective frequency $\omega \pm \omega_{\mathbf{B}}$ depending on the sign of the circular polarization.

Due to the electronic motion there is a current density in the plasma given by

$$\mathbf{J} = n_e e\mathbf{v} = \frac{-in_e e^2}{m(\omega \pm \omega_{\mathbf{B}})} \mathbf{E}.$$

In Maxwell's equation

$$\nabla \times \mathbf{H} = \frac{\partial}{\partial t}\mathbf{D} + \mathbf{J} \qquad (8.5)$$

we may write, in the absence of $\mathbf{J}$, $\mathbf{D} = \varepsilon_0\mathbf{E}$ but the presence of $\mathbf{J}$ will modify this and the right hand side of equation (8.5) becomes

$$\frac{\partial}{\partial t}\mathbf{D} + \mathbf{J} = \frac{\partial}{\partial t}\varepsilon_0 E\,\mathrm{e}^{i\omega t} - \frac{in_e e^2}{m(\omega \pm \omega_{\mathbf{B}})}\mathbf{E}$$

$$= i\omega\varepsilon_0\mathbf{E} - \frac{in_e e^2}{m\varepsilon_0(\omega \pm \omega_{\mathbf{B}})}\varepsilon_0\mathbf{E}$$

$$= i\omega\varepsilon_0\left(1 - \frac{\omega_p^2}{\omega(\omega \pm \omega_{\mathbf{B}})}\right)\mathbf{E} = i\omega\varepsilon\mathbf{E}$$

giving

$$\frac{\varepsilon}{\varepsilon_0} = \varepsilon_r = n_\pm^2 = \left(1 - \frac{\omega_p^2}{\omega(\omega \pm \omega_{\mathbf{B}})}\right)$$

We see that the ionosphere is birefringent with two different values of the refractive index, n_+ for the right handed circularly polarized wave and n_- for the left handed incident polarization. These waves propagate at different velocities and their reception by the ionosphere will depend on their polarization. In its lower D layer the ionosphere has an electron number density $n_e \leq 10^9\,\mathrm{m}^{-3}$ with $\omega_p \approx 10^6\,\mathrm{rad\,s}^{-1}$ and for the upper F_2 layer, $n_e \leq 10^{12}\,\mathrm{m}^{-3}$ with $\omega_p \approx 10^7\,\mathrm{rad\,s}^{-1}$. Taking the value of the earth's magnetic field as 3×10^{-5} T; that is (0.3 G) gives an electron precession frequency $\omega_{\mathbf{B}} \approx 6 \times 10^6\,\mathrm{rad\,s}^{-1}$.

Figure 8.13 shows the behaviour of n_+^2 and n_-^2 versus $\omega/\omega_{\mathbf{B}}$ give for the fixed value of $\omega_p/\omega_{\mathbf{B}} = 2$. Other values of $\omega_p/\omega_{\mathbf{B}}$ give curves of a similar shape. In the wide frequency intervals where n_+^2 and n_-^2 have opposite signs (positive or negative), one state of the circular polarization cannot propagate in the plasma and will be reflected when it strikes the ionosphere. The other wave will be partially transmitted. So, when a linearly polarized wave with $\omega \leq \omega_{\mathbf{B}}$ in Figure 8.14 is incident on the ionosphere, the reflected wave will be elliptically polarized. The electron number densities in the ionosphere are measured by varying the frequency ω of the transmitted electromagnetic waves until reflection occurs. This method is similar to that used on the laboratory plasmas of the previous section. However, the value of n_e varies in an ionospheric layer. It is found to increase with height until it reaches a maximum, only to fall off rapidly with a further increase in height. The height for a particular value of n_e is measured by timing the interval between the transmitted and reflected wave.

The analysis above explains the main features of radio reception which are:

- Very high frequencies (VHF) are received over relatively short distances only.

- Medium wave (MW) reception is possible over longer distances and improves at night.

- Short wave (SW) reception is possible over very long distances.

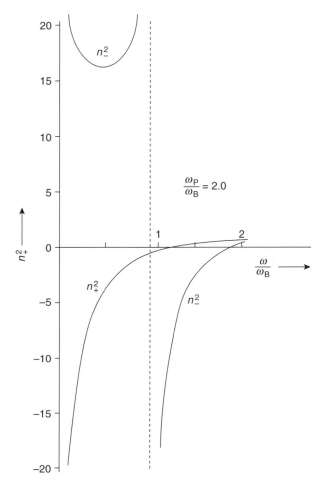

Figure 8.13 The ionospheric plasma is birefringent to electromagnetic waves with different values of the refractive index n_+ for right handed circularly polarized waves and n_- for left handed circularly polarized waves. These values depend upon the ratio of the plasma frequency ω_p to the Larmor frequency ω_B. Graphs of n_+^2 and n_-^2 are shown for a fixed value $\omega_p/\omega_B = 2$ with a horizontal axis ω/ω_B, where ω is the frequency of the propagating e.m. wave

Very high frequencies are greater than ω_p for both the D and F_2 layers; the waves propagate through both layers without reflection (Figure 8.15). The D layer has a plasma frequency $\sim$300 kHz; that is, a wavelength of $\sim$ 1 km and medium waves with $200 < \lambda < 600$ km are attenuated within it. However, the electron number density in the D layer, sustained by ionizing radiation during the day, drops very sharply after sunset and the medium waves are transmitted to the higher F_2 layer where they are reflected and received over longer distances. The D layer is transparent to short waves, $10 < \lambda < 80$ m, but these are reflected by the layer F_2 allowing long-distance radio reception around the earth.

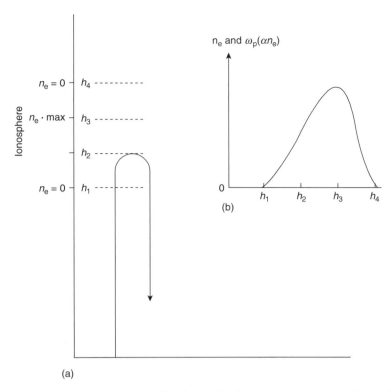

(a)

(b)

Figure 8.14 (a) The number density n_e of a plasma (in this case the ionosphere) may be measured by a probing electromagnetic wave, the frequency of which is varied until reflection occurs. The time of the wave from transmission to reception is a measure of the height at which reflection occurs. The variation of number density n_e with height h in an ionospheric layer is shown in (b)

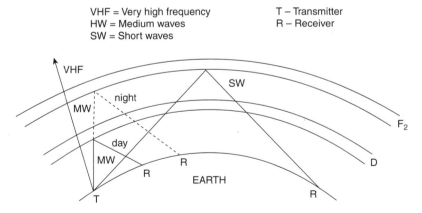

VHF = Very high frequency
HW = Medium waves
SW = Short waves

T – Transmitter
R – Receiver

Figure 8.15 Electron number densities in the ionosphere layers D and F_2 govern the pattern of radio reception. Very high frequencies (VHF) penetrate both layers and are received only over short distances Medium waves (MW) are reflected at the D layer during the daytime but are received over longer distances at night when n_e of the D layer drops and medium waves proceed to the F_2 layer before reflection. Short waves (SW) penetrate the D layer to be reflected at the F_2 layer and are received over very long distances

Problem 8.1

The solutions to the e.m. wave equations are given in Figure 8.3 as

$$E_x = E_0 \sin \frac{2\pi}{\lambda}(vt - z)$$

and

$$H_y = H_0 \sin \frac{2\pi}{\lambda}(vt - z)$$

Use equations (8.1a) and (8.2a) to prove that they have the same wavelength and phase as shown in figure.

Problem 8.2

Show that the concept of $B^2/2\mu$ (magnetic energy per unit volume) as a magnetic pressure accounts for the fact that two parallel wires carrying currents in the same direction are forced together and that reversing one current will force them apart. (Consider a point midway between the two wires.) Show that it also explains the motion of a conductor carrying a current which is situated in a steady externally applied magnetic field.

Problem 8.3

At a distance r from a charge e on a particle of mass m the electric field value is $E = e/4\pi\varepsilon_0 r^2$. Show by integrating the electrostatic energy density over the spherical volume of radius a to infinity and equating it to the value mc^2 that the 'classical' radius of the electron is given by

$$a = 1.41 \times 10^{-15}\,\text{m}$$

Problem 8.4

The rate of generation of heat in a long cylindrical wire carrying a current I is I^2R, where R is the resistance of the wire. Show that this joule heating can be described in terms of the flow of energy into the wire from surrounding space and is equal to the product of the Poynting vector and the surface area of the wire.

Problem 8.5

Show that when a current is increasing in a long uniformly wound solenoid of coil radius r the total inward energy flow rate over a length l (the Poynting vector times the surface area $2\pi r l$) gives the time rate of change of the magnetic energy stored in that length of the solenoid.

Problem 8.6

The plane polarized electromagnetic wave (E_x, H_y) of this chapter travels in free space. Show that its Poynting vector (energy flow in watts per squaremetre) is given by

$$S = E_x H_y = c(\tfrac{1}{2}\varepsilon_0 E_x^2 + \tfrac{1}{2}\mu_0 H_y^2) = c\varepsilon_0 E_x^2$$

where c is the velocity of light. The intensity in such a wave is given by

$$I = \overline{S}_{av} = c\varepsilon_0 \overline{E^2} = \tfrac{1}{2}c\varepsilon_0 E_{max}^2$$

Show that

$$\overline{S} = 1.327 \times 10^{-3} E_{max}^2$$

$$E_{max} = 27.45 \, \overline{S}^{1/2} \, \text{V m}^{-1}$$

$$H_{max} = 7.3 \times 10^{-2} \, \overline{S}^{1/2} \, \text{A m}^{-1}$$

Problem 8.7

A light pulse from a ruby laser consists of a linearly polarized wave train of constant amplitude lasting for 10^{-4} s and carrying energy of 0.3 J. The diameter of the circular cross section of the beam is 5×10^{-3} m. Use the results of Problem 8.6 to calculate the energy density in the beam to show that the root mean square value of the electric field in the wave is

$$2.4 \times 10^5 \, \text{V m}^{-1}$$

Problem 8.8

One square metre of the earth's surface is illuminated by the sun at normal incidence by an energy flux of 1.35 kW. Show that the amplitude of the electric field at the earth's surface is 1010 V m^{-1} and that the associated magnetic field in the wave has an amplitude of 2.7 A m^{-1} (See Problem 8.6). The electric field energy density $\tfrac{1}{2}\varepsilon E^2$ has the dimensions of a pressure. Calculate the **radiation pressure** of sunlight upon the earth.

Problem 8.9

If the total power lost by the sun is equal to the power received per unit area of the earth's surface multiplied by the surface area of a sphere of radius equal to the earth sun distance (15×10^7 km), show that the mass per second converted to radiant energy and lost by the sun is 4.2×10^9 kg. (See Problem 8.6.)

Problem 8.10

A radio station radiates an average power of 10^5 W uniformly over a hemisphere concentric with the station. Find the magnitude of the Poynting vector and the amplitude of the electric and magnetic fields of the plane electromagnetic wave at a point 10 km from the station. (See Problem 8.6)

Problem 8.11

A plane polarized electromagnetic wave propagates along a transmission line consisting of two parallel strips of a perfect conductor containing a medium of permeability μ and permittivity ε. A section of one cubic metre in the figure shows the appropriate field vectors. The electric field E_x generates equal but opposite surface charges on the conductors of magnitude εE_x C m^2. The motion of these surface charges in the direction of wave propagation gives rise to a surface current (as in the discussion associated with Figure 7.1). Show that the magnitude of this current is H_y and that the characteristic impedance of the transmission line is

$$\frac{E_x}{H_y} = \sqrt{\frac{\mu}{\varepsilon}}$$

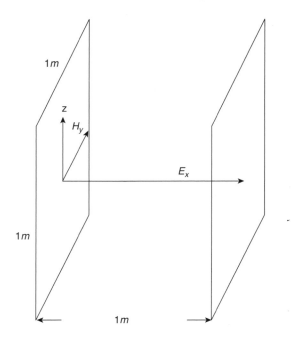

Problem 8.12

Show that equation (8.6) is dimensionally of the form (per unit area)

$$V = L\frac{\mathrm{d}I}{\mathrm{d}t}$$

where V is a voltage, L is an inductance and I is a current.

Problem 8.13

Show that when a group of electromagnetic waves of nearly equal frequencies propagates in a conducting medium the group velocity is twice the wave velocity.

Problem 8.14

A medium has a conductivity $\sigma = 10^{-1}$ S m^{-1} and a relative permittivity $\varepsilon_r = 50$, which is constant with frequency. If the relative permeability $\mu_r = 1$, is the medium a conductor or a dielectric at a frequency of (a) 50 kHz, and (b) 10^4 MHz?

$$[\varepsilon_0 = (36\pi \times 10^9)^{-1}\,\mathrm{F\,m^{-1}}; \ \mu_0 = 4\pi \times 10^{-7}\,\mathrm{H\,m^{-1}}]$$

Answer: (a) $\sigma/\omega\varepsilon = 720$ (conductor)

(b) $\sigma/\omega\varepsilon = 3.6 \times 10^{-3}$ (dielectric).

Problem 8.15

The electrical properties of the Atlantic Ocean are given by

$$\varepsilon_r = 81, \quad \mu_r = 1, \quad \sigma = 4.3\,\mathrm{S\,m^{-1}}$$

Show that it is a conductor up to a frequency of about 10 MHz. What is the longest electromagnetic wavelength you would expect to propagate under water?

Problem 8.16

Show that when a plane electromagnetic wave travelling in air is reflected normally from a plane conducting surface the transmitted magnetic field value $H_t \approx 2H_i$, and that a magnetic standing wave exists in air with a very large standing wave ratio. If the wave is travelling in a conductor and is reflected normally from a plane conductor–air interface, show that $E_t \approx 2E_i$. Show that these two cases are respectively analogous to a short-circuited and an open-circuited transmission line.

Problem 8.17

Show that in a conductor the average value of the Poynting vector is given by

$$S_{av} = \tfrac{1}{2}E_0 H_0 \cos 45°$$
$$= \tfrac{1}{2}H_0^2 \times (\text{real part of } Z_c) \, \text{W m}^2$$

where E_0 and H_0 are the peak field values. A plane 1000 MHz wave travelling in air with $E_0 = 1 \, \text{V m}^{-1}$ is incident normally on a large copper sheet. Show firstly that the real part of the conductor impedance is $8.2 \times 10^{-3} \Omega$ and then (remembering from Problem 8.16 that H_0 doubles in the conductor) show that the average power absorbed by the copper per square metre is 1.6×10^{-7} W.

Problem 8.18

For a good conductor $\varepsilon_r = \mu_r = 1$. Show that when an electromagnetic wave is reflected normally from such a conducting surface its fractional loss of energy ($1 -$ reflection coefficient I_r) is $\approx \sqrt{8\omega\varepsilon/\sigma}$. Note that the ratio of the displacement current density to the conduction current density is therefore a direct measure of the reflectivity of the surface.

Problem 8.19

Using the value of the Poynting vector in the conductor from Problem 8.17, show that the ratio of this value to the value of the Poynting vector in air is $\approx \sqrt{8\omega\varepsilon/\sigma}$, as expected from Problem 8.18.

Problem 8.20

The electromagnetic wave of Problems 8.18 and 8.19 has electric and magnetic field magnitudes in the conductor given by

$$E_x = A e^{-kz} e^{i(\omega t - kz)}$$

and

$$H_y = A \left(\frac{\sigma}{\omega\mu}\right)^{1/2} e^{-kz} e^{i(\omega t - kz)} e^{-i\pi/4}$$

where $k = (\omega\mu\sigma/2)^{1/2}$.

Show that the average value of the Poynting vector in the conductor is given by

$$S_{av} = \tfrac{1}{2}A^2 \left(\frac{\sigma}{2\omega\mu}\right)^{1/2} e^{-2kz} \, (\text{W m}^2)$$

This is the power absorbed per unit area by the conductor. We know, however, that the wave propagates only a distance of the order of the skin depth, so that this power is rapidly transformed. The rate at which it changes with distance is given by $\partial S_{av}/\partial z$, which gives the energy transformed per unit volume in unit time. Show that this quantity is equal to the conductivity σ times the square of the mean value of the electric field vector **E**, that is, the joule heating from currents flowing in the surface of the conductor down to a depth of the order of the skin depth.

Problem 8.21

Show that when light travelling in free space is normally incident on the surface of a dielectric of refractive index n the reflected intensity

$$I_r = \left(\frac{E_r}{E_i}\right)^2 = \left(\frac{1-n}{1+n}\right)^2$$

and the transmitted intensity

$$I_t = \frac{Z_i E_t^2}{Z_t E_i^2} = \frac{4n}{(1+n)^2}$$

(Note $I_r + I_t = 1$.)

Problem 8.22

Show that if the medium of Problem 8.21 is glass ($n = 1.5$) then $I_r = 4\%$ and $I_t = 96\%$. If an electromagnetic wave of 100 MHz is normally incident on water ($\varepsilon_r = 81$) show that $I_r = 65\%$ and $I_t = 35\%$.

Problem 8.23

Light passes normally through a glass plate suffering only one air to glass and one glass to air reflection. What is the loss of intensity?

Problem 8.24

A radiating antenna in simplified form is just a length x_0 of wire in which an oscillating current is maintained. The expression for the radiating power is that used on p. 47 for an oscillating electron

$$P = \frac{\mathrm{d}E}{\mathrm{d}t} = \frac{q^2 \omega^4 x_0^2}{12\pi\varepsilon_0 c^3}$$

where q is the electron charge and ω is the oscillation frequency. The current I in the antenna may be written $I_0 = \omega q$. If $P = \frac{1}{2}RI_0^2$ show that the radiation resistance of the antenna is given by

$$R = \frac{2\pi}{3}\sqrt{\frac{\mu_0}{\varepsilon_0}}\left(\frac{x_0}{\lambda}\right)^2 = 787\left(\frac{x_0}{\lambda}\right)^2 \Omega$$

where λ is the radiated wavelength (an expression valid for $\lambda \gg x_0$).

If the antenna is 30 m long and transmits at a frequency of 5×10^5 H with a root mean square current of 20 A, show that its radiation resistance is $1.97\,\Omega$ and that the power radiated is 400 W. (Verify that $\lambda \gg x_0$.)

Summary of Important Results

Dielectric; μ and $\varepsilon(\sigma = 0)$

Wave equation

$$\frac{\partial^2 E_x}{\partial z^2} = \mu\varepsilon \frac{\partial^2 E_x}{\partial t^2} \quad \left(v^2 = \frac{1}{\mu\varepsilon}\right)$$

$$\frac{\partial^2 H_y}{\partial z^2} = \mu\varepsilon \frac{\partial^2 H_y}{\partial t^2}$$

Impedance

$$\frac{E_x}{H_y} = \sqrt{\frac{\mu}{\varepsilon}} \quad (376.7\Omega \text{ for free space})$$

Energy density $\frac{1}{2}\varepsilon E_x^2 + \frac{1}{2}\mu H_y^2$

$$\text{\textit{Mean energy flow}} = \text{Intensity} = \overline{S} = v(\text{mean energy density})$$
$$= v\left(\tfrac{1}{2}\varepsilon E_x^2 + \tfrac{1}{2}\mu H_y^2\right)_{\text{average}}$$
$$= v\varepsilon\overline{E_x^2} = \tfrac{1}{2}v\varepsilon E_{x(\text{max})}^2$$

Conductor; μ ε *and* σ

Add diffusion equation to wave equation for loss effects from σ

$$\frac{\partial^2 E_x^2}{\partial z^2} = \mu\varepsilon \frac{\partial^2 E_x^2}{\partial t^2} + \mu\sigma \frac{\partial E_x}{\partial t}$$

giving

$$E_x = E_0 \, e^{-kz} \, e^{i(\omega t - kz)}$$

where

$$k^2 = \omega\mu\sigma/2$$

Skin Depth

$$\delta = \frac{1}{k} \quad \text{giving } E_x = E_0 \, e^{-1}$$

Criterion for conductor/dielectric behaviour is ratio

$$\frac{\text{conduction current}}{\text{displacement current}} = \frac{\sigma}{\omega\varepsilon} \quad \text{(note frequency dependence)}$$

Impedance Z_c *(conductor)*

$$\mathbf{Z}_c = \frac{1+i}{\sqrt{2}} \left(\frac{\omega\mu}{\sigma}\right)^{1/2}$$

with magnitude $Z_c = 376.6\sqrt{\mu_r/\varepsilon_r}\,\sqrt{\omega\varepsilon/\sigma}$ ohms

Reflection and Transmission Coefficients (normal incidence),

$$R = \frac{E_r}{E_i} = \frac{Z_2 - Z_1}{Z_2 + Z_1} \quad (E\text{'s and } Z\text{'s may be complex})$$

$$T = \frac{E_t}{E_i} = \frac{2Z_2}{Z_2 + Z_1}$$

Fresnel's Equations (dielectrics)

$$R_\parallel = \frac{\tan{(\phi - \theta)}}{\tan{(\phi + \theta)}}, \quad T_\parallel = \frac{4 \sin \phi \cos \theta}{\sin 2\phi + \sin 2\theta}$$

$$R_\perp = \frac{\sin{(\phi - \theta)}}{\sin{(\phi + \theta)}}, \quad T_\perp = \frac{2 \sin \phi \cos \theta}{\sin{(\phi + \theta)}}$$

Refractive Index

$$n = \frac{c}{v} = \frac{Z\,(\text{free space})}{Z\,(\text{dielectric})}$$

Electromagnetic Waves in a Plasma

Low frequency waves propagate, but a high frequency wave $E_0 \, e^{i\omega t}$ is attenuated or reflected when $\omega < \omega_p$ the plasma frequency, where $\omega_p^2 = n_e e^2 / \varepsilon_0 m_e$. ($n_e$ is the electron number density.)

The plasma has a refractive index n, where

$$n^2 = 1 - \omega_p^2 / \omega^2$$

when $\omega_p \gg \omega_0$, the wave amplitude $E_0 \to E_0 e^{-1}$ in a skin depth distance

$$\delta = \frac{c}{\omega_p} \left(1 - \frac{\omega^2}{\omega_p^2}\right)^{-1/2} \approx \frac{c}{\omega_p}$$

9

Waves in More than One Dimension

Plane Wave Representation in Two and Three Dimensions

Figure 9.1 shows that in two dimensions waves of velocity c may be represented by lines of constant phase propagating in a direction $\mathbf{k}$ which is normal to each line, where the magnitude of $\mathbf{k}$ is the wave number $k = 2\pi/\lambda$.

The direction cosines of $\mathbf{k}$ are given by

$$l = \frac{k_1}{k}, \quad m = \frac{k_2}{k} \quad \text{where} \quad k^2 = k_1^2 + k_2^2$$

and any point $\mathbf{r}(x, y)$ on the line of constant phase satisfies the equation

$$lx + my = p = ct$$

where p is the perpendicular distance from the line to the origin. The displacements at all points $\mathbf{r}(x, y)$ on a given line are in phase and the phase difference ϕ between the origin and a given line is

$$\phi = \frac{2\pi}{\lambda}(\text{path difference}) = \frac{2\pi}{\lambda}p = \mathbf{k} \cdot \mathbf{r} = k_1 x + k_2 y$$
$$= kp$$

Hence, the bracket $(\omega t - \phi) = (\omega t - kx)$ used in a one dimensional wave is replaced by $(\omega t - \mathbf{k} \cdot \mathbf{r})$ in waves of more than one dimension, e.g. we shall use the exponential expression

$$e^{i(\omega t - \mathbf{k} \cdot \mathbf{r})}$$

In three dimensions all points $\mathbf{r}(x, y, z)$ in a given wavefront will lie on *planes* of constant phase satisfying the equation

$$lx + my + nz = p = ct$$

The Physics of Vibrations and Waves, 6th Edition   H. J. Pain
© 2005 John Wiley & Sons, Ltd

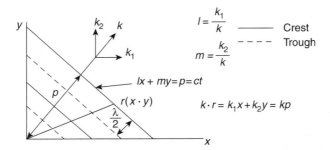

Figure 9.1 Crests and troughs of a two-dimensional plane wave propagating in a general direction **k** (direction cosines l and m). The wave is specified by $lx + my = p = ct$, where p is its perpendicular distance from the origin, travelled in a time t at a velocity c

where the vector **k** which is normal to the plane and in the direction of propagation has direction cosines

$$l = \frac{k_1}{k}, \quad m = \frac{k_2}{k}, \quad n = \frac{k_3}{k}$$

(so that $k^2 = k_1^2 + k_2^2 + k_3^2$) and the perpendicular distance p is given by

$$kp = \mathbf{k} \cdot \mathbf{r} = k_1 x + k_2 y + k_3 z$$

Wave Equation in Two Dimensions

We shall consider waves propagating on a stretched plane membrane of negligible thickness having a mass ρ per unit area and stretched under a uniform tension T. This means that if a line of unit length is drawn in the surface of the membrane, then the material on one side of this line exerts a force T (per unit length) on the material on the other side in a direction perpendicular to that of the line.

If the equilibrium position of the membrane is the xy plane the vibration displacements perpendicular to this plane will be given by z where z depends on the position x, y. In Figure 9.2a where the small rectangular element $ABCD$ of sides δx and δy is vibrating, forces $T\delta x$ and $T\delta y$ are shown acting on the sides in directions which tend to restore the element to its equilibrium position.

In deriving the equation for waves on a string we saw that the tension T along a curved element of string of length dx produced a force perpendicular to x of

$$T\frac{\partial^2 y}{\partial x^2}\, dx$$

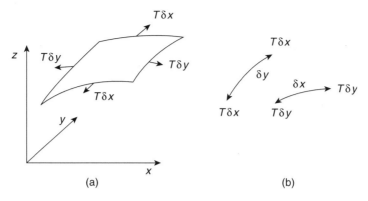

Figure 9.2 Rectangular element of a uniform membrane vibrating in the *z*-direction subject to one restoring force, $T\delta x$, along its sides of length δy and another, $T\delta y$, along its sides of length δx

where *y* was the perpendicular displacement. Here in Figure 9.2b by exactly similar arguments we see that a force $T\delta y$ acting on a membrane element of length δx produces a force

$$T\delta y \frac{\partial^2 z}{\partial x^2} \delta x,$$

where *z* is the perpendicular displacement, whilst another force $T\delta x$ acting on a membrane element of length δy produces a force

$$T\delta x \frac{\partial^2 z}{\partial y^2} \delta y$$

The sum of these restoring forces which act in the *z*-direction is equal to the mass of the element $\rho\,\delta x\,\delta y$ times its perpendicular acceleration in the *z*-direction, so that

$$T\frac{\partial^2 z}{\partial x^2}\delta x\,\delta y + T\frac{\partial^2 z}{\partial y^2}\delta x\,\delta y = \rho\,\delta x\,\delta y\frac{\partial^2 y}{\partial t^2}$$

giving the wave equation in two dimensions as

$$\frac{\partial^2 z}{\partial x^2} + \frac{\partial^2 z}{\partial y^2} = \frac{\rho}{T}\frac{\partial^2 z}{\partial t^2} = \frac{1}{c^2}\frac{\partial^2 z}{\partial t^2}$$

where

$$c^2 = \frac{T}{\rho}$$

The displacement of waves propagating on this membrane will be given by

$$z = A\,\mathrm{e}^{\mathrm{i}(\omega t - \mathbf{k}\cdot\mathbf{r})} = A\,\mathrm{e}^{\mathrm{i}[\omega t - (k_1 x + k_2 y)]}$$

where

$$k^2 = k_1^2 + k_2^2$$

The reader should verify that this expression for z is indeed a solution to the two-dimensional wave equation when $\omega = ck$.

(Problem 9.1)

Wave Guides

Reflection of a 2D Wave at Rigid Boundaries

Let us first consider a 2D wave propagating in a vector direction $\mathbf{k}(k_1, k_2)$ in the xy plane along a membrane of width b stretched under a tension T between two long rigid rods which present an infinite impedance to the wave.

We see from Figure 9.3 that upon reflection from the line $y = b$ the component k_1 remains unaffected whilst k_2 is reversed to $-k_2$. Reflection at $y = 0$ leaves k_1 unaffected whilst $-k_2$ is reversed to its original value k_2. The wave system on the membrane will therefore be given by the superposition of the incident and reflected waves; that is, by

$$\mathbf{z} = A_1 \, e^{i[\omega t - (k_1 x + k_2 y)]} + A_2 \, e^{i[\omega t - (k_1 x - k_2 y)]}$$

subject to the boundary conditions that

$$z = 0 \quad \text{at} \quad y = 0 \quad \text{and} \quad y = b$$

the positions of the frame of infinite impedance.

The condition $z = 0$ at $y = 0$ requires

$$A_2 = -A_1$$

and $z = 0$ at $y = b$ gives

$$\sin k_2 b = 0$$

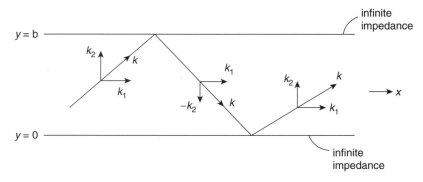

Figure 9.3 Propagation of a two-dimensional wave along a stretched membrane with infinite impedances at $y = 0$ and $y = b$ giving reversal of k_2 at each reflection

or

$$k_2 = \frac{n\pi}{b}$$

(Problem 9.2)

With these values of A_2 and k_2 the displacement of the wave system is given by the real part of **z**, i.e.

$$z = +2A_1 \sin k_2 y \sin (\omega t - k_1 x)$$

which represents a wave travelling along the x direction with a phase velocity

$$v_p = \frac{\omega}{k_1} = \left(\frac{k}{k_1}\right) v$$

where v, the velocity on an infinitely wide membrane, is given by

$$v = \frac{\omega}{k} \quad \text{which is} \quad < v_p$$

because

$$k^2 = k_1^2 + k_2^2$$

Now

$$k^2 = k_1^2 + \frac{n^2\pi^2}{b^2}$$

so

$$k_1 = \left(k^2 - \frac{n^2\pi^2}{b^2}\right)^{1/2} = \left(\frac{\omega^2}{v^2} - \frac{n^2\pi^2}{b^2}\right)^{1/2}$$

and the group velocity for the wave in the x direction

$$v_g = \frac{\partial \omega}{\partial k_1} = \frac{k_1}{\omega} v^2 = \left(\frac{k_1}{k}\right) v$$

giving the product

$$v_p v_g = v^2$$

Since k_1 must be real for the wave to propagate we have, from

$$k_1^2 = k^2 - \frac{n^2\pi^2}{b^2}$$

the condition that

$$k^2 = \frac{\omega^2}{v^2} \geq \frac{n^2\pi^2}{b^2}$$

that is

$$\omega \geq \frac{n\pi v}{b}$$

or

$$\nu \geq \frac{nv}{2b},$$

where n defines the mode number in the y direction. Thus, only waves of certain frequencies ν are allowed to propagate along the membrane which acts as a *wave guide*.

There is a cut-off frequency $n\pi v/b$ for each mode of number n and the wave guide acts as a frequency filter (recall the discussion on similar behaviour in wave propagation on the loaded string in Chapter 4). The presence of the $\sin k_2 y$ term in the expression for the displacement z shows that the amplitude varies across the transverse y direction as shown in Figure 9.4 for the mode values $n = 1, 2, 3$. Thus, along any direction in which the waves meet rigid boundaries a standing wave system will be set up analogous to that on a string of fixed length and we shall discuss the implication of this in the section on normal modes and the method of separation of variables.

Wave guides are used for all wave systems, particularly in those with acoustical and electromagnetic applications. Fibre optics is based on wave guide principles, but the major use of wave guides has been with electromagnetic waves in telecommunications. Here the reflecting surfaces are the sides of a copper tube of circular or rectangular cross section. Note that in this case the free space velocity becomes the velocity of light

$$c = \frac{\omega}{k} < v_{\mathrm{p}}$$

the phase velocity, but the relation $v_{\mathrm{p}} v_{\mathrm{g}} = c^2$ ensures that energy in the wave always travels with a group velocity $v_{\mathrm{g}} < c$.

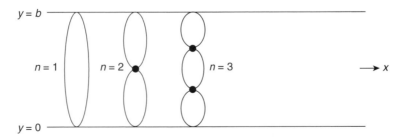

Figure 9.4 Variation of amplitude with y-direction for two-dimensional wave propagating along the membrane of Figure 9.3. Normal modes ($n = 1, 2$ and 3 shown) are set up along any axis bounded by infinite impedances

(Problems 9.3, 9.4, 9.5, 9.6, 9.7, 9.8, 9.9, 9.10, 9.11)

Normal Modes and the Method of Separation of Variables

We have just seen that when waves propagate in more than one dimension a standing wave system will be set up along any axis which is bounded by infinite impedances.

In Chapter 5 we found that standing waves could exist on a string of fixed length l where the displacement was of the form

$$y = A \begin{Bmatrix} \sin \\ \cos \end{Bmatrix} kx \begin{Bmatrix} \sin \\ \cos \end{Bmatrix} \omega_n t,$$

where A is constant and where $\begin{Bmatrix} \sin \\ \cos \end{Bmatrix}$ means that either solution may be used to fit the boundary conditions in space and time. When the string is fixed at both ends, the condition $y = 0$ at $x = 0$ removes the $\cos kx$ solution, and $y = 0$ at $x = l$ requires $k_n l = n\pi$ or $k_n = n\pi/l = 2\pi/\lambda_n$, giving $l = n\lambda_n/2$. Since the wave velocity $c = \nu_n \lambda_n$, this permits frequencies $\omega_n = 2\pi\nu_n = \pi nc/l$, defined as *normal modes of vibration or eigenfrequencies*.

We can obtain this solution in a way which allows us to extend the method to waves in more than one dimension. We have seen that the wave equation

$$\frac{\partial^2 \phi}{\partial x^2} = \frac{1}{c^2} \frac{\partial^2 \phi}{\partial t^2}$$

has a solution which is the product of two terms, one a function of x only and the other a function of t only.

Let us write $\phi = X(x)T(t)$ and apply the method known as separation of variables.

The wave equation then becomes

$$\frac{\partial^2 X}{\partial x^2} \cdot T = \frac{1}{c^2} X \frac{\partial^2 T}{\partial t^2}$$

or

$$X_{xx} T = \frac{1}{c^2} X T_{tt}$$

where the double subscript refers to double differentiation with respect to the variables. Dividing by $\phi = X(x)T(t)$ we have

$$\frac{X_{xx}}{X} = \frac{1}{c^2} \frac{T_{tt}}{T}$$

where the left-hand side depends on x only and the right-hand side depends on t only. However, both x and t are independent variables and the equality between both sides can only be true when both sides are independent of x and t and are equal to a constant, which we shall take, for convenience, as $-k^2$. Thus

$$\frac{X_{xx}}{X} = -k^2, \quad \text{giving} \quad X_{xx} + k^2 X = 0$$

and

$$\frac{1}{c^2}\frac{T_{tt}}{T} = -k^2, \quad \text{giving} \quad T_{tt} + c^2 k^2 T = 0$$

$X(x)$ is therefore of the form $e^{\pm ikx}$ and $T(t)$ is of the form $e^{\pm ickt}$, so that $\phi = A\, e^{\pm ikx}\, e^{\pm ickt}$, where A is constant, and we choose a particular solution in a form already familiar to us by writing

$$\phi = A\, e^{i(ckt - kx)}$$
$$= A\, e^{i(\omega t - kx)} \,,$$

where $\omega = ck$, or we can write

$$\phi = A \left.{\begin{matrix}\sin\\\cos\end{matrix}}\right\} kx \left.{\begin{matrix}\sin\\\cos\end{matrix}}\right\} ckt$$

as above.

Two-Dimensional Case

In extending this method to waves in two dimensions we consider the wave equation in the form

$$\frac{\partial^2 \phi}{\partial x^2} + \frac{\partial^2 \phi}{\partial y^2} = \frac{1}{c^2}\frac{\partial^2 \phi}{\partial t^2}$$

and we write $\phi = X(x)Y(y)T(t)$, where $Y(y)$ is a function of y only.

Differentiating twice and dividing by $\phi = XYT$ gives

$$\frac{X_{xx}}{X} + \frac{Y_{yy}}{Y} = \frac{1}{c^2}\frac{T_{tt}}{T}$$

where the left-hand side depends on x and y only and the right-hand side depends on t only. Since x, y and t are independent variables each side must be equal to a constant, $-k^2$ say. This means that the left-hand side terms in x and y differ by only a constant for all x and y, so that each term is itself equal to a constant. Thus we can write

$$\frac{X_{xx}}{X} = -k_1^2, \quad \frac{Y_{yy}}{Y} = -k_2^2$$

and

$$\frac{1}{c^2}\frac{T_{tt}}{T} = -(k_1^2 + k_2^2) = -k^2$$

giving

$$X_{xx} + k_1^2 X = 0$$
$$Y_{yy} + k_2^2 Y = 0$$
$$T_{tt} + c^2 k^2 T = 0$$

or

$$\phi = A\, e^{\pm ik_1 x}\, e^{\pm ik_2 y}\, e^{\pm ickt}$$

where $k^2 = k_1^2 + k_2^2$. Typically we may write

$$\phi = A \left.{\sin \atop \cos}\right\} k_1 x \left.{\sin \atop \cos}\right\} k_2 y \left.{\sin \atop \cos}\right\} ckt.$$

Three-Dimensional Case

The three-dimensional treatment is merely a further extension. The wave equation is

$$\frac{\partial^2 \phi}{\partial x^2} + \frac{\partial^2 \phi}{\partial y^2} + \frac{\partial^2 \phi}{\partial z^2} = \frac{1}{c^2}\frac{\partial^2 \phi}{\partial t^2}$$

with a solution

$$\phi = X(x)Y(y)Z(z)T(t)$$

yielding

$$\phi = A \left.{\sin \atop \cos}\right\} k_1 x \left.{\sin \atop \cos}\right\} k_2 y \left.{\sin \atop \cos}\right\} k_3 z \left.{\sin \atop \cos}\right\} ckt,$$

where $k_1^2 + k_2^2 + k_3^2 = k^2$.

Using vector notation we may write

$$\phi = A\, e^{i(\omega t - \mathbf{k}\cdot\mathbf{r})}, \quad \text{where} \quad \mathbf{k}\cdot\mathbf{r} = k_1 x + k_2 y + k_3 z$$

Normal Modes in Two Dimensions on a Rectangular Membrane

Suppose waves proceed in a general direction $\mathbf{k}$ on the rectangular membrane of sides a and b shown in Figure 9.5. Each dotted wave line is separated by a distance $\lambda/2$ and a standing wave system will exist whenever $a = n_1\text{AA}'$ and $b = n_2\text{BB}'$, where n_1 and n_2 are integers.

But

$$\text{AA}' = \frac{\lambda}{2\cos\alpha} = \frac{\lambda}{2}\frac{k}{k_1} = \frac{\lambda}{2}\frac{2\pi}{\lambda}\frac{1}{k_1} = \frac{\pi}{k_1}$$

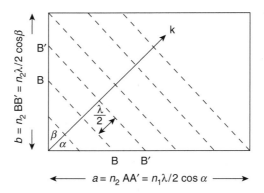

Figure 9.5 Normal modes on a rectangular membrane in a direction **k** satisfying boundary conditions of zero displacement at the edges of length $a = n_1\lambda/2 \cos\alpha$ and $b = n_2\lambda/2 \cos\beta$

so that

$$a = \frac{n_1\pi}{k_1} \quad \text{and} \quad k_1 = \frac{n_1\pi}{a}.$$

Similarly

$$k_2 = \frac{n_2\pi}{b}$$

Hence

$$k^2 = k_1^2 + k_2^2 = \frac{4\pi^2}{\lambda^2} = \pi^2\left(\frac{n_1^2}{a^2} + \frac{n_2^2}{b^2}\right)$$

or

$$\frac{2}{\lambda} = \sqrt{\frac{n_1^2}{a^2} + \frac{n_2^2}{b^2}}$$

defining the frequency of the n_1th mode on the x-axis and the n_2th mode on the y-axis, that is, the $(n_1 n_2)$ normal mode, as

$$\nu = \frac{c}{2}\sqrt{\frac{n_1^2}{a^2} + \frac{n_2^2}{b^2}}, \quad \text{where} \quad c^2 = \frac{T}{\rho}$$

If **k** is not normal to the direction of either a or b we can write the general solution for the waves as

$$z = A \left.\begin{matrix}\sin\\\cos\end{matrix}\right\}k_1x \left.\begin{matrix}\sin\\\cos\end{matrix}\right\}k_2y \left.\begin{matrix}\sin\\\cos\end{matrix}\right\}ckt.$$

with the boundary conditions $z = 0$ at $x = 0$ and a; $z = 0$ at $y = 0$ and b.

The condition $z = 0$ at $x = y = 0$ requires a $\sin k_1 x \sin k_2 y$ term, and the condition $z = 0$ at $x = a$ defines $k_1 = n_1 \pi / a$. The condition $z = 0$ at $y = b$ gives $k_2 = n_2 \pi / b$, so that

$$z = A \sin \frac{n_1 \pi x}{a} \sin \frac{n_2 \pi y}{b} \sin ckt$$

The fundamental vibration is given by $n_1 = 1, n_2 = 1$, so that

$$\nu = \sqrt{\left(\frac{1}{a^2} + \frac{1}{b^2}\right) \frac{T}{4\rho}}$$

In the general mode $(n_1 n_2)$ zero displacement or nodal lines occur at

$$x = 0, \quad \frac{a}{n_1}, \quad \frac{2a}{n_1}, \ldots a$$

and

$$y = 0, \quad \frac{b}{n_2}, \quad \frac{2b}{n_2}, \ldots b$$

Some of these normal modes are shown in Figure 9.6, where the shaded and plain areas have opposite displacements as shown.

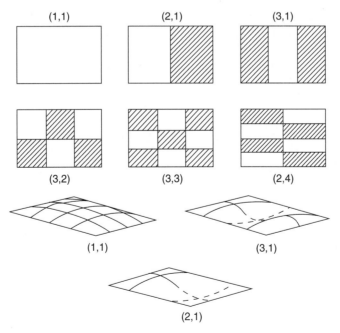

Figure 9.6 Some normal modes on a rectangular membrane with shaded and clear sections having opposite sinusoidal displacements as indicated

The complete solution for a general displacement would be the sum of individual normal modes, as with the simpler case of waves on a string (see the chapter on Fourier Series) where boundary conditions of space and time would have to be met. Several modes of different values $(n_1 n_2)$ may have the same frequency, e.g. in a square membrane the modes (4,7) (7,4) (1,8) and (8,1) all have equal frequencies. If the membrane is rectangular and $a = 3b$, modes (3,3) and (9,1) have equal frequencies.

These modes are then said to be *degenerate*, a term used in describing equal energy levels for electrons in an atom which are described by different quantum numbers.

Normal Modes in Three Dimensions

In three dimensions a normal mode is described by the numbers n_1, n_2, n_3, with a frequency

$$\nu = \frac{c}{2}\sqrt{\frac{n_1^2}{l_1^2} + \frac{n_2^2}{l_2^2} + \frac{n_3^2}{l_3^2}}, \tag{9.1}$$

where l_1, l_2 and l_3 are the lengths of the sides of the rectangular enclosure. If we now form a rectangular lattice with the x-, y- and z-axes marked off in units of

$$\frac{c}{2l_1}, \quad \frac{c}{2l_2} \quad \text{and} \quad \frac{c}{2l_3}$$

respectively (Figure 9.7), we can consider a vector of components n_1 units in the x-direction, n_2 units in the y-direction and n_3 units in the z-direction to have a length

$$\nu = \frac{c}{2}\sqrt{\frac{n_1^2}{l_1^2} + \frac{n_2^2}{l_2^2} + \frac{n_3^2}{l_3^2}}$$

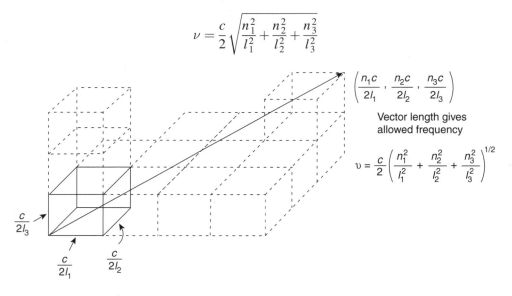

$$\left(\frac{n_1 c}{2l_1}, \frac{n_2 c}{2l_2}, \frac{n_3 c}{2l_3}\right)$$

Vector length gives
allowed frequency

$$\upsilon = \frac{c}{2}\left(\frac{n_1^2}{l_1^2} + \frac{n_2^2}{l_2^2} + \frac{n_3^2}{l_3^2}\right)^{1/2}$$

Figure 9.7 Lattice of rectangular cells in frequency space. The length of the vector joining the origin to any cell corner is the value of the frequency of an allowed normal mode. The vector direction gives the propagation direction of that particular mode

Each frequency may thus be represented by a line joining the origin to a point $cn_1/2l_1, cn_2/2l_2, cn_3/2l_3$ in the rectangular lattice.

The length of the line gives the magnitude of the frequency, and the vector direction gives the direction of the standing waves.

Each point will be at the corner of a rectangular unit cell of sides $c/2l_1, c/2l_2$ and $c/2l_3$ with a volume $c^3/8l_1l_2l_3$. There are as many cells as points (i.e. as frequencies) since each cell has eight points at its corners and each point serves as a corner to eight cells.

A very important question now arises: how many normal modes (stationary states in quantum mechanics) can exist in the frequency range ν to $\nu + d\nu$?

The answer to this question is the total number of all those positive integers n_1, n_2, n_3 for which, from equation (9.1),

$$\nu^2 < \frac{c^2}{4}\left(\frac{n_1^2}{l_1^2} + \frac{n_2^2}{l_2^2} + \frac{n_3^2}{l_3^2}\right) < (\nu + d\nu)^2$$

This total is the number of possible points (n_1, n_2, n_3) lying in the positive octant between two concentric spheres of radii ν and $\nu + d\nu$. The other octants will merely repeat the positive octant values because the n's appear as squared quantities.

Hence the total number of possible points or cells will be

$$\frac{1}{8}\frac{\text{(volume of spherical shell)}}{\text{volume of cell}}$$

$$= \frac{4\pi\nu^2\,d\nu}{8} \cdot \frac{8l_1l_2l_3}{c^3}$$

$$= 4\pi l_1l_2l_3 \cdot \frac{\nu^2\,d\nu}{c^3}$$

so that the number of possible normal modes in the frequency range ν to $\nu + d\nu$ *per unit volume* of the enclosure

$$= \frac{4\pi\nu^2\,d\nu}{c^3}$$

Note that this result, *per unit volume of the enclosure*, is independent of any particular system; we shall consider two very important applications.

Frequency Distribution of Energy Radiated from a Hot Body. Planck's Law

The electromagnetic energy radiated from a hot body at temperature T in the small frequency interval ν to $\nu + d\nu$ may be written $E_\nu\,d\nu$. If this quantity is measured experimentally over a wide range of ν a curve T_1 in Figure 9.8 will result. The general shape of the curve is independent of the temperature, but as T is increased the maximum of the curve increases and shifts towards a higher frequency.

The early attempts to describe the shape of this curve were based on two results we have already used.

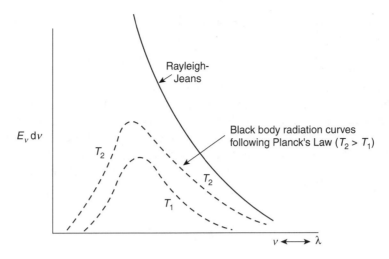

Figure 9.8 Planck's black body radiation curve plotted for two different temperatures $T_2 > T_1$, together with the curve of the classical Rayleigh–0.6-Jeans explanation leading to the 'ultra-violet catastrophe'

In the chapter on coupled oscillations we associated normal modes with 'degrees of freedom', the number of ways in which a system could take up energy. In kinetic theory, assigning an energy $\frac{1}{2}kT$ to each degree of freedom of a monatomic gas at temperature T leads to the gas law $pV = RT = NkT$ where N is Avogadro's number, k is Boltzmann's constant and R is the gas constant.

If we assume that each frequency radiated from a hot body is associated with the normal mode of an oscillator with two degrees of freedom and two transverse planes of polarization, the energy radiated per frequency interval $\mathrm{d}\nu$ may be considered as the product of the number of normal modes or oscillators in the interval $\mathrm{d}\nu$ and an energy contribution of kT from each oscillator for each plane of polarization. This gives

$$E_\nu\, \mathrm{d}\nu = \frac{4\pi\nu^2\, \mathrm{d}\nu\, 2kT}{c^3} = \frac{8\pi\nu^2 kT\, \mathrm{d}\nu}{c^3}$$

a result known as the Rayleigh–Jeans Law.

This, however, gives the energy density proportional to ν^2 which, as the solid curve in Figure 9.8 shows, becomes infinite at very high frequencies, a physically absurd result known as the *ultraviolet catastrophe*.

The correct solution to the problem was a major advance in physics. Planck had introduced the quantum theory, which predicted that the average energy value kT should be replaced by the factor $h\nu/(\mathrm{e}^{h\nu/kT} - 1)$, where h is Planck's constant (the unit of action) as shown in Problem 9.12. The experimental curve is thus accurately described by Planck's Radiation Law

$$E_\nu\, \mathrm{d}\nu = \frac{8\pi\nu^2}{c^3}\, \frac{h\nu}{\mathrm{e}^{h\nu/kT} - 1}\, \mathrm{d}\nu$$

(Problem 9.12)

Debye Theory of Specific Heats

The success of the modern theory of the specific heats of solids owes much to the work of Debye, who considered the thermal vibrations of atoms in a solid lattice in terms of a vast complex of standing waves over a great range of frequencies. This picture corresponds in three dimensions to the problem of atoms spaced along a one dimensional line (Chapter 5). In the specific heat theory each atom was allowed two transverse vibrations (perpendicular planes of polarization) and one longitudinal vibration.

The number of possible modes or oscillations per unit volume in the frequency interval ν to $\nu + d\nu$ is then given by

$$\mathrm{d}n = 4\pi\nu^2\,\mathrm{d}\nu\left(\frac{2}{c_T^3} + \frac{1}{c_L^3}\right) \tag{9.2}$$

where c_T and c_L are respectively the transverse and longitudinal wave velocities.

Problem 9.12 shows that each mode has an average energy (from Planck's Law) of $\bar{\varepsilon} = h\nu/(\mathrm{e}^{h\nu/kT} - 1)$ and the total energy in the frequency range ν to $\nu + d\nu$ for a gram atom of the solid of volume V_A is then

$$V_A\bar{\varepsilon}\,\mathrm{d}n = 4\pi V_A\left(\frac{2}{c_T^3} + \frac{1}{c_L^3}\right)\frac{h\nu^3}{\mathrm{e}^{h\nu/kT} - 1}\,\mathrm{d}\nu$$

The total energy per gram atom over all permitted frequencies is then

$$E_A = \int V_A\bar{\varepsilon}\,\mathrm{d}n = 4\pi V_A\left(\frac{2}{c_T^3} + \frac{1}{c_L^3}\right)\int_0^{\nu_m}\frac{h\nu^3}{\mathrm{e}^{h\nu/kT} - 1}\,\mathrm{d}\nu$$

where ν_m is the maximum frequency of the oscillations.

There are N atoms per gram atom of the solid (N is Avogadro's number) and each atom has three allowed oscillation modes, so an approximation to ν_m is found by writing the integral of equation (9.2) for a gram atom as

$$\int \mathrm{d}n = 3N = 4\pi V_A\left(\frac{2}{c_T^3} + \frac{1}{c_L^3}\right)\int_0^{\nu_m}\nu^2\,\mathrm{d}\nu = \frac{4\pi V_A}{3}\left(\frac{2}{c_T^3} + \frac{1}{c_L^3}\right)\nu_m^3$$

The values of c_T and c_L can be calculated from the elastic constants of the solid (see Chapter 6 on longitudinal waves) and ν_m can then be found.

The values of E_A thus becomes

$$E_A = \frac{9N}{\nu_m^3}\int_0^{\nu_m}\frac{h\nu}{\mathrm{e}^{h\nu/kT} - 1}\nu^2\,\mathrm{d}\nu$$

and the variation of E_A with the temperature T is the molar specific heat of the substance at constant volume. The specific heat of aluminium calculated by this method is compared with experimental results in Figure 9.9.

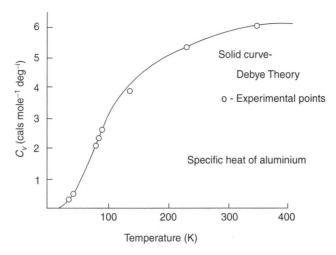

Figure 9.9   Debye theory of specific heat of solids. Experimental values versus theoretical curve for aluminium

(Problems 9.13, 9.14, 9.15, 9.16, 9.17, 9.18, 9.19)

Reflection and Transmission of a Three-Dimensional Wave at a Plane Boundary

To illustrate such an event we choose a physical system of great significance, the passage of a light wave from air to glass. More generally, Figure 9.10 shows a plane polarized electromagnetic wave $\mathbf{E}_i$ incident at an angle θ to the normal of the plane boundary $z = 0$ separating two dielectrics of impedance Z_1 and Z_2, giving reflected and transmitted rays $\mathbf{E}_r$ and $\mathbf{E}_t$, respectively. The boundary condition requires that the tangential electric field E_x is continuous at $z = 0$. The propagation direction $\mathbf{k}_i$ of $\mathbf{E}_i$ lies wholly in the plane of the paper $(y = 0)$ but no assumptions are made about the directions of the reflected and transmitted waves (nor about the planes of oscillation of their electric field vectors).

We write

$$\mathbf{E}_i = A_i\, \mathrm{e}^{\mathrm{i}(\omega t - \mathbf{k}_i \cdot \mathbf{r})} = A_i\, \mathrm{e}^{\mathrm{i}[\omega t - k_i(x \sin\theta + z \cos\theta)]}$$

$$\mathbf{E}_r = A_r\, \mathrm{e}^{\mathrm{i}(\omega t - \mathbf{k}_r \cdot \mathbf{r})} = A_r\, \mathrm{e}^{\mathrm{i}[\omega t - (k_{r1}x + k_{r2}y + k_{r3}z)]}$$

and

$$\mathbf{E}_t = A_t\, \mathrm{e}^{\mathrm{i}(\omega t - \mathbf{k}_t \cdot \mathbf{r})} = A_t\, \mathrm{e}^{\mathrm{i}[\omega t - (k_{t1}x + k_{t2}y + k_{t3}z)]}$$

where $\mathbf{k}_r(k_{r1}, k_{r2}, k_{r3})$ and $\mathbf{k}_t(k_{t1}, k_{t2}, k_{t3})$ are respectively the reflected and transmitted propagation vectors.

Since E_x is continuous at $z = 0$ for all x, y, t we have

$$A_i\, \mathrm{e}^{\mathrm{i}[\omega t - k_i(x \sin\theta)]} + A_r\, \mathrm{e}^{\mathrm{i}[\omega t - (k_{r1}x + k_{r2}y)]}$$
$$= A_t\, \mathrm{e}^{\mathrm{i}[\omega t - (k_{t1}x + k_{t2}y)]}$$

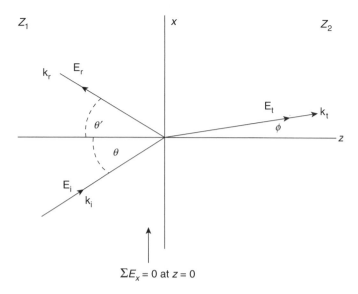

Figure 9.10 Plane-polarized electromagnetic wave propagating in the plane of the paper is represented by vector $\mathbf{E}_i$ and is reflected as vector $\mathbf{E}_r$ and transmitted as vector $\mathbf{E}_t$ at a plane interface between media of impedances Z_1 and Z_2. No assumptions are made about the planes of propagation of $\mathbf{E}_r$ and $\mathbf{E}_t$. From the boundary condition that the electric field component E_x is continuous at the plane $z = 0$ it follows that (1) vectors $\mathbf{E}_i\,\mathbf{E}_r$ and $\mathbf{E}_t$ propagate in the same plane; (2) $\theta = \theta'$ (angle of incidence = angle of reflection); (3) Snell's law $(\sin\theta/\sin\phi) = n_2/n_1$, where n is the refractive index

an identity which is only possible if the indices of all three terms are identical; that is

$$\omega t - k_i x \sin\theta \equiv \omega t - k_{r1}x + k_{r2}y$$
$$\equiv \omega t - k_{t1}x + k_{t2}y$$

Equating the coefficients of x in this identity gives

$$k_i \sin\theta = k_{r1} = k_{t1}$$

whilst equal coefficients of y give

$$0 = k_{r2} = k_{t2}$$

The relation

$$k_{r2} = k_{t2} = 0$$

shows that the reflected and transmitted rays have no component in the y direction and lie wholly in the xz plane of incidence; that is, incident reflected and transmitted (refracted) rays are coplanar.

Now the magnitude

$$k_i = k_r = \frac{2\pi}{\lambda_1}$$

since both incident and reflected waves are travelling in medium Z_1. Hence

$$k_i \sin \theta = k_{rl}$$

gives

$$k_i \sin \theta = k_r \sin \theta'$$

that is

$$\theta = \theta'$$

so the angle of incidence equals the angle of reflection.

The magnitude

$$k_t = \frac{2\pi}{\lambda_2}$$

so that

$$k_i \sin \theta = k_{tl} = k_t \sin \phi$$

gives

$$\frac{2\pi}{\lambda_1} \sin \theta = \frac{2\pi}{\lambda_2} \sin \phi$$

or

$$\frac{\sin \theta}{\sin \phi} = \frac{\lambda_1}{\lambda_2} = \frac{n_2}{n_1} \left[\frac{\text{Refractive Index (medium 2)}}{\text{Refractive Index (medium 1)}} \right]$$

a relationship between the angles of incidence and refraction which is well known as Snell's Law.

Total Internal Reflection and Evanescent Waves

On p. 254 we discussed the propagation of an electromagnetic wave across the boundary between air and a dielectric (glass, say). We now consider the reverse process where a wave in the dielectric crosses the interface into air.

Snell's Law still holds so we have, in Figure 9.11,

$$n_1 \sin \theta = n_2 \sin \phi$$

where

$$n_1 > n_2 \quad \text{and} \quad n_2/n_1 = n_r < 1$$

Thus

$$\sin \theta = (n_2/n_1) \sin \phi = n_r \sin \phi$$

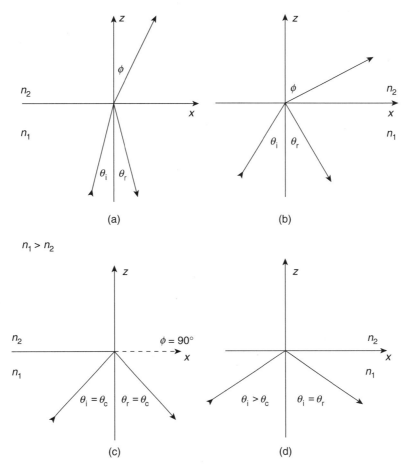

$n_1 > n_2$

Figure 9.11 When light propagates from a dense to a rare medium $(n_1 > n_2)$ Snell's Law defines $\theta = \theta_c$ as that angle of incidence for which $\phi = 90°$ and the refracted ray is tangential to the plane boundary. Total internal reflection can take place but the boundary conditions still require a transmitted wave known as the evanescent or surface wave. It propagates in the x direction but its amplitude decays exponentially with z

with $\phi > \theta$. Eventually a critical angle of incidence θ_c is reached where $\phi = 90°$ and $\sin\theta = n_r$; for $\theta > \theta_c$, $\sin\theta > n_r$. For glass to air $n_r = \frac{1}{1.5}$ and $\theta_c = 42°$.

It is evident that for $\theta \geq \theta_c$ no electromagnetic energy is transmitted into the rarer medium and the incident wave is said to suffer *total internal reflection*.

In the reflection coefficients $R_{\parallel}$ and $R_{\perp}$ on p. 218 we may replace $\cos\phi$ by

$$(1 - \sin^2\phi)^{1/2} = [1 - (\sin\theta/n_r)^2]^{1/2}$$

and rewrite

$$R_{\parallel} = \frac{(n_r^2 - \sin^2\theta)^{1/2} - n_r^2\cos\theta}{(n_r^2 - \sin^2\theta)^{1/2} + n_r^2\cos\theta}$$

and

$$R_{\perp} = \frac{\cos\theta - (n_r^2 - \sin^2\theta)^{1/2}}{\cos\theta + (n_r^2 - \sin^2\theta)^{1/2}}$$

Now for $\theta > \theta_c$, $\sin\theta > n_r$ and the bracketed quantities in $R_{\parallel}$ and $R_{\perp}$ are negative so that $R_{\parallel}$ and $R_{\perp}$ are complex quantities; that is $(E_r)_{\parallel}$ and $(E_r)_{\perp}$ have a phase relation which depends on θ.

It is easily checked that the product of R and R^* is unity so we have $R_{\parallel}R_{\parallel}^* = R_{\perp}R_{\perp}^* = 1$. This means, for both the examples of Figure 8.8, that the incident and reflected intensities I_i and $I_r = 1$. The transmitted intensity $I_t = 0$ so that no energy is carried across the boundary.

However, if there is no transmitted wave we cannot satisfy the boundary condition $E_i + E_r = E_t$ on p. 254, using only incident and reflected waves. We must therefore assert that a transmitted wave does exist but that it cannot on the average carry energy across the boundary.

We now examine the implications of this assertion, using Figure 9.10 above, and we keep the notation of p. 254. This gives a transmitted electric field vector

$$E_t = A_t e^{i[\omega t - (k_{t1}x + k_{t2}y + k_{t3}z)]}$$
$$= A_t e^{i[\omega t - k_t(x\sin\phi + z\cos\phi)]}$$

because $y = 0$ in the xz plane, $k_{t1} = k_t\sin\phi$ and $k_{t3} = k_t\cos\phi$. Now

$$\cos\phi = 1 - \sin^2\phi = 1 - \sin^2\theta/n_r^2$$
$$\therefore k_t\cos\phi = \pm k_t(1 - \sin^2\theta/n_r^2)^{1/2}$$

which for $\theta > \theta_c$ gives $\sin\theta > n_r$ so that

$$k_t\cos\phi = \mp ik_t\left(\frac{\sin^2\theta}{n_r^2} - 1\right)^{1/2} = \mp i\beta$$

We also have

$$k_t\sin\phi = k_t\sin\theta/n_r$$

so

$$E_t = A_t e^{\mp\beta z} e^{i(\omega t - k_r x\sin\theta/n_r)}$$

The alternative factor $e^{+\beta z}$ defines an exponential growth of A_t which is physically untenable and we are left with a wave whose amplitude decays exponentially as it penetrates the less dense medium. The disturbance travels in the x direction along the interface and is known as a *surface* or *evanescent wave*.

It is possible to show from the expressions for $R_{\parallel}$ and $R_{\perp}$ on p. 258 that except at $\theta = 90°$ the incident and the reflected electric field components for $(E)_{\parallel}$ in one case and $(E)_{\perp}$ in the other, do not differ in phase by π rad and cannot therefore cancel each other out. The continuity of the tangential component of $\mathbf{E}$ at the boundary therefore leaves a component parallel to the interface which propagates as the surface wave. This effect has been observed at optical frequencies.

Moreover, if only a very thin air gap exists between two glass blocks it is possible for energy to flow across the gap allowing the wave to propagate in the second glass block. This process is called *frustrated total internal reflection* and has its quantum mechanical analogue in the tunnelling effect discussed on p. 431.

Problem 9.1
Show that

$$z = A\, e^{i\{\omega t - (k_1 x + k_2 y)\}}$$

where $k^2 = \omega^2/c^2 = k_1^2 + k_2^2$ is a solution of the two-dimensional wave equation

$$\frac{\partial^2 z}{\partial x^2} + \frac{\partial^2 z}{\partial y^2} = \frac{1}{c^2}\frac{\partial^2 z}{\partial t^2}$$

Problem 9.2
Show that if the displacement of the waves on the membrane of width b of Figure 9.3 is given by the superposition

$$\mathbf{z} = A_1\, e^{i[\omega t - (k_1 x + k_2 y)]} + A_2\, e^{i[\omega t - (k_1 x - k_2 y)]}$$

with the boundary conditions

$$z = 0 \quad \text{at} \quad y = 0 \quad \text{and} \quad y = b$$

then the real part of $\mathbf{z}$ is

$$z = +2A_1 \sin k_2 y \sin(\omega t - k_1 x)$$

where

$$k_2 = \frac{n\pi}{b}$$

Problem 9.3
An electromagnetic wave loses negligible energy when reflected from a highly conducting surface. With repeated reflections it may travel along a transmission line or wave guide consisting of two parallel, infinitely conducting planes (separation a). If the wave

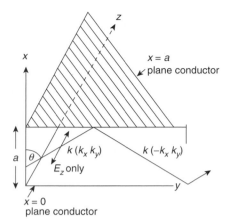

is plane polarized, so that only E_z exists, then the propagating direction **k** lies wholly in the xy plane. The boundary conditions require that the total tangential electric field E_z is zero at the conducting surfaces $x = 0$ and $x = a$. Show that the first boundary condition allows E_z to be written $E_z = E_0(e^{ik_x x} - e^{-ik_x x})e^{i(k_y y - \omega t)}$, where $k_x = k\cos\theta$ and $k_y = k\sin\theta$ and the second boundary condition requires $k_x = n\pi/a$.

If $\lambda_0 = 2\pi c/\omega$, $\lambda_c = 2\pi/k_x$ and $\lambda_g = 2\pi/k_y$ are the wavelengths propagating in the x and y directions respectively show that

$$\frac{1}{\lambda_c^2} + \frac{1}{\lambda_g^2} = \frac{1}{\lambda_0^2}$$

We see that for $n = 1$, $k_x = \pi/a$ and $\lambda_c = 2a$, and that as ω decreases and λ_0 increases, $k_y = k\sin\theta$ becomes imaginary and the wave is damped. Thus, $n = 2(k_x = 2\pi/a)$ gives $\lambda_c = a$, the 'critical wavelength', i.e. the longest wavelength propagated by a waveguide of separation a. Such cut-off wavelengths and frequencies are a feature of wave propagation in periodic structures, transmission lines and wave-guides.

Problem 9.4
Show, from equations (8.1) and (8.2), that the magnetic field in the plane-polarized electromagnetic wave of Problem 9.3 has components in both x- and y-directions. [When an electromagnetic wave propagating in a waveguide has only transverse electric field vectors and no electric field in the direction of propagation it is called a transverse electric (TE) wave. Similarly a transverse magnetic (TM) wave may exist. The plane-polarized wave of Problem 9.3 is a transverse electric wave; the corresponding transverse magnetic wave would have H_z, E_x and E_y components. The values of n in Problem 9.3 satisfying the boundary conditions are written as subscripts to define the exact mode of propagation, e.g. TE_{10}.]

Problem 9.5
Use the value of the inductance and capacitance of a pair of plane parallel conductors of separation a and width b to show that the characteristic impedance of such a waveguide is given by

$$Z_0 = \frac{a}{b}\sqrt{\frac{\mu}{\varepsilon}}\,\Omega$$

where μ and ε are respectively the permeability and permittivity of the medium between the conductors.

Problem 9.6

Consider either the Poynting vector or the energy per unit volume of an electromagnetic wave to show that the power transmitted by a single positive travelling wave in the waveguide of Problem 9.5 is $\frac{1}{2}abE_0^2\sqrt{\varepsilon/\mu}$.

Problem 9.7

An electromagnetic wave $(\mathbf{E}, \mathbf{H})$ propagates in the x-direction down a perfectly conducting hollow tube of arbitrary cross section. The tangential component of $\mathbf{E}$ at the conducting walls must be zero at all times.

Show that the solution $\mathbf{E} = E(y, z)\,\mathbf{n}\cos(\omega t - k_x x)$ substituted in the wave equation yields

$$\frac{\partial^2 E(y, z)}{\partial y^2} + \frac{\partial^2 E(y, z)}{\partial z^2} = -k^2 E(y, z),$$

where $k^2 = \omega^2/c^2 - k_x^2$ and k_x is the wave number appropriate to the x-direction, $\mathbf{n}$ is the unit vector in any direction in the (y, z) plane.

Problem 9.8

If the waveguide of Problem 9.7 is of rectangular cross-section of width a in the y-direction and height b in the z-direction, show that the boundary conditions $E_x = 0$ at $y = 0$ and a and at $z = 0$ and b in the wave equation of Problem 9.7 gives

$$E_x = A\sin\frac{m\pi y}{a}\sin\frac{n\pi z}{b}\cos(\omega t - k_x x),$$

where

$$k^2 = \pi^2\left(\frac{m^2}{a^2} + \frac{n^2}{b^2}\right)$$

Problem 9.9

Show, from Problems 9.7 and 9.8, that the lowest possible value of ω (the cut-off frequency) for k_x to be real is given by $m = n = 1$.

Problem 9.10

Prove that the product of the phase and group velocity ω/k_x, $\partial\omega/\partial k_x$ of the wave of Problems 9.7–9.9 is c^2, where c is the velocity of light.

Problem 9.11

Consider now the extension of Problem 9.2 where the waves are reflected at the rigid edges of the rectangular membrane of sides length a and b as shown in the diagram. The final displacement is the result of the superposition

$$\begin{aligned}
\mathbf{z} = &A_1\,e^{i[\omega t - (k_1 x + k_2 y)]} \\
&+ A_2\,e^{i[\omega t - (k_1 x - k_2 y)]} \\
&+ A_3\,e^{i[\omega t - (-k_1 x - k_2 y)]} \\
&+ A_4\,e^{i[\omega t - (-k_1 x + k_2 y)]}
\end{aligned}$$

with the boundary conditions

$$z = 0 \quad\text{at}\quad x = 0 \quad\text{and}\quad x = a$$

and

$$z = 0 \quad \text{at} \quad y = 0 \quad \text{and} \quad y = b$$

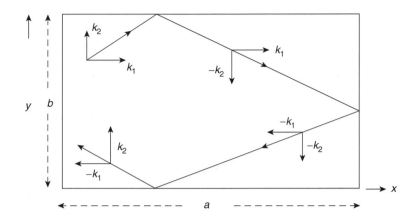

Show that this leads to a displacement

$$z = -4A_1 \sin k_1 x \sin k_2 y \cos \omega t$$

(the real part of $\mathbf{z}$), where

$$k_1 = \frac{n_1 \pi}{a} \quad \text{and} \quad k_2 = \frac{n_2 \pi}{b}$$

Problem 9.12

In deriving the result that the average energy of an oscillator at frequency ν and temperature T is given by

$$\bar{\varepsilon} = \frac{h\nu}{e^{(h\nu/kT)} - 1}$$

Planck assumed that a large number N of oscillators had energies $0, h\nu, 2h\nu \ldots nh\nu$ distributed according to Boltzmann's Law

$$N_n = N_0 \, e^{-nh\nu/kT}$$

where the number of oscillators N_n with energy $nh\nu$ decreases exponentially with increasing n.
Use the geometric progression series

$$N = \sum_n N_n = N_0(1 + e^{-h\nu/kT} + e^{-2h\nu/kT} \ldots)$$

to show that

$$N = \frac{N_0}{1 - e^{-h\nu/kT}}$$

If the total energy of the oscillators in the nth energy state is given by

$$E_n = N_n nh\nu$$

prove that the total energy over all the n energy states is given by

$$E = \sum_n E_n = N_0 \frac{h\nu\, e^{-h\nu/kT}}{(1 - e^{-h\nu/kT})^2}$$

Hence show that the average energy per oscillator

$$\bar{\varepsilon} = \frac{E}{N} = \frac{h\nu}{e^{h\nu/kT} - 1}$$

and expand the denominator to show that for $h\nu \ll kT$, that is low frequencies and long wavelengths. Planck's Law becomes the classical expression of Rayleigh–Jeans.

Problem 9.13

The wave representation of a particle, e.g. an electron, in a rectangular potential well throughout which $V = 0$ is given by Schrödinger's time-independent equation

$$\frac{\partial^2 \Psi}{\partial x^2} + \frac{\partial^2 \Psi}{\partial y^2} + \frac{\partial^2 \Psi}{\partial z^2} = -\frac{8\pi^2 m}{h^2} E\Psi,$$

where E is the particle energy, m is the mass and h is Planck's constant. The boundary conditions to be satisfied are $\psi = 0$ at $x = y = z = 0$ and at $x = L_x, y = L_y, z = L_z$, where L_x, L_y and L_z are the dimensions of the well.

Show that

$$\Psi = A \sin\frac{l\pi x}{L_x} \sin\frac{r\pi y}{L_y} \sin\frac{n\pi z}{L_z}$$

is a solution of Schrödinger's equation, giving

$$E = \frac{h^2}{8m}\left(\frac{l^2}{L_x^2} + \frac{r^2}{L_y^2} + \frac{n^2}{L_z^2}\right)$$

When the potential well is cubical of side L,

$$E = \frac{h^2}{8mL^2}(l^2 + r^2 + n^2)$$

and the lowest value of the quantized energy is given by

$$E = E_0 \quad \text{for} \quad l = 1, \quad r = n = 0$$

Show that the next energy levels are $3E_0, 6E_0$ (three-fold degenerate), $9E_0$ (three-fold degenerate), $11E_0$ (three-fold degenerate), $12E_0$ and $14E_0$ (six-fold degenerate).

Problem 9.14

Show that at low energy levels (long wavelengths) $h\nu \ll kT$, Planck's radiation law is equivalent to the Rayleigh–Jeans expression.

Problem 9.15

Planck's radiation law, expressed in terms of energy per unit range of wavelength instead of frequency, becomes

$$E_\lambda = \frac{8\pi ch}{\lambda^5(e^{ch/\lambda kT} - 1)}$$

Use the variable $x = ch/\lambda kT$ to show that the total energy per unit volume at temperature $T°$ absolute is given by

$$\int_0^\infty E_\lambda \, d\lambda = aT^4 \, \text{J}\,\text{m}^{-3}$$

where

$$a = \frac{8\pi^5 k^4}{15c^3 h^3}$$

(The constant $ca/4 = \sigma$, Stefan's Constant in the Stefan-Boltzmann Law.) Note that

$$\int_0^\infty \frac{x^3 \, dx}{e^x - 1} = \frac{\pi^4}{15}$$

Problem 9.16

Show that the wavelength λ_m at which E_λ in Problem 9.15 is a maximum is given by the solution of

$$\left(1 - \frac{x}{5}\right) e^x = 1, \quad \text{where} \quad x = \frac{ch}{\lambda kT}$$

The solution is $ch/\lambda_m kT = 4.965$.

Problem 9.17

Given that $ch/\lambda_m = 5\,kT$ in Problem 9.16, show that if the sun's temperature is about 6000 K, then $\lambda_m \approx 4.7 \times 10^{-7}$ m, the green region of the visible spectrum where the human eye is most sensitive (evolution ?).

Problem 9.18

The tungsten filament of an electric light bulb has a temperature of $\approx 2000\,\text{K}$. Show that in this case $\lambda_m \approx 14 \times 10^{-7}$ m, well into the infrared. Such a lamp is therefore a good heat source but an inefficient light source.

Problem 9.19

A free electron (travelling in a region of zero potential) has an energy

$$E = \frac{p^2}{2m} = \left(\frac{\hbar^2}{2m}\right) k^2 = E(k)$$

where the wavelength

$$\lambda = h/p = 2\pi/k$$

In a weak periodic potential; for example, in a solid which is a good electrical conductor, $E = (\hbar^2/2m^*) k^2$, where m^* is called the effective mass. (For valence electrons $m^* \approx m$.)

Represented as waves, the electrons in a cubic potential well ($V = 0$) of side L have allowed wave numbers k, where

$$k^2 = k_x^2 + k_y^2 + k_z^2 \quad \text{and} \quad k_i = \frac{n_i \pi}{L}$$

(see Problem 9.13). For each value of k there are two allowed states (each defining the spin state of the single occupying electron–Pauli's principle). Use the arguments in Chapter 9 to show that the total number of states in k space between the values k and $k + dk$ is given by

$$P(k) = 2 \left(\frac{L}{\pi}\right)^3 \frac{4\pi k^2 \, dk}{8}$$

Use the expression $E = (\hbar^2/2m^*) k^2$ to convert this into the number of states $S(E) \, dE$ in the energy interval dE to give

$$S(E) = \frac{A}{2\pi^2} \left(\frac{2m}{\hbar^2}\right)^{3/2} \sqrt{E}$$

where $A = L^3$.

If there are N electrons in the N lowest energy states consistent with Pauli's principle, show that the integral

$$\int_0^{E_f} S(E) \, dE = N$$

gives the Fermi energy level

$$E_f = \frac{\hbar^2}{2m^*} \left(\frac{3\pi^2 N}{A}\right)^{2/3}$$

where E_f is the kinetic energy of the most energetic electron when the solid is in its ground state.

Summary of Important Results

Wave Equation in Two Dimensions

$$\frac{\partial^2 z}{\partial x^2} + \frac{\partial^2 z}{\partial y^2} = \frac{1}{c^2} \frac{\partial^2 z}{\partial t^2}$$

Lines of constant phase $lx + my = ct$ propagate in direction $\mathbf{k}(k_1, k_2)$ where $l = k_1/k$, $m = k_2/k$, $k^2 = k_1^2 + k_2^2$ and $c^2 = \omega^2/k^2$. Solution is

$$z = A \, e^{i(\omega t - \mathbf{k} \cdot \mathbf{r})} \quad \text{for} \quad \mathbf{r}(x, y)$$

where $k \cdot r = k_1 x + k_2 y$.

Wave Equation in Three Dimensions

$$\frac{\partial^2 \phi}{\partial x^2} + \frac{\partial^2 \phi}{\partial y^2} + \frac{\partial^2 \phi}{\partial z^2} = \frac{1}{c^2} \frac{\partial^2 \phi}{\partial t^2}$$

Planes of constant phase $lx + my + nz = ct$ propagate in a direction

$$\mathbf{k}(k_1, k_2, k_3), \quad \text{where} \quad l = k_1/k, \quad m = k_2/k, \quad n = k_3/k$$
$$k^2 = k_1^2 + k_2^2 + k_3^2 \quad \text{and} \quad c^2 = \omega^2/k^2.$$

Solution is

$$\phi = A \, e^{i(\omega t - \mathbf{k} \cdot \mathbf{r})} \quad \text{for} \quad \mathbf{r}(x, y, z).$$

Wave Guides

Reflection from walls $y = 0, y = b$ in a two-dimensional wave guide for a wave of frequency ω and vector direction $\mathbf{k}(k_1, k_2)$ gives normal modes in the y direction with $k_2 = n\pi/b$ and propagation in the x direction with phase velocity

$$v_p = \frac{\omega}{k_1} > \frac{\omega}{k} = v$$

and group velocity

$$v_g = \frac{\partial \omega}{\partial k_1} \quad \text{such that} \quad v_p v_g = v^2$$

Cut-off frequency

Only frequencies $\omega \geq n\pi v/b$ will propagate where n is mode number.

Normal Modes in Three Dimensions

Wave equation separates into three equations (one for each variable x, y, z) to give solution

$$= A \, \frac{\sin}{\cos} k_1 x \, \frac{\sin}{\cos} k_2 y \, \frac{\sin}{\cos} k_3 z \, \frac{\sin}{\cos} \omega t$$

(Boundary conditions determine final form of solution.)

For waves of velocity c, the number of normal modes per unit volume of an enclosure in the frequency range ν to $\nu + d\nu$

$$= \frac{4\pi\nu^2 \, d\nu}{c^3}$$

Directly applicable to

- Planck's Radiation Law

- Debye's theory of specific heats of solids

- Fermi energy level (Problem 9.19)

10

Fourier Methods

Fourier Series

In this chapter we are going to look in more detail at the implications of the principles of superposition which we met at the beginning of the book when we added the two separate solutions of the simple harmonic motion equation. Our discussion of monochromatic waves has led to the idea of repetitive behaviour in a simple form. Now we consider more complicated forms of repetition which arise from superposition.

Any function which repeats itself regularly over a given interval of space or time is called a periodic function. This may be expressed by writing it as $f(x) = f(x \pm \alpha)$ where α is the interval or period.

The simplest examples of a periodic function are sines and cosines of fixed frequency and wavelength, where α represents the period τ, the wavelength λ or the phase angle 2π rad, according to the form of x. Most periodic functions for example the square wave system of Figure 10.1, although quite simple to visualize are more complicated to represent mathematically. Fortunately this can be done for almost all periodic functions of interest in physics using the method of Fourier Series, which states that any periodic function may be represented by the series

$$f(x) = \tfrac{1}{2}a_0 + a_1 \cos x + a_2 \cos 2x \ldots + a_n \cos nx$$
$$+ b_1 \sin x + b_2 \sin 2x \ldots + b_n \sin nx, \tag{10.1}$$

that is, a constant $\tfrac{1}{2}a_0$ plus sine and cosine terms of different amplitudes, having frequencies which increase in discrete steps. Such a series must of course, satisfy certain conditions, chiefly those of convergence. These convergence criteria are met for a function with discontinuities which are not too severe and with first and second differential coefficients which are well behaved. At such discontinuities, for instance in the square wave where $f(x) = \pm h$ at $x = 0, \pm 2\pi$, etc. the series represents the mean of the values of the function just to the left and just to the right of the discontinuity.

The Physics of Vibrations and Waves, 6th Edition   H. J. Pain
© 2005 John Wiley & Sons, Ltd

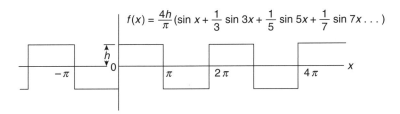

Figure 10.1 Square wave of height h and its Fourier sine series representation (odd function)

We may write the series in several equivalent forms:

$$f(x) = \frac{1}{2}a_0 + \sum_{n=1}^{\infty}(a_n \cos nx + b_n \sin nx)$$

$$= \frac{1}{2}a_0 + \sum_{n=1}^{\infty} c_n \cos(nx - \theta_n)$$

where

$$c_n^2 = a_n^2 + b_n^2$$

and

$$\tan\theta_n = b_n/a_n$$

or

$$f(x) = \sum_{n=-\infty}^{\infty} d_n \,\mathrm{e}^{inx}$$

where

$$2d_n = a_n - ib_n (n \geq 0)$$

and

$$2d_n = a_{-n} + ib_{-n}(n < 0)$$

To find the values of the coefficients a_n and b_n let us multiply both sides of equation (10.1) by $\cos nx$ and integrate with respect to x over the period 0 to 2π (say).

Every term

$$\int_0^{2\pi} \cos mx \cos nx \,\mathrm{d}x = \begin{cases} 0 \text{ if } m \neq n \\ \pi \text{ if } m = n \end{cases}$$

whilst every term

$$\int_0^{2\pi} \sin mx \cos nx \,\mathrm{d}x = 0 \text{ for all } m \text{ and } n.$$

Thus for $m = n$,

$$a_n \int_0^{2\pi} \cos^2 nx \, dx = \pi a_n$$

so that

$$a_n = \frac{1}{\pi} \int_0^{2\pi} f(x) \cos nx \, dx$$

Similarly, by multiplying both sides of equation (10.1) by $\sin nx$ and integrating from 0 to 2π we have, since ·

$$\int_0^{2\pi} \sin mx \sin nx \, dx = \begin{cases} 0 \text{ if } m \neq n \\ \pi \text{ if } m = n \end{cases}$$

that

$$b_n = \frac{1}{\pi} \int_0^{2\pi} f(x) \sin nx \, dx$$

Immediately we see that the constant ($n = 0$), given by $\frac{1}{2}a_0 = 1/2\pi \int_0^{2\pi} f(x) \, dx$, is just the average of the function over the interval 2π. It is, therefore, the steady or 'd.c.' level on which the alternating sine and cosine components of the series are superimposed, and the constant can be varied by moving the function with respect to the x-axis. When a periodic function is symmetric about the x-axis its average value, that is, its steady or d.c. base level, $\frac{1}{2}a_0$, is zero, as in the square wave system of Figure 10.1. If we raise the square waves so that they stand as pulses of height $2h$ on the x-axis, the value of $\frac{1}{2}a_0$ is $h\pi$ (average value over 2π). The values of a_n represent twice the average value of the product $f(x) \cos nx$ over the interval 2π; b_n can be interpreted in a similar way.

We see also that the series representation of the function is the sum of cosine terms which are even functions $[\cos x = \cos(-x)]$ and of sine terms which are odd functions $[\sin x = -\sin(-x)]$. Now every function $f(x) = \frac{1}{2}[f(x) + f(-x)] + \frac{1}{2}[f(x) - f(-x)]$, in which the first bracket is even and the second bracket is odd. Thus, the cosine part of a Fourier series represents the even part of the function and the sine terms represent the odd part of the function. Taking the argument one stage further, a function $f(x)$ which is an even function is represented by a Fourier series having only cosine terms; if $f(x)$ is odd it will have only sine terms in its Fourier representation. Whether a function is completely even or completely odd can often be determined by the position of the y-axis. Our square wave of Figure 10.1 is an odd function $[f(x) = -f(-x)]$; it has no constant and is represented by $f(x) = 4h/\pi(\sin x + 1/3 \sin 3x + 1/5 \sin 5x$, etc. but if we now move the y-axis a half period to the right as in Figure 10.2, then $f(x) = f(-x)$, an even function, and the square wave is represented by

$$f(x) = \frac{4h}{\pi} \left(\cos x - \tfrac{1}{3} \cos 3x + \tfrac{1}{5} \cos 5x - \tfrac{1}{7} \cos 7x + \cdots \right)$$

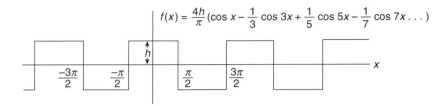

$$f(x) = \frac{4h}{\pi}\left(\cos x - \frac{1}{3}\cos 3x + \frac{1}{5}\cos 5x - \frac{1}{7}\cos 7x \ldots\right)$$

Figure 10.2 The wave of Figure 10.1 is now symmetric about the *y* axis and becomes a cosine series (even function)

If we take the first three or four terms of the series representing the square wave of Figure 10.1 and add them together, the result is Figure 10.3. The fundamental, or first harmonic, has the frequency of the square wave and the higher frequencies build up the squareness of the wave. The highest frequencies are responsible for the sharpness of the vertical sides of the waves; this type of square wave is commonly used to test the frequency response of amplifiers. An amplifier with a square wave input effectively 'Fourier analyses' the input and responds to the individual frequency components. It then puts them together again at its output, and if a perfect square wave emerges from the amplifier it proves that the amplifier can handle the whole range of the frequency components equally well. Loss of sharpness at the edges of the waves shows that the amplifier response is limited at the higher frequency range.

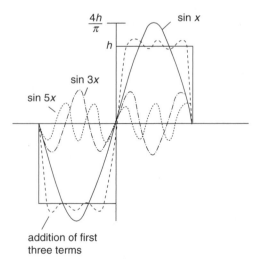

Figure 10.3 Addition of the first three terms of the Fourier series for the square wave of Figure 10.1 shows that the higher frequencies are responsible for sharpening the edges of the pulse

Example of Fourier Series

Consider the square wave of height h in Figure 10.1. The value of the function is given by

$$f(x) = h \quad \text{for} \quad 0 < x < \pi$$

and

$$f(x) = -h \quad \text{for} \quad \pi < x < 2\pi$$

The coefficients of the series representation are given by

$$a_n = \frac{1}{\pi} \left[h \int_0^\pi \cos nx \, dx - h \int_\pi^{2\pi} \cos nx \, dx \right] = 0$$

because

$$\int_0^\pi \cos nx \, dx = \int_\pi^{2\pi} \cos nx \, dx = 0$$

and

$$b_n = \frac{1}{\pi} \left[h \int_0^\pi \sin nx \, dx - h \int_\pi^{2\pi} \sin nx \, dx \right]$$

$$= \frac{h}{n\pi} [[\cos nx]_\pi^0 + [\cos nx]_\pi^{2\pi}]$$

$$= \frac{h}{n\pi} [(1 - \cos n\pi) + (1 - \cos n\pi)]$$

giving $b_n = 0$ for n even and $b_n = 4h/n\pi$ for n odd. Thus, the Fourier series representation of the square wave is given by

$$f(x) = \frac{4h}{\pi} \left(\sin x + \frac{\sin 3x}{3} + \frac{\sin 5x}{5} + \frac{\sin 7x}{7} + \cdots \right)$$

Fourier Series for any Interval

Although we have discussed the Fourier representation in terms of a periodic function its application is much more fundamental, for any section or interval of a well behaved function may be chosen and expressed in terms of a Fourier series. This series will accurately represent the function *only within the chosen interval*. If applied outside that interval it will not follow the function but will periodically repeat the value of the function within the chosen interval. If we represent this interval by a Fourier cosine series the repetition will be that of an even function, if the representation is a Fourier sine series an odd function repetition will follow.

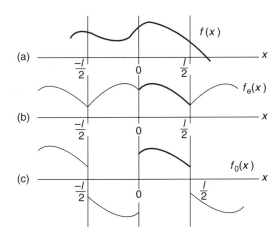

Figure 10.4 A Fourier series may represent a function over a selected half-interval. The general function in (a) is represented in the half-interval $0 < x < l/2$ by f_e, an even function cosine series in (b), and by f_o, an odd function sine series in (c). These representations are valid only in the specified half-interval. Their behaviour outside that half-interval is purely repetitive and departs from the original function

Suppose now that we are interested in the behaviour of a function over only one-half of its full interval and have no interest in its representation outside this restricted region. In Figure 10.4a the function $f(x)$ is shown over its full space interval $-l/2$ to $+l/2$, but $f(x)$ can be represented completely in the interval 0 to $+l/2$ by either a cosine function (which will repeat itself each half-interval as an even function) or it can be represented completely by a sine function, in which case it will repeat itself each half-interval as an odd function. Neither representation will match $f(x)$ outside the region 0 to $+l/2$, but in the half-interval 0 to $+l/2$ we can write

$$f(x) = f_e(x) = f_o(x)$$

where the subscripts e and o are the even (cosine) or odd (sine) Fourier representations, respectively.

The arguments of sines and cosines must, of course, be phase angles, and so far the variable x has been measured in radians. Now, however, the interval is specified as a distance and the variable becomes $2\pi x/l$, so that each time x changes by l the phase angle changes by 2π.

Thus

$$f_e(x) = \frac{a_0}{2} + \sum_{n=1}^{\infty} a_n \cos \frac{2\pi n x}{l}$$

where

$$a_n = \frac{1}{\frac{1}{2}\text{interval}} \int_{-l/2}^{l/2} f(x) \cos \frac{2\pi n x}{l} \, dx$$

$$= \frac{2}{l} \left[\int_{-l/2}^{0} f_e(x) \cos \frac{2\pi n x}{l} \, dx + \int_{0}^{l/2} f_e(x) \cos \frac{2\pi n x}{l} \, dx \right]$$

$$= \frac{4}{l} \int_{0}^{l/2} f(x) \cos \frac{2\pi n x}{l} \, dx$$

because

$$f(x) = f_e(x) \quad \text{from } x = 0 \text{ to } l/2$$

and

$$f(x) = f(-x) = f_e(x) \quad \text{from } x = 0 \text{ to } -l/2$$

Similarly we can represent $f(x)$ by the sine series

$$f(x) = f_o(x) = \sum_{n=1}^{\infty} b_n \sin \frac{2\pi n x}{l}$$

in the range $x = 0$ to $l/2$ with

$$b_n = \frac{1}{\frac{1}{2}\text{interval}} \int_{-l/2}^{l/2} f(x) \sin \frac{2\pi n x}{l} \, dx$$

$$= \frac{2}{l} \left[\int_{-l/2}^{0} f_o(x) \sin \frac{2\pi n x}{l} \, dx + \int_{0}^{l/2} f_o(x) \sin \frac{2\pi n x}{l} \, dx \right]$$

In the second integral $f_o(x) = f(x)$ in the interval 0 to $l/2$ whilst

$$\int_{-l/2}^{0} f_o(x) \sin \frac{2\pi n x}{l} \, dx = \int_{l/2}^{0} f_o(-x) \sin \frac{2\pi n x}{l} \, dx = -\int_{l/2}^{0} f_o(x) \sin \frac{2\pi n x}{l} \, dx$$

$$= \int_{0}^{l/2} f_o(x) \sin \frac{2\pi n x}{l} \, dx = \int_{0}^{l/2} f(x) \sin \frac{2\pi n x}{l} \, dx$$

Hence

$$b_n = \frac{4}{l} \int_{0}^{l/2} f(x) \sin \frac{2\pi n x}{l} \, dx$$

If we follow the behaviour of $f_e(x)$ and $f_o(x)$ outside the half-interval 0 to $l/2$ (Figure 10.4a, b) we see that they no longer represent $f(x)$.

Application of Fourier Sine Series to a Triangular Function

Figure 10.5 shows a function which we are going to describe by a sine series in the half-interval 0 to π. The function is

$$f(x) = x \quad \left(0 < x < \frac{\pi}{2}\right)$$

and

$$f(x) = \pi - x \quad \left(\frac{\pi}{2} < x < \pi\right)$$

Writing $f(x) = \sum b_n \sin nx$ gives

$$b_n = \frac{2}{\pi} \int_0^{\pi/2} x \sin nx \, \mathrm{d}x + \frac{2}{\pi} \int_{\pi/2}^{\pi} (\pi - x) \sin nx \, \mathrm{d}x$$

$$= \frac{4}{n^2 \pi} \sin \frac{n\pi}{2}$$

When n is even $\sin n\pi/2 = 0$, so that only terms with odd values of n are present and

$$f(x) = \frac{4}{\pi} \left(\frac{\sin x}{1^2} - \frac{\sin 3x}{3^2} + \frac{\sin 5x}{5^2} - \frac{\sin 7x}{7^2} + \cdots \right)$$

Note that at $x = \pi/2, f(x) = \pi/2$, giving

$$\frac{\pi^2}{8} = \frac{1}{1^2} + \frac{1}{3^2} + \frac{1}{5^2} + = \sum_{n=0}^{\infty} \frac{1}{(2n+1)^2}$$

We shall use this result a little later.

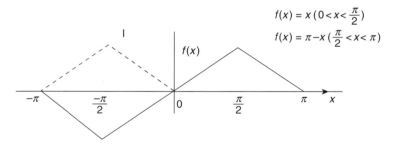

Figure 10.5 Function representing a plucked string and defined over a limited interval. When the string vibrates all the permitted harmonics contribute to the initial configuration

Note that the solid line in the interval 0 to $-\pi$ in Figure 10.5 is the Fourier sine representation for $f(x)$ repeated outside the interval 0 to π whilst the dotted line would result if we had represented $f(x)$ in the interval 0 to π by an even cosine series.

(Problems 10.1, 10.2, 10.3, 10.4, 10.5, 10.6, 10.7, 10.8, 10.9)

Application to the Energy in the Normal Modes of a Vibrating String

If we take a string of length l with fixed ends and pluck its centre a distance d we have the configuration of the half interval 0 to π of Figure 10.5 which we represented as a Fourier sine series. Releasing the string will set up its normal mode or standing wave vibrations, each of which we have shown on p. 126 to have the displacement

$$y_n = (A_n \cos \omega_n t + B_n \sin \omega_n t) \sin \frac{\omega_n x}{c} \tag{5.10}$$

where $\omega_n = n\pi c/l$ is the normal mode frequency.

The total displacement, which represents the shape of the plucked string at $t = 0$ is given by summing the normal modes

$$y = \sum y_n = \sum (A_n \cos \omega_n t + B_n \sin \omega_n t) \sin \frac{\omega_n x}{c}$$

Note that this sum resembles a Fourier series where the fixed ends of the string, $y = 0$ at $x = 0$ and $x = l$ allow only the sine terms in x in the series expansion. If the string remains plucked *at rest* only the terms in x with appropriate coefficients are required to describe it, but its vibrational motion after release has a time dependence which is expressed in each harmonic coefficient as

$$A_n \cos \omega_n t + B_n \sin \omega_n t$$

The significance of these coefficients emerges when we consider the initial or boundary conditions in time.

Let us write the total displacement of the string at time $t = 0$ as

$$y_0(x) = \sum y_n(x) = \sum (A_n \cos \omega_n t + B_n \sin \omega_n t) \sin \frac{\omega_n x}{c}$$

$$= \sum A_n \sin \frac{\omega_n x}{c} \quad \text{at} \quad t = 0$$

Similarly we write the velocity of the string at time $t = 0$ as

$$v_0(x) = \frac{\partial}{\partial t} y_0(x) = \sum \dot{y}_n(x)$$

$$= \sum (-\omega_n A_n \sin \omega_n t + \omega_n B_n \cos \omega_n t) \sin \frac{\omega_n x}{c}$$

$$= \sum \omega_n B_n \sin \frac{\omega_n x}{c} \quad \text{at} \quad t = 0$$

Both $y_0(x)$ and $v_0(x)$ are thus expressed as Fourier sine series, but if the string is at rest at $t = 0$, then $v_0(x) = 0$ and all the B_n coefficients are zero, leaving only the A_n's. If the displacement of the string $y_0(x) = 0$ at time $t = 0$ whilst the string is moving, then all the A_n's are zero and the Fourier coefficients are the $\omega_n B_n$'s.

We can solve for both A_n and $\omega_n B_n$ in the usual way for if

$$y_0(x) = \sum A_n \sin \frac{\omega_n x}{c}$$

and

$$v_0(x) = \sum \omega_n B_n \sin \frac{\omega_n x}{c}$$

for a string of length l then

$$A_n = \frac{2}{l} \int_0^l y_0(x) \sin \frac{\omega_n x}{c} \, dx$$

and

$$\omega_n B_n = \frac{2}{l} \int_0^l v_0(x) \sin \frac{\omega_n x}{c} \, dx$$

If the plucked string of mass m (linear density ρ) is released from rest at $t = 0$ ($v_0(x) = 0$) the energy in each of its normal modes of vibration, given on p. 134 as

$$E_n = \tfrac{1}{4} m \omega_n^2 (A_n^2 + B_n^2)$$

is simply

$$E_n = \tfrac{1}{4} m \omega_n^2 A_n^2$$

because all B_n's are zero.

The total vibrational energy of the released string will be the sum $\sum E_n$ over all the modes present in the vibration.

Let us now solve the problem of the plucked string released from rest. The configuration of Figure 10.5 (string length l, centre plucked a distance d) is given by

$$y_0(x) = \frac{2dx}{l} \qquad 0 \le x \le \frac{l}{2}$$

$$= \frac{2d(l - x)}{l} \quad \frac{l}{2} \le x \le l$$

so

$$A_n = \frac{2}{l} \left[\int_0^{l/2} \frac{2dx}{l} \sin \frac{\omega_n x}{c} \, dx + \int_{l/2}^l \frac{2d(l - x)}{l} \sin \frac{\omega_n x}{c} \, dx \right]$$

$$= \frac{8d}{n^2 \pi^2} \sin \frac{n\pi}{2} \left(\text{for } \omega_n = \frac{n\pi c}{l} \right)$$

We see at once that $A_n = 0$ for n even (when the sine term is zero) so that all even harmonic modes are missing. The physical explanation for this is that the even harmonics would require a node at the centre of the string which is always moving after release.

The displacement of our plucked string is therefore given by the addition of all the permitted (odd) modes as

$$y_0(x) = \sum_{n \text{ odd}} y_n(x) = \sum_{n \text{ odd}} A_n \sin \frac{\omega_n x}{c}$$

where

$$A_n = \frac{8d}{n^2 \pi^2} \sin \frac{n\pi}{2}$$

The energy of the nth mode of oscillation is

$$E_n = \frac{1}{4} m \omega_n^2 A_n^2 = \frac{64 d^2 m \omega_n^2}{4(n^2 \pi^2)^2}$$

and the total vibrational energy of the string is given by

$$E = \sum_{n \text{ odd}} E_n = \frac{16 d^2 m}{\pi^4} \sum_{n \text{ odd}} \frac{\omega_n^2}{n^4} = \frac{16 d^2 c^2 m}{\pi^2 l^2} \sum_{n \text{ odd}} \frac{1}{n^2}$$

for

$$\omega_n = \frac{n\pi c}{l}$$

But we saw in the last section that

$$\sum_{n \text{ odd}} \frac{1}{n^2} = \frac{\pi^2}{8}$$

so

$$E = \sum E_n = \frac{2 m c^2 d^2}{l^2} = \frac{2 T d^2}{l}$$

where $T = \rho c^2$ is the constant tension in the string.

This vibrational energy, in the absence of dissipation, must be equal to the potential energy of the plucked string before release and the reader should prove this by calculating the work done in plucking the centre of the string a small distance d, where $d \ll l$.

To summarize, our plucked string can be represented as a sine series of Fourier components, each giving an allowed normal mode of vibration when it is released. The concept of normal modes allows the energies of each mode to be added to give the total energy of vibration which must equal the potential energy of the plucked string before release. The energy of the nth mode is proportional to n^{-2} and therefore decreases with increasing frequency. Even modes are forbidden by the initial boundary conditions.

The boundary conditions determine which modes are allowed. If the string were struck by a hammer those harmonics having a node at the point of impact would be absent, as in the case of the plucked string. Pianos are commonly designed with the hammer striking a point one seventh of the way along the string, thus eliminating the seventh harmonic which combines to produce discordant effects.

Fourier Series Analysis of a Rectangular Velocity Pulse on a String

Let us now consider a problem similar to that of the last section except that now the displacement $y_0(x)$ of the string is zero at time $t = 0$ whilst the velocity $v_0(x)$ is non-zero. A string of length l, fixed at both ends, is struck by a mallet of width a about its centre point. At the moment of impact the displacement

$$y_0(x) = 0$$

but the velocity

$$v_0(x) = \frac{\partial y_0(x)}{\partial t} = 0 \quad \text{for} \quad \left| x - \frac{l}{2} \right| \geq \frac{a}{2}$$

$$= v \quad \text{for} \quad \left| x - \frac{l}{2} \right| < \frac{a}{2}$$

This situation is shown in Figure 10.6.
The Fourier series is given by

$$v_0(x) = \sum_n \dot{y}_n = \sum_n \omega_n B_n \sin \frac{\omega_n x}{c}$$

where

$$\omega_n B_n = \frac{2}{l} \int_{+l/2-a/2}^{l/2+a/2} v \sin \frac{\omega_n x}{c} \, dx$$

$$= \frac{4v}{n\pi} \sin \frac{n\pi}{2} \sin \frac{n\pi a}{2l}$$

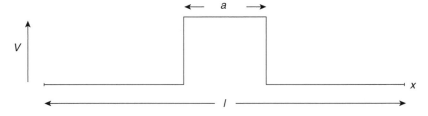

Figure 10.6 Velocity distribution at time $t = 0$ of a string length l, fixed at both ends and struck about its centre point by a mallet of width a. Displacement $y_0(x) = 0$; velocity $v_0(x) = v$ for $|x - l/2| < a/2$ and zero outside this region

Again we see that $\omega_n B_n = 0$ for n even ($\sin n\pi/2 = 0$) because the centre point of the string is never stationary, as is required in an even harmonic.

Thus

$$v_0(x) = \sum_{n \text{ odd}} \frac{4v}{n\pi} \sin \frac{n\pi a}{2l} \sin \frac{\omega_n x}{c}$$

The energy per mode of oscillation

$$\begin{aligned}
E_n &= \tfrac{1}{4} m \omega_n^2 (A_n^2 + B_n^2) \\
&= \tfrac{1}{4} m \omega_n^2 B_n^2 \quad \text{(All } A_n\text{'s} = 0) \\
&= \tfrac{1}{4} m \frac{16 v^2}{n^2 \pi^2} \sin^2 \frac{n\pi a}{2l} \\
&= \frac{4 m v^2}{n^2 \pi^2} \sin^2 \frac{n\pi a}{2l}
\end{aligned}$$

Now

$$n = \frac{\omega_n}{\omega_1} = \frac{\omega_n l}{\pi c}$$

for the fundamental frequency

$$\omega_1 = \frac{\pi c}{l}$$

So

$$E_n = \frac{4 m v^2 c^2}{l^2 \omega_n^2} \sin^2 \frac{\omega_n a}{2c}$$

Again we see, since $\omega_n \propto n$ that the energy of the nth mode $\propto n^{-2}$ and decreases with increasing harmonic frequency. We may show this by rewriting

$$\begin{aligned}
E_n(\omega) &= \frac{m v^2 a^2}{l^2} \frac{\sin^2(\omega_n a/2c)}{(\omega_n a/2c)^2} \\
&= \frac{m v^2 a^2}{l^2} \frac{\sin^2 \alpha}{\alpha^2}
\end{aligned}$$

where

$$\alpha = \omega_n a/2c$$

and plotting this expression as an energy-frequency spectrum in Figure 10.7.

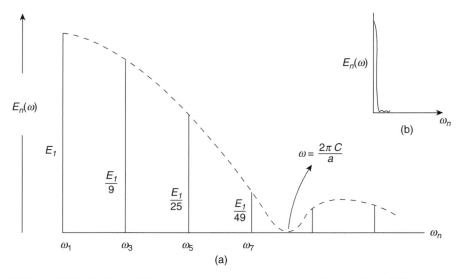

Figure 10.7 (a) Distribution of the energy in the harmonics ω_n of the string of Figure 10.6. The spectrum $E_n(\omega) \propto \sin^2\alpha/\alpha^2$ where $\alpha = \omega_n a/2c$. Most of the energy in the string is contained in the frequency range $\Delta\omega \approx 2\pi c/a$, and for $a = \Delta x$ (the spatial width of the pulse), $\Delta x/c = \Delta t$ and $\Delta\omega\Delta t \approx 2\pi$ (Bandwidth Theorem). Note that the values of $E_n(\omega)$ for $\omega_3, \omega_5, \omega_7$, etc. are magnified for clarity. (b) The true shape of the pulse

The familiar curve of $\sin^2\alpha/\alpha^2$ again appears as the envelope of the energy values for each ω_n.

If the energy at ω_1 is E_1 then $E_3 = E_1/9$ and $E_5 = E_1/25$ so the major portion of the energy in the velocity pulse is to be found in the low frequencies. The first zero of the envelope $\sin^2\alpha/\alpha^2$ occurs when

$$\alpha = \frac{\omega a}{2c} = \pi$$

so the width of the central frequency pulse containing most of the energy is given by

$$\omega \approx \frac{2\pi c}{a}$$

This range of energy-bearing harmonics is known as the 'spectral width' of the pulse written

$$\Delta\omega \approx \frac{2\pi c}{a}$$

The 'spatial width' a of the pulse may be written as Δx so we have

$$\Delta x\Delta\omega \approx 2\pi c$$

Reducing the width Δx of the mallet will increase the range of frequencies $\Delta\omega$ required to take up the energy in the rectangular velocity pulse. Now c is the velocity of waves on the string so a wave travels a distance Δx along the string in a time

$$\Delta t = \Delta x/c$$

which defines the duration of the pulse giving

$$\Delta\omega\,\Delta t \approx 2\pi$$

or

$$\Delta\nu\,\Delta t \approx 1$$

the Bandwidth Theorem we first met on p. 134.

Note that the harmonics have frequencies

$$\omega_n = \frac{n\pi c}{l}$$

so $\pi c/l$ is the harmonic interval. When the length l of the string becomes very long and $l \to \infty$ so that the pulse is isolated and non-periodic, the harmonic interval becomes so small that it becomes differential and the Fourier series summation becomes the Fourier Integral discussed on p. 283.

The Spectrum of a Fourier Series

The Fourier series can always be represented as a frequency spectrum. In Figure 10.8 a the relative amplitudes of the frequency components of the square wave of Figure 10.1 are plotted, each sine term giving a single spectral line. In a similar manner, the distribution of energy with frequency may be displayed for the plucked string of the earlier section. The frequency of the r th mode of vibration is given by $\omega_r = r\pi c/l$, and the energy in each mode varies inversely with r^2, where r is odd. The spectrum of energy distribution is therefore given by Figure 10.8 b.

Suppose now that the length of this string is halved but that the total energy remains constant. The frequency of the fundamental is now increased to $\omega'_r = 2r\pi c/l$ and the frequency interval between consecutive spectral lines is doubled (Figure 10.8 c). Again, the smaller the region in which a given amount of energy is concentrated the wider the frequency spectrum required to represent it.

Frequently, as in the next section, a Fourier series is expressed in its complex or exponential form

$$f(t) = \sum_{n=-\infty}^{\infty} d_n\, e^{in\omega t}$$

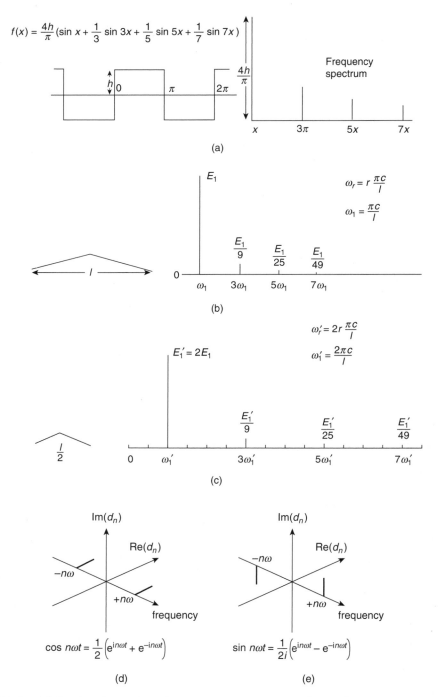

Figure 10.8 (a) Fourier sine series of a square wave represented as a frequency spectrum; (b) energy spectrum of a plucked string of length l; and (c) the energy spectrum of a plucked string of length $l/2$ with the same total energy as (b), demonstrating the Bandwidth Theorem that the greater the concentration of the energy in space or time the wider its frequency spectrum. Complex exponential frequency spectrum of (d) $\cos \omega t$ and (e) $\sin \omega t$

where $2d_n = a_n - \mathrm{i}b_n(n \geq 0)$ and $2d_n = a_{-n} + \mathrm{i}b_{-n}(n < 0)$.
Because

$$\cos n\omega t = \tfrac{1}{2}(\mathrm{e}^{\mathrm{i}n\omega t} + \mathrm{e}^{-\mathrm{i}n\omega t})$$

and

$$\sin n\omega t = \frac{1}{2\mathrm{i}}(\mathrm{e}^{\mathrm{i}n\omega t} - \mathrm{e}^{-\mathrm{i}n\omega t})$$

a frequency spectrum in the complex plane produces two spectral lines for each frequency component $n\omega$, one at $+n\omega$ and the other at $-n\omega$. Figure 10.8 d shows the cosine representation, which lies wholly in the real plane, and Figure 10.8 e shows the sine representation, which is wholly imaginary. The amplitudes of the lines in the positive and negative frequency ranges are, of course, complex conjugates, and the modulus of their product gives the square of the true amplitude. The concept of a negative frequency is seen to arise because the $\mathrm{e}^{-\mathrm{i}n\omega t}$ term increases its phase in the opposite sense to that of the positive term $\mathrm{e}^{\mathrm{i}n\omega t}$. The negative amplitude of the negative frequency in the sine representation indicates that it is in antiphase with respect to that of the positive term.

Fourier Integral

At the beginning of this chapter we saw that one Fourier representation of the function could be written

$$f(x) = \sum_{n=-\infty}^{\infty} d_n \mathrm{e}^{\mathrm{i}nx}$$

where $2d_n = a_n - \mathrm{i}b_n(n \geq 0)$ and $2d_n = a_{-n} + \mathrm{i}b_{-n}(n < 0)$.

If we use the time as a variable we may rewrite this as

$$f(t) = \sum_{n=-\infty}^{\infty} d_n \mathrm{e}^{\mathrm{i}n\omega t}$$

where, if T is the period,

$$d_n = \frac{1}{T} \int_{-T/2}^{T/2} f(t) \mathrm{e}^{-\mathrm{i}n\omega t} \, \mathrm{d}t$$

(for $n = -2, -1, 0, 1, 2$, etc.).

If we write $\omega = 2\pi\nu_1$, where ν_1 is the fundamental frequency, we can write

$$f(t) = \sum_{n=-\infty}^{\infty} \left[\int_{-T/2}^{T/2} f(t') \mathrm{e}^{-\mathrm{i}2\pi n\nu_1 t'} \, \mathrm{d}t' \right] \mathrm{e}^{\mathrm{i}2\pi n\nu_1 t} \cdot \frac{1}{T}$$

If we now let the period T approach infinity we are isolating a single pulse by saying that it will not be repeated for an infinite period; the frequency $\nu_1 = 1/T \to 0$, and $1/T$ becomes infinitesimal and may be written $d\nu$.

Furthermore, n times ν_1, when n becomes as large as we please and $1/T = \nu_1 \to 0$, may be written as $n\nu_1 = \nu$, and the sum over n now becomes an integral, since unit change in n produces an infinitesimal change in $n/T = n\nu_1$.

Hence, for an infinite period, that is for a single non-periodic pulse, we may write

$$f(t) = \int_{-\infty}^{\infty} \left[\int_{-\infty}^{\infty} f(t')\, e^{-i2\pi\nu t'}\, dt' \right] e^{i2\pi\nu t}\, d\nu$$

which is called the *Fourier Integral*.

We may express this as

$$f(t) = \int_{-\infty}^{\infty} F(\nu)\, e^{i2\pi\nu t}\, d\nu$$

where

$$F(\nu) = \int_{-\infty}^{\infty} f(t')\, e^{-i2\pi\nu t'}\, dt'$$

is called the *Fourier Transform* of $f(t)$. We shall discuss the transform in more detail in a later section of this chapter.

We see that when the period is finite and $f(t)$ is periodic, the expression

$$f(t) = \sum_{n=-\infty}^{\infty} d_n\, e^{in\omega t}$$

tells us that the representation is in terms of an infinite number of different frequencies, each frequency separated by a finite amount from its nearest neighbour, but when $f(t)$ is not periodic and has an infinite period then

$$f(t) = \int_{-\infty}^{\infty} F(\nu)\, e^{i2\pi\nu t}\, d\nu$$

and this expression is the integral (not the sum) of an infinite number of frequency components of amplitude $F(\nu)\, d\nu$ infinitely close together, since ν varies continuously instead of in discrete steps.

For a periodic function the amplitude of the Fourier series coefficient

$$d_n = \frac{1}{T} \int_{-T/2}^{T/2} f(t)\, e^{-in\omega t}\, dt$$

whereas the corresponding amplitude in the Fourier integral is

$$F(\nu)\,\mathrm{d}\nu = \left(\frac{1}{T}\right)\int_{-\infty}^{\infty} f(t')\,\mathrm{e}^{-in\omega t'}\,\mathrm{d}t'$$

This corroborates the statement we made when discussing the frequency spectrum that the narrower or less extended the pulse the wider the range of frequency components required to represent it. A truly monochromatic wave of one frequency and wavelength (or wave number) requires a wave train of infinite length before it is properly defined.

No wave train of finite length can be defined in terms of *one* unique wavelength.

Since a monochromatic wave, infinitely long, of single frequency and constant amplitude transmits no information, its amplitude must be modified by adding other frequencies (as we have seen in Chapter 5) before the variation in amplitude can convey information. These ideas are expressed in terms of the Bandwidth Theorem.

Fourier Transforms

We have just seen that the Fourier integral representing a non-periodic wave group can be written

$$f(t) = \int_{-\infty}^{\infty} F(\nu)\,\mathrm{e}^{i2\pi\nu t}\,\mathrm{d}\nu$$

where its Fourier transform

$$F(\nu) = \int_{-\infty}^{\infty} f(t')\,\mathrm{e}^{-i2\pi\nu t'}\,\mathrm{d}t'$$

so that integration with respect to one variable produces a function of the other. Both variables appear as a product in the index of an exponential, and this product must be non-dimensional. Any pair of variables which satisfy this criterion forms a Fourier pair of transforms, since from the symmetry of the expressions we see immediately that if

$$F(\nu) \text{ is the Fourier transform of } f(t)$$

then

$$f(-\nu) \text{ is the Fourier transform of } F(t)$$

If we are given the distribution in time of a function we can immediately express it as a spectrum of frequency, and vice versa. In the same way, a given distribution in space can be expressed as a function of wave numbers (this merely involves a factor, $1/2\pi$, in front of the transform because $k = 2\pi/\lambda$).

A similar factor appears if ω is used instead of ν. If the function of $f(t)$ is even only the cosine of the exponential is operative, and we have a Fourier cosine transform

$$f(t) = \int_{0}^{\infty} F(\nu)\cos 2\pi\nu t\,\mathrm{d}\nu$$

and

$$F(\nu) = \int_0^\infty f(t)\cos 2\pi\nu t\,\mathrm{d}t$$

If $f(t)$ is odd only the sine terms operate, and sine terms replace the cosines above. Note that only positive frequencies appear. The Fourier transform of an even function is real and even, whilst that of an odd function is imaginary and odd.

Examples of Fourier Transforms

The two examples of Fourier transforms chosen to illustrate the method are of great physical significance. They are

1. The 'slit' function of Figure 10.9a,

2. The Gaussian function of Figure 10.11.

As shown, they are both even functions and their transforms are therefore real; the physical significance of this is that all the frequency components have the same phase at zero time.

The Slit Function

This is a function having height h over the time range $\pm d/2$. Thus, $f(t) = h$ for $|t| < d/2$ and zero for $|t| > d/2$, so that

$$F(\nu) = \int_{-\infty}^{\infty} f(t)\,\mathrm{e}^{-\mathrm{i}2\pi\nu t}\,\mathrm{d}t = \int_{-d/2}^{d/2} h\,\mathrm{e}^{-\mathrm{i}2\pi\nu t}\,\mathrm{d}t$$

$$= \frac{-h}{\mathrm{i}2\pi\nu}[\mathrm{e}^{-\mathrm{i}2\pi\nu d/2} - \mathrm{e}^{+\mathrm{i}2\pi\nu d/2}] = hd\frac{\sin\alpha}{\alpha}$$

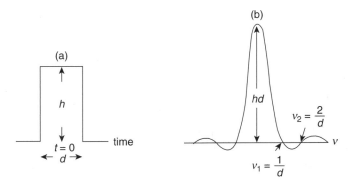

Figure 10.9 (a) Narrow slit function of extent d in time and of height h, and (b) its Fourier transform

where

$$\alpha = \frac{2\pi\nu d}{2}$$

Again we see the Fourier transformation of a rectangular pulse in time to a $\sin\alpha/\alpha$ pattern in frequency. The Fourier transform of the same pulse in space will give the same distribution as a function of wavelength. Figure 10.9b shows that as the pulse width decreases in time the separation between the zeros of the transform is increased. The negative values in the spectrum of the transform indicate a phase reversal for the amplitude of the corresponding frequency component.

The Fourier Transform Applied to Optical Diffraction from a Single Slit

This topic belongs more properly to the next chapter where it will be treated by another method, but here we derive the fundamental result as an example of the Fourier Transform. The elegance of this method is seen in problems more complicated than the one-dimensional example considered here. We shall see its extension to two dimensions in Chapter 12 when we consider the diffraction patterns produced by rectangular and circular apertures.

The amplitude of light passing through a single slit may be represented in space by the rectangular pulse of Figure 10.9a where d is now the width of the slit. A plane wave of monochromatic light, wavelength λ, falling normally on a screen which contains the narrow slit of width $d \sim \lambda$, forms a secondary system of plane waves diffracted in all directions with respect to the screen. When these diffracted waves are focused on to a second screen the intensity distribution (square of the amplitude) may be determined in terms of the aperture dimension d, the wavelength λ and the angle of diffraction θ.

In Figure 10.10 the light diffracted through an angle θ is brought to focus at a point P on the screen PP_0. Finding the amplitude of the light at P is the simple problem of adding all the small contributions in the diffracted wavefront taking account of all the phase differences which arise with variation of path length from P to the points in the slit aperture from which the contributions originate. The diffraction amplitude in k or wave number space is the Fourier transform of the pulse, width d, in x space in Figure 10.9b. The conjugate parameters ν and t are exactly reciprocal but the product of x and k involves the term 2π which requires either a constant factor $1/2\pi$ in front of one of the transform integrals or a common factor $1/\sqrt{2\pi}$ in front of each. This factor is however absorbed into the constant value of the maximum intensity and all other intensities are measured relative to it.

The constant pulse height now measures the amplitude h of the small wave sources across the slit width d and the Fourier transform method is the addition by integration of their contributions.

In Figure 10.10 we see that the path difference between the contribution at the centre of the slit and that at a point x in the slit is given by $x\sin\theta$, so that the phase difference is

$$\phi = \frac{2\pi}{\lambda}x\sin\theta = kx\sin\theta$$

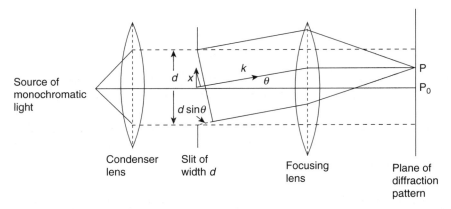

Figure 10.10 A monochromatic plane wave normally incident on a narrow slit of width d is diffracted an angle θ, and the light in this direction is focused at a point P. The amplitude at P is the superposition of all contributions with their appropriate phases with respect to the central point in the slit. The contribution from a point x in the slit has phase $\phi = 2\pi x \sin\theta/\lambda$ with respect to the central contribution. The phase difference from contributing points on opposite edges of the slit is $\phi = 2\pi d \sin\theta/\lambda = 2\alpha$

The product $kx\sin\theta$ can, however be expressed in a form more suitable for extension to two- and three-dimensional examples by writing it as $\mathbf{k}\cdot\mathbf{x} = klx$, the scalar product of the vector $\mathbf{k}$, giving the wave propagation direction, and the vector $\mathbf{x}$, l being the direction cosine

$$l = \cos(\pi/2 - \theta)$$
$$= \sin\theta$$

of $\mathbf{k}$ with respect to the x-axis.

Adding all the small contributions across the slit to obtain the amplitude at P by the Fourier transform method gives

$$F(k) = \frac{1}{2\pi}\int f(x)\,e^{-i\phi}\,dx$$
$$= \frac{1}{2\pi}\int_{-d/2}^{+d/2} h\,e^{-iklx}\,dx$$
$$= \frac{h}{-ikl}\frac{1}{2\pi}\left(e^{-ikld/2} - e^{+ikld/2}\right)$$
$$= \frac{-2ih}{-ikl2\pi}\sin\frac{kld}{2}$$
$$= \frac{dh}{2\pi}\frac{\sin\alpha}{\alpha}$$

where

$$\alpha = \frac{kld}{2} = \frac{\pi}{\lambda} d \sin \theta$$

The intensity I at P is given by the square of the amplitude; that is, by the product of $F(k)$ and its complex conjugate $F^*(k)$, so that

$$I = \frac{d^2 h^2}{4\pi^2} \frac{\sin^2 \alpha}{\alpha^2}$$

where I_0, the principal maximum intensity at $\alpha = 0$, (P_0 in Figure 10.10) is now

$$I_0 = \frac{d^2 h^2}{4\pi^2}$$

The Gaussian Curve

This curve often appears as the wave group description of a particle in wave mechanics. The Fourier transform of a Guassian distribution is another Gaussian distribution.

In Figure 10.11a the Gaussian function of height h is symmetrically centred at time $t = 0$, and is given by $f(t) = h e^{-t^2/\sigma^2}$, where the width parameter or standard deviation σ is that value of t at which the height of the curve has a value equal to e^{-1} of its maximum.

Its transform is

$$F(\nu) = \int_{-\infty}^{\infty} h e^{-t/\sigma^2} e^{-i2\pi\nu t} \, dt$$

$$= \int_{-\infty}^{\infty} h e^{(-t/\sigma^2 - i2\pi\nu t + \pi^2\nu^2\sigma^2)} e^{-\pi^2\nu^2\sigma^2} \, dt$$

$$= h e^{(-\pi^2\nu^2\sigma^2)} \int_{-\infty}^{\infty} e^{-(t/\sigma + i\pi\nu\sigma)^2} \, dt$$

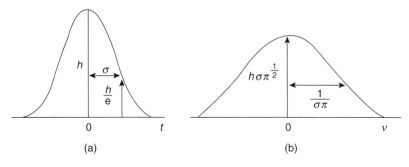

(a) (b)

Figure 10.11 (a) A Gaussian function Fourier transforms (b) into another Gaussian function

The integral

$$\int_{-\infty}^{\infty} e^{-x^2}\, dx = \sqrt{\pi}$$

and substituting, with $x = (t/\sigma + i\pi\nu\sigma)$ and $dt = \sigma\, dx$, gives

$$F(\nu) = h\sigma\pi^{1/2}e^{-\pi^2\nu^2\sigma^2}$$

another Gaussian distribution in frequency space (Figure 10.11b) with a new height $h\sigma\pi^{1/2}$ and a new width parameter $(\sigma\pi)^{-1}$.

As in the case of the slit and the diffraction pattern, we see again that a narrow pulse in time (width σ) leads to a wide frequency distribution [width $(\sigma\pi)^{-1}$].

When the curve is normalized so that the area under it is unity, h takes the value $(\sigma\pi)^{1/2}$ because

$$\frac{1}{(\sigma\pi^{1/2})}\int_{-\infty}^{\infty} e^{-t^2/\sigma^2}\, dt = 1$$

Thus, the height of a normalized curve transforms into a pulse of unit height whereas a pulse of unit height transforms to a pulse of width $(\sigma\pi)^{-1}$.

If we consider a family of functions with progressively increasing h values and decreasing σ values, each satisfying the condition of unit area under their curves, we are led in the limit as the height $h \to \infty$ and the width $\sigma \to 0$ to an infinitely narrow pulse of finite area unity which defines the Dirac delta (δ) function. The transform of such a function is the constant unity, and Figures. 10.12a and b show the family of normalized Gaussian distributions and their transforms. Figure 10.13 shows a number of common Fourier transform pairs.

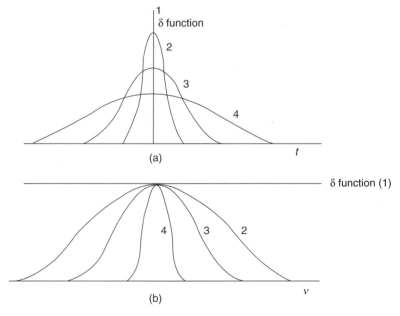

Figure 10.12 (a) A family of normalized Gaussian functions narrowed in the limit to Dirac's delta function; (b) the family of their Fourier transforms

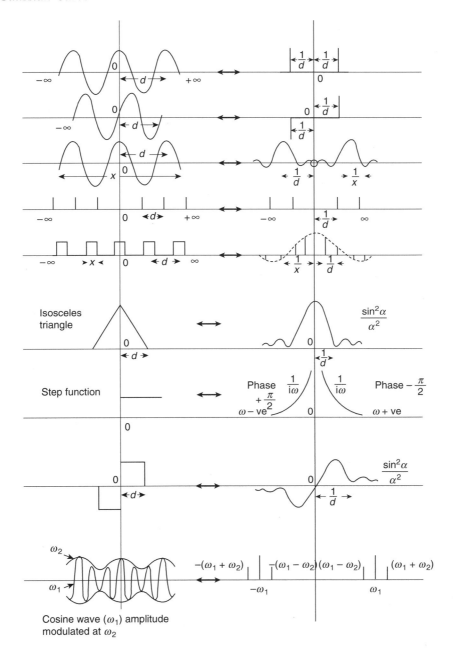

Figure 10.13   Some common Fourier transform pairs

In wave mechanics the position x of a particle and its momentum p_x are conjugate parameters and its Gaussian wave group representation may be Fourier transformed from x to p_x space and vice versa. The Fourier Transform gives the amplitude of the wave function but the probability of finding the particle at x or its having a given momentum p_x is proportional to the square of the amplitude.

The Dirac Delta Function, its Sifting Property and its Fourier Transform

The Dirac δ function is defined by

$$\delta(x) = 0 \text{ at } x \neq 0$$
$$= \infty \text{ at } x = 0$$

and

$$\int_{-\infty}^{\infty} \delta(x)\mathrm{d}x = 1$$

i.e., an infinitely narrow pulse centred on $x = 0$. It is also known as the unit impulse function.

A valuable characteristic is its sifting property, that is

$$\int_{-\infty}^{\infty} \delta(x - x_0)f(x)\mathrm{d}x = f(x_0)$$

The Fourier Transform of $\delta(x - x_0) = \mathrm{e}^{-ikx_0}$ because by definition

$$F(\delta(x - x_0)) = \int_{-\infty}^{\infty} \delta(x - x_0)\mathrm{e}^{-ikx}\mathrm{d}x$$

so writing $f(x) = \mathrm{e}^{-ikx}$ and applying the sifting property gives $f(x_0) = \mathrm{e}^{-ikx_0}$. Note that $\mathrm{e}^{-ikx_0} = \mathrm{e}^{ikx_0} = 1$ for $x_0 = 0$.

From the form of the transform we see that if a function $f(x)$ is a sum of individual functions then the Fourier Transform $F(f(x))$ is the sum of their individual transforms. Thus, if

$$f(x) = \sum_j \delta(x - x_j)$$

then

$$Ff(x) = \sum_j \mathrm{e}^{-ikx_j}$$

Figure 10.14 shows two Dirac δ functions situated at $x = \pm\frac{a}{2}$ so that $f(x) = \delta(x - \frac{a}{2}) + \delta(x + \frac{a}{2})$ giving $F(f(x)) = \mathrm{e}^{\frac{ika}{2}} + \mathrm{e}^{-\frac{ika}{2}} = 2\cos ka/2$.

Convolution

Given two functions $f(x)$ and $h(x)$, their convolution, written

$$f(x) \otimes h(x) = \int_{-\infty}^{\infty} f(x)h(x)\mathrm{d}x$$

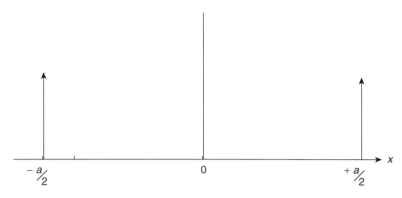

Figure 10.14 The Fourier transform of two Dirac δ functions located at $x = \pm a/2$ is $2\cos ka/2$

is the overlap area under the product of the two functions as one function scans across the other. It the functions are two dimensional, $f(x, y)$ and $h(x, y)$, their convolution is the volume overlap under their product.

To illustrate a one-dimensional convolution consider the rectangular pulse of length D in Figure 10.15 convolved with an identical pulse. This is known as self-convolution. The convolution will be the sum of the shaded areas such as that of Figure 10.15a as one pulse slides over the other. We can see that the base length of the resulting convolved pulse will be $2D$ and that it will be symmetric about its peak, that is, when the two pulses completely overlap. If we consider the left-hand pulse as an infinite series of δ functions, of which we show a few, then Figure 10.15b shows that the integrated sum is an isosceles triangle of base length $2D$.

Another example is the convolution of a small triangular pulse with a rectangular pulse length D, Figure 10.16. Again, we use the series of δ functions to show the sum of the components of the resulting convolution and its integrated form for an infinite series of δ functions. The length of the final pulse is again the sum of the lengths of the two pulses.

Such a pulse would result in the convolution of a rectangular pulse with an exponential time function, for example, when a rectangular pulse is passed into an integrating network formed by a series resistance and parallel condenser, Figure 10.17. Here, the exponential time function of the network may be considered as fixed in time while the pulse performs the scanning operation. Note in Figures 10.15, 10.16 and 10.17 that the component contributions of the left hand pulses are summed in reverse order. This is explained in the discussion following eq. 10.2.

A convolution $f(x) \otimes h(x)$ is generally written in the form

$$g(x') = \int_{-\infty}^{\infty} f(x)h(x' - x)\mathrm{d}x \tag{10.2}$$

This a particularly relevant form when we consider the Optical Transfer Function on page 391. There, x is an object space coordinate and x' is an image space coordinate so the convolution relates image to object. If the function $h(x' - x)$ is a localized pulse in the object space and x' lies within it on the object axis x then the pulse $h(x' - x)$ is reversed

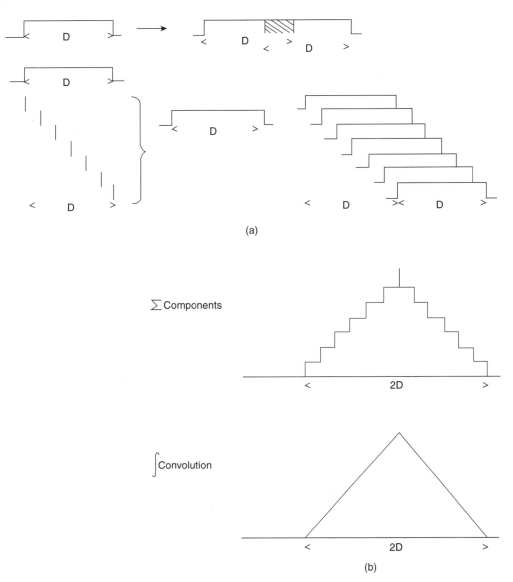

(a)

$\sum$ Components

$<$ 2D $>$

$\int$ Convolution

$<$ 2D $>$

(b)

Figure 10.15 (a) A convolution is the integral of all overlapping areas as one function scans another. A rectangular pulse length D scans an identical pulse and the overlap area is shaded at one point of the scanning. (b) The scanning pulse is represented by several Dirac δ (impulse) functions and the component overlap areas are summed. When the number of impulse functions is large the sum of the components is integrated to become the triangular pulse

in image space (axis x') so that its trailing edge becomes its leading edge. Figure 10.18(a) shows the pulse on the object axis and Figure 10.18(b) shows the reversed pulse on the image axis.

The product $f(x)\,h(x'-x)$ exists only where the functions overlap and in Figure 10.18(b) $g(x_1')$ is the superposition of all the individual overlapping contributions that

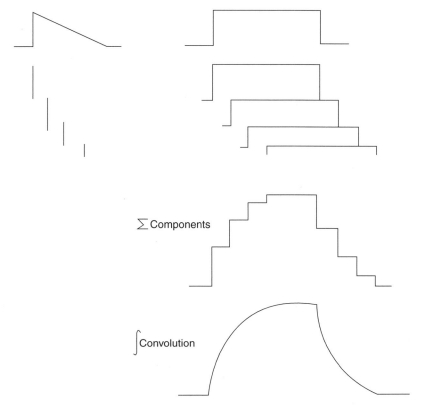

Figure 10.16 The convolution of a triangular with a rectangular pulse using the method of Figure 10.15

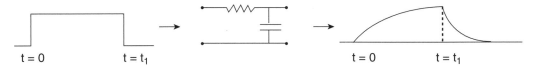

t = 0   t = t₁   t = 0   t = t₁

Figure 10.17 The convolution of Figure 10.16 is the same as that of a rectangular electrical pulse passing through an integrating circuit formed by a series resistance and a parallel condenser

exist at x'_1. The contribution to $g(x'_1)$ at x'_1 by x_1 and dx at x_1 is $f(x_1)h(x'_1 - x_1)\mathrm{d}x$ where $f(x_1)$ is a number which magnifies the pulse of Figure 10.18(b) to become the pulse of Figure 10.18(c). Each value of x in the overlap region makes a contribution to $g(x'_1)$; x values beyond the overlap make no contribution. The contributions begin when the leading edge of $h(x' - x)$ reaches x'_1 and they cease when its trailing edge passes x'_1.

Note that by changing the variable $x'' = x' - x$ in Equation (10.2).

$$f \otimes h = h \otimes f$$

This result is also evident when we consider the Convolution Theorem in the next section.

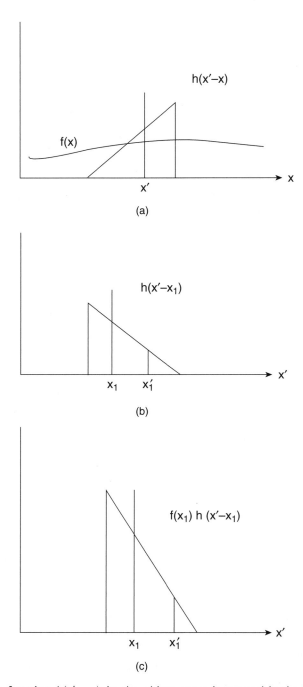

Figure 10.18   The function $h(x' - x)$ in the object space is reversed in the image space in Figure 10.18(b). (b) The convolution $g(x'_1)$ is the superposition of all individual overlapping contributions to $f(x)h(x' - x)$ that exist at x'_1. (c) The contribution made by $f(x_1)\mathrm{d}x$ to $g(x'_1)$ where $f(x_1)$ is a number which magnifies $h(x'_1 - x)$

Returning to the convolution of the rectangular pulses in Figure 10.15 and taking the left-hand pulse as $f(x)$ each impulse x_i of the infinite series sweeps across the right-hand pulse $h(x' - x)$ to give the triangular convolution $g(x')$. If the left-hand pulse is now $h(x' - x)$ sweeping across the right-hand pulse $f(x)$ with x_i' as a fixed location in $h(x' - x)$, the series of overlaps, as x_i' moves across $f(x)$, gives the same triangular convolution.

The Convolution Theorem

The importance of the convolution process may be seen by considering the following.

When a signal, electrical or optical, passes through a system such as an amplifier or a lens, the resulting output is a function of the original signal and the system response. We have seen that a slit, in passing light from an optical source, may act as an angular filter, restricting the amount of information it passes and superimposing its own transform on the radiation passing through. An electrical filter can behave in a similar fashion.

Effectively there are two transformations, one into the intermediate system and one out again.

A convolution reduces this to a single transformation. The transform of the intermediate system is applied to the orginal function or signal and the resulting output is the integrated product of each point operating on the transformed response.

The convolution theorem states that the Fourier transform of the convolution of two functions is the product of the Fourier transforms of the individual functions, that is, if

$$g(x') = f(x) \otimes h(x)$$

then

$$F(g) = F(f \otimes h) = F(f) \cdot F(h)$$

The proof is straightforward.

The convolution $g(x')$ is a function of k, so its transform is

$$F(g) = G(k) = \int_{-\infty}^{\infty} g(x') e^{-ihx'} dx'$$

$$= \int_{-\infty}^{\infty} \left[\int_{-\infty}^{\infty} f(x) h(x' - x) dx \right] e^{-ikx'} dx'$$

$$= \int_{-\infty}^{\infty} \left[\int_{-\infty}^{\infty} h(x' - x) e^{-ikx'} dx' \right] f(x) dx$$

Putting $x' - x = y$ gives $dy = dx'$ and $e^{-ikx'} = e^{-iky} e^{-ikx}$ and so

$$F(g) = G(k) = \int_{-\infty}^{\infty} f(x) e^{-ikx} dx \int_{-\infty}^{\infty} h(y) e^{-iky} dy$$

$$= F(f) \cdot F(h) = F(h) \cdot F(f)$$

We can use this result to find the Fourier Transform of the resulting triangular pulse in Figure 10.15(b). The slit may be seen as a rectangular pulse of width d and its Fourier

Transform on page 288 gave its diffraction pattern as $\propto \sin\alpha/\alpha$ where $\alpha = kld/2$. Each of the pulses in Figure 10.15(b) contributes a Fourier Transform $\propto \sin\beta/\beta$ where

$$\beta = \frac{klD}{2}$$

so the Fourier Transform of the isosceles triangular pulse is $\propto \sin^2\beta/\beta^2$.

Note that the analysis above is equally true if the arguments of the two functions are exchanged under the convolution process so that we have $f(x' - x)$ and $h(x)$. We use this in the discussion on the Optical Transfer Function on page 393.

(Problems 10.10, 10.11, 10.12, 10.13, 10.14, 10.15, 10.16, 10.17, 10.18, 10.19)

Problem 10.1

After inspection of the two wave forms in the diagram what can you say about the values of the constant, absence or presence of sine terms, cosine terms, odd or even harmonics, and range of harmonics required in their Fourier series representation? (Do not use any mathematics.)

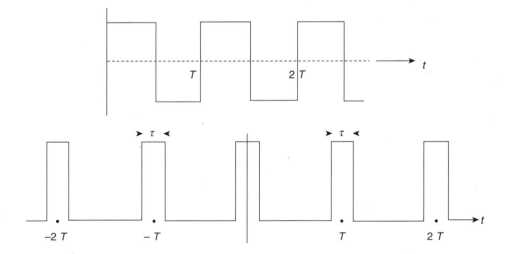

Problem 10.2

Show that if a periodic waveform is such that each half-cycle is identical except in sign with the previous one, its Fourier spectrum contains no even order frequency components. Examine the result physically.

Problem 10.3

A half-wave rectifier removes the negative half-cycles of a pure sinusoidal wave $y = h\sin x$. Show that the Fourier series is given by

$$y = \frac{h}{\pi}\left(1 + \frac{\pi}{1\cdot 2}\sin x - \frac{2}{1\cdot 3}\cos 2x - \frac{2}{3\cdot 5}\cos 4x - \frac{2}{5\cdot 7}\cos 6x \ldots\right)$$

Problem 10.4
A full-wave rectifier merely inverts the negative half-cycle in Problem 10.3. Show that this doubles the output and removes the undesirable modulating ripple of the first harmonic.

Problem 10.5
Show that $f(x) = x^2$ may be represented in the interval $\pm \pi$ by

$$f(x) = \frac{2}{3}\pi^2 + \sum(-1)^n \frac{4}{n^2}\cos nx$$

Problem 10.6
Use the square wave sine series of unit height $f(x) = 4/\pi(\sin x + \frac{1}{3}\sin 3x + \frac{1}{5}\sin 5x)$ to show that

$$1 - \tfrac{1}{3} + \tfrac{1}{5} - \tfrac{1}{7} = \pi/4$$

Problem 10.7
An infinite train of pulses of unit height, with pulse duration 2τ and a period between pulses of T, is expressed as

$$\begin{aligned} f(t) &= 0 \quad \text{for } -\tfrac{1}{2}T < t < -\tau \\ &= 1 \quad \text{for } -\tau < t < \tau \\ &= 0 \quad \text{for } \tau < t < \tfrac{1}{2}T \end{aligned}$$

and

$$f(t+T) = f(t)$$

Show that this is an even function with the cosine coefficients given by

$$a_n = \frac{2}{n\pi}\sin\frac{2\pi}{T}n\tau$$

Problem 10.8
Show, in Problem 10.7, that as τ becomes very small the values of $a_n \to 4\tau/T$ and are independent of n, so that the spectrum consists of an infinite set of lines of constant height and spacing. The representation now has the same form in both time and frequency; such a function is called 'self reciprocal'. What is the physical significance of the fact that as $\tau \to 0$, $a_n \to 0$?

Problem 10.9
The pulses of Problems 10.7 and 10.8 now have amplitude $1/2\tau$ with unit area under each pulse. Show that as $\tau \to 0$ the infinite series of pulses is given by

$$f(t) = \frac{1}{T} + \frac{2}{T}\sum_{n=1}^{\infty}\cos 2\pi nt/T$$

Under these conditions the amplitude of the original pulses becomes infinite, the energy per pulse remains finite and for an infinity of pulses in the train the total energy in the waveform is also infinite. The amplitude of the individual components in the frequency representation is finite, representing finite energy, but again, an infinity of components gives an infinite energy.

Problem 10.10

The unit step function is defined by the relation

$$f(t) = 1 \ (t > 0)$$
$$= 0 \ (t < 0)$$

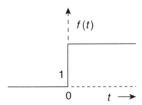

This is a very important function in physics and engineering, but it does not satisfy the criteria for Fourier representation because its integral is not finite. A similar function of finite period will satisfy the criteria. If this function is defined

$$f(t) = 1(0 < t < T)$$
$$= 0 \text{ elsewhere}$$

show that if the transform

$$F(\omega) = \int_{-\infty}^{\infty} f(t) \, e^{-i\omega t} \, dt = \int_{0}^{T} e^{-i\omega t} \, dt$$
$$= \frac{1}{i\omega} [1 - e^{i\omega T}]$$

then

$$f(t) = \frac{1}{2\pi} \int_{-\infty}^{\infty} F(\omega) \, e^{i\omega t} \, d\omega$$
$$= \frac{1}{2} + \frac{1}{2\pi} \int_{-\infty}^{\infty} \frac{1}{i\omega} \, e^{i\omega t} \, d\omega$$

(use the fact that for T very large

$$\int_{-\infty}^{\infty} \frac{1}{i\omega} \, e^{i\omega(t-T)} \, d\omega = \int_{-\infty}^{\infty} \frac{1}{i\omega} \, e^{-i\omega T} \, d\omega = -\pi$$

Note that the integral for the second term of $f(t)$ gives $-\pi$ for $t < 0$ and $+\pi$ for $t > 0$. This spectral representation is shown in Figure 10.13.)

Problem 10.11

Optical wave trains emitted by radiating atoms are of finite length and only an infinite wave train may be defined in terms of one frequency. The radiation from atoms therefore has a frequency bandwidth which contributes to the spectral linewidth. The random phase relationships between these wave trains create incoherence and produce the difficulties in obtaining interference effects from separate sources.

Let a finite length monochromatic wave train of wavelength λ_0 be represented by

$$f(t) = f_0 \, e^{i2\pi\nu_0 t}$$

and be a cosine of constant amplitude f_0 extending in time between $\pm\tau/2$. The distance $l = c\tau$ is called the coherence length. This finite train is the superposition of frequency components of amplitude $F(\nu)$ where the transform gives

$$f(t) = \int_{-\infty}^{\infty} F(\nu) \, e^{i2\pi\nu t} \, d\nu$$

so that

$$F(\nu) = \int_{-\infty}^{\infty} f(t') \, e^{-i2\pi\nu t'} \, dt'$$
$$= \int_{-\tau/2}^{+\tau/2} f_0 \, e^{-i2\pi(\nu-\nu_0)t'} \, dt'$$

Show that

$$F(\nu) = f_0\tau \frac{\sin[\pi(\nu - \nu_0)\tau]}{\pi(\nu - \nu_0)\tau}$$

and that the relative energy distribution in the spectrum follows the intensity distribution curve in a single slit diffraction pattern.

Problem 10.12

Show that the total width of the first maximum of the energy spectrum of Problem 10.11 has a frequency range $2\Delta\nu$ which defines the coherence length l of Problem 10.11 as $\lambda_0^2/\Delta\lambda$.

Problem 10.13

For a ruby beam the value of $\Delta\nu$ in Problem 10.12 is found to be 10^4 Hz and $\lambda_0 = 6.936 \times 10^{-7}$ m. Show that $\Delta\lambda = 1.6 \times 10^{-17}$ m and that the coherence length l of the beam is 3×10^4 m.

Problem 10.14

The energy of the finite wave train of the damped simple harmonic vibrations of the radiating atom in Chapter 2 was described by $E = E_0 \, e^{-\omega_0 t/Q}$. Show from physical arguments that this defines a frequency bandwidth in this train of $\Delta\omega$ about the frequency ω_0, where the quality factor $Q = \omega_0/\Delta\omega$. (Suggested line of argument—at the maximum amplitude all frequency components are in phase. After a time τ the frequency component ω_0 has changed phase by $\omega_0\tau$. Other components have a phase change which interfere destructively. What bandwidth and phase change is acceptable?)

Problem 10.15

Consider Problem 10.14 more formally. Let the damped wave be represented as a function of time by

$$f(t) = f_0 \, e^{i2\pi\nu_0 t} \, e^{-t/\tau}$$

where f_0 is constant and τ is the decay constant.

Use the Fourier transform to show that the amplitudes in the frequency spectrum are given by

$$F(\nu) = \frac{f_0}{1/\tau + i2\pi(\nu - \nu_0)}$$

Write the denominator of $F(\nu)$ as $r\,e^{i\theta}$ to show that the energy distribution of frequencies in the region of $\nu - \nu_0$ is given by

$$|F(\nu)|^2 = \frac{f_0^2}{r^2} = \frac{f_0^2}{(1/\tau)^2 + (\omega - \omega_0)^2}$$

Problem 10.16
Show that the expression $|F(\nu)|^2$ of Problem 10.15 is the resonance power curve of Chapter 3; show that it has a width at half the maximum value $(f_0\tau)^2$ which gives $\Delta\nu = 1/\pi\tau$, and show that a spectral line which has a value of $\Delta\lambda$ in Problem 10.12 equal to 3×10^{-9} m has a finite wave train of coherence length equal to 32×10^{-6} m (32 μm) if $\lambda_0 = 5.46 \times 10^{-7}$ m.

Problem 10.17
Sketch the self-convolution of the double slit function shown in Figure Q 10.17.

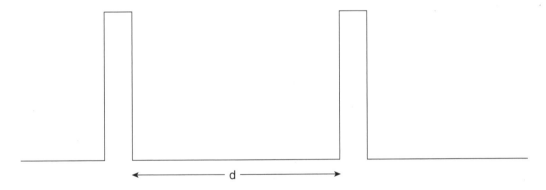

Figure Q.10.17

Problem 10.18
Sketch the convolution of the two functions in Figure Q 10.18 and use the convolution theorem to find its Fourier transform.

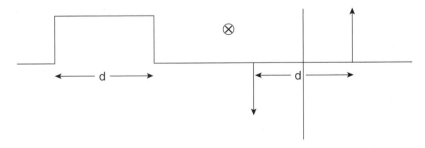

Figure Q.10.18

Problem 10.19

The convolution of two identical circles of radius r is very important in the modern method of testing lenses against an ideal diffraction limited criterion.

In Figure Q 10.19 show that the area of overlap is

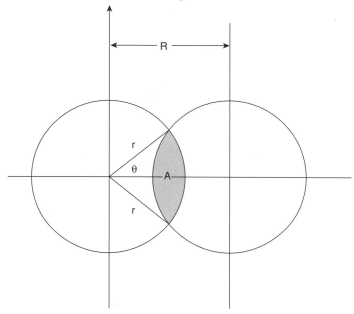

Figure Q.10.19

$$A = r^2(2\theta - 2\sin\theta\cos\theta)$$

and show for

$$R \leq 2r$$

that the convolution

$$O(R) = r^2\left[2\cos^{-1}\frac{R}{2r} - 2\left(1 - \frac{R^2}{4r^2}\right)^{\frac{1}{2}}\frac{R}{2r}\right]$$

Sketch $O(R)$ for $O \leq R \leq 2r$

Apart from a constant the linear operator $\hat{O}$ is known as the modulation factor of the optical transfer function.

Summary of Important Results

Fourier Series

Any function may be represented in the interval $\pm\pi$ by

$$f(x) = \frac{1}{2}a_0 + \sum_1^n a_n\cos nx + \sum_1^n b_n\sin nx$$

where

$$a_n = \frac{1}{\pi} \int_0^{2\pi} f(x) \cos nx \, dx$$

and

$$b_n = \frac{1}{\pi} \int_0^{2\pi} f(x) \sin nx \, dx$$

Fourier Integral

A single non-periodic pulse may be represented as

$$f(t) = \int_{-\infty}^{+\infty} \left[\int_{-\infty}^{+\infty} f(t') e^{-i2\pi\nu t'} \, dt' \right] e^{i2\pi\nu t} \, d\nu$$

or as

$$f(t) = \int_{-\infty}^{+\infty} F(\nu) e^{i2\pi\nu t} \, d\nu$$

where

$$F(\nu) = \int_{-\infty}^{+\infty} f(t') e^{-i2\pi\nu t} \, dt'$$

$f(t)$ and $F(\nu)$ are *Fourier Transforms* of each other. When t is replaced by x and ν by k the right hand side of each transform has a factor $1/\sqrt{2\pi}$. The Fourier Transform of a rectangular pulse has the shape of $\sin\alpha/\alpha$. (Important in optical diffraction.)

11

Waves in Optical Systems

Light. Waves or Rays?

Light exhibits a dual nature. In practice, its passage through optical instruments such as telescopes and microscopes is most easily shown by geometrical ray diagrams but the fine detail of the images formed by these instruments is governed by diffraction which, together with interference, requires light to propagate as waves. This chapter will correlate the geometrical optics of these instruments with wavefront propagation. In Chapter 12 we shall consider the effects of interference and diffraction.

The electromagnetic wave nature of light was convincingly settled by Clerk–Maxwell in 1864 but as early as 1690 Huygens was trying to reconcile waves and rays. He proposed that light be represented as a wavefront, each point on this front acting as a source of secondary wavelets whose envelope became the new position of the wavefront, Figure 11.1(a). Light propagation was seen as the progressive development of such a process. In this way, reflection and refraction at a plane boundary separating two optical media could be explained as shown in Figure 11.1(b) and (c).

Huygens' theory was explicit only on those contributions to the new wavefront directly ahead of each point source of secondary waves. No statement was made about propagation in the backward direction nor about contributions in the oblique forward direction. Each of these difficulties is resolved in the more rigorous development of the theory by Kirchhoff which uses the fact that light waves are oscillatory (see Appendix 2, p. 547).

The way in which rays may represent the propagation of wavefronts is shown in Figure 11.2 where spherically diverging, plane and spherically converging wavefronts are moving from left to right. All parts of the wavefront (a surface of constant phase) take the same time to travel from the source and all points on the wavefront are the same *optical distance* from the source. This optical distance must take account of the changes of refractive index met by the wavefront as it propagates. If the physical path length is measured as x in a medium of refractive index n then the *optical path length* in the medium is the product nx. In travelling from one point to another light chooses a unique optical path which may always be defined in terms of Fermat's Principle.

The Physics of Vibrations and Waves, 6th Edition   H. J. Pain
© 2005 John Wiley & Sons, Ltd

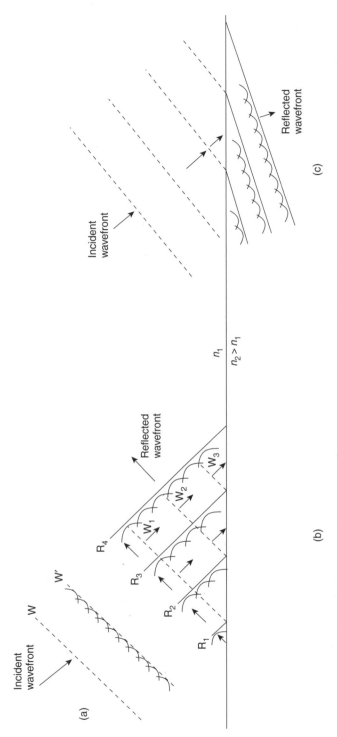

Figure 11.1 (a) Incident plane wavefront W propagates via Huygens wavelets to W′. (b) At the plane boundary between the media (refractive index $n_2 > n_1$) the incident wavefront W_1 has a reflected section R_1. Increasing sections R_2 and R_3 are reflected until the whole wavefront is *reflected* as R_4. (c) An increasing section of the incident wavefront is refracted. Incident wavefronts are shown dashed, and reflected and refracted wavefronts as a continuous line

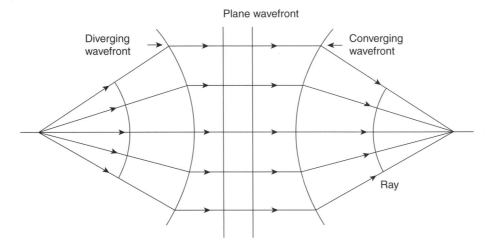

Figure 11.2 Ray representation of spherically diverging, plane and spherically converging wavefronts

Fermat's Principle

Fermat's Principle states that the optical path length has a stationary value; its first order variation or first derivative in a Taylor series expansion is zero. This means that when an optical path lies wholly within a medium of constant refractive index the path is a straight line, the shortest distance between its end points, and the light travels between these points in the minimum possible time. When the medium has a varying refractive index or the path crosses the boundary between media of different refractive indices the direction of the path always adjusts itself so that the time taken between its end points is a minimum. Fermat's Principle is therefore sometimes known as the Principle of Least Time. Figure 11.3 shows examples of light paths in a medium of varying refractive index. As examples of light meeting a boundary between two media we use Fermat's Principle to derive the laws of reflection and refraction.

The Laws of Reflection

In Figure 11.4a Fermat's Principle requires that the optical path length OSI should be a minimum where O is the object, S lies on the plane reflecting surface and I is the point on the reflected ray at which the image of O is viewed. The plane OSI must be perpendicular to the reflecting surface for, if reflection takes place at any other point S′ on the reflecting surface where OSS′ and ISS′ are right angles then evidently OS′ > OS and IS′ > IS, giving OS′I > OSI.

The laws of reflection also require, in Figure 11.4a that the angle of incidence i equals the angle of reflection r. If the coordinates of O, S and I are those shown and the velocity of light propagation is c then the time taken to traverse OS is

$$t = (x^2 + y^2)^{1/2}/c$$

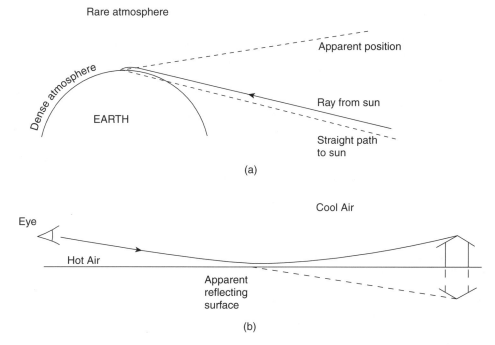

(a)

(b)

Figure 11.3 Light takes the shortest optical path in a medium of varying refractive index. (a) A light ray from the sun bends towards the earth in order to shorten its path in the denser atmosphere. The sun remains visible after it has passed below the horizon. (b) A light ray avoids the denser atmosphere and the road immediately below warm air produces an apparent reflection

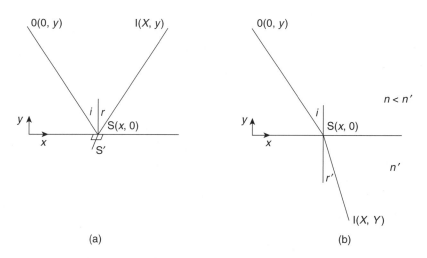

(a)   (b)

Figure 11.4 The time for light to follow the path OSI is a minimum (a) in reflection, when OSI forms a plane perpendicular to the reflecting surface and $\hat{i} = \hat{r}$; and (b) in refraction, when $n \sin i = n' \sin r'$ (Snell's Law)

and the time taken to traverse SI is

$$t' = [(X - x)^2 + y^2]^{1/2}/c$$

so that the total time taken to travel the path OSI is

$$T = t + t'$$

The position of S is now varied along the x axis and we seek, via Fermat's Principle of Least Time, that value of x which minimizes T, so that

$$\frac{dT}{dx} = \frac{x}{c(x^2 + y^2)^{1/2}} - \frac{X - x}{c[(X - x)^2 + y^2]^{1/2}} = 0$$

But

$$\frac{x}{(x^2 + y^2)^{1/2}} = \sin i$$

and

$$\frac{X - x}{[(X - x)^2 + y^2]^{1/2}} = \sin r$$

Hence

$$\sin i = \sin r$$

and

$$\hat{i} = \hat{r}$$

The Law of Refraction

Exactly similar arguments lead to Snell's Law, already derived on p. 256.

Here we express it as

$$n \sin i = n' \sin r'$$

where i is the angle of incidence in the medium of refractive index n and r' is the angle of refraction in the medium of refractive index $n'(n' > n)$. In Figure 11.4b a plane boundary separates the media and light from O $(0, y)$ is refracted at S $(x, 0)$ and viewed at I (X, Y) on the refracted ray. If v and v' are respectively the velocities of light propagation in the media n and n' then OS is traversed in the time

$$t = (x^2 + y^2)^{1/2}/v$$

and SI is traversed in the time

$$t' = [(X - x)^2 + Y^2]^{1/2}/v'$$

The total time to travel from O to I is $T = t + t'$ and we vary the position of S along the x axis which lies on the plane boundary between n and n', seeking that value of x which minimizes T. So

$$\frac{\mathrm{d}T}{\mathrm{d}x} = \frac{1}{v} \frac{x}{(x^2 + y^2)^{1/2}} - \frac{1}{v'} \frac{(X - x)}{[(X - x)^2 + Y^2]^{1/2}} = 0$$

where

$$\frac{x}{(x^2 + y^2)^{1/2}} = \sin i$$

and

$$\frac{(X - x)}{[(X - x)^2 + Y^2]^{1/2}} = \sin r'$$

But

$$\frac{1}{v} = \frac{n}{c}$$

and

$$\frac{1}{v'} = \frac{n'}{c}$$

Hence

$$n \sin i = n' \sin r'$$

Rays and Wavefronts

Figure 11.2 showed the ray representation of various wavefronts. In order to reinforce the concept that rays trace the history of wavefronts we consider the examples of a thin lens and a prism.

The Thin Lens

In Figure 11.5 a plane wave in air is incident normally on the plane face of a plano convex glass lens of refractive index n and thickness d at its central axis. Its spherical face has a radius of curvature $R \gg d$. The power of a lens to change the curvature of a wavefront is the inverse of its focal length f. A lens of positive power converges a wavefront, negative power diverges the wavefront.

Simple rays optics gives the power of the plano convex lens as

$$\mathscr{P} = \frac{1}{f} = (n - 1)\frac{1}{R}$$

but we derive this result from first principles that is, by considering the way in which the lens modifies the wavefront.

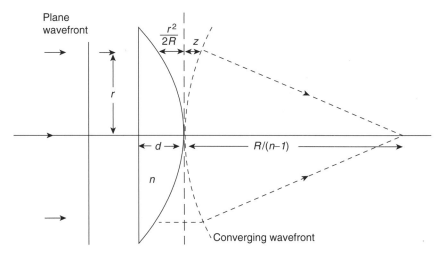

Figure 11.5 A plane wavefront is normally incident on a plano-convex lens of refractive index n and thickness d at the central axis. The radius of the curved surface $R \gg d$. The wavefront is a surface of constant phase and the optical path length is the same for each section of the wavefront. At a radius r from the central axis the wavefront travels a shorter distance in the denser medium and the lens curves the incident wavefront which converges at a distance $R/(n-1)$ from the lens

At the central axis the wavefront takes a time $t = nd/c$ to traverse the thickness d. At a distance r from the axis the lens is thinner by an amount $r^2/2R$ (using the elementary relation between the sagitta, arc and radius of a circle) so that, in the time $t = nd/c$, points on the wavefront at a distance r from the axis travel a distance

$$(d - r^2/2R)$$

in the lens plus a distance $(r^2/2R + z)$ in air as shown in the figure. Equating the times taken by the two parts of the wave front we have

$$nd/c = (n/c)(d - r^2/2R) + (1/c)(z + r^2/2R)$$

which yields

$$z = (n - 1)r^2/2R$$

But this is again the relation between the sagitta z, its arc and a circle of radius $R/(n-1)$ so, in three dimensions, the locus of z is a sphere of radius $R/(n-1)$ and the emerging spherical wavefront converges to a focus at a distance

$$f = R/(n-1)$$

(Problems 11.1, 11.2, 11.3)

The Prism

In Figure 11.6 a section, height y, of a plane wavefront in air is deviated through an angle θ when it is refracted through an isosceles glass prism, base l, vertex angle β and refractive

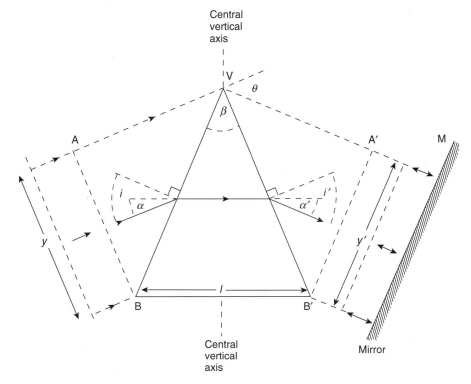

Figure 11.6 A plane wavefront suffers minimum deviation $(\theta_{\min})$ when its passage through a prism is symmetric with respect to the central vertical axis $(i = i')$. The wavefront obeys the Optical Helmholtz Condition that $ny \tan \alpha$ is a constant where n is the refractive index, y is the width of the wavefront and α is shown. (Here $\alpha = \alpha'$)

index n. Experiment shows that there is *one*, and only one, value of the incident angle i for which the angle of deviation is a minimum $= \theta_{\min}$. It is easily shown *using ray optics* that this unique value of i requires the passage of the wavefront through the prism to be symmetric about the central vertical axis as shown in the figure so that the incident angle i equals the emerging angle i'. Equating the lengths of the *optical* paths AVA$'$ and BB$'(= nl)$ followed by the edges of the wavefront section gives the familiar result

$$\sin \left(\frac{\theta_{\min} + \beta}{2} \right) = n \sin \frac{\beta}{2}$$

which is used in the standard experiment to determine n, the refractive index of the prism.

Now there is only *one* value of i which produces minimum deviation and this leads us to expect that the passage of the wavefront will be symmetric about the central vertical axis for if a plane mirror (M in the figure) is placed parallel to the emerging wavefront the wavefront is reflected back along its original path, and if $i \neq i'$ there would be *two* values of incidence, each producing minimum deviation. At i for minimum deviation any rotation increases i'.

However, the real argument for symmetry from a wavefront point of view depends on the optical Helmholtz equation which we shall derive on p. 321. This states that for a plane wavefront the product $ny \tan \alpha$ remains constant as it passes through an optical system irrespective of the local variations of the factors n, y and $\tan \alpha$. Now the wavefront has the same width on entry into and exit from the prism so $y = y'$ and although n changes at the prism faces the initial and final medium for the wavefront is air where $n = 1$.

Hence, from the optical Helmholtz equation $\tan \alpha = \tan \alpha'$ in Figure 11.6. It is evident that as long as its width $y = y'$ the wavefront section will turn through a minimum angle when the physical path length BB′ followed by its lower edge is a maximum with respect to AVA′, the physical path length of its upper edge.

Ray Optics and Optical Systems

An optical system changes the curvature of a wavefront. It is formed by one or more optical surfaces separating media of different refractive indices. In Fig. 11.7 rays from the object point L_0 pass through the optical system to form an image point L'. When the optical surfaces are spherical the line joining L_0 and L', which passes through the centres of curvature of the surfaces, is called the *optical axis*. This axis cuts each optical surface at its *pole*. If the object lies in a plane normal to the optical axis its perfect image lies in a *conjugate* plane normal to the optical axis. Conjugate planes cut the optical axis at conjugate points, e.g. L_0 and L'. In Figure 11.7 the plane at $+\infty$ has a conjugate focal plane cutting the optical axis at the focal point F. The plane at $-\infty$ has a conjugate focal plane cutting the optical axis at the focal point F'.

Paraxial Rays

Perfect geometrical images require perfect plane and spherical optical surfaces and in a real optical system a perfect spherical optical surface is obtained by using only that part of the wavefront close to the optical axis. This means that all angles between the axis and rays are very small. Such rays are called paraxial rays.

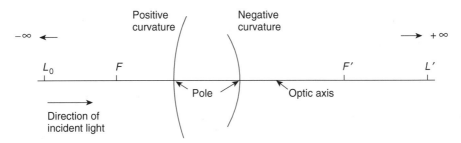

Figure 11.7 Optical system showing direction of incident light from left to right and optical surfaces of positive and negative curvature. Rays from L_0 pass through L' and this defines L_0 and L' as conjugate points. The conjugate point of F is at $+\infty$, the conjugate point of F' is at $-\infty$

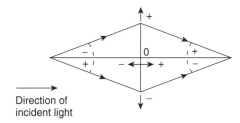

Direction of
incident light

Figure 11.8 Sign convention for lengths is Cartesian measured from the right angles at 0. Angles take the sign of their tangents. 0 is origin of measurements

Sign Convention

The convention used here involves only signs of lengths and angles. The direction of incident light is positive and is always taken from left to right. Signs for horizontal and vertical directions are Cartesian. If AB $= l$ then BA $= -l$. The radius of curvature of a surface is measured from its pole to its centre so that, in Figure 11.7, the convex surface presented to the incident light has a positive radius of curvature and the concave surface has a negative radius of curvature.

The Cartesian convention with origin O at the right angles of Figure 11.8 gives the angle between a ray and the optical axis the sign of its tangent.

If the angle between a ray and the axis is α then, for paraxial rays

$$\sin \alpha = \tan \alpha = \alpha$$

and

$$\cos \alpha = 1$$

so that Snell's Law of Refraction

$$n \sin i = n' \sin r'$$

becomes

$$ni = n'r'$$

Power of a Spherical Surface

In Figure 11.9(a) and (b) a spherical surface separates media of refractive indices n and n'. Any ray through L_0 is refracted to pass through its conjugate point L'. The angles are exaggerated so that the base of h is very close to the pole of the optical surface which is taken as the origin. In Figure 11.9(a) the signs of R, l' and α' are positive with l and α negative. In Figure 11.9(b) R, l, l', α and α' are all positive quantities. In both figures Snell's Law gives

$$ni = n'r'$$

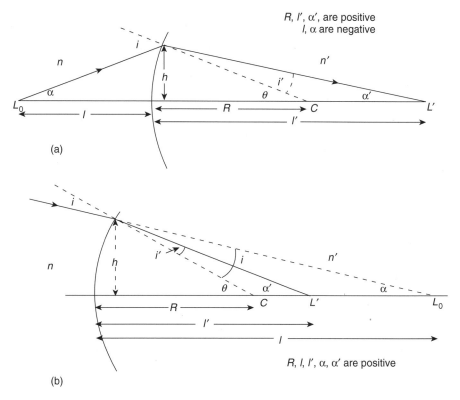

Figure 11.9 Spherical surface separating media of refractive indices n and n'. Rays from L_0 pass through L'. Snell's Law gives the power of the surface as

$$\mathscr{P} = \frac{n'}{l'} - \frac{n}{l} = \frac{n' - n}{R}$$

i.e.

$$n(\theta - \alpha) = n'(\theta - \alpha')$$

or

$$n'\alpha' - n\alpha = (n' - n)\theta = \left(\frac{n' - n}{R}\right)h = \mathscr{P}h \tag{11.1}$$

Thus

$$\frac{n'}{l'} - \frac{n}{l} = \frac{n' - n}{R} = \mathscr{P} \tag{11.2}$$

where $\mathscr{P}$ is the power of the surface. For $n' > n$ the power $\mathscr{P}$ is positive and the surface converges the wavefront. For $n' < n$, $\mathscr{P}$ is negative and the wavefront diverges. When the radius of curvature R is measured in metres the units of $\mathscr{P}$ are *dioptres*.

Magnification by the Spherical Surface

In Figure 11.10 the points QQ' form a conjugate pair, as do L_0L'. The ray QQ' passes through C the centre of curvature, L_0Q is the object height y, $L'Q'$ is the image height y' so

$$ni = n'r'$$

gives

$$ny/l = n'y'/l'$$

or

$$nyh/l = n'y'h/l'$$

that is

$$ny\alpha = n'y'\alpha' \tag{11.3}$$

This is the paraxial form of the optical Helmholtz equation.

The Transverse Magnification is defined as

$$M_T = y'/y = nl'/n'l.$$

The image y' is inverted so y and y' (and l and l') have opposite signs.

The Angular Magnification is defined as

$$M_\alpha = \alpha'/\alpha$$

Note that

$$M_T = n/n'M_\alpha$$

which, being independent of y, applies to any point on the object so that the object in the plane L_0Q is similar to the image in the plane $L'Q'$.

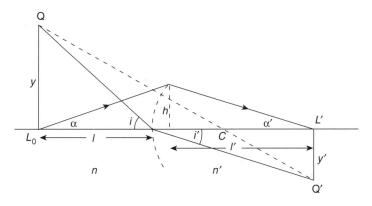

Figure 11.10 Magnification by a spherical surface. The paraxial form of the optical Helmholtz equation is $ny\alpha = n'y'\alpha'$ so Transverse Magnification $M_T = y'/y = nl'/ln'$ Angular Magnification $M_\alpha = \alpha'/\alpha$. Note that the image is inverted so y and y' (and l and l') have opposite signs

A series of optical surfaces separating media of refractive indices n, $n'n''$ yields the expression

$$ny\alpha = n'y'\alpha' = n''y''\alpha''$$

which is the paraxial form of the optical Helmholtz equation.

Power of Two Optically Refracting Surfaces

If Figure 11.11 the refracting surfaces have powers $\mathscr{P}_1$ and $\mathscr{P}_2$, respectively. At the first surface equation (11.1) gives

$$n_1\alpha_1 - n\alpha = \mathscr{P}_1 h_1$$

and at the second surface

$$n'\alpha' - n_1\alpha_1 = \mathscr{P}_2 h_2$$

Adding these equations gives

$$n'\alpha' - n\alpha = \mathscr{P}_1 h_1 + \mathscr{P}_2 h_2$$

If the object is located at $-\infty$ so that $\alpha = 0$ we have

$$n'\alpha' = \mathscr{P}_1 h_1 + \mathscr{P}_2 h_2$$

or

$$\alpha' = \frac{1}{n'}(\mathscr{P}_1 h_1 + \mathscr{P}_2 h_2)$$

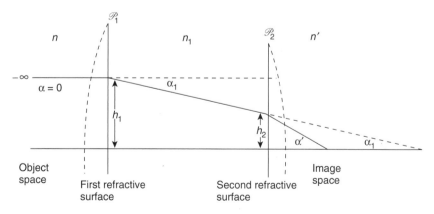

Figure 11.11 Two optically refracting surfaces of power $\mathscr{P}_1$ and $\mathscr{P}_2$ have a combined power of

$$\mathscr{P} = \frac{1}{h_1}(\mathscr{P}_1 h_1 + \mathscr{P}_2 h_2)$$

This gives the same image as a single element of power $\mathscr{P}$ if

$$\alpha' = \frac{1}{n'}(\mathscr{P}_1 h_1 + \mathscr{P}_2 h_2) = \frac{1}{n'}\mathscr{P} h_1$$

where

$$\mathscr{P} = \frac{1}{h_1}(\mathscr{P}_1 h_1 + \mathscr{P}_2 h_2) \tag{11.4}$$

is the total power of the system. *This is our basic equation* and we use it first to find the power of a thin lens in air.

Power of a Thin Lens in Air (Figure 11.12)

Equation (11.2) gives

$$\frac{n'}{l'} - \frac{n}{l} = \frac{n' - n}{R} = \mathscr{P}$$

for each surface, so that in Figure 11.12

$$\mathscr{P}_1 = (n_1 - 1)/R_1$$

and

$$\mathscr{P}_2 = (1 - n_1)/R_2$$

From equation (11.4)

$$\mathscr{P} = \frac{1}{h_1}(\mathscr{P}_1 h_1 + \mathscr{P}_2 h_2)$$

with

$$h_1 = h_2$$

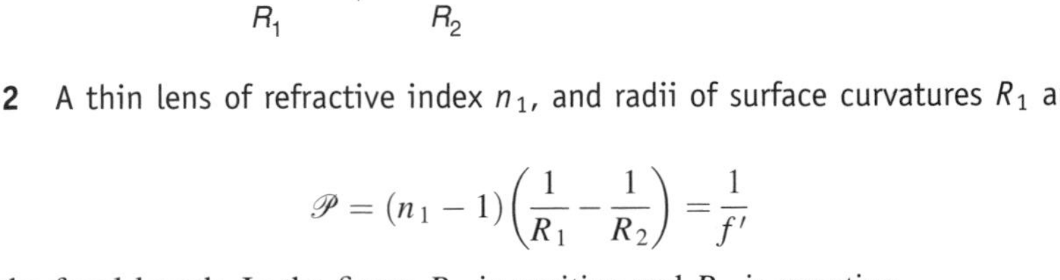

Figure 11.12 A thin lens of refractive index n_1, and radii of surface curvatures R_1 and R_2 has a power

$$\mathscr{P} = (n_1 - 1)\left(\frac{1}{R_1} - \frac{1}{R_2}\right) = \frac{1}{f'}$$

where f' is the focal length. In the figure R_1 is positive and R_2 is negative we have

$$\mathscr{P} = \mathscr{P}_1 + \mathscr{P}_2$$

so the expression for the thin lens in air with surfaces of power $\mathscr{P}_1$ and $\mathscr{P}_2$ becomes

$$\mathscr{P} = \frac{1}{l'} - \frac{1}{l} = (n_1 - 1)\left(\frac{1}{R_1} - \frac{1}{R_2}\right) = \frac{1}{f'}$$

where f' is the focal length.

Applying this result to the plano convex lens of p. 311 we have $R_1 = \infty$ and R_2 negative from our sign convention. This gives a positive power which we expect for a converging lens.

Effect of Refractive Index on the Power of a Lens

Suppose, in Figure 11.13, that the object space of the lens remains in air $(n = 1)$ but that the image space is a medium of refractive index $n_2' \neq 1$. How does this affect the focal length in the medium n_2'?

If $\mathscr{P}$ is the power of the lens in air we have

$$n_2'\alpha' - n\alpha = \mathscr{P}h_1 \tag{11.5}$$

and for

$$\alpha = 0$$

we have

$$\alpha' = \mathscr{P}h_1/n_2' = h_1/n_2'f'$$

where f' is the focal length in air.

If f_2' is the focal length in the medium n_2' then

$$f_2'\alpha' = h_1$$

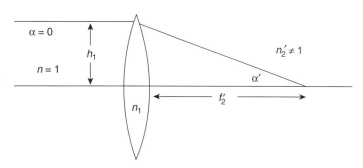

Figure 11.13 The focal length of a thin lens measured in the medium n_2' is given by $f_2' = n_2'f'$ where f' is the focal length of the lens measured in air

so

$$\alpha' = h_1/f_2' = h_1/n_2'f'$$

giving

$$f_2' = n_2'f'$$

Thus, the focal length changes by a factor equal to the refractive index of the medium in which it is measured and the power is affected by the same factor.

If the lens has a medium n_0 in its object space and a medium n_i in its image space then the respective focal lengths f_0 and f_i in these spaces are related by the expression

$$f_i/f_0 = -n_i/n_0 \qquad (11.6)$$

where the negative signs shows that f_0 and f_i are measured in opposite directions (f_0 is negative and f_i is positive).

Principal Planes and Newton's Equation

There are two particular planes normal to the optic axis associated with every lens element of an optical system. These planes are called principal planes or unit planes because between these planes there is unit transverse magnification so the path of every ray between them is parallel to the optic axis. Moreover, any complex optical system has two principal planes of its own. In a thin lens the principal planes coincide.

The principal planes of a single lens do not, in general, coincide with its optical surfaces; focal lengths, object and image distances are measured from the principal planes and not from the optical surfaces. In Figure 11.14, PH and $P'H'$ define the first and second

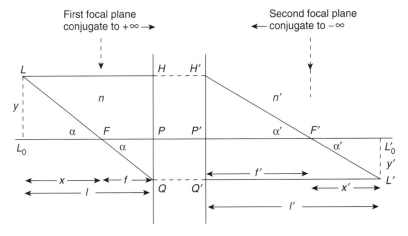

Figure 11.14 Between the principal planes PH and $P'H'$ of a lens or lens system there is unit magnification and rays between these planes are parallel to the optic axis. Newton's equation defines $xx' = ff'$. The optical Helmholtz equation is $ny\alpha =$ constant for paraxial rays and $ny \tan \alpha =$ constant for rays from ∞

principal planes, respectively, of a lens or optical system and PF and $P'F'$ are respectively the first and second focal lengths. The object and image planes cut the optic axis in L_0 and L_0', respectively.

The ray LH parallel to the optic axis proceeds to H' and thence through F' the second focal point. The rays LH and $H'F'$ meet at H' and therefore define the position of the second principal plane, $P'H'$. The position of the first principal plane may be found in a similar way.

If Figure 11.14, the similar triangles FL_0L and FPQ give $y/y' = x/f$ where, measured from P, only y is algebraically positive. The similar triangles $F'L_0'L'$ and $F'P''H'$ give

$$y/y' = f'/x',$$

where, measured from P', only y' is algebraically negative.

We have, therefore,

$$x/f = f'/x',$$

where x and f are negative and x' and f' are positive.

Thus,

$$xx' = ff'$$

This is known as Newton's equation.

If l, the object distance, and l', the image distance, are measured from the principal planes as in Figure 11.14, then

$$l = f + x \quad \text{and} \quad l' = f' + x'$$

and Newton's equation gives

$$xx' = (l - f)(l' - f') = ll' - l'f - lf' + ff' = ff'$$

so that

$$\frac{f'}{l'} + \frac{f}{l} = 1$$

But from $nf' = -n'f$ (equation (11.6)) we have

$$\frac{n'}{l'} - \frac{n}{l} = \frac{n'}{f'} = \frac{-n}{f} = \mathscr{P}$$

the power of the lens.

Optical Helmholtz Equation for a Conjugate Plane at Infinity

Suppose now that the object is no longer located at L_0L but at infinity so that the ray LH originates at one point from the distant object while the ray LFQ comes from a point on the object much more distant from the optic axis.

We still have from triangle $F'P'H'$ that

$$y = f' \tan \alpha'$$

and from triangle FPQ that

$$y' = f \tan \alpha$$

so

$$\frac{f \tan \alpha}{f' \tan \alpha'} = \frac{y}{y'} \quad \text{and} \quad \frac{f}{f'} y \tan \alpha = y' \tan \alpha'$$

But

$$\frac{f}{f'} = \frac{-n}{n'}$$

so

$$ny \tan \alpha = -n'y' \tan \alpha'$$

(Note that α, α' and y' are negative.)

This form of the Helmholtz equation applies when one of the conjugate planes is at infinity and is to be compared with the general unrestricted form of the Helmholtz equation for paraxial rays

$$ny\alpha = n'y'\alpha'$$

The infinite conjugate form $ny \tan \alpha = $ constant is valid when applied to the prism of p. 312 because the plane wavefront originated at infinity.

(Problems 11.4, 11.5, 11.6, 11.7, 11.8)

The Deviation Method for (a) Two Lenses and (b) a Thick Lens

Figure 11.11 illustrated how the deviation of a ray through two optically refracting surfaces could be used to find the power of a thin lens. We now apply this process to (a) a combination of two lenses and (b) a thick lens in order to find the power of these systems and the location of their principal planes. We have already seen in equation (11.5), which may be written

$$n'_1\alpha' - n_1\alpha = \mathscr{P}_1 y \tag{11.7}$$

where $\mathscr{P}_1$ is the power of the first lens in Figure 11.15a or the power of the first refracting surface in Figure 11.15b. If the incident ray is parallel to the optic axis, then $\alpha = 0$ and we have

$$n'_1\alpha' = \mathscr{P}_1 y_1 \tag{11.8}$$

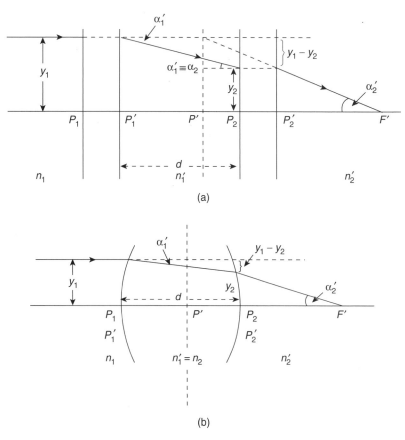

Figure 11.15 Deviation of a ray through (a) a system of two lenses and (b) a single thick lens. P' is a principal plane of the system. All the significant optical properties may be derived via this method

At the second lens or refracting surface

$$n_2\alpha_2 \equiv n_1'\alpha_1'$$

so

$$n_2'\alpha_2' - n_1'\alpha_1' = \mathscr{P}_2 y_2 \qquad (11.9)$$

Equation (11.8) plus equation (11.9) gives

$$n_2'\alpha_2' = \mathscr{P}_1 y_1 + \mathscr{P}_2 y_2 \qquad (11.10)$$

Now the incident ray strikes the principal plane P' at a height y_1 so, extrapolating the ray from F', the focal point of the system, through the plane P_2' to the plane P', we have

$$n_2'\alpha_2' = \mathscr{P} y_1 \qquad (11.11)$$

where $\mathscr{P}$ is the power of the complete system.

From equations (11.10) and (11.11) we have

$$\mathscr{P}y_1 = \mathscr{P}_1 y_1 + \mathscr{P}_2 y_2 \tag{11.12}$$

Moreover, Figure 11.15 shows that, algebraically

$$y_2 = y_1 - d\alpha_1'$$

which, with equation (11.8) gives

$$y_2 = y_1 - \frac{d}{n_1'}\mathscr{P}_1 y_1 = y_1 - \bar{d}\mathscr{P}_1 y_1, \tag{11.13}$$

where

$$\bar{d} = d/n_1'$$

This, with equation (11.12), gives

$$\mathscr{P} = \mathscr{P}_1 + \mathscr{P}_2 - \bar{d}\mathscr{P}_1\mathscr{P}_2 \tag{11.14}$$

where $\mathscr{P}$ is the power of the whole system.

From Figure 11.15 we have algebraically

$$P_2'P' = -\frac{y_1 - y_2}{\alpha_2'}$$

which with equations (11.11) and (11.13) gives

$$P_2'P' = \frac{-n_2'\bar{d}\mathscr{P}_1}{\mathscr{P}} \tag{11.15}$$

For a similar ray incident from the right we can find

$$P_1P = \frac{n_1\bar{d}\mathscr{P}_2}{\mathscr{P}}$$

where P is the first principal plane (not shown in the figures).

A more significant distance for the thick lens of Figure 11.15(b) is P_2F' the distance between the second refracting surface and the focal point F'.

Now

$$P_2F' = P'F' - P'P_2'$$

which with

$$P'F' = n_2'/\mathscr{P} \tag{11.16}$$

gives

$$P_2 F' = \frac{n_2'}{\mathscr{P}} - \frac{n_2' \bar{d} \mathscr{P}_1}{\mathscr{P}}$$
$$= \frac{n_2'}{\mathscr{P}} (1 - \bar{d} \mathscr{P}_1) \tag{11.17}$$

We shall see in the following section that the factor $1 - \bar{d}\mathscr{P}_1$ and the power $\mathscr{P}$ of the system arise quite naturally in the matrix treatment of this problem.

The Matrix Method

Tracing paraxial rays through an optical system involves the constant repetition of two consecutive processes and is particularly suited to matrix methods.

A refracting R process carries the ray from one medium across a refracting surface into a second medium from where it is taken by a transmitting T process through the second medium to the next refracting surface for R to be repeated. Both R and T processes and their products are represented by 2×2 matrices.

An R process is characterized by

$$n'\alpha' - n\alpha = \mathscr{P}_1 y \tag{11.7}$$

which changes $n\alpha$ but which leaves y unaffected.

We write this in the form

$$\bar{\alpha}' - \bar{\alpha} = \mathscr{P}_1 y \tag{11.18}$$

where

$$\bar{\alpha}_i = n_i \alpha_i$$

The reader should review Figure 11.8 for the sign convention for angles.

A T process is characterized by

$$y' = y - \bar{d}' \bar{\alpha}' \tag{11.19}$$

which changes y but leaves $\bar{\alpha}$ unaffected. The thick lens of the last section demonstrates the method particularly well and reproduces the results we have already found.

In Figure 11.16 note that

$$n_2 \alpha_2 \equiv n_1' \alpha_1'$$

that is

$$\bar{\alpha}_2 = \bar{\alpha}_1'$$

We express equations (11.18) and (11.19) in a suitable 2×2 matrix form by writing them as separate pairs.

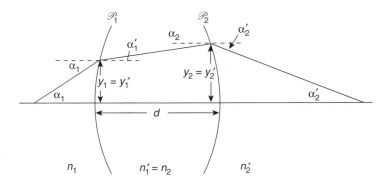

Figure 11.16 The single lens of Figure 11.15 is used to demonstrate the equivalence of the deviation and matrix methods for determining the important properties of a lens system. The matrix method is easily extended to a system of many optical elements

For R we have

$$\bar{\alpha}'_1 = \bar{\alpha}_1 + \mathscr{P}_1 y_1$$

where $\mathscr{P}_1$ is the power of the first refracting surface and

$$y'_1 = 0\bar{\alpha}_1 + 1 y_1$$

so, in matrix form we have

$$\begin{bmatrix} \bar{\alpha}'_1 \\ y'_1 \end{bmatrix} = \begin{bmatrix} 1 & \mathscr{P}_1 \\ 0 & 1 \end{bmatrix} \begin{bmatrix} \bar{\alpha}_1 \\ y_1 \end{bmatrix} = R_1 \begin{bmatrix} \bar{\alpha}_1 \\ y_1 \end{bmatrix}$$

This carries the ray across the first refracting surface.

For T we have

$$\bar{\alpha}_2 = 1\bar{\alpha}'_1 + 0 y'_1$$
$$y_2 = -\bar{d}'_1 \bar{\alpha}'_1 + 1 y'_1$$

where $\bar{\alpha}_2 = \bar{\alpha}'_1$, so

$$\begin{bmatrix} \bar{\alpha}_2 \\ y_2 \end{bmatrix} = \begin{bmatrix} 1 & 0 \\ -\bar{d}'_1 & 1 \end{bmatrix} \begin{bmatrix} \bar{\alpha}'_1 \\ y'_1 \end{bmatrix} = T_{12} \begin{bmatrix} \bar{\alpha}'_1 \\ y'_1 \end{bmatrix}$$

This carries the ray through the lens between its two refracting surfaces.

At the second refracting surface we repeat R to give

$$\bar{\alpha}'_2 = 1\bar{\alpha}_2 + \mathscr{P}_2 y_2$$
$$y'_2 = 0\bar{\alpha}_2 + 1 y_2$$

or

$$\begin{bmatrix} \bar{\alpha}'_2 \\ y'_2 \end{bmatrix} = \begin{bmatrix} 1 & \mathscr{P}_2 \\ 0 & y_2 \end{bmatrix} = R_2 \begin{bmatrix} \bar{\alpha}_2 \\ y_2 \end{bmatrix}$$

Therefore

$$\begin{bmatrix} \bar{\alpha}'_2 \\ y'_2 \end{bmatrix} = R_2 \begin{bmatrix} \bar{\alpha}_2 \\ y_2 \end{bmatrix} = R_{12}T_{12} \begin{bmatrix} \bar{\alpha}'_1 \\ y'_1 \end{bmatrix} = R_2 T_{12} R_1 \begin{bmatrix} \bar{\alpha}_1 \\ y_1 \end{bmatrix}$$

$$= \begin{bmatrix} 1 & \mathscr{P}_2 \\ 0 & 1 \end{bmatrix} \begin{bmatrix} 1 & 0 \\ -\bar{d}'_1 & 1 \end{bmatrix} \begin{bmatrix} 1 & \mathscr{P}_1 \\ 0 & 1 \end{bmatrix} \begin{bmatrix} \bar{\alpha}_1 \\ y_1 \end{bmatrix}$$

which, after matrix multiplication, gives

$$\begin{bmatrix} \bar{\alpha}'_2 \\ y'_2 \end{bmatrix} = \begin{bmatrix} 1 - \bar{d}'_1 \mathscr{P}_2 & \mathscr{P}_1 + \mathscr{P}_2 - \bar{d}'_1 \mathscr{P}_1 \mathscr{P}_2 \\ -\bar{d}'_1 & 1 - \bar{d}'_1 \mathscr{P}_1 \end{bmatrix} \begin{bmatrix} \bar{\alpha}_1 \\ y_1 \end{bmatrix}$$

Writing

$$R_2 T_{12} R_1 = \begin{bmatrix} a_{11} & a_{12} \\ a_{21} & a_{22} \end{bmatrix}$$

we see that a_{12} is the power $\mathscr{P}$ of the thick lens (equation (11.14)) and that a_{22} apart from the factor $n'_2/\mathscr{P}$ is the distance between the second refracting surface and the second focal point. The product of the coefficient a_{11} and $n_1/\mathscr{P}$ gives the separation between the first focal point and the first refracting surface. Note, too, that a_{11} and a_{22} enable us to locate the principal planes with respect to the refracting surfaces.

The order of the matrices for multiplication purposes is the reverse of the progress of the ray through $R_1 T_{12} R_2$, etc.

If the ray experiences a number J of such transformations, the general result is

$$\begin{bmatrix} \bar{\alpha}'_J \\ y'_J \end{bmatrix} = R_J T_{J-1,J} R_{J-1} \dots R_2 T_{12} R_1 \begin{bmatrix} \bar{\alpha}_1 \\ y_1 \end{bmatrix}$$

The product of all these 2×2 matrices is itself a 2×2 matrix.

It is important to note that the determinant of each matrix and of their products is unity, which implies that the column vector represents a property which is invariant in its passage through the system.

The components of the column vector are, of course, $\bar{\alpha}_1 y_1$; that is, $n\alpha$ and y and we already know that for paraxial rays the Helmholtz equation states that the product $ny\alpha$ remains constant throughout the system.

(Problems 11.9, 11.10, 11.11)

Problem 11.1
Apply the principle of p. 311 to a thin bi-convex lens of refractive index n to show that its power is

$$\mathscr{P} = (n-1)\left(\frac{1}{R_1} - \frac{1}{R_2}\right)$$

where R_1 and R_2, the radii of curvature of the convex faces, are both much greater than the thickness of the lens.

Problem 11.2

A plane parallel plate of glass of thickness d has a non-uniform refractive index n given by $n = n_0 - \alpha r^2$ where n_0 and α are constants and r is the distance from a certain line perpendicular to the sides of the plate. Show that this plate behaves as a converging lens of focal length $1/2\alpha d$.

Problem 11.3

For oscillatory waves the focal point F of the converging wavefront of Figure 11.17 is located where Huygens secondary waves all arrive in phase: the point F' vertically above F receives waves whose total phase range $\Delta\phi$ depends on the path difference AF'−BF'. When F' is such that $\Delta\phi$ is 2π the resultant amplitude tends to zero. Thus,

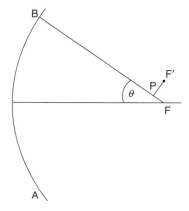

Figure 11.17

the focus is not a point but a region whose width x depends on the wavelength λ and the angle θ subtended by the spherical wave. If PF' is perpendicular to BF the phase at F' and P may be considered the same. Show that the width of the focal spot is given by $x = \lambda/\sin\theta$. Note that $\sin\theta$ is directly related to the f/d ratio for a lens (focal length/diameter) so that x defines the minimum size of the image for a given wavelength and a given lens.

Problem 11.4

As an object moves closer to the eye its apparent size grows with the increasing angle it subtends at the eye. A healthy eye can accommodate (that is, focus) objects from infinity to about 25 cm, the closest 'distance of distinct vision'. Beyond this 'near point' the eye can no longer focus and a magnifying glass is required. A healthy eye has a range of accommodation of 4 dioptres ($1/\infty$ to $1/0.25$ m). If a man's near point is 40 cm from his eye, show that he needs spectacles of power equal to 1.5 dioptres. If another man is unable to focus at distances greater than 2 m, show that he needs diverging spectacles with a power of -0.5 dioptres.

Problem 11.5

Figure 11.18 shows a magnifying glass of power P with an erect and virtual image at l'. The angular magnification

$$M_\alpha = \beta/\gamma$$

$$= \frac{\text{angular size of image seen through the glass at distance } l'}{\text{angular size of object seen by the unaided eye at } d_o}$$

where d_o is the distance of distinct vision. Show that the transverse magnification $M_T = l'/l$ where l is the actual distance (not d_o) at which the object O is held. Hence show that $M_\alpha = d_o/l$ and use the thin lens power equation, p. 318, to show that

$$M_\alpha = d_o(P + 1/l') = Pd_o + 1$$

when $l' = d_o$. Note that M_α reduces to the value Pd_o when the eye relaxes by viewing the image at ∞.

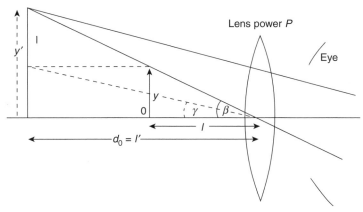

Figure 11.18

Problem 11.6

A telescope resolves details of a distant object by accepting plane wavefronts from individual points on the object and amplifying the very small angles which separate them. In Figure 11.19, α is the angle between two such wavefronts one of which propagates along the optical axis. In normal adjustment the astronomical telescope has both object and image at ∞ so that the total power of the system is zero. Use equation (11.14) to show that the separation of the lenses must be $f_o + f_e$ where f_o and f_e are respectively the focal lengths of the object and eye lenses.

If $2y$ is the width of the wavefront at the objective and $2y'$ is the width of the wavefront at the eye ring show that

$$M_\alpha = \left| \frac{\alpha'}{\alpha} \right| = \left| \frac{f_o}{f_e} \right| = \frac{D}{d}$$

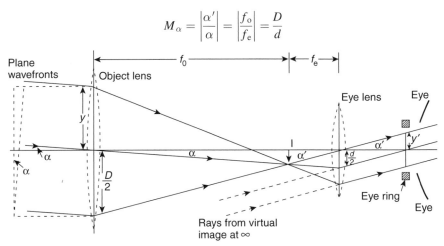

Figure 11.19

where D is the effective diameter of the object lens and d is the effective diameter of the eye lens. Note that the image is inverted.

Problem 11.7

The two lens microscope system of Figure 11.20 has a short focus objective lens of power P_o and a magnifying glass eyepiece of power P_e. The image is formed at the near point of the eye (the distance d_o of Problems 11.4 and 11.5). Show that the magnification by the object lens is $M_o = -P_o x'$ where the minus sign shows that the image is inverted. Hence use the expression for the magnifying glass in Problem 11.5 to show that the total magnification is

$$M = M_o M_e = -P_o P_e d_o x'$$

The length x' is called the optical tube length and is standardized for many microscopes at 0.14 m.

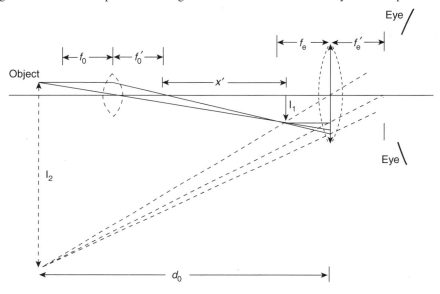

Figure 11.20

Problem 11.8

Microscope objectives are complex systems of more than one lens but the principle of the oil immersion objective is illustrated by the following problem. In Figure 11.21 the object O is embedded a distance R/n from the centre C of a glass sphere of radius

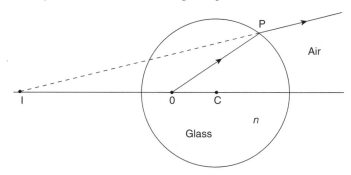

Figure 11.21

R and refractive index n. Any ray OP entering the microscope is refracted at the surface of the sphere and, when projected back, will always meet the axis CO at the point I. Use Snell's Law to show that the distance $IC = nR$.

Problems 11.9, 11.10, 11.11

Using the matrix method or otherwise, find the focal lengths and the location of the principal plane for the following lens systems (a), (b) and (c). The glass in all lenses has a refractive index of $n = 1.5$ and all measurements have the same units. R_i is a radius of curvature.

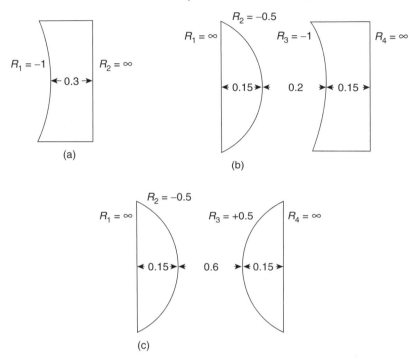

(a)

(b)

(c)

Summary of Important Results

Power of a Thin Lens

$$\mathscr{P} = (n-1)\left(\frac{1}{R_1} - \frac{1}{R_2}\right) = \frac{1}{f}$$

where n is the refractive index of the lens material, R_1 and R_2 are the radii of curvature of the lens surfaces and f is the focal length.

Power of two thin lenses separated a distance d in Air

$$\mathscr{P} = \mathscr{P}_1 + \mathscr{P}_2 - d\mathscr{P}_1\mathscr{P}_2$$

where $\mathscr{P}_1$ and $\mathscr{P}_2$ are the powers of the thin lenses.

Power of a thick lens of thickness d and refractive index n

$$\mathscr{P} = \mathscr{P}_1 + \mathscr{P}_2 - d/n\,\mathscr{P}_1\,\mathscr{P}_2$$

where $\mathscr{P}_1$ and $\mathscr{P}_2$ are the powers of the refracting surfaces of the lens.

Optical Helmholtz Equation

For a plane wavefront (source at ∞) passing through an optical system the product

$$ny\tan\alpha = \text{constant}$$

where n is the refractive index, y is the width of the wavefront section and α is the angle between the optical axis and the normal to the wavefront.

For a source at a finite distance, this equation becomes, for paraxial rays,

$$ny\alpha = \text{constant}$$

12

Interference and Diffraction

All waves display the phenomena of interference and diffraction which arise from the superposition of more than one wave. At each point of observation within the interference or diffraction pattern the phase difference between any two component waves of the same frequency will depend on the different paths they have followed and the resulting amplitude may be greater or less than that of any single component. Although we speak of separate waves the waves contributing to the interference and diffraction pattern must ultimately derive from the same single source. This avoids random phase effects from separate sources and guarantees coherence. However, even a single source has a finite size and spatial coherence of the light from different parts of the source imposes certain restrictions if interference effects are to be observed. This is discussed in the section on spatial coherence on p. 360. The superposition of waves involves the addition of two or more harmonic components with different phases and the basis of our approach is that laid down in the vector addition of Figure 1.11. More formally in the case of diffraction we have shown the equivalence of the Fourier transform method on p. 287 of Chapter 10.

Interference

Interference effects may be classified in two ways:

1. Division of amplitude

2. Division of wavefront

1. *Division of amplitude.* Here a beam of light or ray is reflected and transmitted at a boundary between media of different refractive indices. The incident, reflected and transmitted components form separate waves and follow different optical paths. They interfere when they are recombined.

2. *Division of wavefront.* Here the wavefront from a single source passes simultaneously through two or more apertures each of which contributes a wave at the point of superposition. Diffraction also occurs at each aperture.

The Physics of Vibrations and Waves, 6th Edition   H. J. Pain
© 2005 John Wiley & Sons, Ltd

The difference between interference and diffraction is merely one of scale: in optical diffraction from a narrow slit (or source) the aperture is of the order of the wavelength of the diffracted light. According to Huygens Principle every point on the wavefront in the plane of the slit may be considered as a source of secondary wavelets and the further development of the diffracted wave system may be obtained by superposing these wavelets.

In the interference pattern arising from two or more such narrow slits each slit may be seen as the source of a single wave so the number of superposed components in the final interference pattern equals the number of slits (or sources). This suggests that the complete pattern for more than one slit will display both interference and diffraction effects and we shall see that this is indeed the case.

Division of Amplitude

First of all we consider interference effects produced by division of amplitude. In Figure 12.1 a ray of monochromatic light of wavelength λ in air is incident at an angle i on a plane parallel slab of material thickness t and refractive index $n > 1$. It suffers partial reflection and transmission at the upper surface, some of the transmitted light is reflected at the lower surface and emerges parallel to the first reflection with a phase difference determined by the extra optical path it has travelled in the material. These parallel beams meet and interfere at infinity but they may be brought to focus by a lens. Their optical path difference is seen to be

$$n(\mathrm{AB} + \mathrm{BD}) - \mathrm{AC} = 2n\mathrm{AB} - \mathrm{AC}$$
$$= 2nt/\cos\theta - 2t\tan\theta\sin i$$
$$= \frac{2nt}{\cos\theta}(1 - \sin^2\theta) = 2nt\cos\theta$$

(because $\sin i = n\sin\theta$).

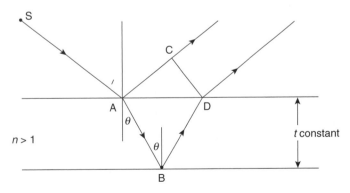

Figure 12.1 Fringes of constant inclination. Interference fringes formed at infinity by division of amplitude when the material thickness t is constant. The mth order bright fringe is a circle centred at S and occurs for the constant θ value in $2nt\cos\theta = (m + \frac{1}{2})\lambda$

This path difference introduces a phase difference

$$\delta = \frac{2\pi}{\lambda} 2nt \cos \theta$$

but an additional phase change of π rad occurs at the upper surface.

The phase difference δ between the two interfering beams is achieved by writing the beam amplitudes as

$$y_1 = a(\sin \omega t + \delta/2) \quad \text{and} \quad y_2 = a \sin (\omega t - \delta/2)$$

with a resultant amplitude

$$R = a[\sin (\omega t + \delta/2) + \sin (\omega t - \delta/2)$$
$$= 2a \sin \omega t \cos \delta/2$$

and an intensity

$$I = R^2 = 4a^2 \sin^2 \omega t \cos^2 \delta/2$$

Figure 12.2 shows the familiar $\cos^2 \delta/2$ intensity fringe pattern for the spatial part of I.

Thus, if $2nt \cos \theta = m\lambda$ (m an integer) the two beams are anti-phase and cancel to give zero intensity, a minimum of interference. If $2nt \cos \theta = (m + \frac{1}{2})\lambda$ the amplitudes will reinforce to give an interference maximum.

Since t is constant the locus of each interference fringe is determined by a constant value of θ which depends on a constant angle i. This gives a circular fringe centred on S. An extended source produces a range of constant θ values at one viewing position so the complete pattern is obviously a set of concentric circular fringes centred on S and formed at infinity. They are fringes of *equal inclination* and are called Haidinger fringes. They are observed to high orders of interference, that is values of m, so that t may be relatively large.

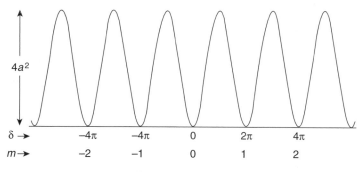

Figure 12.2 Interference fringes of $\cos^2$ intensity produced by the division of amplitude in Figure 12.1. The phase difference $\delta = 2\pi nt \cos \theta/\lambda$ and m is the order of interference

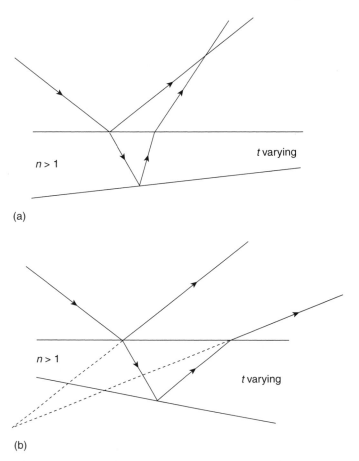

(a)

(b)

Figure 12.3 Fringes of constant thickness. When the thickness t of the material is not constant the fringes are localized where the interfering beams meet (a) in a real position and (b) in a virtual position. These fringes are almost parallel to the line where $t = 0$ and each fringe defines a locus of constant t

When the thickness t is not constant and the faces of the slab form a wedge, Figure 12.3a and b the interfering rays are not parallel but meet at points (real or virtual) near the wedge. The resulting interference fringes are localized near the wedge and are almost parallel to the thin end of the wedge. When observations are made at or near the normal to the wedge $\cos \theta \sim 1$ and changes slowly in this region so that $2nt \cos \theta \approx 2nt$. The condition for bright fringes then becomes

$$2nt = (m + \tfrac{1}{2})\lambda$$

and *each fringe locates a particular value of the thickness t of the wedge* and this defines the patterns as *fringes of equal thickness*. As the value of m increases to $m + 1$ the thickness of the wedge increases by $\lambda/2n$ so the fringes allow measurements to be made to within a fraction of a wavelength and are of great practical importance.

The spectral colours of a thin film of oil floating on water are fringes of constant thickness. Each frequency component of white light produces an interference fringe at that film thickness appropriate to its own particular wavelength.

In the laboratory the most familiar fringes of constant thickness are Newton's Rings.

Newton's Rings

Here the wedge of varying thickness is the air gap between two spherical surfaces of different curvature. A constant value of t yields a circular fringe and the pattern is one of concentric fringes alternately dark and bright. The simplest example, Figure 12.4, is a plano convex lens resting on a plane reflecting surface where the system is illuminated from above using a partially reflecting glass plate tilted at 45°. Each downward ray is partially reflected at each surface of the lens and at the plane surface. Interference takes

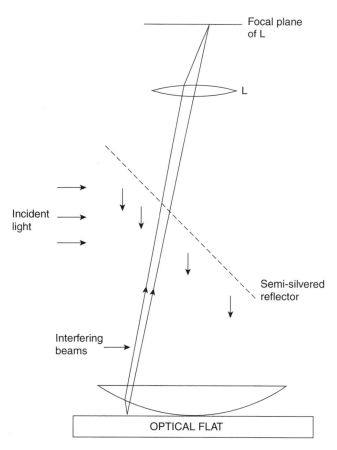

Figure 12.4 Newton's rings of interference formed by an air film of varying thickness between the lens and the optical flat. The fringes are circular, each fringe defining a constant value of the air film thickness

place between the light beams reflected at each surface of the air gap. At the lower (air to glass) surface of the gap there is a π rad phase change upon reflection and the centre of the interference fringe pattern, at the point of contact, is dark. Moving out from the centre, successive rings are light and dark as the air gap thickness increases in units of $\lambda/2$. If R is the radius of curvature of the spherical face of the lens, the thickness t of the air gap at a radius r from the centre is given approximately by $t \approx r^2/2R$. In the mth order of interference a bright ring requires

$$2t = (m + \tfrac{1}{2})\lambda = r^2/R$$

and because $t \propto r^2$ the fringes become more crowded with increasing r. Rings may be observed with very simple equipment and good quality apparatus can produce fringes for $m > 100$.

(Problem 12.1)

Michelson's Spectral Interferometer

This instrument can produce both types of interference fringes, that is, *circular fringes of equal inclination at infinity* and *localized fringes of equal thickness.* At the end of the nineteenth century it was one of the most important instruments for measuring the structure of spectral lines.

As shown in Figure 12.5 it consists of two identical plane parallel glass plates G_1 and G_2 and two highly reflecting plane mirrors M_1 and M_2. G_1 has a partially silvered back face, G_2 does not. In the figure G_1 and G_2 are parallel and M_1 and M_2 are perpendicular. Slow, accurately monitored motion of M_1 is allowed in the direction of the arrows but the mounting of M_2 is fixed although the angle of the mirror plane may be tilted so that M_1 and M_2 are no longer perpendicular.

The incident beam from an extended source divides at the back face of G_1. A part of it is reflected back through G_1 to M_1 where it is returned through G_1 into the eye or detector. The remainder of the incident beam reaches M_2 via G_2 and returns through G_2 to be reflected at the back face of G_1 into the eye or detector where it interferes with the beam from the M_1 arm of the interferometer. The presence of G_2 assures that each of the two interfering beams has the same optical path in glass. This condition is not essential for fringes with monochromatic light but it is required with a white light source where dispersion in glass becomes important.

An observer at the detector looking into G_1 will see M_1, a reflected image of M_2 (M_2', say) and the images S_1 and S_2' of the source provided by M_1 and M_2. This may be represented by the linear configuration of Figure 12.6 which shows how interference takes place and what type of firnges are produced.

When the optical paths in the interferometer arms are equal and M_1 and M_2 are perpendicular the planes of M_1 and the image M_2' are coincident. However a small optical path difference t between the arms becomes a difference of $2t$ between the mirrored images of the source as shown in Figure 12.6. The divided ray from a single point P on the extended source is reflected at M_1 and M_2 (shown as M_2') but these reflections appear to

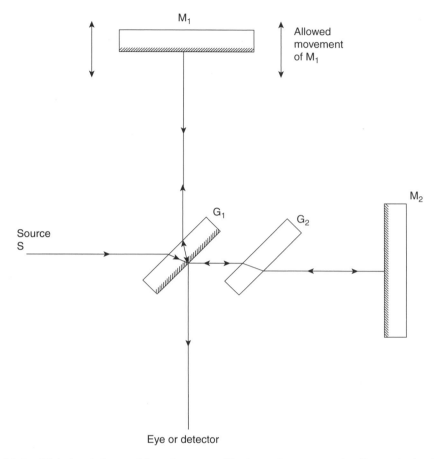

Figure 12.5 Michelson's Spectral Interferometer. The beam from source S splits at the back face of G_1, and the two parts are reflected at mirrors M_1 and M_2 to recombine and interfere at the eye or detector. G_2 is not necessary with monochromatic light but is required to produce fringes when S is a white light source

come from P_1 and P_2' in the image planes of the mirrors. The path difference between the rays from P_1 and P_2' is evidently $2t \cos \theta$. When $2t \cos \theta = m\lambda$ a maximum of interference occurs and for constant θ the interference fringe is a circle. The extended source produces a range of constant θ values and a pattern of concentric circular fringes of constant inclination.

If the path difference t is very small and the plane of M_2 is now tilted, a wedge is formed and straight localized fringes may be observed at the narrowest part of the wedge. As the wedge thickens the fringes begin to curve because the path difference becomes more strongly dependent upon the angle of observation. These curved fringes are always convex towards the thin end of the wedge.

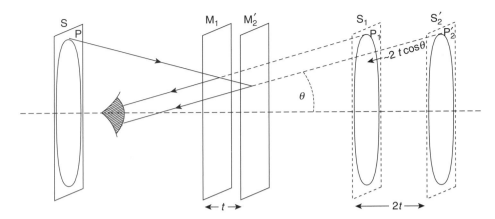

Figure 12.6 Linear configuration to show fringe formation by a Michelson interferometer. A ray from point P on the extended source S reflects at M_1, and appears to come from P_1 in the reflected plane S_1. The ray is reflected from M_2 (shown here as M_2') and appears to come from P_2' in the reflected plane S_2'. The path difference at the detector between the interfering beams is effectively $2t\cos\theta$ where t is the difference between the path lengths from the source S to the separate mirrors M_1 and M_2

The Structure of Spectral Lines

The discussion on spatial coherence, p. 362, will show that two close identical sources emitting the same wavelength λ produce interference fringe systems slightly displaced from each other (Figure 12.17).

The same effect is produced by a *single* source, such as sodium, emitting two wavelengths, λ and $\lambda - \Delta\lambda$ so that the maxima and minima of the $\cos^2$ fringes for λ are slightly displaced from those for $\lambda - \Delta\lambda$. This displacement increases with the order of interference m until a value of m is reached when the maximum for λ coincides with a minimum for $\lambda - \Delta\lambda$ and the fringes disappear as their visibility is reduced to zero.

In 1862, Fizeau, using a sodium source to produce Newton's Rings, found that the fringes disappeared at the order $m = 490$ but returned to maximum visibility at $m = 980$. He correctly identified the presence of two components in the spectral line.

The visibility

$$(I_{max} - I_{min})/(I_{max} + I_{min})$$

equals zero when

$$m\lambda = (m + \tfrac{1}{2})(\lambda - \Delta\lambda)$$

and for $\lambda = 0.5893\,\mu m$ and $m = 490$ we have $\Delta\lambda = 0.0006\,\mu m$ (6 Å), which are the accepted values for the D lines of the sodium doublet.

Using his spectral interferometer, Michelson extended this work between the years 1890 and 1900, plotting the visibility of various fringe systems and building a mechanical harmonic analyser into which he fed different component frequencies in an attempt to reproduce his visibility curves. The sodium doublet with angular frequency components ω and $\omega + \Delta\omega$ produced a visibility curve similar to that of Figures 1.7 and 4.4 and was easy to interpret. More complicated visibility patterns were not easy to reproduce and the modern method of Fourier transform spectroscopy reverses the procedure by extracting the frequency components from the observed pattern.

Michelson did however confirm that the cadmium red line, $\lambda = 0.6438\,\mu\text{m}$ was highly monochromatic. The visibility had still to reach a minimum when the path difference in his interferometer arms was 0.2 m.

Fabry – Perot Interferometer

The interference fringes produced by division of amplitude which we have discussed so far have been observed as reflected light and have been produced by only two interfering beams. We now consider fringes which are observed in transmission and which require multiple reflections. They are fringes of constant inclination formed in a pattern of concentric circles by the Fabry–Perot interferometer. The fringes are particularly narrow and sharply defined so that a beam consisting of two wavelengths λ and $\lambda - \Delta\lambda$ forms two patterns of rings which are easily separated for small $\Delta\lambda$. This instrument therefore has an extremely high resolving power. The main component of the interferometer is an etalon Figure 12.7 which consists of two plane parallel glass plates with identical highly reflecting inner surfaces S_1 and S_2 which are separated by a distance d.

Suppose a monochromatic beam of unit amplitude, angular frequency ω and wavelength (in air) of λ strikes the surface S_1 as shown. A fraction t of this beam is transmitted in passing from glass to air. At S_2 a further fraction t' is transmitted in passing from air to glass to give an emerging beam of amplitude $tt' = T$. The reflection coefficient at the air–S_1 and air–S_2 surfaces is r so each subsequent emerging beam is parallel but has an amplitude factor $r^2 = R$ with respect to its predecessor. Other reflection and transmission losses are common to all beams and do not affect the analysis. Each emerging beam has a phase lag $\delta = 4\pi d \cos\theta / \lambda$ with respect to its predecessor and these parallel beams interfere when they are brought to focus via a lens.

The vector sum of the transmitted interfering amplitudes together with their appropriate phases may be written

$$A = T\,e^{i\omega t} + TR\,e^{i(\omega t - \delta)} + TR^2\,e^{i(\omega t - 2\delta)} \cdots$$
$$= T\,e^{i\omega t}[1 + R\,e^{-i\delta} + R^2\,e^{-i2\delta} \cdots$$

which is an infinite geometric progression with the sum

$$A = T\,e^{i\omega t}/(1 - R\,e^{-i\delta})$$

This has a complex conjugate

$$A^* = T\,e^{-i\omega t}/(1 - R\,e^{i\delta})$$

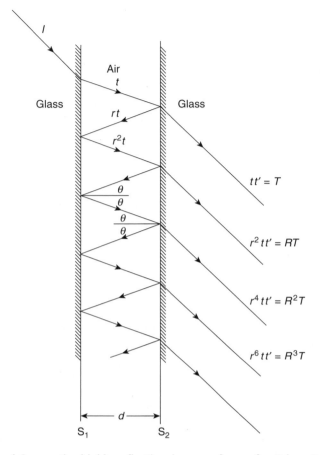

Figure 12.7 S_1 and S_2 are the highly reflecting inner surfaces of a Fabry–Perot etalon with a constant air gap thickness d. Multiple reflections produce parallel interfering beams with amplitudes T, RT, R^2T, etc. each beam having a phase difference

$$\delta = 4\pi d \cos\theta/\lambda$$

with respect to its neighbour

If the incident unit intensity is I_0 the fraction of this intensity in the transmitted beam may be written

$$\frac{I_t}{I_0} = \frac{AA^*}{I_0} = \frac{T^2}{(1 - R\,e^{-i\delta})(1 - R\,e^{i\delta})} = \frac{T^2}{(1 + R^2 - 2R\cos\delta)}$$

or, with

$$\cos\delta = 1 - 2\sin^2\delta/2$$

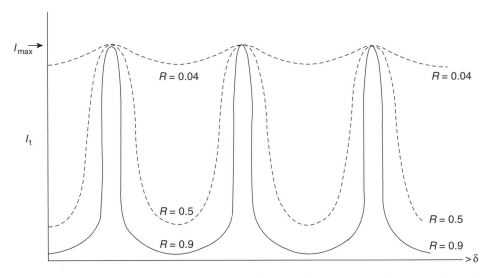

Figure 12.8 Observed intensity of fringes produced by a Fabry–Perot interferometer. Transmitted intensity I_t versus δ. $R = r^2$ where r is the reflection coefficient of the inner surfaces of the etalon. As R increases the fringes become narrower and more sharply defined

as

$$\frac{I_t}{I_0} = \frac{T^2}{(1-R)^2 + 4R\sin^2 \delta/2} = \frac{T^2}{(1-R)^2} \frac{1}{1 + [4R\sin^2 \delta/2/(1-R)^2]}$$

But the factor $T^2/(1-R)^2$ is a constant, written C so

$$\frac{I_t}{I_0} = C \cdot \frac{1}{1 + [4R\sin^2 \delta/2/(1-R)^2]}$$

Writing $CI_0 = I_{max}$, the graph of I_t versus δ in Figure 12.8 shows that as the reflection coefficient of the inner surfaces is increased, the interference fringes become narrow and more sharply defined. Values of $R > 0.9$ may be reached using the special techniques of multilayer dielectric coating. In one of these techniques a glass plate is coated with alternate layers of high and low refractive index materials so that each boundary presents a large change of refractive index and hence a large reflection. If the *optical* thickness of each layer is $\lambda/4$ the emerging beams are all in phase and the reflected intensity is high.

Resolving Power of the Fabry – Perot Interferometer

Figure 12.8 shows that a value of $R = 0.9$ produces such narrow and sharply defined fringes that if the incident beam has two components λ and $\lambda - \Delta\lambda$ the two sets of fringes should be easily separated. The criterion for separation depends on the shape of the fringes:

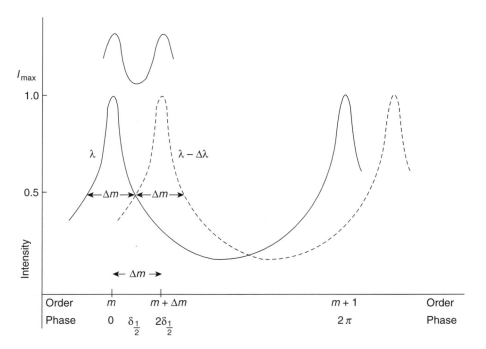

Figure 12.9 Fabry–Perot interference fringes for two wavelength λ and $\lambda - \Delta\lambda$ are resolved at order m when they cross at half their maximum intensity. Moving from order m to $m + 1$ changes the phase δ by 2π rad and the full 'half-value' width of each maximum is given by $\Delta m = 2\delta_{1/2}$ which is also the separation between the maxima of λ and $\lambda - \Delta\lambda$ when the fringes are just resolved

the diffraction grating of p. 373 uses the Rayleigh criterion, but the fringes here are so sharp that they are resolved at a much smaller separation than that required by Rayleigh.

Here the fringes of the two wavelengths may be resolved when they cross at half their maximum intensities; that is, at $I_t = I_{max}/2$ in Figure 12.9.

Using the expression

$$I_t = I_{max} \cdot \cfrac{1}{1 + \cfrac{4R \sin^2 \delta/2}{(1 - R)^2}}$$

we see that $I_t = I_{max}$ when $\delta = 0$ and $I_t = I_{max}/2$ when the factor

$$4R \sin^2 \delta/2 / (1 - R)^2 = 1$$

The fringes are so narrow that they are visible only for very small values of δ so we may replace $\sin \delta/2$ by $\delta/2$ in the expression

$$4R \sin^2 \delta/2 / (1 - R)^2 = 1$$

to give the value

$$\delta_{1/2} = \frac{(1 - R)}{R^{1/2}}$$

as the phase departure from the maximum, $\delta = 0$, which produces the intensity $I_t = I_{max}/2$ for wavelength λ. Our criterion for resolution means, therefore, that the maximum intensity for $\lambda - \Delta\lambda$ is removed an extra amount $\delta_{1/2}$ along the phase axis of Figure 12.9. This axis also shows the order of interference m at which the wavelengths are resolved, together with the order $m + 1$ which represents a phase shift of $\delta = 2\pi$ along the phase axis.

In the mth order of interference we have

$$2d \cos\theta = m\lambda$$

and for fringes of equal inclination (θ constant), logarithmic differentiation gives

$$\lambda/\Delta\lambda = -m/\Delta m$$

Now $\Delta m = 1$ represents a phase change of $\delta = 2\pi$ and the phase difference of $2.\delta_{1/2}$ which just resolves the two wavelengths corresponds to a change of order

$$\Delta m = 2.\delta_{1/2}/2\pi$$

Thus, the resolving power, defined as

$$\frac{\lambda}{\Delta\lambda} = \left| \frac{m}{\Delta m} \right| = \frac{m\pi}{\delta_{1/2}} = \frac{m\pi R^{1/2}}{(1 - R)}$$

The equivalent expression for the resolving power in the mth order for a diffracting grating of N lines (interfering beams) is shown on p. 376 to be

$$\frac{\lambda}{\Delta\lambda} = mN$$

so we may express

$$N' = \pi R^{1/2}/(1 - R)$$

as the effective number of interfering beams in the Fabry–Perot interferometer.

This quantity N' is called the *finesse* of the etalon and is a measure of its quality. We see that

$$N' = \frac{2\pi}{2\delta_{1/2}} = \frac{1}{\Delta m} = \frac{\text{separation between orders } m \text{ and } m + 1}{\text{'half value' width of } m\text{th order}}$$

Thus, using one wavelength only, the ratio of the separation between successive fringes to the narrowness of each fringe measures the quality of the etalon. A typical value of $N' \sim 30$.

Free Spectral Range

There is a limit to the wavelength difference $\Delta\lambda$ which can be resolved with the Fabry–Perot interferometer. This limit is reached when $\Delta\lambda$ is such that the circular fringe for λ in the mth order coincides with that for $\lambda - \Delta\lambda$ in the $m + 1$th order. The pattern then loses its unique definition and this value of $\Delta\lambda$ is called the *free spectral range*.

From the preceding section we have the expression

$$\frac{\lambda}{\Delta\lambda} = -\frac{m}{\Delta m}$$

and in the limit when $\Delta\lambda$ represents the free spectral range then

$$\Delta m = 1$$

and

$$\Delta\lambda = -\lambda/m$$

But $m\lambda = 2d$ when $\theta \simeq 0$ so the free spectral range

$$\Delta\lambda = -\lambda^2/2d$$

Typically $d \sim 10^{-2}$ m and for λ (cadmium red) $= 0.6438$ microns we have, from $2d = m\lambda$, a value of

$$m \approx 3 \times 10^4$$

Now the resolving power

$$\frac{\lambda}{\Delta\lambda} = mN'$$

so, for

$$N' \approx 30$$

the resolving power can be as high as 1 part in 10^6.

Central Spot Scanning

Early interferometers recorded flux densities on photographic plates but the non-linear response of such a technique made accurate resolution between two wavelengths tedious and more difficult. This method has now been superseded by the use of photoelectronic detectors which have the advantage of a superior and more reliable linearity. Moreover, the response of such a device with controlled vibration of one mirror of the etalon allows the variation of the intensity across the free spectral range to be monitored continuously.

The vibration of the mirror, originally electro-mechanical, is now most often produced by using a piezoelectric material on which to mount one of the etalon mirrors. When a voltage is applied to such a material it changes its length and the distance d between the etalon mirrors can be varied. The voltage across the piezoelectric mount is tailored to produce the desired motion.

Changing d by $\lambda/2$ is equivalent to changing Δm by 1, which corresponds to a scan of the free spectral range, $\Delta\lambda$, when $\lambda/\Delta\lambda = |m/\Delta m|$ (Figure 12.9).

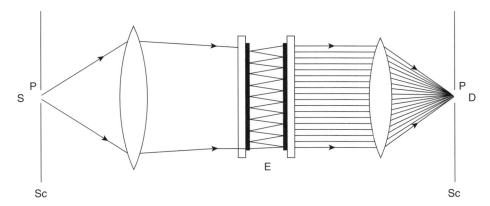

Figure 12.10 Fabry–Perot etalon central spot scanning. The distance between the etalon mirrors changes when one mirror vibrates on its piezoelectric mount. The free spectral range is scanned over many vibration cycles at a central spot and a stationary trace is obtained on the oscilloscope screen

One of the most common experimental arrangements is that of central spot scanning (Figure 12.10). Where the earlier photographic technique recorded the flux density over a wide region for a short period, central spot scanning focuses on a single point in space for a long period over many cycles of the etalon vibration. Matching the time base of the oscilloscope to the vibration period of the etalon produces a stationary trace on the screen which can be measured directly in addition to being filmed for record purposes.

The Laser Cavity

The laser cavity is in effect an extended Fabry–Perot etalon. Mirrors coated with multi-dielectric films described in the next section can produce reflection coefficients $R \approx 0.99$ and the amplified stimulated emission in the laser produces a beam which is continuously reflected between the mirror ends of the cavity. The high value of R allows the amplitudes of the beam in opposing directions to be taken as equal, so a standing wave system is generated (Figure 12.11) to form a longitudinal mode in the cavity.

The superposed amplitudes after a return journey from one mirror to the other and back are written for a wave number k and a frequency $\omega = 2\pi\nu$ as

$$E = A_1(e^{i(\omega t - kx)} - e^{i(\omega t + kx)})$$
$$= A_1(e^{-ikx} - e^{ikx}) e^{i\omega t} = -2iA_1 \sin kx\, e^{i\omega t}$$

of which the real part is $E = 2A_1 \sin kx \sin \omega t$.

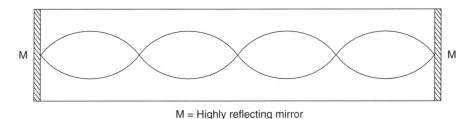

M = Highly reflecting mirror

Figure 12.11 A longitudinal mode in a laser cavity which behaves as an extended Fabry–Perot etalon with highly reflecting mirrors at each end. The standing wave system acquires an extra $\lambda/2$ for unit change in the mode number m. A typical output is shown in Figure 12.12

If the cavity length is L, one round trip between the mirrors creates a phase change of

$$\phi = -2Lk + 2\alpha = -\frac{4\pi L}{c}\nu + 2\alpha$$

where α is the phase change on reflection at each mirror.

For this standing wave mode to be maintained, the phase change must be a multiple of 2π, so for m an integer

$$\phi = m2\pi = \frac{4\pi L}{c}\nu - 2\alpha$$

or

$$\nu = \frac{mc}{2L} + \frac{\alpha c}{2\pi L}$$

When m changes to $m + 1$, the phase change of 2π corresponds to an extra wavelength λ for the return journey; that is, an extra $\lambda/2$ in the standing wave mode. A series of longitudinal modes can therefore exist with frequency intervals $\Delta\nu = c/2L$ determined by a unit change in m.

The intensity profile for each mode and the separation $\Delta\nu$ is best seen by reference to Figure 12.9, where $\phi \equiv \delta$ is given by the unit change in the order of interference from m to $m + 1$.

The intensity profile for each cavity mode is that of Figure 12.9, where the full width at half maximum intensity is given by the phase change

$$2\delta_{1/2} = \frac{2(1 - R)}{R^{1/2}}$$

where R is the reflection coefficient. This corresponds to a full width intensity change over a frequency $d\nu$ generated by the phase change

$$d\phi = \frac{4\pi L}{c}\,d\nu \quad \text{in the expression for } \phi \text{ above}$$

The width at half maximum intensity for each longitudinal mode is therefore given by

$$\frac{4\pi L}{c} d\nu = \frac{2(1-R)}{R^{1/2}}$$

or

$$d\nu = \frac{(1-R)c}{R^{1/2}2\pi L}$$

For a laser 1 m long with $R = 0.99$, the longitudinal modes are separated by frequency intervals

$$\Delta\nu = \frac{c}{2L} = 1.5 \times 10^8 \, \text{Hz}$$

Each mode intensity profile has a full width at half maximum of

$$d\nu = 10^{-2}\frac{c}{2\pi} \approx 4.5 \times 10^5 \, \text{Hz}$$

For a He−Ne laser the mean frequency of the output at 632.8 nm is 4.74×10^{14} Hz. The pattern for $\Delta\nu$ and $d\nu$ is shown in Figure 12.12, where the intensities are reduced under the dotted envelope as the frequency difference from the mean is increased.

The finesse of the laser cavity is given by

$$\frac{\Delta\nu}{d\nu} = \frac{1.5 \times 10^8}{4.5 \times 10^5} \approx 300$$

for the example quoted.

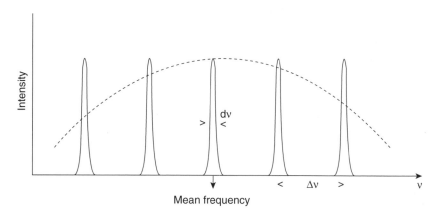

Figure 12.12 Output of a laser cavity. A series of longitudinal modes separated by frequency intervals $\Delta\nu = c/2L$, where c is the velocity of light and L is the cavity length. The modes are centred about the mean output frequency and are modulated under the dotted envelope. For a He−Ne laser 1 m long the separation $\Delta\nu$ between the modes ≈ 300 full widths of a mode intensity profile at half its maximum value

The intensity of each longitudinal mode is of course, amplified by each passage of the stimulated emission. Radiation allowed from out of one end represents the laser output but the amplification process is dominant and the laser produces a continuous beam.

Multilayer Dielectric Films

We have just seen that in the mth order of interference the resolving power of a Fabry–Perot interferometer is given by

$$\lambda/\Delta\lambda \equiv mN'$$

where the finesse or number of interfering beams

$$N' = \pi R^{1/2}/(1-R) = \pi r/(1-r^2)$$

and r is the reflection coefficient of the inner surfaces of the etalon.

It is evident that as $r \to 1$ the values of N' and the resolving power become much larger. The value of r can be increased to more than 99% by using a metallic coating on the inner surfaces of the etalon or by depositing on them a multilayer of dielectric films with alternating high and low refractive indices. For a given monochromatic electromagnetic wave each layer or film has an optical thickness of $\lambda/4$.

The reflection coefficient r for such a wave incident on the surface of a higher refractive index film is increased because the externally and internally reflected waves are in phase; a phase change of π occurs only on the outer surface and is reinforced by the π phase change of the wave reflected at the inner surface which travels an extra $\lambda/2$ optical distance.

High values of r result from films of alternating high and low values of the refractive index because reflections from successive boundaries are in phase on return to the front surface of the first film. Those retarded an odd multiple of π by the extra optical path length per film also have a π phase change on reflection to make a total of 2π rad.

We consider the simplest case of a monochromatic electromagnetic wave in a medium of refractive index n_1, *normally incident* on a single film of refractive index n_1', and thickness d_1'. This film is deposited on the surface of a material of refractive index n_2', which is called a substrate (Figure 12.13). The phase lag for a single journey across the film is written δ.

The boundary conditions are that the components of the E and H fields parallel to a surface are continuous across that surface. We write these field amplitudes as E_f and $H_f = nE_f$ for the forward-going wave to the right in Figure 12.13 and E_r and $H_r = nE_r$ for the reflected wave going to the left.

We see that at surface 1 the boundary conditions for the electric field E are

$$E_{f1} + E_{r1} = E_{f1}' + E_{r1}' \tag{12.1a}$$

and for the magnetic field

$$n_1 E_{f1} - n_1 E_{r1} = n_1' E_{f1}' - n_1' E_{r1}' \tag{12.1b}$$

where the negative sign for the reflected amplitude arises when the $\mathbf{E} \times \mathbf{H}$ direction of the wave is reversed (see Figure 8.7).

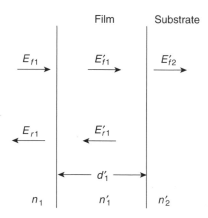

Figure 12.13 A thin dielectric film is deposited on a substrate base. At each surface an electromagnetic wave is normally incident, as E_{fi}, in a medium of refractive index n_i and is reflected as E_{ri}. A multilayer stack of such films, each of optical thickness $\lambda/4$ and of alternating high and low refractive indices can produce reflection coefficients >99%

At surface 2 in Figure 12.13, E'_{f1} arrives with a phase lag of δ with respect to E'_{f1} at surface 1 but the E'_{r1} wave at surface 2 has a phase δ in advance of E'_{r1} at surface 1, so we have the boundary conditions

$$E'_{f1}\,e^{-i\delta} + E'_{r1}\,e^{i\delta} = E'_{f2} \tag{12.1c}$$

and

$$n'_1 E'_{f1}\,e^{-i\delta} - n'_1 E'_{r1}\,e^{i\delta} = n'_2 E'_{f2} \tag{12.1d}$$

We can eliminate E'_{f1} and E'_{r1} from equations (12.1a)–(12.1d) to give

$$1 + \frac{E_{r1}}{E_{f1}} = \left(\cos\delta + i\frac{n'_2}{n'_1}\sin\delta\right)\frac{E'_{f2}}{E_{f1}} \tag{12.2}$$

and

$$n_1 - n_1\frac{E_{r1}}{E_{f1}} = (in'_1\sin\delta + n'_2\cos\delta)\frac{E'_{f2}}{E_{f1}} \tag{12.3}$$

which we can express in matrix form

$$\begin{bmatrix} 1 \\ n_1 \end{bmatrix} + \begin{bmatrix} 1 \\ -n_1 \end{bmatrix}\frac{E_{r1}}{E_{f1}} = \begin{bmatrix} \cos\delta & i\sin\delta/n'_1 \\ in'_1\sin\delta & \cos\delta \end{bmatrix}\begin{bmatrix} 1 \\ n'_2 \end{bmatrix}\frac{E'_{f2}}{E_{f1}}$$

We write this as

$$\begin{bmatrix} 1 \\ n_1 \end{bmatrix} + \begin{bmatrix} 1 \\ -n_1 \end{bmatrix} r = M_1 \begin{bmatrix} 1 \\ n_2' \end{bmatrix} t$$

where $r = E_{r1}/E_{f1}$ is the reflection coefficient at the first surface and $t = E_{f2}'/E_{f1}$ is the transmitted coefficient into medium n_2' (a quantity we shall not evaluate).

The 2×2 matrix

$$M_1 = \begin{bmatrix} \cos \delta & i \sin \delta / n_1' \\ in_1' \sin \delta & \cos \delta \end{bmatrix}$$

relates r and t across the first film and is repeated with appropriate values of n_i' for each successive film. The product of these 2×2 matrices is itself a 2×2 matrix as with the repetitive process we found in the optical case of p. 325.

Thus, for N films we have

$$\begin{bmatrix} 1 \\ n_1 \end{bmatrix} + \begin{bmatrix} 1 \\ -n_1 \end{bmatrix} R = M_1 M_2 M_3 \cdots M_N \begin{bmatrix} 1 \\ n_{N+1}' \end{bmatrix} T, \tag{12.4}$$

where $R = E_{r1}/E_{f1}$ as before and $T = E_{f(N+1)}'/E_{f1}$ the transmitted coefficient after the final film. *Note, however, that E_{r1} in R is now the result of reflection from all the film surfaces and that these are in phase.*

The typical matrix M_3 relates r to t across the third film and the product of the matrices

$$M_1 M_2 M_3 \cdots M_N = M = \begin{bmatrix} M_{11} & M_{12} \\ M_{21} & M_{22} \end{bmatrix}$$

is a 2×2 matrix.

Eliminating T from the two simultaneous equations (12.4) we have, in terms of the coefficients of M

$$R = \frac{A - B}{A + B} \tag{12.5}$$

where

$$A = n_1(M_{11} + M_{12}n_{N+1}')$$

and

$$B = (M_{21} + M_{22}n_{N+1}')$$

If we now consider a system of two films, the first of higher refractive index n_H and the second of lower refractive index n_L, where each has an optical thickness $d = \lambda/4$, then the phase $\delta = \pi/2$ for each film and

$$M_1 M_2 = \begin{bmatrix} 0 & i/n_H \\ in_H & 0 \end{bmatrix} \begin{bmatrix} 0 & i/n_L \\ in_L & 0 \end{bmatrix} = \begin{bmatrix} -n_L/n_H & 0 \\ 0 & -n_H/n_L \end{bmatrix}$$

A stack of N such pairs, $2N$ films in all with alternating n_H and n_L, produces

$$M_1 M_2 \cdots M_{2N} = [M_1 M_2]^N = \begin{bmatrix} \left(\dfrac{-n_L}{n_H}\right)^N & 0 \\ 0 & \left(\dfrac{-n_H}{n_L}\right)^N \end{bmatrix}$$

giving R the total reflection coefficient from equation (12.5) equal to

$$R = \frac{\left(\dfrac{-n_L}{n_H}\right)^N - \left(\dfrac{-n_H}{n_L}\right)^N}{\left(\dfrac{-n_L}{n_H}\right)^N + \left(\dfrac{-n_H}{n_L}\right)^N}$$

We see that as long as $n_H \neq n_L$, then as $N \to \infty$, $R \to 1$ and this value may be used in our derivation of the expressions for the finesse and resolving power of the Fabry–Perot interferometer.

Multilayer stacks using zinc sulphate ($n_H = 2.3$) and cryolite ($n_L = 1.35$) have achieved R values $> 99.5\%$.

Note that all the 2×2 matrices and their products have determinants equal to unity which states that the column vectors represent a quantity which remains invariant throughout the matrix transformations.

(Problem 12.2)

The Thin Film Optical Wave Guide

Figure 12.14 shows a thin film of width d and refractive index n along which light of frequency ν and wave number k is guided by multiple internal reflections. The extent of the

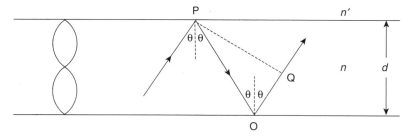

Figure 12.14 A thin dielectric film or fibre acts as an optical wave guide. The reflection angle θ must satisfy the relation $n \sin \theta \geq n'$, where n' is the refractive index of the coating over the film of refractive index n. Propagating modes have standing wave systems across the film as shown and constructive interference occurs on the standing wave axis where the amplitude is a maximum. Destructive interference occurs at the nodes

wave guide is infinite in the direction normal to the page. The internal reflection angle θ must satisfy

$$n \sin \theta \geq n'$$

where n' is the refractive index of the medium bounding the thin film surfaces. Each reflected ray is normal to a number of wave fronts of constant phase separated by λ, where $k = 2\pi/\lambda$ and constructive interference is necessary for any mode to propagate. Reflections may take place at any pair of points P and O along the film and we examine the condition for constructive interference by considering the phase difference along the path POQ, taking into account a phase difference α introduced by reflection at each of P and Q.

Now

$$PO = \cos \theta / d$$

and

$$OQ = PO \cos 2\theta$$

so with

$$\cos 2\theta = 2 \cos^2 \theta - 1$$

we have

$$PO + OQ = 2d \cos \theta$$

giving a phase difference

$$\Delta\phi = \phi_Q - \phi_P = -\frac{2\pi\nu}{c}(n\, 2d \cos \theta) + 2\alpha$$

Constructive interference requires

$$\Delta\phi = m\, 2\pi$$

where m is an integer, so we write

$$m\, 2\pi = \frac{2\pi\nu}{c}\, n\, 2d \cos \theta - 2\pi\Delta m$$

where

$$\Delta m = 2\alpha/2\pi$$

represents the phase change on reflection.

Radiation will therefore propagate only when

$$\cos \theta = \frac{c(m + \Delta m)}{\nu\, 2nd}$$

for $m = 0, 1, 2, 3$.

The condition $n \sin \theta \geq n'$ restricts the values of the frequency ν which can propagate. If $\theta = \theta_m$ for mode m and

$$\cos \theta_m = (1 - \sin^2 \theta_m)^{1/2}$$

then

$$n \sin \theta_m \geq n'$$

becomes

$$\cos \theta_m \leq \left[1 - \left(\frac{n'}{n} \right)^2 \right]^{1/2}$$

and ν must satisfy

$$\nu \geq \frac{c(m + \Delta m)}{2d(n^2 - n'^2)^{1/2}}$$

The mode $m = 0$ is the mode below which ν will not propagate, while Δm is a constant for a given wave guide. Each mode, Figure 12.14, is represented by a standing wave system across the wave guide normal to the direction of propagation. Constructive interference occurs on the axis of this wave system where the amplitude is a maximum and destructive interference is indicated by the nodes.

Division of Wavefront

Interference Between Waves from Two Slits or Sources

In Figure 12.15 let S_1 and S_2 be two equal sources separated by a distance f, each generating a wave of angular frequency ω and amplitude a. At a point P sufficiently distant from S_1 and S_2 only plane wavefronts arrive with displacements

$$y_1 = a \sin (\omega t - kx_1) \quad \text{from } S_1$$

and

$$y_2 = a \sin (\omega t - kx_2) \quad \text{from } S_2$$

so that the phase difference between the two signals at P is given by

$$\delta = k(x_2 - x_1) = \frac{2\pi}{\lambda}(x_2 - x_1)$$

This phase difference δ, which arises from the path difference $x_2 - x_1$, depends only on x_1, x_2 and the wavelength λ and not on any variation in the source behaviour. This requires that there shall be no sudden changes of phase in the signal generated at either source – such sources are called *coherent*.

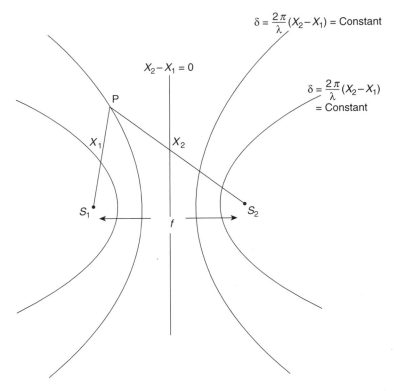

Figure 12.15 Interference at P between waves from equal sources S_1 and S_2, separation f, depends only on the path difference $x_2 - x_1$. Loci of points with constant phase difference $\delta = (2\pi/\lambda)$ $(x_2 - x_1)$ are the family of hyperbolas with S_1 and S_2 as foci

The superposition of displacements at P gives a resultant

$$R = y_1 + y_2 = a[\sin(\omega t - kx_1) + \sin(\omega t - kx_2)]$$

Writing $X \equiv (x_1 + x_2)/2$ as the average distance from the two sources to point P we obtain

$$kx_1 = kX - \delta/2 \quad \text{and} \quad kx_2 = kX + \delta/2$$

to give

$$R = a[\sin(\omega t - kX + \delta/2) + \sin(\omega t - kX - \delta/2)]$$
$$= 2a \sin(\omega t - kX) \cos \delta/2$$

and an intensity

$$I = R^2 = 4a^2 \sin^2(\omega t - kX) \cos^2 \delta/2$$

When

$$\cos \frac{\delta}{2} = \pm 1$$

the spatial intensity is a maximum,

$$I = 4a^2$$

and the component displacements reinforce each other to give *constructive interference*. This occurs when

$$\frac{\delta}{2} = \frac{\pi}{\lambda}(x_2 - x_1) = n\pi$$

that is, when the path difference

$$x_2 - x_1 = n\lambda$$

When

$$\cos \frac{\delta}{2} = 0$$

the intensity is zero and the components cancel to give *destructive interference*. This requires that

$$\frac{\delta}{2} = (2n + 1)\frac{\pi}{2} = \frac{\pi}{\lambda}(x_2 - x_1)$$

or, the path difference

$$x_2 - x_1 = (n + \tfrac{1}{2})\lambda$$

The loci or sets of points for which $x_2 - x_1$ (or δ) is constant are shown in Figure 12.15 to form hyperbolas about the foci S_1 and S_2 (in three dimensions the loci would be the hyperbolic surfaces of revolution).

Interference from Two Equal Sources of Separation *f*

Separation $f \gg \lambda$. Young's Slit Experiment

One of the best known methods for producing optical interference effects is the Young's slit experiment. Here the two coherent sources, Figure 12.16, are two identical slits S_1 and S_2 illuminated by a monochromatic wave system from a single source equidistant from S_1 and S_2. The observation point P lies on a screen which is set at a distance l from the plane of the slits.

The intensity at P is given by

$$I = R^2 = 4a^2 \cos^2 \frac{\delta}{2}$$

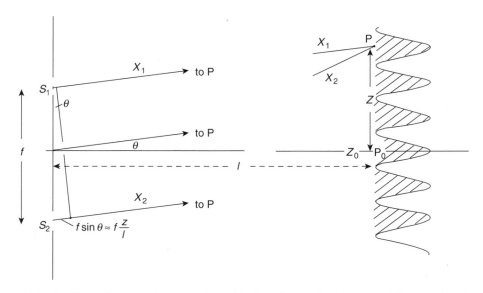

Figure 12.16 Waves from equal sources S_1 and S_2 interfere at P with phase difference $\delta = (2\pi/\lambda)$ $(x_2 - x_1) = (2\pi/\lambda) f \sin\theta \approx (2\pi/\lambda) f(z/l)$. The distance $l \gg z$ and f so S_1P and S_2P are effectively parallel. Interference fringes of intensity $I = I_0 \cos^2 \delta/2$ are formed in the plane PP_0

and the distances $PP_0 = z$ and slit separation f are both very much less than l (experimentally $\approx 10^{-3} l$). This is indicated by the break in the lines x_1 and x_2 in Figure 12.16 where S_1P and S_2P may be considered as sufficiently parallel for the path difference to be written as

$$x_2 - x_1 = f \sin\theta = f\frac{z}{l}$$

to a very close approximation.

Thus

$$\delta = \frac{2\pi}{\lambda}(x_2 - x_1) = \frac{2\pi}{\lambda} f \sin\theta = \frac{2\pi}{\lambda} f \frac{z}{l}$$

If

$$I = 4a^2 \cos^2\frac{\delta}{2}$$

then

$$I = I_0 = 4a^2 \quad \text{when} \quad \cos\frac{\delta}{2} = 1$$

that is, when the path difference

$$f\frac{z}{l} = 0, \quad \pm\lambda, \quad \pm2\lambda, \quad \dots \pm n\lambda$$

and

$$I = 0 \quad \text{when} \quad \cos\frac{\delta}{2} = 0$$

that is, when

$$f\frac{z}{l} = \pm\frac{\lambda}{2}, \quad \pm\frac{3\lambda}{2}, \quad \pm(n+\tfrac{1}{2})\lambda$$

Taking the point P_0 as $z = 0$, the variation of intensity with z on the screen P_0P will be that of Figure 12.16, a series of alternating straight bright and dark fringes parallel to the slit directions, the bright fringes having $I = 4a^2$ whenever $z = n\lambda l/f$ and the dark fringes $I = 0$, occurring when $z = (n+\tfrac{1}{2})\lambda l/f$, n being called the *order of interference* of the fringes. The zero order $n = 0$ at the point P_0 is the central bright fringe. The distance on the screen between two bright fringes of orders n and $n + 1$ is given by

$$z_{n+1} - z_n = [(n+1) - n]\frac{\lambda l}{f} = \frac{\lambda l}{f}$$

which is also the physical separation between two consecutive dark fringes. The spacing between the fringes is therefore constant and independent of n, and a measurement of the spacing, l and f determines λ.

The intensity distribution curve (Figure 12.17) shows that when the two wave trains arrive at P exactly out of phase they interfere destructively and the resulting intensity or energy flux is zero. Energy conservation requires that the energy must be redistributed, and that lost at zero intensity is found in the intensity peaks. The average value of $\cos^2\delta/2$ is $\tfrac{1}{2}$, and the dotted line at $I = 2a^2$ is the average intensity value over the interference system which is equal to the sum of the separate intensities from each slit.

There are two important points to remember about the intensity interference fringes when discussing diffraction phenomena; these are

- The intensity varies with $\cos^2\delta/2$.

- The maxima occur for path differences of zero or integral numbers of the wavelength, whilst the minima represent path differences of odd numbers of the half-wavelength.

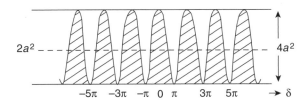

Figure 12.17 Intensity of interference fringes is proportional to $\cos^2\delta/2$, where δ is the phase difference between the interfering waves. The energy which is lost in destructive interference (minima) is redistributed into regions of constructive interference (maxima)

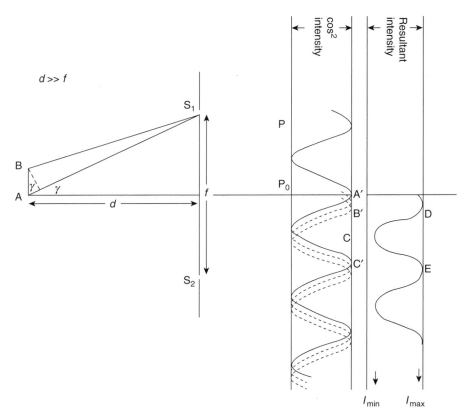

Figure 12.18 The point source A produces the cos^2 interference fringes represented by the solid curve A'C'. Other points on the line source AB produce cos^2 fringes (the displaced broken curves B') and the observed total intensity is the curve DE. When the points on AB extend A'B' to C the fringes disappear and the field is uniformly illuminated

Spatial Coherence   In the preceding section nothing has been said about the size of the source producing the plane wave which falls on S_1 and S_2. If this source is an ideal *point* source A equidistant from S_1 and S_2, Figure 12.18, then a single set of cos^2 fringes is produced. But every source has a finite size, given by AB in Figure 12.18, and each point on the line source AB produces its own set of interference fringes in the plane PP$_0$; the eye observing the sum of their intensities.

If the solid curve A'C' is the intensity distribution for the point A of the source and the broken curves up to B' represent the corresponding fringes for points along AB the resulting intensity curve is DE. Unless A'B' extends to C the variations of DE will be seen as faint interference bands. These intensity variations were quantified by Michelson, who defined the

$$\text{Visibility} = \frac{I_{\max} - I_{\min}}{I_{\max} + I_{\min}}$$

The $\cos^2$ fringes from a point source obviously have a visibility of unity because the minimum intensity $I_{min} = 0$.

When $A'B'$ of Figure 12.18 $= A'C$, the point source fringe separation (or a multiple of it) the field is uniformly illuminated, fringe visibility $= 0$ and the fringes disappear.

This occurs when the path difference

$$AS_2 - BS_1 \approx AB \sin\gamma = \lambda/2 \quad \text{where } AS_2 = AS_1.$$

Thus, the requirement for fringes of good visibility imposes a limit on the finite size of the source. Light from points on the source must be *spatially coherent* in the sense that $AB \sin\gamma \ll \lambda/2$ in Figure 12.18.

But for $f \ll d$,

$$\sin\gamma \approx f/2d$$

so the coherence condition becomes

$$\sin\gamma = f/2d \ll \lambda/2AB$$

or

$$AB/d \ll \lambda/f$$

where AB/d measures the angle subtended by the source at the plane S_1S_2.

Spatial coherence therefore requires that the angle subtended by the source

$$\ll \lambda/f$$

where f is the linear size of the diffracting system. (Note also that λ/f measures $\theta(\sim z/l)$ the angular separation of the fringes in Figure 12.16.)

As an example of spatial coherence we may consider the production of Young's interference fringes using the sun as a source.

The sun subtends an angle of 0.018 rad at the earth and if we accept the approximation

$$\frac{AB}{d} \ll \frac{\lambda}{f} \approx \frac{\lambda}{4f}$$

with $\lambda = 0.5\,\mu\text{m}$,
we have

$$f \sim \frac{0.5}{4(0.018)} \sim 14\,\mu\text{m}$$

This small value of slit separation is required to meet the spatial coherence condition.

Separation $f \ll \lambda(kf \ll 1$ where $k = 2\pi/\lambda)$

If there is a zero phase difference between the signals leaving the sources S_1 and S_2 of Figure 12.16 then the intensity at some distant point P may be written

$$I = 4a^2 \cos^2 \frac{\delta}{2} = 4I_s \cos^2 \frac{kf \sin\theta}{2} \approx 4I_s,$$

where the path difference $S_2P - S_1P = f \sin\theta$ and $I_s = a^2$ is the intensity from each source.

We note that, since $f \ll \lambda (kf \ll 1)$, the intensity has a very small θ dependence and the two sources may be effectively replaced by a single source of amplitude $2a$.

Dipole Radiation $(f \ll \lambda)$

Suppose, however, that the signals leaving the sources S_1 and S_2 are anti-phase so that their total phase difference at some distant point P is

$$\delta = (\delta_0 + kf \sin\theta)$$

where $\delta_0 = \pi$ is the phase difference introduced at source.

The intensity at P is given by

$$I = 4 I_s \cos^2 \frac{\delta}{2} = 4 I_s \cos^2 \left(\frac{\pi}{2} + \frac{kf \sin\theta}{2} \right)$$

$$= 4 I_s \sin^2 \left(\frac{kf \sin\theta}{2} \right)$$

$$\approx I_s (kf \sin\theta)^2$$

because

$$kf \ll 1$$

Two anti-phase sources of this kind constitute a *dipole* whose radiation intensity $I \ll I_s$ the radiation from a single source, when $kf \ll 1$. The efficiency of radiation is seen to depend on the product kf and, for a fixed separation f the dipole is a less efficient radiator at low frequencies (small k) than at higher frequencies. Figure 12.19 shows the radiation intensity I plotted against the polar angle θ and we see that for the dipole axis along the direction $\theta = \pi/2$, completely destructive interference occurs only on the perpendicular axis $\theta = 0$ and $\theta = \pi$. There is no direction (value of θ) giving completely constructive interference. The highest value of the radiated intensity occurs along the axis $\theta = \pi/2$ and $\theta = 3\pi/2$ but even this is only

$$I = (kf)^2 I_s,$$

where

$$kf \ll 1$$

The directional properties of a radiating dipole are incorporated in the design of transmitting aerials. In acoustics a loudspeaker may be considered as a multi dipole source, the face of the loudspeaker generating compression waves whilst its rear propagates rarefactions. Acoustic reflections from surrounding walls give rise to undesirable interference effects which are avoided by enclosing the speaker in a cabinet. Bass reflex or phase inverter cabinets incorporate a vent on the same side as the speaker face at an acoustic distance of half a wavelength from the rear of the speaker. The vent thus acts as a second source *in phase* with the speaker face and radiation is improved.

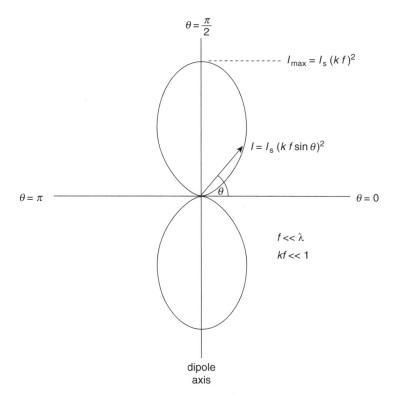

Figure 12.19 Intensity I versus direction θ for interference pattern between waves from two equal sources, π rad out of phase (dipole) with separation $f \ll \lambda$. The dipole axis is along the direction $\theta = \pm\pi/2$

(Problems 12.3, 12.4, 12.5)

Interference from Linear Array of *N* Equal Sources

Figure 12.20 shows a linear array of N equal sources with constant separation f generating signals which are all in phase $(\delta_0 = 0)$. At a distant point P in a direction θ from the sources the phase difference between the signals from two successive sources is given by

$$\delta = \frac{2\pi}{\lambda}f \sin \theta$$

and the resultant at P is found by superposing the equal contribution from each source with the constant phase difference δ between successive contributions.

But we found from Figure 1.11 that the resultant of such a superposition was given by

$$R = a\frac{\sin (N\delta/2)}{\sin (\delta/2)}$$

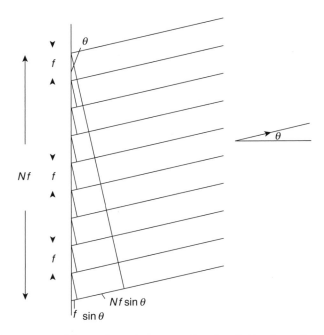

Figure 12.20   Linear array of N equal sources separation f radiating in a direction θ to a distant point P. The resulting amplitude at P (see Figure 1.11) is given by

$$R = a[\sin N(\delta/2)/\sin(\delta/2)]$$

where a is the amplitude from each source and

$$\delta = (2\pi/\lambda)f\sin\theta$$

is the common phase difference between successive sources

where a is the signal amplitude at each source, so the intensity may be written

$$I = R^2 = a^2\frac{\sin^2(N\delta/2)}{\sin^2(\delta/2)} = I_s\frac{\sin^2(N\pi f\sin\theta/\lambda)}{\sin^2(\pi f\sin\theta/\lambda)}$$

$$= I_s\frac{\sin^2 N\beta}{\sin^2\beta}$$

where I_s is the intensity from each source and $\beta = \pi f\sin\theta/\lambda$.

If we take the case of $N = 2$, then

$$I = I_s\frac{\sin^2 2\beta}{\sin^2\beta} = 4I_s\cos^2\beta = 4I_s\cos^2\frac{\delta}{2}$$

which gives us the Young's Slit Interference pattern.

We can follow the intensity pattern for N sources by considering the behaviour of the term $\sin^2 N\beta/\sin^2\beta$.

We see that when

$$\beta = \frac{\pi}{\lambda} \sin \theta = 0 \pm \pi \pm 2\pi, \text{etc.}$$

i.e. when

$$f \sin \theta = 0, \ \pm \lambda, \ \pm 2\lambda \ldots \pm n\lambda$$

constructive interference of the order n takes place, and

$$\frac{\sin^2 N\beta}{\sin^2 \beta} \to \frac{N^2 \beta^2}{\beta^2} \to N^2$$

giving

$$I = N^2 I_s$$

that is, a very strong intensity at the *Principal Maximum* condition of

$$f \sin \theta = n\lambda$$

We can display the behaviour of the $\sin^2 N\beta / \sin^2 \beta$ term as follows

Numerator $\sin^2 N\beta$ is zero for $N\beta \to 0\pi \ldots N\pi \ldots 2N\pi$

$$\downarrow \quad \downarrow \quad \downarrow$$

Denominator $\sin^2 \beta$ is zero for $\beta \to 0 \ \ldots \ \pi \ \ldots \ 2\pi$

The coincidence of zeros for both numerator and denominator determine the Principal Maxima with the factor N^2 in the intensity, i.e. whenever $f \sin \theta = n\lambda$.

Between these principal maxima are $N - 1$ points of zero intensity which occur whenever the numerator $\sin^2 N\beta = 0$ but where $\sin^2 \beta$ remains finite.

These occur when

$$f \sin \theta = \frac{\lambda}{N}, \ \frac{2\lambda}{N} \ldots (n - 1)\frac{\lambda}{N}$$

The $N - 2$ subsidiary maxima which occur between the principal maxima have much lower intensities because none of them contains the factor N^2. Figure 12.21 shows the intensity curves for $N = 2, 4, 8$ and $N \to \infty$.

Two scales are given on the horizontal axis. One shows how the maxima occur at the order of interference $n = f \sin \theta / \lambda$. The other, using units of $\sin \theta$ as the ordinate displays two features. It shows that the separation between the principal maxima in units of $\sin \theta$ is λ/f and that the width of half the base of the principal maxima in these units is λ/Nf (the same value as the width of the base of subsidiary maxima). As N increases not only does the principal intensity increase as N^2 but the width of the principal maximum becomes very small.

As N becomes very large, the interference pattern becomes highly directional, very sharply defined peaks of high intensity occurring whenever $\sin \theta$ changes by λ/f.

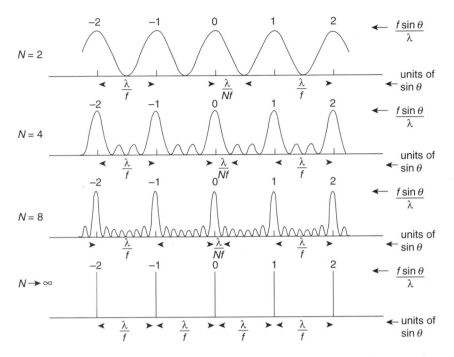

Figure 12.21 Intensity of interference patterns from linear arrays of N equal sources of separation f. The horizontal axis in units of $f \sin\theta / \lambda$ gives the spectral order n of interference. The axis in units of $\sin\theta$ shows that the separation between principal maxima is given by $\sin\theta = \lambda/f$ and the half-width of the principal maximum is given by $\sin\theta = \lambda/Nf$

The directional properties of such a linear array are widely used in both transmitting and receiving aerials and the polar plot for $N = 4$ (Figure 12.22) displays these features. For N large, such an array, used as a receiver, forms the basis of a radio telescope where the receivers (sources) are set at a constant (but adjustable) separation f and tuned to receive a fixed wavelength. Each receiver takes the form of a parabolic reflector, the axes of which are kept parallel as the reflectors are oriented in different directions. The angular separation between the directions of incidence for which the received signal is a maximum is given by $\sin\theta = \lambda/f$.

(Problems 12.6, 12.7)

Diffraction

Diffraction is classified as Fraunhofer or Fresnel. In Fraunhofer diffraction the pattern is formed at such a distance from the diffracting system that the waves generating the pattern may be considered as plane. A Fresnel diffraction pattern is formed so close to the diffracting system that the waves generating the pattern still retain their curved characteristics.

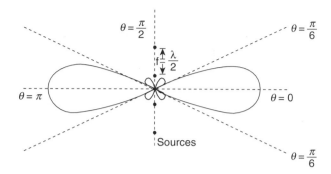

Figure 12.22 Polar plot of the intensity of the interference pattern from a linear array of four sources with common separation $f = \lambda/2$. Note that the half-width of the principal maximum is $\theta = \pi/6$ satisfying the relation $\sin\theta = \lambda/Nf$ and that the separation between principal maxima satisfies the relation that the change in $\sin\theta = \lambda/f$

Fraunhofer Diffraction

The single narrow slit. Earlier in this chapter it was stated that the difference between interference and diffraction is merely one of scale and not of physical behaviour.

Suppose we contract the scale of the N equal sources separation f of Figure 12.20 until the separation between the first and the last source, originally Nf, becomes equal to a distance d where d is now assumed to be the width of a narrow slit on which falls a monochromatic wavefront of wavelength λ where $d \sim \lambda$. Each of the large number N equal sources may now be considered as the origin of secondary wavelets generated across the plane of the slit on the basis of Huygens' Principle to form a system of waves diffracted in all directions.

When these diffracted waves are focused on a screen as shown in Figure 12.23 the intensity distribution of the diffracted waves may be found in terms of the aperture of the slit, the wavelength λ and the angle of diffraction θ. In Figure 12.23 a plane light wave falls normally on the slit aperture of width d and the waves diffracted at an angle θ are brought to focus at a point P on the screen PP_0. The point P is sufficiently distant from the slit for all wavefronts reaching it to be plane and we limit our discussion to *Fraunhofer Diffraction*.

Finding the amplitude of the light at P is the simple problem of superposing all the small contributions from the N equals sources in the plane of the slit, taking into account the phase differences which arise from the variation in path length from P to these different sources. We have already solved this problem several times. In Chapter 10 we took it as an example of the Fourier transform method but here we reapply the result already used in this chapter on p. 364, namely that the intensity at P is given by

$$I = I_s \frac{\sin^2 N\beta}{\sin^2 \beta} \quad \text{where} \quad N\beta = \frac{\pi}{\lambda} Nf \sin\theta$$

is half the phase difference between the contributions from the first and last sources. But now $Nf = d$ the slit width, and if we replace β by α where $\alpha = (\pi/\lambda)\,d\sin\theta$ is now half

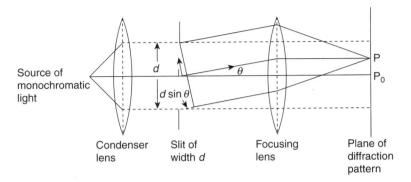

Figure 12.23 A monochromatic wave normally incident on a narrow slit of width d is diffracted through an angle θ and the light in this direction is focused at a point P. The amplitude at P is the superposition of all the secondary waves in the plane of the slit with their appropriate phases. The extreme phase difference from contributing waves at opposite edges of the slit is $\phi = 2\pi d \sin\theta/\lambda = 2\alpha$

the phase difference between the contributions from the opposite edges of the slit, the intensity of the diffracted light at P is given by

$$I = I_s = \frac{\sin^2(\pi/\lambda)d\sin\theta}{\sin^2(\pi/\lambda N)d\sin\theta} = I_s \frac{\sin^2\alpha}{\sin^2(\alpha/N)}$$

For N large

$$\sin^2\frac{\alpha}{N} \rightarrow \left(\frac{\alpha}{N}\right)^2$$

and we have

$$I = N^2 I_s \frac{\sin^2\alpha}{\alpha^2} = I_0 \frac{\sin^2\alpha}{\alpha^2}$$

(recall that in the Fourier Transform derivation on p. 289,

$$I_0 = \frac{d^2 h^2}{4\pi^2}$$

where h was the amplitude from each source).

Plotting $I = I_0(\sin^2\alpha/\alpha^2)$ with $\alpha = (\pi/\lambda)d\sin\theta$ in Figure 12.24 we see that its pattern is symmetrical about the value

$$\alpha = \theta = 0$$

where $I = I_0$ because $\sin\alpha/\alpha \rightarrow 1$ as $\alpha \rightarrow 0$. The intensity $I = 0$ whenever $\sin\alpha = 0$ that is, whenever α is a multiple of π or

$$\alpha = \frac{\pi}{\lambda}d\sin\theta = \pm\pi \ \pm 2\pi \ \pm 3\pi, \text{etc.}$$

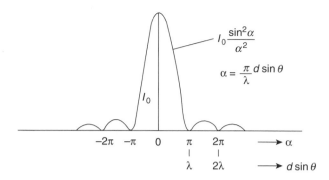

Figure 12.24 Diffraction pattern from a single narrow slit of width d has an intensity $I = I_0 \sin^2 \alpha / \alpha^2$ where $\alpha = \pi d \sin \theta / \lambda$

giving

$$d \sin \theta = \pm \lambda \pm 2\lambda \pm 3\lambda, \text{etc.}$$

This condition for *diffraction minima* is the same as that for *interference maxima* between two slits of separation d, and this is important when we consider the problem of light transmission through more than one slit.

The intensity distribution maxima occur whenever the factor $\sin^2 \alpha / \alpha^2$ has a maximum; that is, when

$$\frac{\mathrm{d}}{\mathrm{d}\alpha}\left(\frac{\sin \alpha}{\alpha}\right)^2 = \frac{\mathrm{d}}{\mathrm{d}\alpha}\left(\frac{\sin \alpha}{\alpha}\right) = 0$$

or

$$\frac{\cos \alpha}{\alpha} - \frac{\sin \alpha}{\alpha^2} = 0$$

This occurs whenever $\alpha = \tan \alpha$, and Figure 12.25 shows that the roots of this equation are closely approximated by $\alpha = \pm 3\pi/2, \pm 5\pi/2$, etc. (see problem at end of chapter on exact values).

Table 12.1 shows the relative intensities of the subsidiary maxima with respect to the principal maximum I_0.

The rapid decrease in intensity as we move from the centre of the pattern explains why only the first two or three subsidiary maxima are normally visible.

Scale of the Intensity Distribution

The width of the principal maximum is governed by the condition $d \sin \theta = \pm \lambda$. A constant wavelength λ means that a decrease in the slit width d will increase the value of $\sin \theta$ and will widen the principal maximum and the separation between subsidiary maxima. The narrower the slit the wider the diffraction pattern; that is, in terms of a Fourier transform the narrower the pulse in x-space the greater the region in k- or wave number space required to represent it.

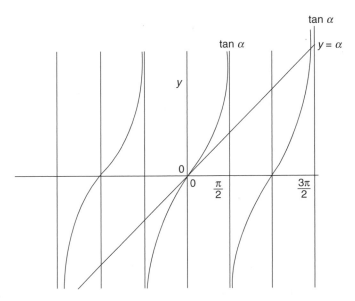

Figure 12.25   Position of principal and subsidiary maxima of single slit diffraction pattern is given by the intersections of $y = \alpha$ and $y = \tan \alpha$

Table 12.1

α	$\dfrac{\sin^2 \alpha}{\alpha^2}$	$\dfrac{I_0 \sin^2 \alpha}{\alpha^2}$
0	1	I_0
$\dfrac{3\pi}{2}$	$\dfrac{4}{9\pi^2}$	$\dfrac{I_0}{22.2}$
$\dfrac{5\pi}{2}$	$\dfrac{4}{25\pi^2}$	$\dfrac{I_0}{61.7}$
$\dfrac{7\pi}{2}$	$\dfrac{4}{49\pi^2}$	$\dfrac{I_0}{121}$

(Problems 12.8, 12.9)

Intensity Distribution for Interference with Diffraction from *N* Identical Slits

The extension of the analysis from the example of one slit to that of N equal slits of width d and common spacing f, Figure 12.26, is very simple.

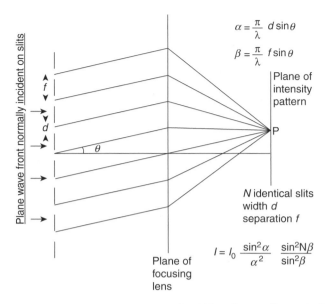

Figure 12.26 Intensity distribution for diffraction by N equal slits is

$$I = I_0 \frac{\sin^2 \alpha}{\alpha^2} \frac{\sin^2 N\beta}{\sin^2 \beta}$$

the product of the diffraction intensity for one slit, $I_0 \sin^2 \alpha / \alpha^2$ and the interference intensity between N sources $\sin^2 N\beta / \sin^2 \beta$, where $\alpha = (\pi/\lambda) d \sin \theta$ and $\beta = (\pi/\lambda) f \sin \theta$

To obtain the expression for the intensity at a point P of diffracted light from a single slit we considered the contributions from the multiple equal sources across the plane of the slit.

We obtained the result

$$I = I_0 \frac{\sin^2 \alpha}{\alpha^2}$$

by contracting the original linear array of N sources of spacing f on p. 364. If we expand the system again to recover the linear array, where each source is *now* a slit giving us the diffraction contribution

$$I_s = I_0 \frac{\sin^2 \alpha}{\alpha^2}$$

we need only insert this value at I_s in the original expression for the interference intensity,

$$I = I_s \frac{\sin^2 N\beta}{\sin^2 \beta}$$

on p. 364 where

$$\beta = \frac{\pi}{\lambda} f \sin\theta$$

to obtain, for the intensity at P in Figure 12.26, the value

$$I = I_0 \frac{\sin^2\alpha}{\alpha^2} \frac{\sin^2 N\beta}{\sin^2\beta},$$

where

$$\alpha = \frac{\pi}{\lambda} d \sin\theta$$

Note that this expression combines the diffraction term $\sin^2\alpha/\alpha^2$ for each slit (source) and the interference term $\sin^2 N\beta/\sin^2\beta$ from N sources (which confirms what we expected from the opening paragraphs on interference). The diffraction pattern for any number of slits will always have an envelope

$$\frac{\sin^2\alpha}{\alpha^2} \quad \text{(single slit diffraction)}$$

modifying the intensity of the multiple slit (source) interference pattern

$$\frac{\sin^2 N\beta}{\sin^2\beta}$$

Fraunhofer Diffraction for Two Equal Slits ($N = 2$)

When $N = 2$ the factor

$$\frac{\sin^2 N\beta}{\sin^2\beta} = 4\cos^2\beta$$

so that the intensity

$$I = 4I_0 \frac{\sin^2\alpha}{\alpha^2} \cos^2\beta$$

the factor 4 arising from N^2 whilst the $\cos^2\beta$ term is familiar from the double source interference discussion. The intensity distribution for $N = 2, f = 2d$, is shown in Figure 12.27. The intensity is zero at the diffraction minima when $d \sin\theta = n\lambda$. It is also zero at the interference minima when $f \sin\theta = (n + \frac{1}{2})\lambda$.

At some value of θ an *interference maximum* occurs for $f \sin\theta = n\lambda$ at the same position as a *diffraction minimum* occurs for $d \sin\theta = m\lambda$.

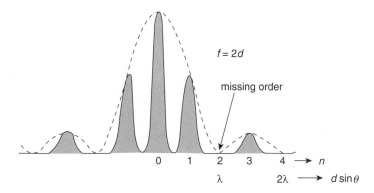

Figure 12.27 Diffraction pattern for two equal slits, showing interference fringes modified by the envelope of a single slit diffraction pattern. Whenever diffraction minima coincide with interference maxima a fringe is suppressed to give a 'missing order' of interference

In this case the diffraction minimum suppresses the interference maximum and the order n of interference is called a *missing order*.

The value of n depends upon the ratio of the slit spacing to the slit width for

$$\frac{n\lambda}{m\lambda} = \frac{f\sin\theta}{d\sin\theta}$$

i.e.

$$\frac{n}{m} = \frac{f}{d} = \frac{\beta}{\alpha}$$

Thus, if

$$\frac{f}{d} = 2$$

the missing orders will be $n = 2, 4, 6, 8$, etc. for $m = 1, 2, 3, 4$, etc.

The ratio

$$\frac{f}{d} = \frac{\beta}{\alpha}$$

governs the scale of the diffraction pattern since this determines the number of interference fringes between diffraction minima and the scale of the diffraction envelope is governed by α.

(Problem 12.10)

Transmission Diffraction Grating (*N* Large)

A large number N of equivalent slits forms a transmission diffraction grating where the common separation f between successive slits is called the *grating space*.

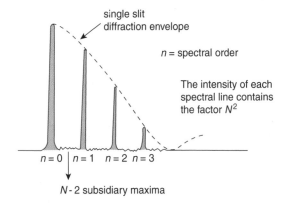

Figure 12.28 Spectral line of a given wavelength produced by a diffraction grating loses intensity with increasing order *n* as it is modified by the single slit diffraction envelope. At the principal maxima each spectral line has an intensity factor N^2 where *N* is the number of lines in the grating

Again, in the expression for the intensity

$$I = I_0 \frac{\sin^2 \alpha}{\alpha^2} \frac{\sin^2 N\beta}{\sin^2 \beta}$$

the pattern lies under the single slit diffraction term (Figure 12.28).

$$\frac{\sin^2 \alpha}{\alpha^2}$$

The principal interference maxima occur at

$$f \sin \theta = n\lambda$$

having the factor N^2 in their intensity and these are observed as *spectral lines* of order *n*. We see, however, that the intensities of the spectral lines of a given wavelength decrease with increasing spectral order because of the modifying $\sin^2 \alpha / \alpha^2$ envelope.

Resolving Power of Diffraction Grating

The importance of the diffraction grating as an optical instrument lies in its ability to resolve the spectral lines of two wavelengths which are too close to be separated by the naked eye. If these two wavelengths are λ and $\lambda + d\lambda$ where $d\lambda/\lambda$ is very small the *Resolving Power* for any optical instrument is given by the ratio $\lambda/d\lambda$.

Two such lines are just resolved, according to Rayleigh's Criterion, when the maximum of one falls upon the first minimum of the other. If the lines are closer than this their separate intensities cannot be distinguished.

If we recall that the spectral lines are the principal maxima of the interference pattern from many slits we may display Rayleigh's Criterion in Figure 12.29 where the nth order spectral lines of the two wavelengths are plotted on an axis measured in units of $\sin\theta$. We have already seen in Figure 12.21 that the half width of the spectral lines (principal maxima) measured in such units is given by λ/Nf where N is now the number of grating lines (slits) and f is the grating space. In Figure 12.29 the nth order of wavelength λ occurs when

$$f\sin\theta = n\lambda$$

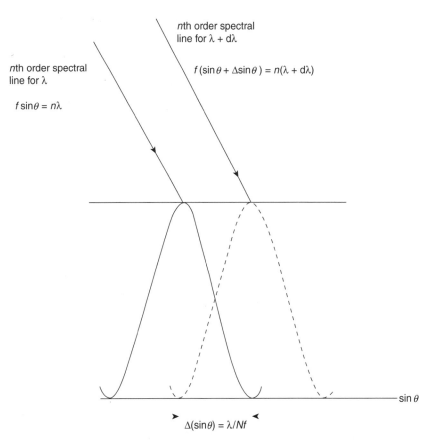

Figure 12.29 Rayleigh's criterion states that the two wavelengths λ and $\lambda + d\lambda$ are just resolved in the nth spectral order when the maximum of one line falls upon the first minimum of the other as shown. This separation, in units of $\sin\theta$, is given by λ/Nf where N is the number of diffraction lines in the grating and f is the grating space. This leads to the result that the resolving power of the grating $\lambda/d\lambda = nN$

whilst the *n*th order for $\lambda + d\lambda$ satisfies the condition

$$f[\sin \theta + \Delta(\sin \theta)] = n(\lambda + d\lambda)$$

so that

$$f\Delta(\sin \theta) = n \, d\lambda$$

Rayleigh's Criterion requires that the fractional change

$$\Delta(\sin \theta) = \frac{\lambda}{Nf}$$

so that

$$f\Delta(\sin \theta) = n \, d\lambda = \frac{\lambda}{N}$$

Hence the Resolving Power of the diffraction grating in the *n*th order is given by

$$\frac{\lambda}{d\lambda} = Nn$$

Note that the Resolving Power increases with the number of grating lines N and the spectral order n. A limitation is placed on the useful range of n by the decrease of intensity with increasing n due to the modifying diffraction envelope

$$\frac{\sin^2 \alpha}{\alpha^2} \quad \text{(Fig. 12.28)}$$

Resolving Power in Terms of the Bandwidth Theorem

A spectral line in the *n*th order is formed when $f \sin \theta = n\lambda$ where $f \sin \theta$ is the path difference between light coming from two successive slits in the grating. The extreme path difference between light coming from opposite ends of the grating of N lines is therefore given by

$$Nf \sin \theta = Nn\lambda$$

and the time difference between signals travelling these extreme paths is

$$\Delta t = \frac{Nn\lambda}{c}$$

where c is the velocity of light.

The light frequency $\nu = c/\lambda$ has a resolvable differential change

$$|\Delta\nu| = c\frac{|\Delta\lambda|}{\lambda^2} = \frac{c}{Nn\lambda}$$

because $\Delta\lambda/\lambda = 1/Nn$ (from the inverse of the Resolving Power).

Hence

$$\Delta\nu = \frac{c}{Nn\lambda} = \frac{1}{\Delta t}$$

or $\Delta\nu\,\Delta t = 1$ (the Bandwidth Theorem).

Thus, the frequency difference which can be resolved is the inverse of the time difference between signals following the extreme paths

$$(\Delta\nu\,\Delta t = 1 \quad \text{is equivalent of course to } \Delta\omega\,\Delta t = 2\pi)$$

If we now write the extreme path difference as

$$Nn\lambda = \Delta x$$

we have, from the inverse of the Resolving Power, that

$$\frac{\Delta\lambda}{\lambda} = \frac{1}{Nn}$$

so

$$\frac{|\Delta\lambda|}{\lambda^2} = \Delta\left(\frac{1}{\lambda}\right) = \frac{\Delta k}{2\pi} = \frac{1}{Nn\lambda} = \frac{1}{\Delta x}$$

where the wave number $k = 2\pi/\lambda$.

Hence we also have

$$\Delta x\,\Delta k = 2\pi$$

where Δk is a measure of the resolvable wavelength difference expressed in terms of the difference Δx between the extreme paths.

On pp. 70 and 71 we discussed the quality factor Q of an oscillatory system. Note that the resolving power may be considered as the Q of an instrument such as the diffraction grating or a Fabry–Perot cavity for

$$\frac{\lambda}{\Delta\lambda} = \left|\frac{\nu}{\Delta\nu}\right| = \frac{\omega}{\Delta\omega} = Q$$

(Problems 12.11, 12.12, 12.13, 12.14)

Fraunhofer Diffraction from a Rectangular Aperture

The value of the Fourier transform method of Chapter 10 becomes apparent when we consider plane wave diffraction from an aperture which is finite in two dimensions.

Although Chapter 10 carried through the transform analysis for the case of only one variable it is equally applicable to functions of more than one variable.

In two dimensions, the function $f(x)$ becomes the function $f(x, y)$, giving a transform $F(k_x, k_y)$ where the subscripts give the directions with which the wave numbers are associated.

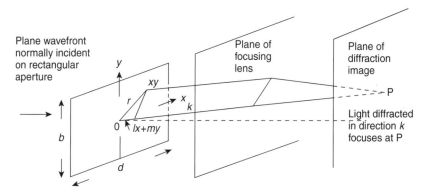

Figure 12.30 Plane waves of monochromatic light incident normally on a rectangular aperture are diffracted in a direction **k**. All light in this direction is brought to focus at P in the image plane. The amplitude at P is the superposition of contributions from all the typical points, x, y in the aperture plane with their appropriate phase relationships

In Figure 12.30 a plane wavefront is diffracted as it passes through the rectangular aperture of dimensions d in the x-direction and b in the y-direction. The vector **k**, which is normal to the diffracted wavefront, has direction cosines l and m with respect to the x- and y-axes respectively. This wavefront is brought to a focus at point P, and the amplitude at P is the superposition of the contributions from all points (x, y) in the aperture with their appropriate phases.

A typical point (x, y) in the aperture may be denoted by the vector **r**; the difference in phase between the contribution from this point and the central point O of the aperture is, of course, $(2\pi/\lambda)$ (path difference). But the path difference is merely the projection of the vector **r** upon the vector **k**, and the phase difference is $\mathbf{k} \cdot \mathbf{r} = (2\pi/\lambda)(lx + my)$, where $lx + my$ is the projection of **r** on **k**.

If we write

$$\frac{2\pi l}{\lambda} = k_x \quad \text{and} \quad \frac{2\pi m}{\lambda} = k_y$$

we have the Fourier transform in two dimensions

$$F(k_x, k_y) = \frac{1}{(2\pi)^2} \int_{-\infty}^{\infty} \int_{-\infty}^{\infty} f(x, y)\, e^{-i(k_x x + k_y y)}\, dx\, dy$$

where $f(x, y)$ is the amplitude of the small contributions from the points in the aperture.

Taking $f(x, y)$ equal to a constant a, we have $F(k_x, k_y)$ the amplitude in k-space at P

$$= \frac{a}{(2\pi)^2} \int_{-d/2}^{+d/2} \int_{-b/2}^{+b/2} e^{-ik_x x}\, e^{-ik_y y}\, dx\, dy$$

$$= \frac{a}{4\pi^2} bd \frac{\sin \alpha}{\alpha} \frac{\sin \beta}{\beta}$$

where

$$\alpha = \frac{\pi l d}{\lambda} = \frac{k_x d}{2}$$

and

$$\beta = \frac{\pi m b}{\lambda} = \frac{k_y b}{2}$$

Physically the integration with respect to y evaluates the contribution of a strip of the aperture along the y direction, and integrating with respect to x then adds the contributions of all these strips with their appropriate phase relationships.

The intensity distribution of the rectangular aperture is given by

$$I = I_0 \frac{\sin^2 \alpha}{\alpha^2} \frac{\sin^2 \beta}{\beta^2}$$

and relative intensities of the subsidiary maxima depend upon the product of the two diffraction terms $\sin^2 \alpha / \alpha^2$ and $\sin^2 \beta / \beta^2$.

These relative values will therefore be numerically equal to the product of any two terms of the series

$$\frac{4}{9\pi^2}, \quad \frac{4}{25\pi^2}, \quad \frac{4}{49\pi^2}, \quad \text{etc.}$$

The diffraction pattern from such an aperture together with a plan showing the relative intensities is given in Figure 12.31.

Fraunhofer Diffraction from a Circular Aperture

Diffraction through a circular aperture presents another two-dimensional problem to which the Fourier transform technique may be applied.

As in the case of the rectangular aperture, the diffracted plane wave propagates in a direction **k** with direction cosines l and m with respect to the x- and y-axes (Figure 12.32a). The circular aperture has a radius r_0 and any point in it is specified by polar coordinates (r, θ) where $x = r \cos \theta$ and $y = r \sin \theta$. This plane wavefront in direction **k** is focused at a point P in the plane of the diffraction pattern and the amplitude at P is the superposition of the contributions from all points (r, θ) in the aperture with their appropriate phase relationships. The phase difference between the contribution from a point defined (x, y) and that from the central point of the aperture is

$$\frac{2\pi}{\lambda} \text{ (path difference)} = \frac{2\pi}{\lambda} (lx + my) = k_x x + k_y y \tag{12.6}$$

as with the rectangular aperture, so that the Fourier transform becomes

$$F(k_x k_y) = \frac{1}{(2\pi)^2} \int_{-\infty}^{\infty} \int_{-\infty}^{\infty} f(x, y) \, e^{-i(k_y x + k_y y)} \, dx \, dy \tag{12.7}$$

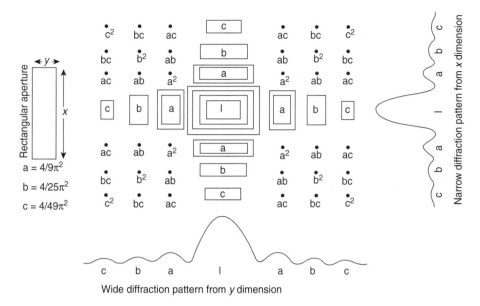

Figure 12.31 The distribution of intensity in the diffraction pattern from a rectangular aperture is seen as the product of two single-slit diffraction patterns, a wide diffraction pattern from the narrow dimension of the slit and a narrow diffraction pattern from the wide dimension of the slit. This 'rotates' the diffraction pattern through 90° with respect to the aperture

If we use polar coordinates, $f(x, y)$ becomes $f(r, \theta)$ and $dx\, dy$ becomes $r\, dr\, d\theta$, where the limits of θ are from 0 to 2π. Moreover, because of the circular symmetry we may simplify the problem. The amplitude or intensity distribution along any radius of the diffraction pattern is sufficient to define the whole of the pattern, and we may choose this single radial direction conveniently by restricting **k** to lie wholly in the xz plane (Figure 12.32b) so that $m = k_y = 0$ and the phase difference is simply

$$\frac{2\pi}{\lambda} lx = k_x x = k_x r \cos \theta$$

Assuming that $f(r, \theta)$ is a constant amplitude a at all points in the circular aperture, the transform becomes

$$F(k_x) = \frac{a}{2\pi} \int_0^{2\pi} d\theta \int_0^{r_0} e^{-ik_x r \cos \theta} r\, dr \qquad (12.8)$$

This can be integrated by parts with respect to r and then term by term in a power series for $\cos \theta$, but the result is well known and conveniently expressed in terms of a *Bessel function* as

$$F(k_x) = \frac{ar_0}{k_x} J_1(k_x r_0)$$

where $J_1(k_x r_0)$ is called a Bessel function of the first order.

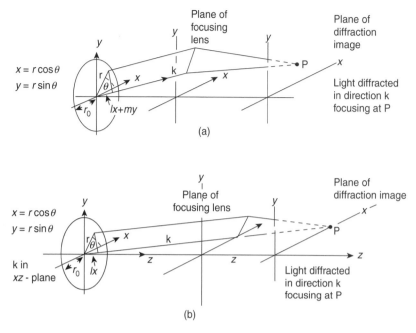

Figure 12.32 (a) A plane monochromatic wave diffracted in a direction **k** from a circular aperture is focused at a point P in the image plane. Contributions from all points *x, y* in the aperture superpose at P with appropriate phase relationships. (b) The direction **k** of (a) is chosen to lie wholly in the *xz*-plane to simplify the analysis. No generality is lost because of circular symmetry. The variation of the amplitude of diffracted light along any one radius determines the complete pattern

Bessel functions are series expansions which are analogous to sine and cosine functions. Where sines and cosines are those functions which satisfy rectangular boundary conditions defined in Cartesian coordinates, Bessel functions satisfy circular or cylindrical boundary conditions requiring polar coordinates.

Standing waves on a circular membrane, e.g. a drum, would require definition in terms of Bessel functions.

The Bessel function of order n is written

$$J_n(x) = \frac{x^n}{2^n n!} \left(1 - \frac{x^2}{2 \cdot 2n + 2} + \frac{x^4}{2 \cdot 4 \cdot 2n + 2 \cdot 2n + 4} \cdots \right)$$

so that

$$J_1(x) = \frac{x}{2} - \frac{x^3}{2^2 4} + \frac{x^5}{2^2 4^2 6} - \frac{x^7}{2^2 4^2 6^2 8}$$

The expression $a^2 r_0^2 [J_1(k_x r_0)/k_x r_0]^2$, which measures the intensity along any radius of the diffraction pattern due to a circular aperture is normalized and plotted in Figure 12.33.

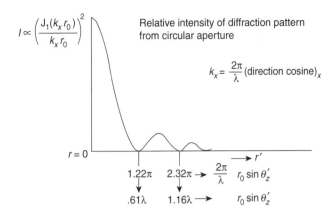

Figure 12.33 Intensity of the diffraction pattern from a circular aperture of radius r_0 versus r', the radius of the pattern. The intensity is proportional to $[J_1(k_x r_0)/k_x r_0]^2$, where J_1 is Bessel's function of order 1. The pattern consists of a central circular principal maximum surrounded by a series of concentric rings of minima and subsidiary maxima of rapidly diminishing intensity

$J_1(k_x r_0)$ has an infinite number of zeros, and the diffraction pattern is formed by an infinite number of light and dark concentric rings. The first dark band will occur at the first zero of $J_1(k_x r_0)$ which is given by $k_x r_0 = 1.219\pi$.

However,

$$k_x r_0 = \frac{2\pi}{\lambda} l r_0 = \frac{2\pi}{\lambda} r_0 \sin \theta_z'$$

where θ_z' is the angle between the vector **k** and the z-axis and defines the angle of diffraction. The first minimum therefore occurs at $r_0 \sin \theta_z' = 0.61\lambda$ and the next minimum at $r_0 \sin \theta_z' = 1.16\lambda$.

If the aperture were square with a side length $2r_0$ (the diameter of the circle) the first dark fringe would be at $r_0 \sin \theta_z' = 0.5\lambda$ and the second at $r_0 \sin \theta_z' = \lambda$.

As the radius of the circular aperture is reduced the value of θ_z' for the first minimum is increased and the whole pattern expands. This reminds us that a reduction of the pulse in x-space requires an increase in wave number or k-space to represent it.

We may write equation (12.8) as

$$F(k_x) = \frac{a}{2\pi} \int_0^{r_o} \int_0^{2\pi} e^{-ik_x \cdot r \cos \theta} r \, dr d\theta$$

where $\int_0^{2\pi} e^{-ik_x \cdot r \cos \theta} d\theta = 2\pi J_0(k_x r)$ and J_0 is the Bessel function of order zero.

Then

$$F(k_x) = a \int_0^{r_o} J_0(k_x r) r dr$$

Now $J_1(k_x r)$ and $J_0(k_x r)rdr$ are related by

$$\int_0^{k_x r_0} J_0(k_x r)k_x rd(k_x r) = k_x r_0 J_1(k_x r_0)$$

giving

$$F(k_x) = a\pi r_0^2 \left[\frac{2J_1(k_x r_0)}{k_x r_0}\right]$$

where r_0 is the radius of the aperture.

The Intensity

$$I = I_0 \left[\frac{J_1(k_x r_0)}{k_x r_0}\right]^2$$

with the curve shown in Figure 12.33.

Fraunhofer Far Field Diffraction

If we remove the focusing lens in Figure 12.32 and leave the aperture open or place the lens within it we have the conditions for far field diffraction, Figure 12.34, where R_0' the distance from $\tilde{O}$ to P' is $\gg$ distances in the aperture and image planes from the optic axis. The aperture is uniformly illuminated by a distant monochromatic source and a small area $d\tilde{s} = d\tilde{x}d\tilde{y}$ in the aperture is $\ll \lambda^2$, where λ is the wavelength.

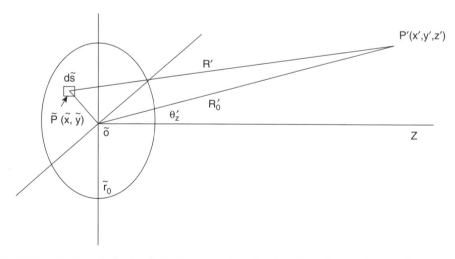

Figure 12.34 In Fraunhofer far field diffraction the distance from the aperture to the image point P' is $\gg$ distances in the aperture and image planes from the optic axis. The electric field at P' is the integral of the spherical waves from small areas $d\tilde{s}$ in the aperture plane and the resulting intensity pattern is that of Figure 12.33. It is known as the Airy disc

The electric field at P' due to the spherical wave from $d\tilde{s}$ is

$$dE_{P'} = \frac{\tilde{E}}{R'} e^{i\omega t - kR'} d\tilde{s}$$

Where $\tilde{E}e^{i\omega t}$ is the field at $d\tilde{s}$
  Now

$$R'^2 = z'^2 + (x' - \tilde{x})^2 + (y' - \tilde{y})^2$$

and

$$R_0'^2 = z'^2 + x'^2 + y'^2$$

which combine to give

$$R' = R_0'[1 + (\tilde{x}^2 + \tilde{y}^2)/R_0'^2 - 2(x'\tilde{x} + y'\tilde{y})/R_0'^2]^{1/2}$$

and $R_0'^2 \gg (\tilde{x}^2 + \tilde{y}^2)$
  so we write

$$R' = R_0'[1 - 2(x'\tilde{x} + y'\tilde{y})/R_0'^2]^{1/2}$$

and if we neglect higher terms

$$R' = R_0'[1 - (x'\tilde{x} + y'\tilde{y})/R_0'^2]$$
$$= R_0' - \frac{x'\tilde{x}}{R_0'} - \frac{y'\tilde{y}}{R_0'}$$

We use this value for R' in the expression for $dE_{p'}$ to give the total field at P' as

$$E_{P'} = \frac{\tilde{E}e^{i\omega t - kR_0'}}{R_0'} \int\int_{\text{aperture}} e^{ik\frac{(x'\tilde{x} + y'\tilde{y})}{R_0'}} d\tilde{s}$$

Comparison with equation (12.6) shows that $k\tilde{x}/R_0' = kl$ and $k\tilde{y}/R_0' = km$ of that equation and proceeding via polar co-ordinates we obtain the same value for the intensity of the diffraction pattern,
  i.e.

$$I = I_0 \left(\frac{J_1(kr_0 \sin \theta_z')}{kr_0 \sin \theta_z'} \right)^2 \qquad \text{in Figure 12.33}$$

This far field diffraction pattern is known as the Airy disc, Figure 12.35, and its size places a limit on the resolving power of a telescope. When the two components of a double star with an angular separation $\Delta\phi$ are viewed through a telescope with an objective lens of focal length l and diameter d their images will appear as two Airy discs separated by the angle $\Delta\phi$. The two diffraction patterns will be resolved if $\Delta\phi$ is much wider than the

Figure 12.35 Photograph of an Airy disc showing the central bright disc, the first dark ring and the first subsidiary maximum. Compare this with Figure 12.33

angluar width of a disc but not if it is much less. Lord Rayleigh's criterion (Figure 12.29) gives the critical angle $\Delta\phi$ for resolution as that when the maximum of one disc falls on the first minimum of the other, Figure 12.36. Figure 12.33 then gives

$$\Delta\phi = \frac{0.61\lambda}{r_0} = \frac{1.22\lambda}{d}$$
$$(\Delta\phi = \Theta'_z \quad \text{in Figure 12.33})$$

where λ is the radiated wavelength.

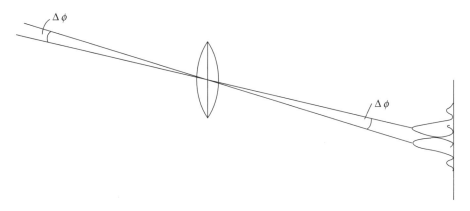

Figure 12.36 Two stars with angular separation $\Delta\phi$ form separate Airy disc images when viewed through a telescope. Rayleigh's criterion (Figure 12.29) states that the these images are resolved when the central maximum of one falls upon the first minimum of the other

This condition is known as diffraction-limited resolution. A poor quality lens will introduce aberrations and will not meet this criterion.

The Michelson Stellar Interferometer

In the discussion on Spatial Coherence (p. 360) we saw that the relative displacement of the interference fringes from separate sources 1 and 2 led to a partial loss of the visibility of the fringes defined as

$$V = \frac{I_{\max} - I_{\min}}{I_{\max} + I_{\min}}$$

and eventually when the displacement was equal to half a fringe width $V = 0$ and there was a complete loss of contrast.

Michelson's Stellar Interferomenter (1920) used this to measure the angular separation between the two components of a double star or, alternatively, the angular width of a single star.

Initially, we take the simplest case to illustrate the principle and then discuss the practical problems which arise. We assume in the first instance that light from the stars is monochromatic with a wavelenght λ_0. Michelson used four mirros $M_1 M_2 M_3 M_4$ mounted on a girder with two slits S_1 and S_2 in front of the lens of an astronomical telescope, Figure 12.37. The slits were perpendicular to the line joining the two stars. The separation h of the outer pair of mirrors ($\sim$meters) was increased until the fringes observed in the focal plane of the objective just disappeared. Assuming zero path difference between $M_1 M_2 \, P_0$ and $M_4 M_3 \, P_0$ the light from star A will form its zero order fringe maximum at P_0 and its first order fringe maximum at P_1, due to a path difference $S_2 N = d \sin \theta = \lambda_0$ so the fringe spacing is determined by d, the separation between the inner mirrors M_2 and M_3.

The condition for fringe disappearance is that rays from star B will form a first order maximum fringe midway between P_0 and P_1, that is, when

$$CM_1 M_2 S_1 P_0 - M_4 M_3 S_2 P_0 = CM_1 = h \sin \phi = \lambda_0/2$$

The condition for fringe disappearnce is therefore determined by h while the angular size of the fringes depends on d so there is an effective magnification of h/d over a fringe system produced by the slits alone.

The angles θ and ϕ are small and the minimum value of h is found which produces $V = 0$ so that the fringes disappear at

$$h\phi = \lambda_0/2 \qquad \text{or} \qquad h = \frac{\lambda}{2\phi}$$

Measurement of h thus determines the double-star angular separation.

Several assumptions have been made in this simple case presentation. First, that the intensities of the light radiated by the stars are equal and that they are coherent soruces. In

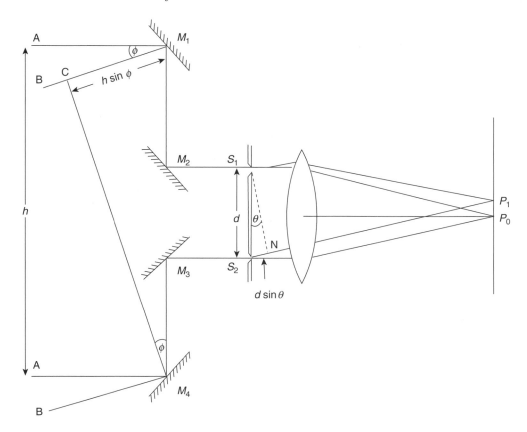

Figure 12.37 In the Michelson stellar interferometer light from stars A and B strike the movable outer mirrors M_1 and M_4 to be reflected via fixed mirrors M_2 and M_3 through two slits S_1 and S_2 and a lens to form interference fringes. Light from Star A forms its zero order fringe at P_0 and its first order fringe at P_1 when $S_2N = d\sin\theta = \lambda_0$. The minimum separation h of M_1M_4 is found for light from B to reduce the fringe visibility to zero, that is, when the path difference $h = \sin\phi = \lambda_0/2$. The angles are so small that θ and ϕ replace their sines. Note that the fringe separation depends on d, but the fringe visibility is governed by h

fact, even if the sources are incoherent their radiation is essentially coherent at the interferometer. Second, the radiation is not monochromatic and only a few fringes around the zero order were visible so λ_0 must be taken as a mean wavelength. Finally, the introduction of a lens into the system inevitably creates Airy discs and the visibility must be expressed in terms of the Airy disc intensity distribution. This results in

$$V = 2\left(\frac{J_1(u)}{u}\right)$$

where

$$u = \pi h\phi/\lambda_0$$

If this visibility is plotted against $h\phi/\lambda_0$ its first zero occurs at 1.22 so the fringes disappear when $h = 1.22\,\lambda_0/\phi$.

In fact, Michelson first used his interferometer in 1920 to measure the angular diameter of the star Betelgeuse the colour of which is orange. His astronomical telescope was the 2.54 m (100 in.) telescope of the Mt. Wilson Observatory. A mean wavelength $\lambda_0 = 570 \times 10^{-9}$ m was used and the fringes vanished when $h = 3.07$ m to give an angular diameter $\phi = 22.6 \times 10^{-8}$ radians or 0.047 arc seconds. The distance of Betelgeuse from the Earth was known and its diameter was calculated to be about 384×10^6 km, roughly 280 times that of the Sun. This magnitude is greater than that of the orbital diameter of Mars around the Sun.

The Convolution Array Theorem

This is a very useful application of the Convolution Theorem when one of the members is the sum of a series of δ functions.

e.g.

$$g(x) = f_1(x) \otimes \sum_m \delta(x - x_m)$$

$$= \int_{-\infty}^{\infty} f_1(x') \sum_m \delta(x - x' - x_m)\,dx'$$

$$= \sum_m f_1(x - x_m)$$

This is a linear addition of functions each of the form $f_1(x)$ but shifted to new origins at $x_m(m = 1, 2, 3 \ldots)$, Figure 12.38.

The convolution theorem gives the Fourier Transform of $g(x)$ as

$$F[g(x)] = F[f_1(x)]F\left[\sum_m \delta(x - x_m)\right]$$

i.e.

$$F(k_x) = F_1[f_1(x)] \sum_m e^{-ik_x x_m}$$

so the transform of the spatially shifted local function is just the product of the transform of the local function and a phase factor.

This is the Array Theorem which we now apply in a more rigorous approach to the effect of diffraction on the interference fringes in Young's slit experiment (p. 358) where the illuminating source is equidistant from both slits.

The Array Theorem may be applied to any combination of identical apertures but Young's experiment involves only the two rectangular (slits) pulses in Figure 12.39a. Here, $f_1(x)$ is a rectangular pulse of width d and the x_m values above are $x_m = \pm a/2$.

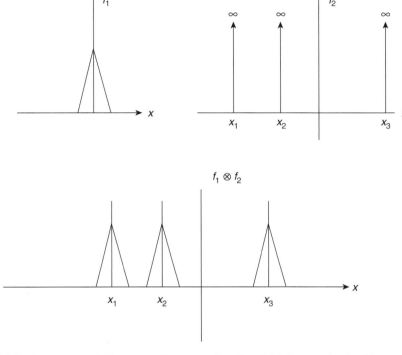

Figure 12.38 In the convolution array theorem a function $f_1(x)$ is convolved with a series of Dirac functions which shift it to new origins

Thus, we have the transform amplitude

$$F(k_x) = F_1(k_x) \sum_m e^{-ik_x x_m}$$

where $k_x = \mathbf{k} \cdot \mathbf{x} = kx \sin \theta$ and $\mathbf{k}$ in Figure 13.39b is the vector direction from $x = -a/2$ to a point P on the diffraction-interference pattern. p. 288 gives

$$F_1(k_x) \propto \frac{\sin \alpha}{\alpha}$$

where

$$\alpha = \frac{\pi}{\lambda} d \sin \theta$$

The second term, a phase factor, is

$$\sum_m e^{-ik_x x_m} = [e^{ik_x a/2} + e^{-ik_x a/2}] = 2 \cos k_x a/2$$

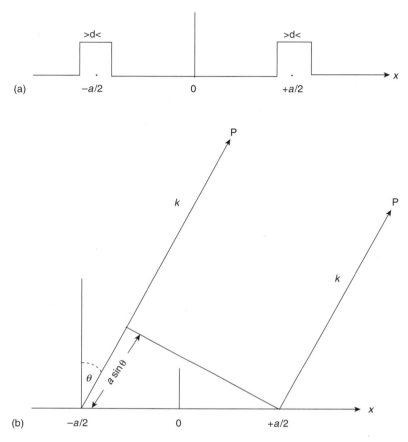

Figure 12.39 Young's double slit experiment represented in convolution array theorem (a) by two reactangular pulses and (b) with a path difference in the direction **k** of $d \sin \phi$ where a is the separation between the pulse centres

We may equate $k_x a/2$ with $\delta/2$ on p. 358 where $\delta = \frac{2\pi}{\lambda}(x_2 - x_1)$ is the phase difference at point P due to the path difference from the two sources. Here, $k_x a/2 = ka \sin \theta/2 = \pi a \sin \theta/\lambda$ (Figure 13.39b). When $\cos k_x a/2 = 1$ for maximum constructive interference

$$ka \sin \theta/2 = \frac{\pi}{\lambda} a \sin \theta = n\pi$$

i.e.

$$a \sin \theta = n\lambda$$

The amplitude squared or intensity is, therefore

$$I \propto \frac{\sin^2 \alpha}{\alpha^2} 4 \cos^2(\delta/2)$$

a cos^2 interference system modulated by a diffraction envelope as shown in Figure 12.27

This method can be extended to produce the pattern for a diffraction grating of N identical slits.

The Optical Transfer Function

The modern method of testing an optical system, e.g. a lens, is to consider the object as a series of Fourier frequency components and to find the response of the system to these frequencies. A test chart with a sinusoidal distribution of intensity would make a suitable object for this purpose. The function of the lens or optical system is considered to be that of a linear operator which transforms a sinusoidal input into an undistorted sinusoidal output.

The linear operator is defined in terms of the Optical Transfer Function (OTF) which may be real or complex. The real part, the Modulation Transfer Function (MTF), measures the effect of the lens on the amplitude of the sinusoidal input; the complex element is the Phase Transfer Function (PTF), a shift in phase when aberrations are present. If there are no aberrations and the effect on the image is limited to diffraction the PTF is zero.

Changing the amplitude of the object frequency components affects the contrast between different parts of the image compared with the corresponding parts of the object. We shall evaluate this effect at the end of the analysis.

We shall assume that the object is space invariant and incoherent. Space invariance means that the only effect of moving a point source over the object is to change the location of the image. When an object is incoherent its intensity or irradiance varies from point to point and all contributions to the final image are added under the integral sign.

Over a small area $dx\,dy$ of the object the radiated flux will be $I_0(x,y)dx\,dy$ and this makes its contribution to the image intensity. In addition, every point source on the object creates a circular diffraction pattern (Airy disc) around the corresponding image point so the resulting intensity of the image at (x',y') will be

$$d\,I'(x',y') = I_0(x,y)O(x,y,x'y')dx\,dy$$

where $O(x,y,x'y')$ is the radially symmetric intensity distribution of the diffraction pattern (Airy disc). In this context it is called the Point Spread Function (PSF).

Adding all contributions gives the image intensity

$$I'(x',y') = \int_{-\infty}^{\infty}\int_{-\infty}^{\infty} I_0(x,y)O(x,y,x'y')dx\,dy$$

If, as we shall assume for simplicity, the magnification is unity, there is a one-to-one correspondence between the point (x,y) on the object and the centre of its diffraction pattern in the image plane. Using (x,y) as the coordinate of this centre the value of $O(x,y,x',y')$ at any other point (x',y') in the diffraction pattern is given by

$$O(x'-x,\ y'-y)$$

Thus, the intensity or irradiance at any image point may be written

$$I'(x', y') = \int_{-\infty}^{\infty} \int_{-\infty}^{\infty} I_0(x, y) O(x' - x, y' - y) \mathrm{d}x \, \mathrm{d}y$$

This is merely the two-dimensional form of the convolution we met on p. 293 and we reduce it to one dimension by writing

$$I'(x') = \int_{-\infty}^{\infty} I_0(x) O(x' - x) \mathrm{d}x = \int_{-\infty}^{\infty} I_0(x' - x) O(x) \mathrm{d}x$$

because the convolution theorem of p. 297 allows us to exchange the variables of the functions under the convolution integral.

This is evidently of the form

$$I' = I_0 \otimes O$$

with Fourier Transforms

$$F(I') = F(I_0) \cdot F(O)$$

The choice of one dimension which adds clarity to the following analysis tranforms the PSF to a Line Spread Function (LSF) by cutting a narrow slice from the three-dimensional PSF. This is achieved by using a line source represented by a Dirac δ function, the sifting property of which isolates an infinitesimally narrow section of the PSF.

The shape of the three-dimensional PSF may be imagined by rotating Figure 12.33 about its vertical axis for a complete revolution. The profile of a slice along the diameter through the centre of the PSF is then the intensity of Figure 12.33 together with its reflection about the vertical axis. Any other slice, not through the centre, will have a similar profile but will differ in some details, e.g. its minimum values will not be zero, Figure 12.40.

Thus, in one dimension, replacing $O(x)$ by $L(x)$ the LSF, we have

$$I'(x') = \int_{-\infty}^{\infty} I_0(x' - x) L(x) \mathrm{d}x$$

or

$$I' = I_0 \otimes L = L \otimes I_0$$

with

$$F(I') = F(I_0) \cdot F(L) = F(L) \cdot F(I_0)$$

Let us write the intensity distribution of an object frequency component in one dimension as $a + b\cos k_x x$, where b modulates the cosine and a is a positive d.c. bias greater than b so

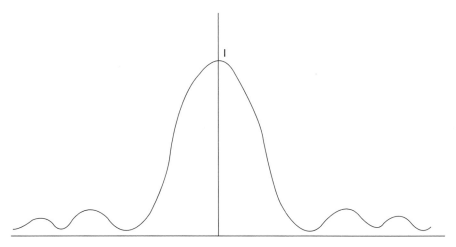

Figure 12.40 The profile of the Line Spread Function $L(x)$ is formed by cutting an off-centre slice from the three-dimensional Point Spread Function: $L(x)$ is the area under the curve. Note that the minimum values of $L(x)$ are non-zero, unlike the curve of Figure 12.33

that the intensity is always positive. Then, in the convolution above

$$I_0 = a + b\cos k_x(x' - x)$$

and the image intensity at x' is

$$I'(x') = \int_{-\infty}^{\infty} [a + b\cos k_x(x' - x)]L(x)\,\mathrm{d}x$$

$$= \int_{-\infty}^{\infty} L(x)[a + b\cos k_x(x' - x)\,\mathrm{d}x$$

We remove the x' terms from the integral by expanding the cosine term to give

$$I'(x') = a\int_{-\infty}^{\infty} L(x)\mathrm{d}x + b\cos k_x x' \int_{-\infty}^{\infty} L(x)\cos k_x x\mathrm{d}x + b\sin k_x x' \int_{-\infty}^{\infty} L(x)\sin k_x x\,\mathrm{d}x$$

$$(12.9)$$

The integrals in the second and third terms on right-hand side of this equation are, repectively, the cosine and sine Fourier transforms from pp. 285, 286.

If we write

$$C(k_x) = \int_{-\infty}^{\infty} L(x)\cos k_x x\mathrm{d}x$$

and

$$S(k_x) = \int_{-\infty}^{\infty} L(x)\sin k_x x\mathrm{d}x$$

we have

$$C(k_x) - i\,S(k_x) = \int_{-\infty}^{\infty} L(x)e^{-ik_xx}dx = F(L_x) = M(k_x)e^{-i\phi(k_x)}$$

where

$$M(k_x) = [C(k_x)^2 + S(k_x)^2]^{1/2}$$

is the MTF and $e^{-i\phi(k_x)}$ is the PTF with

$$\tan\phi = S(k_x)/C(k_x)$$

The OTF is, therefore, the Fourier transform of the LSF.

If the LSF is symmetrical, as in the case of the diffraction pattern, the odd terms in $S(k_x)$ are zero, so the phase change $\phi = 0$ and the OTF is real.

For a given frequency component n we can normalize $L(x)$ to give

$$L_n(x) = \frac{L(x)}{\int_{-\infty}^{\infty} L_n(x)dx} = 1$$

so that equation (12.9) becomes

$$\begin{aligned}I'(x') &= a + M(k_x)b(\cos k_xx'\cos\phi - \sin k_xx'\sin\phi)\\ &= a + M(k_x)b(\cos k_xx' + \phi)\end{aligned}$$

In the absence of aberrations, that is, in the symmetric diffraction limited case, $\phi = 0$. I_0 is shown in Figure 12.41(a) and $I'(x')$ in Figure 12.41(b) where $\phi \neq 0$ due to aberrations.

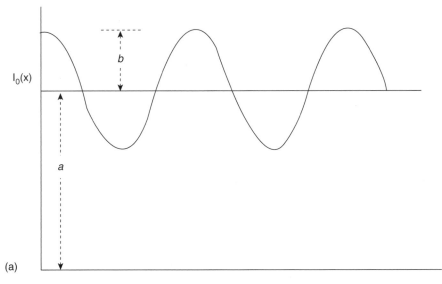

Figure 12.41   (a) The object frequency component $a + b\cos k_xx$ is modified by the Optical Transfer Function

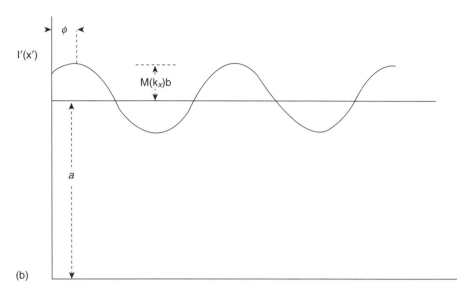

Figure 12.41 (b) In the image component $a + M(k)b\cos(k_xx' + \phi)$, $M(k)$ is the Modulation Transfer Function, which is < 1 and the phase change ϕ results from aberrations. The contrast in the image is less than that in the object. Note that in (b) ϕ is negative in the expression $\cos(k_xx' + \phi)$

The effect of the MTF on the amplitude of the frequency components is to reduce the contrast between parts of the image compared with corresponding parts of the object.

We have already met an expression for the contrast which we called Visibility on p. 360. Thus, we can write

$$\text{Contrast} = \frac{I_{\max} - I_{\min}}{I_{\text{mix}} + I_{\min}} = \frac{(a + b) - (a - b)}{(a + b) + (a - b)} = \frac{b}{a} \qquad \text{for the object}$$

The image contrast $M(k_x)b/a < b/a$ so the image contrast is less than that of the object.

Fresnel Diffraction

The Straight Edge and Slit

Our discussion of Fraunhofer diffraction considered a plane wave normally incident upon a slit in a plane screen so that waves at each point in the plane of the slit were in phase. Each point in the plane became the source of a new wavefront and the superposition of these wavefronts generated a diffraction pattern. At a sufficient distance from the slit the superposed wavefronts were plane and this defined the condition for Fraunhofer diffraction. Its pattern followed from summing the contributions from these waves together with their relative phases and on p. 21 we saw that these formed an arc of constant length. When the contributions were all in phase the arc was a straight line but as the relative phases increased the arc curved to form *closed* circles of decreasing radii. The length of the chord joining the ends of the arc measured the resulting amplitude of the superposition and the square of that length measured the light intensity within the pattern.

Nearer the slit where the superposed wavefronts are not yet plane but retain their curved character the diffraction pattern is that of Fresnel. There is no sharp division between Fresnel and Fraunhofer diffraction, the pattern changes continuously from Fresnel to Fraunhofer as the distance from the slit increases.

The Fresnel pattern is determined by a procedure exactly similar to that in Fraunhofer diffraction, an arc of constant length is obtained but now it convolutes around the arms of a pair of joined spirals, Figure 12.42, and not around closed circles.

An understanding of Fresnel diffraction is most easily gained by first considering, not the slit, but a straight edge formed by covering the lower half of the incident plane wavefront with an infinite plane screen. The undisturbed upper half of the wavefront will contribute one half of the total spiral pattern, that part in the first quadrant.

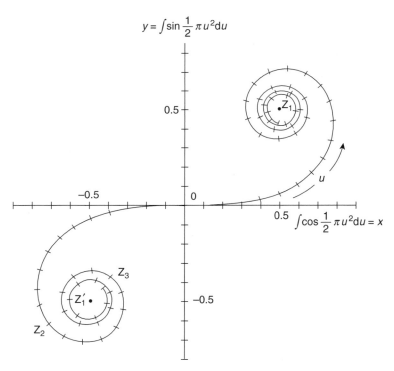

Figure 12.42 Cornu spiral associated with Fresnel diffraction. The spiral in the first quadrant represents the contribution from the upper half of an infinite plane wavefront above an infinite straight edge. The third quadrant spiral results from the downward withdrawal of the straight edge. The width of the wavefront contributing to the diffraction pattern is correlated with the length u along the spiral. The upper half of the wavefront above the straight edge contributes an intensity $(OZ_1)^2$ which is the square of the length of the chord from the origin to the spiral eye. This intensity is 0.25 of the intensity $(Z_1Z_1')^2$ due to the whole wavefront

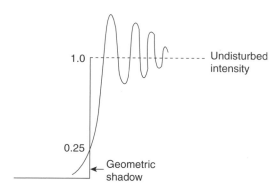

Figure 12.43 Fresnel diffraction pattern from a straight edge. Light is found within the geometric shadow and fringes of varying intensity form the observed pattern. The intensity at the geometric shadow is 0.25 of that due to the undisturbed wavefront

The Fresnel diffraction pattern from a straight edge, Figure 12.43, has several significant features. In the first place light is found beyond the geometric shadow; this confirms its wave nature and requires a Huygens wavelet to contribute to points not directly ahead of it (see the discussion on p. 305). Also, near the edge there are fringes of intensity greater and less than that of the normal undisturbed intensity (taken here as unity). On this scale the intensity at the geometric shadow is exactly 0.25.

To explain the origin of this pattern we consider the point O at the straight edge of Figure 12.44 and the point P directly ahead of O. The line OP defines the geometric shadow. Below O the screen cuts off the wavefront. The phase difference between the contributions to the disturbance at P from O and from a point H, height h above O is given by

$$\Delta(h) = \frac{2\pi}{\lambda}(\text{HP} - \text{OP}) \simeq \frac{2\pi}{\lambda}\frac{1}{2}\frac{h^2}{l}$$

where OP $= l$ and higher powers of h^2/l^2 are neglected.

We now divide the wavefront above O into strips which are parallel to the infinite straight edge and we call these strips 'half period zones'. This name derives from the fact that the width of each strip is chosen so that the contributions to the disturbance at P from the lower and upper edges of a given strip differ in phase by π radians.

Since the phase $\Delta(h) \propto h^2$ we shall not expect these strips or half period zones to be of equal width and Figure 12.45 shows how the width of each strip decreases as h increases. The total contribution from a strip will depend upon its area; that is, upon its width. The amplitude and phase of the contribution at P from a narrow strip of width $\mathrm{d}h$ at a height h above O may be written as $(\mathrm{d}h)\,\mathrm{e}^{\mathrm{i}\Delta}$ where $\Delta = \pi h^2/\lambda l$.

This contribution may be resolved into two perpendicular components

$$\mathrm{d}x = \mathrm{d}h\cos\Delta$$

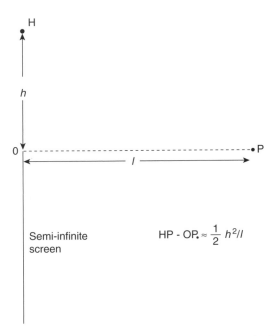

Figure 12.44 Line OP normal to the straight edge defines the geometric shadow. The wavefront at height *h* above 0 makes a contribution to the disturbance at P which has a phase lag of $\pi h^2 / \lambda l$ with respect to that from 0. The total disturbance at P is the vector sum (amplitude and phase) of all contributions from the wavefront section above 0

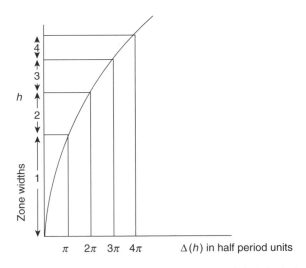

Figure 12.45 Variation of the width of each half period zone with height *h* above the straight edge

and

$$dy = dh \sin \Delta$$

If we now plot the vector sum of these contributions the total disturbance at P from that section of the wavefront measured from O to a height h will have the component values $x = \int dx$ and $y = \int dy$. These integrals are usually expressed in terms of the dimensionless variable $u = h(2/\lambda l)^{1/2}$, the physical significance of which we shall see shortly.

We then have $\Delta = \pi u^2/2$ and $dh = (\lambda l/2)^{1/2} du$ and the integrals become

$$x = \int dx = \int_0^u \cos(\pi u^2/2)\, du$$

and

$$y = \int dy = \int_0^u \sin(\pi u^2/2)\, du$$

These integrals are called Fresnel's Integrals and the locus of the coordinates x and y with variation of u (that is, of h) is the spiral in the first quadrant of Figure 12.42. The complete figure is known as Cornu's spiral.

As h, the width of the contributing wavefront above the straight edge, increases, we measure the increasing length u from 0 along the curve of the spiral in the first quadrant unit, as h and $u \to \infty$ we reach Z_1 the centre of the spiral eye with coordinate $x = \frac{1}{2}, y = \frac{1}{2}$.

The tangent to the spiral curve is

$$\frac{dy}{dx} = \tan \frac{\pi u^2}{2}$$

and this is zero when the phase

$$\Delta(h) = \pi h^2/\lambda l = \pi u^2/2 = m\pi$$

where m is an integer so that $u = \sqrt{(2m)}$ relates u, the distance measured along the spiral to m the number of half period zones contributing to the disturbance at P. The total intensity at P due to all the half period zones above the straight edge is given by the square of the length of the 'chord' OZ_1. This is the intensity at the geometric shadow.

Suppose now that we keep P fixed as we slowly withdraw the screen vertically downwards from O. This begins to reveal contributions to P from the lower part of the wavefront; that is, the part which contributes to the Cornu spiral in the third quadrant. The length u now includes not only the whole of the upper spiral arm but an increasing part of the lower spiral until, when u has extended to Z_2 the 'chord' $Z_1 Z_2$ has its maximum value and this corresponds to the fringe of maximum intensity nearest the straight edge. Further withdrawal of the screen lengthens u to the position Z_3 which corresponds to the first minimum of the fringe pattern and the convolutions of an increasing length u around the spiral eye will produce further intensity oscillations of decreasing magnitude until, with the final removal of the screen, u is now the total length of the spiral and the square of the 'chord' length Z_1Z_1' gives the undisturbed intensity of unit value. But $Z_1Z_1' = 2Z_1O$ so that the undisturbed intensity $(Z_1Z_1')^2$ is a factor of four greater than $(Z_1O)^2$ the intensity at the geometric shadow.

The Fresnel diffraction pattern from a slit may now be seen as that due to a fixed height h of the wavefront equal to that of the slit width. This defines a fixed length u of the spiral between the end points of which the 'chord' is drawn and its length measured and squared to give the intensity. At a given distance from the slit the intensity at a point P in the diffraction pattern will correlate with the precise location of the fixed length u along the spiral. At the centre of the pattern P is symmetric with respect to the upper and lower edges of the slit and the fixed length u is centred about O (Figure 12.46). As P moves across the pattern towards the geometric shadow the length u moves around the convolutions of the spiral. In the geometric shadow this length is located entirely within the first or third quadrant of the spiral and the magnitude of the 'chord' between its ends is less than OZ_1. When the slit is wide enough to produce the central minimum of the diffraction pattern in Figure 12.47 the length u is centred at O with its ends at Z_3 and Z_4 in Figure 12.46.

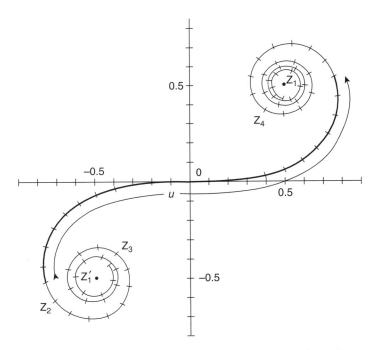

Figure 12.46 The slit width h defines a fixed length u of the spiral. The intensity at a point P in the diffraction pattern is correlated with the precise location of u on the spiral. When P is at the centre of the pattern u is centred on O and moves along the spiral as P moves towards the geometric shadow. Within the geometric shadow the chord joining the ends of u is less than OZ_1

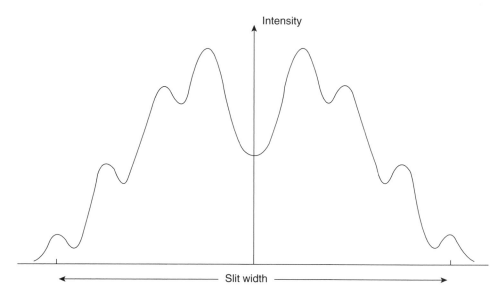

Figure 12.47 Fresnel diffraction pattern from a slit which is wide enough for the spiral length u to be centred at 0 and to end on points Z_3 and Z_4 of Figure 12.46. This produces the intensity minimum at the centre of the pattern

Circular Aperture (Fresnel Diffraction)

In this case the half period zones become annuli of decreasing width. If r_n is the mean radius of the half period zone whose phase lag is $n\pi$ with respect to the contribution from the central ring the path difference in Figure 12.48 is given by

$$\mathrm{NP} - \mathrm{OP} = \Delta = n\lambda/2 = \tfrac{1}{2}r_n^2/l$$

Unlike the rectangular example of the straight edge where the area of the half period zone was proportional to its width dh each zone here has the same area equal to $\pi\lambda l$.

Each zone thus contributes equally to the disturbance at P except for a factor arising from the rigorous Kirchhoff theory which, until now, we have been able to ignore. This is the so-called obliquity factor $\cos\chi$ where χ is shown in the figure. This factor is negligible for small values of n but its effect is to reduce a zone contribution as n increases. A large circular aperture with many zones produces, in the limit, an undisturbed normal intensity on the axis and from Figure 12.49 where we show the magnitude and phase from successive half zones we see that the sum of these vectors which represents the amplitude produced by an undisturbed wave is only half of that from the innermost zone.

It is evident that if alternate zones transmit no light then the contributions from the remaining zones would all be in phase and combine to produce a high intensity at P similar

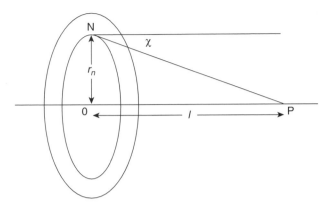

Figure 12.48 Fresnel diffraction from a circular aperture. The mean radius r_n defines the half period zone with a phase lag of $n\pi$ at P with respect to the contribution from the central zone. The obliquity angle χ which reduces the zone contribution at P increases with n

to the focusing effect of a lens. Such circular 'zone plates' are made by blacking out the appropriate areas of a glass slide, Figure 12.50. A further refinement increases the intensity still more. If the alternate zone areas are not blacked out but become areas where the *optical* thickness of the glass is reduced, via etching, by $\lambda/2$ the light transmitted through these zones is advanced in phase by π rad so that the contributions from all the zones are now in phase.

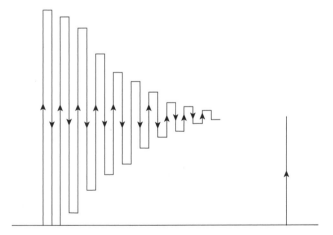

Figure 12.49 The vector contributions from successive zones in the circular aperture. The amplitude produced by an undisturbed wave is seen to be only half of that from the central zone. Removing the contributions from alternate zones leaves the remainder in phase and produces a very high intensity. This is the principle of the zone plate of Figure 12.50

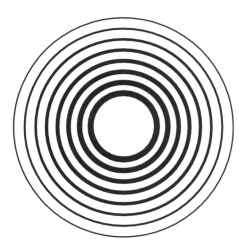

Figure 12.50 Zone plate produced by removing alternate half zones from a circular aperture to leave the remaining contributions in phase

Holography

Why is it that when we observe an object we see it in three dimensions but when we photograph it we obtain only a flat two dimensional distribution of light intensity? The answer is that the photograph has lost the information contained in the *phase* of the incident light. Holographic processes retain this information and a hologram reconstructs a three-dimensional image.

The principle of holography was proposed by Gabor in 1948 but its full development needed the intense beams of laser light. A hologram requires two coherent beams and the holographic plate records their interference pattern. In practice both beams derive from the same source, one serves as a direct reference beam the other is the wavefront scattered from the object.

Figure 12.51 shows one possible arrangement where a partly silvered beam splitter passes the direct reference beam and reflects light on to the object which scatters it on to the photographic plate. Mirrors or deviating prisms are also used to split the incident beam.

In Figure 12.51 let the reference beam amplitude be $A_0\,\mathrm{e}^{i\omega t}$. If the holographic plate lies in the yz plane both the amplitude and phase of scattered light which strikes a given point (y, z) on the plate will depend on these co-ordinates. We simplify the analysis by considering only the y co-ordinate shown in the plane of the paper and we represent the scattered light in amplitude and phase as a function of y, namely

$$A(y)\,\mathrm{e}^{i(\omega t + \phi(y))}$$

It is this information we shall wish to recover.

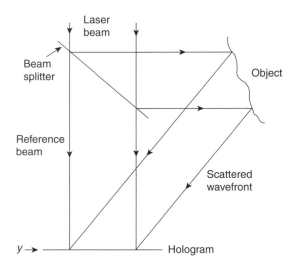

Figure 12.51 The hologram records the interference between two parts of the same laser beam. The original beam is divided by the partially silvered beam splitter to form a direct reference beam and a wavefront scattered from the object. The amplitude and phase information contained in the scattered wavefront must be preserved and recovered

We may now write the resulting amplitude at y (after removing the common $e^{i\omega t}$ factor) as

$$A = A_0 + A(y)\,e^{i\phi(y)}$$

The intensity is therefore

$$
\begin{aligned}
I = AA^* &= [A_0 + A(y)\,e^{i\phi(y)}][A_0 + A(y)\,e^{-i\phi(y)}] \\
&= A_0^2 + A(y)^2 + A_0 A(y)[e^{i\phi(y)} + e^{-i\phi(y)}]
\end{aligned}
$$

The holographic plate records this intensity as shown in Figure 12.52 where the reference intensity A_0^2 is modulated by the terms which contain $A(y)$ and $\phi(y)$, the original scattered amplitude and phase information. This modulation shows of course as contrasting interference fringes whose local intensity is governed by the amplitude $A(y)$ and whose distribution along the y axis is determined by the phase $\phi(y)$. The wavefront scattered by the object is now reconstructed to form the holographic image. This is done by shining the reference beam through the processed hologram which acts as a diffraction grating. The greater the recorded intensity the lower the transmitted amplitude. If the developed photographic emulsion possessed idealized characteristics the relation between the transmitted amplitude of the reference beam and the *exposure* would be linear.

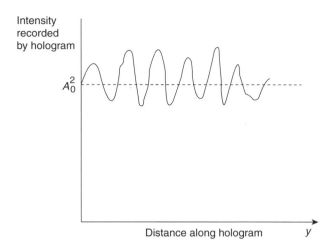

Figure 12.52 Total intensity recorded as a function of y by the holographic plate. The direct reference beam intensity A_0^2 is modulated by information from the scattered wavefront. This shows as variations in the intensity of an interference fringe pattern

Exposure defines the product of incident intensity and exposure time. The curve relating the characteristics for a real holographic emulsion is shown in Figure 12.53 and this is linear only over a limited range near the centre indicated by the dotted lines. This imposes several conditions on practical holography.

In the first place the exposure must be correctly chosen at the value E_C. Secondly, the value of the reference beam intensity A_0^2 must be chosen to produce the correct transmitted amplitude T_0 on the vertical axis of Figure 12.53. This value of T_0 is at the centre of the linear range. Finally, the modulation of A_0^2 by the scattered intensity $A(y)^2$ in Figure 12.53 must be small enough for the transmission of the modulated signal to remain within the linear range of the characteristic curve. Excursions outside this range introduce non-linear distortions by generating extra Fourier frequency components (the situation is similar to that for characteristic curves in electronic amplifiers).

Experimentally this final restriction requires $A(y) \ll A_0$.

Shining the reference beam through the processed hologram produces a transmitted *amplitude*

$$A_0 T = A_0^3 + A_0^2 A(y)\, e^{i\phi(y)} + A_0^2 A(y)\, e^{-i\phi(y)}$$
$$= A_0^2 [A_0 + A(y)\, e^{i\phi(y)} + A(y)\, e^{-i\phi(y)}]$$

where we have neglected the $A(y)^2$ term as $\ll A_0^2$ and have written the negative and positive exponential terms separately. This has a profound physical significance for we see that apart from the common constant factor A_0^2, the observed transmitted beam has three components $A_0, A(y)\, e^{i\phi(y)}$ and $A(y)\, e^{-i\phi(y)}$.

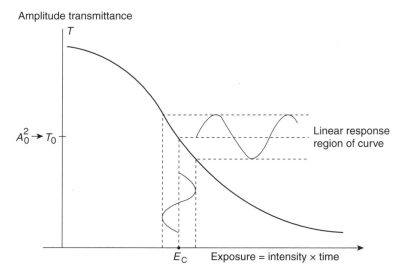

Amplitude transmittance

$A_0^2 \rightarrow T_0$

E_C Exposure = intensity × time

Linear response
region of curve

Figure 12.53 Characteristic curve of a real holographic emulsion (transmittance versus exposure). Only the central linear section of the curve is used. The transmittance T_0 (governed by the reference beam intensity A_0^2) is chosen with the critical exposure E_C to produce the central point on the linear part of the curve. Information from the scattered wavefront must keep the modulations within the linear range for faithful reproduction free from distortion

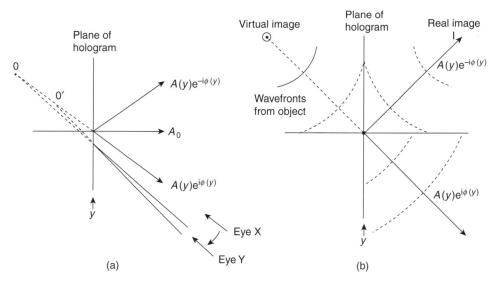

Figure 12.54 (a) Shining the reference beam through the processed hologram produces three components $A_0, A(y)\,e^{i\phi(y)}$ and $A(y)\,e^{-i\phi(y)}$ in the directions shown. Movement of the eye from X to Y about the component $A(y)\,e^{i\phi(y)}$ resolves the separate points 0 and 0' on the image of the object to reveal its three dimensional nature. (b) This image at 0 is seen to be virtual while the image associated with the component $A(y)\,e^{-i\phi(y)}$ is real. This real image is 'phase reversed' and the object appears 'inside out'

The first term, A_0, shows that the incident reference beam has continued beyond the hologram to form the central beam of Figure 12.54a. The second component $A(y)\,e^{i\phi(y)}$ has the same form in amplitude and phase as the original wavefront scattered from the object. As shown in Figure 12.54b it is seen to be a wavefront diverging from a virtual image of the object having the same size and three dimensional distribution as the object itself. Moving the eye across this beam in 12.54a exposes a different section OO' of the virtual image to produce a three dimensional effect.

The third component of the transmitted beam is identical with the second except for the phase reversal; it has a negative exponential index. It forms another image, in this case a real image often referred to as 'pseudoscopic'. It is an image of the original object turned inside out. All contours are reversed, bumps become dents and the closest point on the original object when viewed directly by the observer now becomes the most distant.

Problem 12.1

Suppose that Newton's Rings are formed by the system of Figure 12.4 except that the plano convex lens now rests centrally in a concave surface of radius of curvature R_1 and not on an optical flat. Show that the radius r_n of the nth dark ring is given by

$$r_n^2 = R_1 R_2 n\lambda/(R_1 - R_2)$$

where R_2 is the radius of curvature of the lens and $R_1 > R_2$ (note that R_1 and R_2 have the same sign).

Problem 12.2

Light of wavelength λ in a medium of refractive index n_1 is normally incident on a thin film of refractive index n_2 and optical thickness $\lambda/4$ which coats a plane substrate of refractive index n_3. Show that the film is a perfect anti-reflector $(r = 0)$ if $n_2^2 = n_1 n_3$.

Problem 12.3

Two identical radio masts transmit at a frequency of 1500 kc s^{-1} and are 400 m apart. Show that the intensity of the interference pattern between these radiators is given by $I = 2I_0[1 + \cos(4\pi \sin\theta)]$, where I_0 is the radiated intensity of each. Plot this intensity distribution on a polar diagram in which the masts lie on the $90°$–$270°$ axis to show that there are two major cones of radiation in opposite directions along this axis and 6 minor cones at $0°$, $30°$, $150°$, $180°$, $210°$ and $330°$.

Problem 12.4

(a) Two equal sources radiate a wavelength λ and are separated a distance $\lambda/2$. There is a phase difference $\delta_0 = \pi$ between the signals at source. If the intensity of each source is I_s, show that the intensity of the radiation pattern is given by

$$I = 4I_s \sin^2\left(\frac{\pi}{2}\sin\theta\right)$$

where the sources lie on the axis $\pm\pi/2$.

    Plot I versus θ.

(b) If the sources in (a) are now $\lambda/4$ apart and $\delta_0 = \pi/2$ show that

$$I = 4I_s\left[\cos^2\frac{\pi}{4}(1 + \sin\theta)\right]$$

Plot I versus θ.

Problem 12.5

(a) A large number of identical radiators is arranged in rows and columns to form a lattice of which the unit cell is a square of side d. Show that all the radiation from the lattice in the direction θ will be in phase at a large distance if $\tan \theta = m/n$, where m and n are integers.

(b) If the lattice of section (a) consists of atoms in a crystal where the rows are parallel to the crystal face, show that radiation of wavelength λ incident on the crystal face at a grazing angle of θ is scattered to give interference maxima when $2d \sin \theta = n\lambda$ (Bragg reflection).

Problem 12.6

Show that the separation of equal sources in a linear array producing a principal maximum along the line of the sources ($\theta = \pm \pi/2$) is equal to the wavelength being radiated. Such a pattern is called 'end fire'. Determine the positions (values of θ) of the secondary maxima for $N = 4$ and plot the angular distribution of the intensity.

Problem 12.7

The first multiple radio astronomical interferometer was equivalent to a linear array of $N = 32$ sources (receivers) with a separation $f = 7$ m working at a wavelength $\lambda = 0.21$ m. Show that the angular width of the central maximum is 6 min of arc and that the angular separation between successive principal maxima is $1°42'$.

Problem 12.8

Monochromatic light is normally incident on a single slit, and the intensity of the diffracted light at an angle θ is represented in magnitude and direction by a vector **I**, the tip of which traces a polar diagram. Sketch several polar diagrams to show that as the ratio of slit width to the wavelength gradually increases the polar diagram becomes concentrated along the direction $\theta = 0$.

Problem 12.9

The condition for the maxima of the intensity of light of wavelength λ diffracted by a single slit of width d is given by $\alpha = \tan \alpha$, where $\alpha = \pi d \sin \theta/\lambda$. The approximate values of α which satisfy this equation are $\alpha = 0, +3\pi/2, +5\pi/2$, etc. Writing $\alpha = 3\pi/2 - \delta, 5\pi/2 - \delta$, etc. where δ is small, show that the real solutions for α are $\alpha = 0, \pm 1.43\pi, \pm 2.459\pi, \pm 3.471\pi$, etc.

Problem 12.10

Prove that the intensity of the secondary maximum for a grating of three slits is $\frac{1}{9}$ of that of the principal maximum if only interference effects are considered.

Problem 12.11

A diffraction grating has N slits and a grating space f. If $\beta = \pi f \sin \theta/\lambda$, where θ is the angle of diffraction, calculate the phase change $d\beta$ required to move the diffracted light from the principal maximum to the first minimum to show that the half width of the spectral line produced by the grating is given by $d\theta = (nN \cot \theta)^{-1}$, where n is the spectral order. (For $N = 14,000, n = 1$ and $\theta = 19°$, $d\theta \approx 5$ s of arc.)

Problem 12.12

(a) Dispersion is the separation of spectral lines of different wavelengths by a diffraction grating and increases with the spectral order n. Show that the dispersion of the lines when projected by a lens of focal length F on a screen is given by

$$\frac{dl}{d\lambda} = F \frac{d\theta}{d\lambda} = \frac{nF}{f}$$

for a diffraction angle θ and the nth order, where l is the linear spacing on the screen and f is the grating space.

(b) Show that the change in linear separation per unit increase in spectral order for two wavelengths $\lambda = 5 \times 10^{-7}$ m and $\lambda_2 = 5.2 \times 10^{-7}$ m in a system where $F = 2$ m and $f = 2 \times 10^{-6}$ m is 2×10^{-2} m.

Problem 12.13

(a) A sodium doublet consists of two wavelength $\lambda_1 = 5.890 \times 10^{-7}$ m and $\lambda_2 = 5.896 \times 10^{-7}$ m. Show that the minimum number of lines a grating must have to resolve this doublet in the third spectral order is ≈ 328.

(b) A red spectral line of wavelength $\lambda = 6.5 \times 10^{-7}$ m is observed to be a close doublet. If the two lines are just resolved in the third spectral order by a grating of 9×10^4 lines show that the doublet separation is 2.4×10^{-2} m.

Problem 12.14

Optical instruments have circular apertures, so that the Rayleigh criterion for resolution is given by $\sin \theta = 1.22\lambda/a$, where a is the diameter of the aperture.

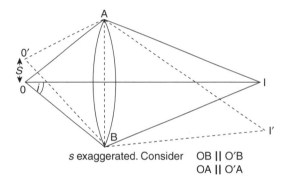

s exaggerated. Consider   OB || O′B
OA || O′A

Two points O and O′ of a specimen in the object plane of a microscope are separated by a distance s. The angle subtended by each at the objective aperture is $2i$ and their images I and I′ are just resolved. By considering the path difference between O′A and O′B show that the separation $s = 1.22\lambda/2 \sin i$.

Summary of Important Results

Interference: Division of Wavefront (Two Equal Sources)

Intensity

$$I = 4I_s \cos^2 \delta/2$$

where

$$I_s = \text{source intensity}$$

and

$$\delta = \left[\frac{2\pi}{\lambda}(\text{path difference})\right] \text{is phase difference}$$

Interference (N Equal Sources – Separation f)

$$I = I_s \frac{\sin^2 N\beta}{\sin^2 \beta} \quad \text{where} \quad \beta = \frac{\pi}{\lambda} f \sin \theta$$

Principal Maxima

$$I = N^2 I_s \quad \text{at } f \sin \theta = n\lambda$$

Fraunhofer Diffraction (Single Slit – Width d)

Intensity

$$I = I_0 \frac{\sin^2 \alpha}{\alpha^2} \quad \text{where} \quad \alpha = \frac{\pi}{\lambda} d \sin \theta$$

Intensity Distribution from N Slits (Width d – Separation f)

$$I = I_0 \frac{\sin^2 \alpha}{\alpha^2} \frac{\sin^2 N\beta}{\sin^2 \beta}$$

(interference pattern modified by single slit diffraction envelope).

Resolving Power of Transmission Grating

$$\frac{\lambda}{d\lambda} = nN$$

where n is spectral order and N is number of grating lines:
Expressible in terms of Bandwidth Theorem as

$$\Delta\nu\Delta t = 1$$

where $\Delta\nu$ is resolvable frequency difference and Δt is the time difference between extreme optical paths.

   Resolving power

$$\frac{\lambda}{\Delta\lambda} = \left|\frac{\nu}{\Delta\nu}\right| = \frac{\omega}{\Delta\omega} = Q$$

where Q is the quality factor of the system.

13

Wave Mechanics

The wave mechanics of Schrödinger (1926) and the equivalent matrix formulation by Heisenberg (1926) are the basis of what is known as 'modern physics'. Without exception they have been successful in replacing or including classical mechanics over the whole range of physics at atomic and molecular levels; these in turn govern the larger scale macroscopic properties. Very high energy phenomena in the physics of elementary particles still, however, present many problems.

In this chapter we shall be concerned only with Schrödinger's wave mechanics and in the way it displays the dual wave–particle nature of matter. This dual nature was first established for electromagnetic radiation but the parallel attempt to establish the wave nature of material particles is the basic history of twentieth century physics.

Origins of Modern Quantum Theory

In the nineteenth century interference and diffraction experiments together with classical electromagnetic theory had confirmed the wave nature of light beyond all doubt but in 1901, in order to explain the experimental curves of black body radiation, Planck postulated that electromagnetic oscillators of frequency ν had discrete energies $nh\nu$ where n was an integer and h was a constant (p. 252). A quarter of a century was to elapse before this was formally derived from the new quantum mechanics.

X-rays had been found by Roentgen in 1895, their wave-like properties were displayed by the diffraction experiments of von Laue in 1912, and their electromagnetic nature was soon proved. A much longer time was required to reconcile a wave nature with the negatively charged particles which J. J. Thomson found in his cathode ray experiments of 1897. It was not until 1927 that interference effects from reflected or scattered electrons were obtained by Davisson and Germer whilst in 1928 G. P. Thomson (the son of J. J.) produced concentric ring diffraction patterns by firing electrons through a thin foil.

In the meantime, in 1906, Einstein had used Planck's idea to explain the photoelectric effect where light falling on a given surface caused electrons to be ejected. Einstein considered the light beam as a stream of individual photons, or quanta of light, each of

The Physics of Vibrations and Waves, 6th Edition   H. J. Pain
© 2005 John Wiley & Sons, Ltd

energy $h\nu$. Collisions between these quanta and electrons in the target material gave the electrons sufficient energy to escape.

In 1912 the alpha particle scattering experiments of Rutherford led to his proposal that the atom consisted of a small positively charged nucleus surrounded by enough negative electrons to leave the atom electrically neutral. This atom was the model used by Bohr and Sommerfeld in their 'old quantum theory', a mixture of classical mechanics and quantum postulates, attempting to explain, amongst other things, the regularity of spectroscopic series from radiating atoms. Electrons were required to orbit the nucleus at definite energy levels (like planets round the Sun), and radiation at a fixed frequency ν was given out when an electron moved from a higher to a lower energy orbit with an energy difference $\Delta E = h\nu$. These orbits were required to be stable or 'stationary' orbits with quantized, that is, allowed values of energy and angular momentum. The fact that classical electromagnetic theory had shown that an accelerating charge (electron in an orbit) was itself a source of radiation remained an unresolved difficulty.

By 1920 Einstein had provided two of the vital tools necessary for further progress (a) that a quantum of radiation has energy $E = h\nu$, and (b) that a particle of momentum $p = mv$ and rest mass m_0 has a relativistic energy E where $E^2 = p^2 c^2 + (m_0 c^2)^2$.

This relation established the equivalence of matter and energy; a stationary particle $v = 0$ has an energy $E = m_0 c^2$ where c is the velocity of light.

The time was now ripe for the final steps leading to the modern quantum theory. The first of these was provided by Compton (1922–23) and the second by de Broglie in 1924.

Compton fired X-rays of a known frequency at a thin foil and observed that the frequency ν of the scattered radiation was independent of the foil material. This implied that the scattering was the result of collisions between X-ray quanta of energy $h\nu$ and the electrons in the target material. In addition to scattering at the incident frequency a lower frequency of scattered radiation was always found which depended only on the mass of the scattering particles (electrons) and the angle of scattering. Compton showed that these results were consistent if momentum and energy were conserved in an elastic collision between two 'particles', the electron and an X-ray of energy $h\nu$, a rest mass $m_0 = 0$ and (from Einstein's relativistic energy equation), a momentum

$$p = \frac{E}{c} = \frac{h\nu}{c} = \frac{h}{\lambda},$$

where $c = \nu\lambda$.

De Broglie in 1924 proposed that if the dual wave-particle nature of electromagnetic fields required a particle momentum of $p = h/\lambda$, it was possible that a wavelength λ of a 'matter' field could be associated with **any** particle of momentum $p = mv$ to give the relation $p = h/\lambda$. His argument was as follows.

If the phase velocity of such a 'matter' wave obeys the usual relation

$$v_{\mathrm{p}} = \nu\lambda$$

where ν is the frequency, the assumption that any particle has a momentum $p = h/\lambda$ together with Einstein's expression $E = h\nu$ yields $v_{\mathrm{p}} = E/p$.

The theory of relativity gives, for a particle of rest mass m_0 and velocity v an energy $E = mc^2$ and a momentum $p = mv$, where

$$m = m_0 \left(1 - \frac{v^2}{c^2} \right)^{-1/2}$$

is the particle mass at velocity v. For such a particle the phase velocity

$$v_p = \frac{E}{p} = \frac{c^2}{v}$$

that is,

$$v v_p = c^2$$

(an expression we met earlier for the wave guides of p. 243).

This gives a phase velocity $v_p > c$ for a particle velocity $v < c$. However, the energy in the de Broglie wave (or particle) travels with the group velocity

$$v_g = \frac{\partial \omega}{\partial k}$$

which, for

$$E = h\nu = \frac{h}{2\pi} \omega$$

and

$$p = \frac{h}{\lambda} = \frac{h}{2\pi} k$$

gives

$$v_g = \frac{\partial \omega}{\partial k} = \frac{\partial E}{\partial p}$$

Such a particle with relativistic energy E where

$$E^2 = p^2 c^2 + (m_0 c^2)^2$$

has

$$2E \frac{\partial E}{\partial p} = 2p c^2$$

or

$$v_{\mathrm{g}} = \frac{\partial E}{\partial p} = \frac{pc^2}{E} = \frac{vc^2}{c^2} = v$$

so that *the group velocity of de Broglie matter wave corresponds to the particle velocity v.*

Even the 'old quantum theory' of Bohr–Sommerfeld gained something from de Broglie's hypothesis. Their postulate that the angular momentum of stationary orbits was restricted to integral (quantum) numbers of the unit angular momentum h was shown, for the circular orbit of radius r, to yield

$$2\pi r p = nh$$

or

$$2\pi r = \frac{nh}{p} = n\lambda$$

so that the circumference of a stationary orbit was a standing wave system and contained an integral number n of λ, the de Broglie wavelength.

Within three years, however, such quantum numbers ceased to be assumptions. They were the natural outcome of the new quantum theory of Schrödinger and Heisenberg.

Heisenberg's Uncertainty Principle

Although, as we shall see, Schrödinger's equation takes the form of a standing wave equation, the fitting of an integral number of de Broglie standing waves around a Bohr orbit presents a fundamental difficulty. The azimuthal symmetry of such a pattern, Figure 13.1,

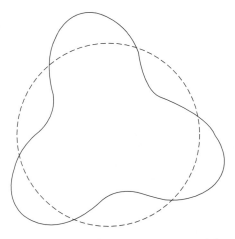

Figure 13.1 Integral number of de Broglie standing waves $\lambda = h/p$ around a circular Bohr orbit does not allow the exact position of the electron to be specified at a particular time

representing an electron in an orbit, does not allow the exact position of the electron to be specified at a particular time. This dilemma was resolved by Heisenberg on the basis of the Bandwidth Theorem we first met on p. 134.

There, a group of waves with a group velocity v_g and a frequency range $\Delta\nu$ superposed effectively only for a time Δt where

$$\Delta\nu\,\Delta t \approx 1$$

Similarly, a group in the wave number range Δk superposed in space over a distance Δx where

$$\Delta x\,\Delta k \approx 2\pi$$

But the velocity of the de Broglie matter wave is essentially a group velocity with a momentum

$$p = \frac{h}{\lambda} = \frac{h}{2\pi}k = \hbar k$$

where

$$\hbar = \frac{h}{2\pi}$$

so

$$\Delta p = \hbar\,\Delta k$$

and the Bandwidth Theorem becomes Heisenberg's Uncertainty Principle

$$\Delta x\,\Delta p \approx h$$

Since

$$E = h\nu = \frac{h}{2\pi}\omega = \hbar\omega$$

it follows that

$$\frac{\Delta E}{\Delta\nu} = \Delta E\,\Delta t \approx h$$

and

$$\Delta E \approx \hbar\Delta\omega$$

are also expressions of Heisenberg's Uncertainty Principle.

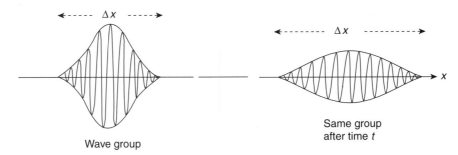

Wave group

Same group
after time t

Figure 13.2   A wave group representing a particle showing dispersion after time t. The square of the wave amplitude at any point represents the probability of the particle being in that position, and the dispersion represents the increasing uncertainty of the particle position with time (Heisenberg's Uncertainty Principle)

This relation sets a fundamental limit on the ultimate precision with which we can know the position x of a particle and the x component of its momentum. If Figure 13.2 shows a wave group representing the particle, the range Δx shows the uncertainty of the position of the particle in the range of space over which it could be found, with the probability of its being at a particular place given by the square of the wave amplitude of that position. The relation

$$\Delta x \, \Delta p \approx h$$

means that the velocity of the particle ($p = mv$) is also uncertain, the more accurate the knowledge of the particle position, the less certain that of the value of its velocity. If the particle is 'observed' at some later time, dispersion of the group will have increased the range Δx and decreased the amplitude. The uncertainty of the position has increased and the probability of its being at any one place has become less. But this is because of the original uncertainty of its velocity, through Δp, which makes an accurate forecast of its position after time t even more unlikely.

The shape of the wave group above is often taken as a Gaussian curve written $\Psi(x, t)$ with a width Δx at $t = 0$ where the value $\Psi(x, t)$ is e^{-1} of its maximum value (see p. 289).

$P(xt)$ defines the probability density of finding the particle at a position Δx, i.e. within the range x and $x + \Delta x$.

The position x and momentum p_x of a particle are conjugate parameters, so the representation of the particle in momentum space $\Phi(p_x, t)$ is the Fourier transform of $\Psi(x, t)$ and $\Phi(p_x, t)$ is also a Gaussian curve with a width Δp_x where $\Phi(p_x t)$ is e^{-1} of its maximum value.

If the group velocity of the wave packet is $v_g = p_0/m$ a rigorous treatment of the time development of these functions leads to the conclusion that $P(xt)$ falls to e^{-1} of its maximum value at the points where

$$x - v_g t = \pm \Delta x$$

where

$$\Delta x(t) = \frac{\hbar}{p_x} \left[1 + \frac{(\Delta p_x)^4}{m^2 \hbar^2} t^2 \right]^{\frac{1}{2}}$$

and hence increases with time.

If the time is sufficiently small so that

$$t \ll t_1 = \frac{m\hbar}{(\Delta p_x)^2}$$

the second term in the bracket is negligible and the wave packet propagates with only a very small change in its width.

As an example, a Gaussian wave packet for an electron localized at time $t = 0$ to within a distance of $10^{-10} m$ (atomic dimensions) with $\Delta p_x = \hbar / \Delta x \approx 10^{-24} \mathrm{kg \cdot m \cdot s^{-1}}$ will have spread to twice its size at time $t = t_1 \sqrt{3} \approx 10^{-16} \mathrm{s}$.

An example of the relation

$$\Delta E \, \Delta t \approx h$$

may be found in considering the time spent by an electron in an atomic orbit. In a stable orbit this time Δt is long and the energy uncertainty ΔE is small so the energy levels of stable orbits are well defined. When an electron changes energy levels and radiation is emitted the time in the orbit may be short and the energy levels ill defined so that the term ΔE contributes to the breadth of a spectral line.

(Problems 13.1, 13.2, 13.3, 13.4, 13.5, 13.6, 13.7, 13.8, 13.9, 13.10)

Schrödinger's Wave Equation

The old quantum theory had sought to establish rules for the existence of discrete frequencies and energy levels. An integral number of de Broglie half wavelengths could be fitted around a circular Bohr orbit. Both of these facts are consistent with the classical standing wave systems we examined in Chapters 5 and 9 when waves travelling between rigid boundaries were perfectly reflected.

In Chapter 5 we saw that the transverse displacement $y(xt)$ of a string of length l with both ends fixed obeys the wave equation

$$\frac{\partial^2 y}{\partial x^2} - \frac{1}{v_p^2} \frac{\partial^2 y}{\partial t^2} = 0$$

where v_p is the wave velocity.

The x and t dependence could be separated in the solution for standing waves to give

$$y(x, t) = A \sin \frac{\omega_n x}{v_p} \sin \omega_n t$$

where n could take the integral values $n = 1, 2, 3$, etc. to give the discrete *eigenfrequencies*,

$$\omega_n = \frac{n\pi v_{\mathrm{p}}}{l}$$

The solution $y(x, t)$ corresponding to a given ω_n is called an *eigenfunction* or a *wave function*.

In developing the Schrödinger wave equation which applies to particle behaviour we use arguments below which in no way constitute a proof because wave mechanics cannot be derived from classical mechanics. Wave mechanics is based on certain postulates the validity of which can be confirmed only by the accuracy of the predicted results.

From the preceding sections we have the representation of a particle as a matter wave with energy $E = \hbar\omega$, momentum $\mathbf{p} = \hbar\mathbf{k}$ and velocity $v_{\mathrm{g}} = \partial\omega/\partial k$.

Wave mechanics uses the notation

$$\Psi(x, t) = \Psi_0\, \mathrm{e}^{-\mathrm{i}(\omega t - kx)} = \Psi_0\, \mathrm{e}^{\mathrm{i}(px - Et)/\hbar}$$

to define the amplitude of a matter wave at a point x at time t. The physical significance of ψ is amplified on p. 422 but for the moment we note the reversed sign of the exponential index which follows the convention used in all books on quantum mechanics. This merely introduces a π rad phase difference from the notation consistently used in the earlier chapters of this book but the new convention will be used throughout this chapter to avoid confusion with other texts and attention will be carefully drawn to any possible ambiguity.

In classical mechanics the total energy E of a particle of mass m and momentum $\mathbf{p}$ in a conservative field of potential V is given by

$$E = p^2/2m + V$$

Differentiating $\Psi(x, t)$ gives

$$\frac{\partial^2}{\partial x^2}\Psi(x, t) = \frac{-p^2}{\hbar^2}\Psi(x, t)$$

and inserting this value of p^2 in the classical energy equation above gives

$$\frac{\hbar^2}{2m}\frac{\partial^2}{\partial x^2}\Psi(x, t) + (E - V)\Psi(x, t) = 0$$

If we now express $\Psi(x, t) = \psi(x)\, \mathrm{e}^{-\mathrm{i}\omega t}$ we may cancel the common $\mathrm{e}^{-\mathrm{i}\omega t}$ factor from the equation above to obtain the *time independent* Schrödinger wave equation

$$\frac{\hbar^2}{2m}\frac{\partial^2}{\partial x^2}\psi(x) + (E - V)\psi(x) = 0$$

This *time independent* wave equation will give states of *constant frequency*; that is, of *constant energy*, and these are the only states we shall consider in this book.

Note that this equation has the same form as the standing wave equation we first met on p. 124.

States which are not of constant energy require the time dependence to be retained in Schrödinger's equation. We do this by using the fact that

$$\frac{\partial}{\partial t}\Psi(x,t) = \frac{-iE}{\hbar}\Psi(x,t)$$

and inserting this value of E in the classical energy equation. This gives the *time dependent* Schrödinger wave equation

$$\frac{-\hbar^2}{2m}\frac{\partial^2}{\partial x^2}\Psi(x,t) + V\Psi(x,t) = i\hbar\frac{\partial}{\partial t}\Psi(x,t).$$

One-dimensional Infinite Potential Well

Consider as a first example the case of a particle constrained to move in a region between $x = 0$ and $x = a$ where the potential $V = 0$. At $x = 0$ and $x = a$ the potential walls are infinitely high as shown in Figure 13.3. This is an idealized form of the potential seen by an electron in the low energy levels near the nucleus of an atom.

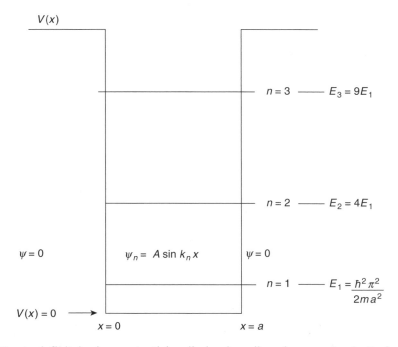

Figure 13.3 An infinitely deep potential well showing allowed energy levels E_n for a particle constrained to move within it with wave function $\psi_n = A\sin k_n x$ where $k_n^2 = 2mE/\hbar^2$ and m is the particle mass

Since $V(x) = 0$ for $0 < x < a$ Schrödinger's equation becomes

$$\frac{\partial^2 \psi(x)}{\partial x^2} + \frac{2mE}{\hbar^2} \psi = 0$$

which may be written, as on p. 124, in the form

$$\frac{\partial^2 \psi}{\partial x^2} + k^2 \psi = 0$$

with

$$k^2 = \frac{2mE}{\hbar^2}$$

The boundary conditions are that $\psi(x) = 0$ at $x = 0$ and $x = a$ where $V(x)$ becomes infinite, whilst the other terms in the equation remain finite. The particle must lie within the well and classically, whatever the value of its energy E it will rebound elastically off the potential 'walls'. When moving to the right the particle behaviour may be represented by a wave function of the form e^{+ikx} which satisfies Schrödinger's equation, and when moving to the left by a wave function of the form

$$e^{-ikx}$$

But, as with the waves on the string, perfect reflection which reverses the amplitude allows $\psi_n(x)$, the solution of Schrödinger's equation, to represent a standing wave system at ω_n; expressed in the form

$$\psi_n(x) = C e^{ik_n x} - C e^{-ik_n x}$$
$$= A \sin k_n x$$

where

$$A = \frac{C}{2i}$$

The boundary condition $\psi_n(x) = 0$ at $x = a$ gives $k_n a = n\pi$ for $n = 1, 2, 3$, etc. i.e. $k_n = n\pi/a$.

Hence

$$k_n^2 = \frac{2mE_n}{\hbar^2} = \frac{n^2 \pi^2}{a^2}$$

giving energy eigenvalues

$$E_n = \frac{n^2 \pi^2 \hbar^2}{a^2 2m}$$

Thus, we see that discrete energy values governed by the quantum number n arise naturally from the application of boundary conditions to the wave function solutions of Schrödinger's equation. Values of the particle momentum are also quantized since

$$p = \frac{h}{\lambda} = \hbar k = \frac{n\pi\hbar}{a}$$

It is evident that in an infinite potential well, an electron or particle cannot have an arbitrary energy but must take only the quantized values E_n. This restriction will hold whenever Schrödinger's equation is solved for a potential $V(x)$ which imposes boundary conditions constraining the particle to move in a limited region.

The wave functions $\psi_n(x)$ for $n = 1, 2, 3$ are plotted in Figure 13.4 showing them to be identical with the allowed amplitude functions for standing waves on a vibrating string with fixed ends. Note that the interval between allowed energy states decreases as either the mass of the particle or the dimensions of the potential box increase relative to h. For particles of large mass and systems of large dimensions the allowed energy states form, for all practical purposes, a continuum and are no longer quantized. Thus, in passing from atomic to much larger dimensions the results of quantum mechanics approach those of classical physics.

We see that the minimum value of the energy of the particle in the potential well is not zero but

$$E_1 = \frac{\hbar^2 \pi^2}{2ma^2}$$

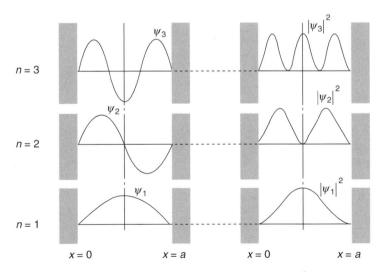

Figure 13.4 Wave functions $\psi_n(x)$ and probability densities $|\psi_n(x)|^2$ for the first three allowed energy levels in an infinitely deep potential well of width a

This minimum energy is related to Heisenberg's Uncertainty Principle

$$\Delta x \, \Delta p \approx h$$

The uncertainty in the position of the particle is obviously $\Delta x = a$ and the particle momentum p may be in either the positive or negative x direction giving an uncertainty

$$\Delta p = 2p$$

Thus

$$\Delta x \, \Delta p = a \cdot 2p \approx h$$

or

$$p \approx \frac{h}{2a}$$

Now, for $V(x) = 0$

$$E = \frac{p^2}{2m} \approx \frac{h^2}{8ma^2} \approx \frac{\hbar^2 \pi^2}{2ma^2}$$

This is an example of the so-called *zero point* energy. We shall meet others.

(Problem 13.11)

Significance of the Amplitude $\psi_n(x)$ of the Wave Function

In Figure 13.4 the amplitude $\psi_n(x)$ of the wave function is plotted for the values $n = 1, 2, 3$ together with the values

$$|\psi_n(x)|^2$$

In the waves we have met so far, the amplitude, or rather the amplitude squared, has been a measure of the intensity of the wave. At a position of high amplitude, the wave was more intense—more energy was localized there. Here we have expressed the motion of a particle confined to a small region of space in terms of its associated matter wave. The amplitude of the wave function $\psi(x)$ varies from point to point within the small region in which the particle is to be found. Outside the infinite well $\psi(x)$ is zero. The intensity of the matter wave is written

$$|\psi(x)|^2 = \psi^*(x)\psi(x)$$

where the complex conjugate $\psi^*(x)$ indicates that $\psi(x)$ may sometimes be complex. Since the matter field describes the motion of the particle we may say that the regions of space in which the particle is more likely to be found are those in which the intensity $|\psi(x)|^2$ is large, or, more formally

'the probability of finding the particle described by the wave function $\psi(x)$ in the interval dx around the point x is $|\psi(x)|^2\,dx$'.

The probability per unit length of finding the particle at x is

$$P(x) = |\psi(x)|^2$$

In three dimensions a wave function would be of the form $\psi(x, y, z)$ and the probability of finding the particle in the unit volume element surrounding the point xyz is

$$P(xyz) = |\psi(xyz)|^2$$

The probability of finding the particle within a finite volume V is obviously

$$P_V = \int_V |\psi(xyz)|^2\,dx\,dy\,dz$$

Now the particle must always be somewhere in space so, in extending the integral over all space, the probability becomes a certainty; that is, it equals unity, or

$$\int_{\text{all space}} |\psi(xyz)|^2\,dx\,dy\,dz = 1$$

This process of integrating over all possible locations to give unity is called *normalization* and it always imposes restrictions on the form of $\psi(x, y, z)$ which must tend to zero as x, y or z tends to infinity.

Normalization determines the value of the constant A in our wave function

$$\psi_n(x) = A \sin \frac{n\pi x}{a}$$

for the case of the infinite potential well.

There

$$\int_{-\infty}^{\infty} |\psi_n(x)|^2\,dx = \int_0^a |\psi_n(x)|^2\,dx$$

$$= A^2 \int_0^a \sin^2 \frac{n\pi x}{a}\,dx = A^2 \frac{a}{2} = 1$$

Hence

$$A = \sqrt{\frac{2}{a}}$$

and the *normalized* wave function

$$\psi_n(x) = \sqrt{\frac{2}{a}} \sin \frac{n\pi x}{a}$$

(Problem 13.12)

Particle in a Three-dimensional Box

Suppose the particle is confined to a rectangular volume abc at the bottom of an infinitely deep potential well $(V = 0)$ where a, b and c are the lengths of the sides of the rectangular box.

The energy of the particle is then

$$E = \frac{p^2}{2m} = \frac{1}{2m}(p_x^2 + p_y^2 + p_z^2)$$

where the momentum components are

$$p_x = n_1 \frac{\pi\hbar}{a}$$
$$p_y = n_2 \frac{\pi\hbar}{b}$$
$$p_z = n_3 \frac{\pi\hbar}{c}$$

and n_1, n_2 and n_3 are integers.

The energy levels allowed in the box are therefore given by

$$E = \frac{\pi^2 \hbar^2}{2m} \left(\frac{n_1^2}{a^2} + \frac{n_2^2}{b^2} + \frac{n_3^2}{c^2} \right)$$

and solutions for the space part of the wave function may be written

$$\psi(x, y, z) = A \sin \frac{n_1 \pi x}{a} \sin \frac{n_2 \pi y}{b} \sin \frac{n_3 \pi z}{c}$$

in accordance with the three-dimensional normal mode solution of p. 249.

If the box is cubical so that $a = b = c$ the allowed energy levels become

$$E = \frac{\pi^2 \hbar^2}{2ma^2}(n_1^2 + n_2^2 + n_3^2) = \frac{\pi^2 \hbar^2}{2ma^2} k^2$$

where $k^2 = n_1^2 + n_2^2 + n_3^2$ with wave functions

$$\psi(xyz) = A \sin \frac{n_1 \pi x}{a} \sin \frac{n_2 \pi y}{a} \sin \frac{n_3 \pi z}{a}$$

Table 13.1

Energy	n_1, n_2, n_3 Combinations	Degeneracy
$3E_1$	(1, 1, 1)	1
$6E_1$	(2, 1, 1) (1, 2, 1) (1, 1, 2)	3
$9E_1$	(2, 2, 1) (2, 1, 2) (1, 2, 2)	3
$11E_1$	(3, 1, 1) (1, 3, 1) (1, 1, 3)	3
$12E_1$	(2, 2, 2)	1
$14E_1$	(1, 2, 3) (3, 2, 1) (2, 3, 1) (1, 3, 2) (2, 1, 3) (3, 1, 2)	6

We saw, however, on p. 250 that combinations of different n values can give the same k value; that is, the same energy value. When n_1, n_2 and n_3 are permuted without changing the k value, the wave function is also changed so that a certain energy level may be associated with several different wave functions or dynamical states. The energy level is said to be *degenerate*, the *order of degeneracy* being defined by the number of different or independent wave functions associated with the given energy.

In the case of the cubic potential box, the lowest energy level is $3E_1$, i.e.

$$(n_1 = n_2 = n_3 = 1)$$

where

$$E_1 = \frac{\pi^2 \hbar^2}{2ma^2}$$

The next energy level is given by $6E_1$, with a degeneracy of 3 where the n values are given by (2, 1, 1) (1, 2, 1) and (1, 1, 2). Higher energy values with degeneracy orders are shown in Table 13.1 above.

(Problem 13.13)

Number of Energy States in Interval *E* to *E* + d*E*

As long as the dimensions of the cubical box above are small the energy levels remain distinct. However, when the volume increases, as is the case for free electrons in a metal, successive energy levels become so close that an almost continuous spectrum is formed.

If we wish to find how many energy levels may be contained in the small energy range dE when the potential box is very large, we have only to apply the result of p. 251 where we found that the number of possible normal modes of oscillation *per unit volume* of an enclosure in the frequency range ν to $\nu + d\nu$ is given by

$$dn = \frac{4\pi \nu^2 d\nu}{c^3}$$

There we stressed that the result was independent of any particular system and we applied it to Planck's Radiation Law and Debye's Theory of Specific Heats. Here we use it with

$$E = \frac{p^2}{2m} = h\nu \quad \text{and} \quad p = \frac{E}{c} = \frac{h\nu}{c}$$

$\Big($ so that

$$dE = \frac{p}{m}\,dp = h\,d\nu$$

and

$$dp = \frac{h\,d\nu}{c}\Big)$$

to give the number of states *per unit volume* in the energy interval dE as

$$dn(E) = \frac{4\pi(2m^3)^{1/2}E^{1/2}}{h^3}\,dE$$

This may be applied directly to determine how free electrons in a metal may distribute themselves in a band of energies from zero to some value E. Each energy level can accommodate two electrons (with opposing spins) according to Pauli's Principle so the total number of electrons *per unit volume* in the energy range zero to E is

$$n = \int dn(E) = \frac{2 \cdot 4\pi(2m_e^3)^{1/2}}{h^3} \int_0^E E^{1/2}\,dE$$
$$= \frac{16\pi(2m_e^3)^{1/2}}{3h^3}\,E^{3/2}$$

where m_e is the electron mass.

If the metal is in its ground state the available electrons will occupy the lowest possible energy levels, and if the total number of electrons *per unit volume* n_0 is less than the total number of energy levels in the band, then the electrons will occupy all energy states up to a maximum energy E_F called the Fermi Energy which is given by

$$n_0 = \frac{16\pi(2m_e^3)^{1/2}\,E_F^{3/2}}{3h^3}$$

Typical values of E_F are of the order of 5 eV $(1\,\text{eV} = 1.6 \times 10^{-19}\,\text{J})$.

(Problems 13.14, 13.15)

The Potential Step

The standing wave system of the infinite potential well where the wave function

$$\psi_n(x)$$

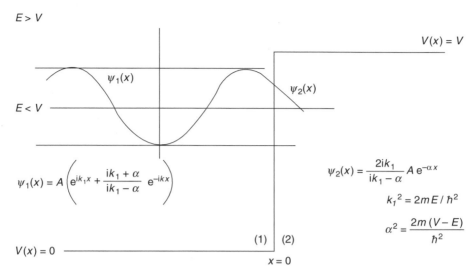

Figure 13.5 Wave functions $\psi_1(x)$ and $\psi_2(x)$ for a particle mass m, energy $E < V$ at a potential step $V(x) = V$

is finite in the region $V(x) = 0$ but zero at all other points is unique in the formal correspondence it presents between classical and quantum mechanical results. The quantum effects become evident when we consider the general case of the potential step of finite height V in Figure 13.5 which is an idealized form of the very steep potential gradient of a conservative force

$$F(x) = -\frac{\partial V}{\partial x}$$

Such a potential step would be seen by a free electron near the surface of a metal.

It is necessary to consider separately the two cases where the total particle energy E is (a) less than the potential energy V, and (b) greater than V, where

$$E = \frac{p^2}{2m} + V(x)$$

(a) $E < V$

When E is less than V, the region $x > 0$ of Figure 13.5 is forbidden to the particle by classical mechanics for the kinetic energy

$$\frac{p^2}{2m}$$

would then have a negative value.

In finding the complete solution for $\psi(x)$ for the potential step we must solve Schrödinger's equation for the separate regions of Figure 13.5, $x < 0$ (region 1) and $x > 0$ (region 2).

In region 1, $V(x) = 0$ and we have

$$\frac{\partial^2 \psi_1(x)}{\partial x^2} + \frac{2mE}{\hbar^2} \psi_1(x) = 0$$

with a solution

$$\psi_1(x) = A\,e^{ik_1x} + B\,e^{-ik_1x}$$

where

$$k_1^2 = \frac{2mE}{\hbar^2}$$

The term $A\,e^{ik_1x}$ (with the sign convention of this chapter) is the wave representation of an incident particle moving to the right, and $B\,e^{-ik_1x}$ represents a reflected particle moving to the left.

In region 2, $V(x) = V$ and Schrödinger's equation becomes

$$\frac{\partial^2 \psi_2(x)}{\partial x^2} + \frac{2m(E-V)}{\hbar^2} \psi_2(x) = 0$$

or

$$\frac{\partial^2 \psi_2(x)}{\partial x^2} - \alpha^2 \psi_2(x) = 0$$

where

$$\alpha^2 = \frac{2m(V-E)}{\hbar^2}$$

This equation has the solution

$$\psi_2(x) = C\,e^{-\alpha x} + D\,e^{\alpha x}$$

Now the probability of finding the particle in region 2 where it is classically forbidden depends on the square of the wave function amplitude $|\psi_2(x)|^2$ with the condition that for any wave function to be normalized $\left(\text{i.e. for}\right.$

$$\int |\psi_2(x)|^2\,dx = 1 \Bigg)$$

the wave function $\psi_2(x) \to 0$ as $x \to \infty$.

This forbids the second term $D\,e^{\alpha x}$ which increases with x but still leaves

$$\psi_2(x) = C\,e^{-\alpha x}$$

to give a finite probability of finding the particle beyond the potential step, a probability which decreases exponentially with distance. This is a profound departure from classical behaviour.

At the boundary $x = 0$, $\psi(x)$ must be finite to give a finite probability of finding the particle there, but there is a finite discontinuity in $V(x)$. In these circumstances Schrödinger's equation asserts that the second derivative

$$\frac{\partial^2 \psi(x)}{\partial x^2}$$

at $x = 0$ is finite, which means that both $\psi(x)$ and $(\partial \psi(x)/\partial x)$ are continuous at $x = 0$.

These are the boundary conditions which allow the separate solutions

$$\psi_1(x) \quad \text{and} \quad \psi_2(x)$$

for the wave function, to be matched across the boundary of the two regions.

The continuity of $\psi(x)$ at $x = 0$ gives $\psi_1(x) = \psi_2(x)$ or $A + B = C$ whilst

$$\frac{\partial \psi_1(x)}{\partial x} = \frac{\partial \psi_2(x)}{\partial x}$$

at $x = 0$ gives

$$ik_1(A - B) = -\alpha C = -\alpha(A + B)$$

Thus

$$B = \left(\frac{ik_1 + \alpha}{ik_1 - \alpha}\right) A$$

and

$$C = \frac{2ik_1}{ik_1 - \alpha} A$$

The wave functions for the separate regions then become

$$\psi_1(x) = A\left(e^{ik_1 x} + \frac{ik_1 + \alpha}{ik_1 - \alpha} e^{-ik_1 x}\right)$$

and

$$\psi_2(x) = \frac{2ik_1}{ik_1 - \alpha} A e^{-\alpha x}$$

and these are shown in Figure 13.5. Note particularly that the intensity of the incident part of the wave function

$$|\psi_1(x)|^2 = |A|^2$$

whilst the reflected intensity is

$$|B|^2 = \left| \frac{ik_1 + \alpha}{ik_1 - \alpha} A \right|^2 = |A|^2$$

Thus, for any energy $E < V$ we have total reflection as in the classical case, even for those particles which penetrate the classically forbidden region $x > 0$ where $\psi_2(x)$ is finite.

In region 2 the probability of finding the particle is

$$P(x) = |\psi_2(x)|^2 = |C e^{-\alpha x}|^2$$
$$= \left| \frac{2ik_1}{ik_1 - \alpha} A e^{-\alpha x} \right|^2 = \frac{4k_1^2}{k_1^2 + \alpha^2} A^2 e^{-2\alpha x}$$

Since the exponential coefficient α depends on $V(x)$ the greater the value $V(x)$ the faster the wave function $\psi_2(x)$ goes to zero in region 2 for a given total energy $E < V$.

When $V(x) \to \infty$, as in the case of the infinite potential well, $\psi_2(x)$ becomes zero, as we have seen; and there is no penetration into the classically forbidden region.

Several important physical phenomena may be explained on the assumption that a particle with $E < V$ meeting a potential step of finite height V and *finite width b* has a wave function $\psi_2(x)$ which is still finite at $x = b$, making it possible for the particle to tunnel through the potential barrier (Figure 13.6). The probability that the particle will penetrate the barrier to $x = b$ is given by

$$P(x) = |\psi_2(x)|^2 \propto e^{-2\alpha x}$$

and beyond this barrier the particle will propagate in region 3 with a wave function $\psi_3(x)$ of reduced amplitude. The boundary conditions must then be applied at $x = b$ to match $\psi_2(x)$ to $\psi_3(x)$.

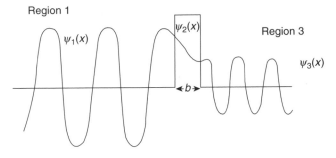

Figure 13.6 Narrow potential barrier of width b penetrated by a particle represented by $\psi_1(x)$ leaving a finite amplitude $\psi_3(x)$ as a measure of the reduced probability of finding the particle in region 3

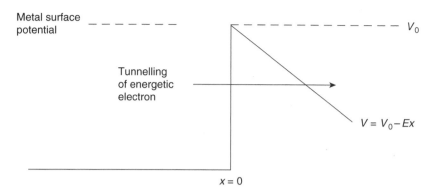

Figure 13.7 Application of an electric field E to the surface of a metal at potential V_0 reduces the potential to $V = V_0 - Ex$ forming a barrier of finite width which may be penetrated by an energetic electron near the metal surface

This quantum 'tunnel effect' is the basis of the explanation of the radioactive decay of the nucleus. In addition the potential step seen by a free electron near the surface of a metal may be distorted, as shown in Figure 13.7, by the application of an external electric field, to form a barrier of finite width. The most energetic electrons near the surface of the metal can leak through the barrier in a process known as *field electron* emission.

Another example results from the two possible positions of the single nitrogen atom with respect to the three hydrogen atoms in the ammonia molecule NH_3. These positions are shown as N and N' in Figure 13.8 together with the potential barrier presented to the nitrogen atom as it moves to and fro between N and N'. This penetration occurs at a frequency of 2.3786×10^{10} Hz for the ground state of NH_3 and its high definition is used as an atomic clock to fix standards of time.

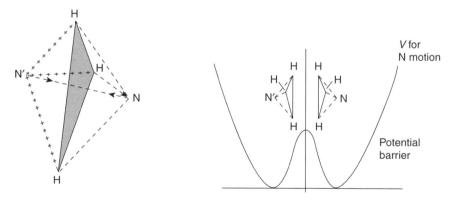

Figure 13.8 The two possible configurations N and N' of the nitrogen atom with respect to the triangular hydrogen base in the ammonia molecule NH_3 and the finite potential barrier penetrated by the nitrogen atom at a frequency of $>10^{10}$ Hz in the NH_3 ground state

(Problem 13.16)

(b) $E > V$

In the region $x < 0$ in Figure 13.5 $V(x) = 0$ and Schrödinger's equation is

$$\frac{\partial^2 \psi_1(x)}{\partial x^2} + \frac{2mE}{\hbar^2} \psi_1(x) = 0$$

or

$$\frac{\partial^2 \psi_1}{\partial x^2} + k_1^2 \psi_1 = 0$$

with

$$k_1^2 = \frac{2mE}{\hbar^2}$$

having a solution

$$\psi_1(x) = A\,e^{ik_1 x} + B\,e^{-ik_1 x}$$

with both incident and reflected terms.

The momentum of the particle is p_1 where $p_1^2/2m = E$.

In the region $x > 0$, $V(x) = V$ and Schrödinger's equation is

$$\frac{\partial^2 \psi_2(x)}{\partial x^2} + \frac{2m(E - V)}{\hbar^2} \psi_2(x) = 0$$

or

$$\frac{\partial^2 \psi_2}{\partial x^2} + k_2^2 \psi_2 = 0$$

where

$$k_2^2 = \frac{2m(E - V)}{\hbar^2}$$

and the particle momentum p_2 is given by $p_2^2/2m = (E - V)$.

In the wave function solution for this region we consider only the right-going or transmitted term since there is nothing beyond $x = 0$ to cause a reflection, so

$$\psi_2(x) = C\,e^{ik_2 x}$$

Now the wave number k is related to the de Broglie wavelength of the particle and we see that k changes when the potential V changes; that is, when the particle experiences a change in the force acting on it. Such a particle therefore reacts to a changing potential as light reacts to changing refractive index. As the potential V increases for $E > V$ the momentum p and wave number $k(p = \hbar k)$ decrease and the wavelength λ increases.

At $x = 0$ the conditions for continuity give

$$\psi_1(x) = \psi_2(x)$$

or

$$A + B = C$$

and

$$\frac{\partial \psi_1(x)}{\partial x} = \frac{\partial \psi_2(x)}{\partial x}$$

or

$$k_1(A - B) = k_2 C$$

These two equations give

$$B = \frac{(k_1 - k_2)}{(k_1 + k_2)} A$$

and

$$C = \frac{2k_1}{k_1 + k_2} A$$

Since B is not zero, some reflection takes place at $x = 0$ even though the energy $E > V$. This is clearly not classical behaviour. If many particles form an incident beam at $x = 0$ and each particle has velocity

$$v_1 = \frac{p_1}{m} = \frac{\hbar k_1}{m}$$

then the velocity of transmitted particles will be

$$v_2 = \frac{p_2}{m} = \frac{\hbar k_2}{m}$$

The incident flux of particles; that is, the number crossing unit area per unit time, may be seen as the product of the velocity and the intensity; that is

$$v_1 |A|^2$$

The reflected flux is

$$v_1|B|^2$$

and the transmitted flux is

$$v_2|C|^2$$

Thus, the reflection coefficient, the ratio of reflected to incident flux is

$$R = \frac{v_1|B|^2}{v_1|A|^2} = \frac{(k_1 - k_2)^2}{(k_1 + k_2)^2}$$

and the transmission coefficient, the ratio of transmitted to incident flux is

$$T = \frac{v_2|C|^2}{v_1|A|^2} = \frac{k_2}{k_1}\frac{(2k_1)^2}{(k_1 + k_2)^2} = \frac{4k_1k_2}{(k_1 + k_2)^2}$$

results which are similar to those for our classical waves in earlier chapters.

Note that $R + T = 1$ showing that the number of particles is conserved.

We have chosen here to apply R and T to a number of particles forming a beam. These coefficients, when applied to identical particles forming the beam, measure the average probability that an individual particle will be reflected or transmitted.

(Problem 13.17)

The Square Potential Well

Let us consider a particle with energy $E < V$ moving in the square potential well of width a in Figure 13.9. Within the well the potential is zero, and the value V of the height of the well

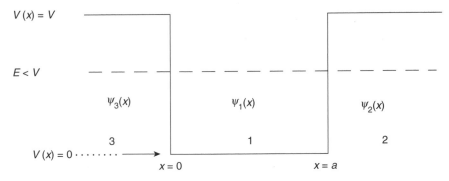

Figure 13.9 A particle with energy $E < V$ (V = the finite height of a square potential well of width a) may take only the energy values E satisfying the equation

$$\tan a\sqrt{\frac{2mE}{\hbar^2}} = \frac{2\sqrt{E(V - E)}}{2E - V}$$

The wave functions in the three regions are matched at the boundaries $x = 0$ and $x = a$ by the conditions that $\psi(x)$ and $\partial\psi(x)/\partial x$ are continuous

is finite. This potential approximates that of a finite range force which has no influence beyond a limited distance. Outside the range of the force the potential may be considered constant. From our discussion of the infinitely deep potential well $(V = \infty)$ and of the potential step we can expect our wave function representation to have the form of an integral number of de Broglie half wavelengths within the well, plus an exponentially decaying penetration into the wall on either side.

Writing Schrödinger's equation for each of the three regions, we have for region $1 (0 < x \leq a)$

$$\frac{\partial^2 \psi_1(x)}{\partial x^2} + \frac{2mE}{\hbar^2} \psi_1(x) = 0$$

with a solution, for $k_1^2 = 2mE/\hbar^2$ of

$$\begin{aligned}
\psi_1(x) &= A\,e^{ik_1 x} + B\,e^{-ik_1 x} \\
&= A(\cos k_1 x + i \sin k_1 x) + B(\cos k_1 x - i \sin k_1 x) \\
&= A_1 \cos k_1 x + B_1 \sin k_1 x
\end{aligned}$$

where $A_1 = A + B$ and $B_1 = i(A - B)$.

In region $2 (x \geq a)$

$$\frac{\partial^2 \psi_2(x)}{\partial x^2} + \frac{2m(E - V)}{\hbar^2} \psi_2(x) = 0$$

has the solution

$$\psi_2(x) = A_2\,e^{\alpha x} + B_2\,e^{-\alpha x}$$

where

$$\alpha^2 = \frac{2m}{\hbar^2}(V - E)$$

In region 3, $(x < 0)$

$$\frac{\partial^2 \psi_3(x)}{\partial x^2} + \frac{2m(E - V)}{\hbar^2} \psi_3(x) = 0$$

has the solution

$$\psi_3(x) = A_3\,e^{\alpha x} + B_3\,e^{-\alpha x}$$

For $\psi(x)$ to remain finite as $x \to \pm\infty$ (normalization condition) A_2 and B_3 must be zero, and the boundary conditions $\psi(x)$ and $\partial\psi(x)/\partial x$ continuous, must be satisfied at $x = 0$ and $x = a$.

At $x = 0$,

$$\psi_1(x) = \psi_3(x) \quad \text{and} \quad \frac{\partial \psi_1(x)}{\partial x} = \frac{\partial \psi_3(x)}{\partial x}$$

give

$$A_1 = A_3 \tag{13.1}$$

and

$$k_1 B_1 = \alpha A_3 \tag{13.2}$$

whilst at $x = a$

$$\psi_1(x) = \psi_2(x) \quad \text{and} \quad \frac{\partial \psi_1(x)}{\partial x} = \frac{\partial \psi_2(x)}{\partial x}$$

give

$$A_1 \cos k_1 a + B_1 \sin k_1 a = B_2 e^{-\alpha a} \tag{13.3}$$

and

$$-k_1 A_1 \sin k_1 a + k_1 B_1 \cos k_1 a = -\alpha B_2 e^{-\alpha a} \tag{13.4}$$

In order to satisfy equations (13.1), (13.2), (13.3) and (13.4) some conditions must be imposed on k and α; that is, on the value of E, so only certain values of E are allowed.

Equations (13.1) and (13.2) give

$$\frac{A_1}{B_1} = \frac{k_1}{\alpha}$$

and this equation with equations (13.3) and (13.4) yields

$$\tan k_1 a = \frac{2 k_1 \alpha}{k_1^2 - \alpha^2}$$

or

$$\tan a \sqrt{\frac{2mE}{\hbar^2}} = \frac{2\sqrt{E(V-E)}}{2E - V}$$

Only those values of E which satisfy this relation are allowed energy states, but these values must be found by numerical or graphical methods.

The wave functions for the first three allowed energy values are shown in Figure 13.10 and their general behaviour may be clarified by considering Schrödinger's equation in the form

$$\frac{\partial^2 \psi}{\partial x^2} \bigg/ \psi = -(+\text{ve constant})(E - V)$$

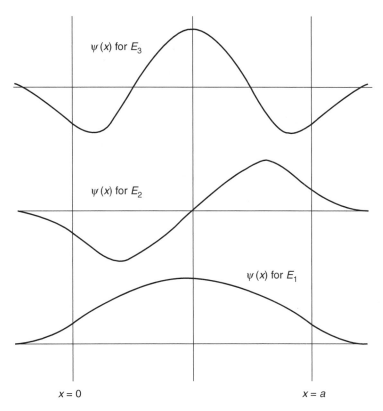

$\psi(x)$ for E_3

$\psi(x)$ for E_2

$\psi(x)$ for E_1

$x = 0$ $x = a$

Figure 13.10 Wave functions for a particle in a square potential well with the lowest three allowed energies E_1, E_2, E_3. Note the exponential decay of $\psi(x)$ outside the box

Now $\partial^2\psi/\partial x^2$ is the rate of change of the slope; that is, the curvature of the wave function and when $E > V$ both sides of the equation are negative and the ψ curve must everywhere keep its concave side towards the x axis as it always does, for example, in sine and cosine curves. The curvature increases with E so we shall expect more de Broglie half wavelengths in the higher energy levels. This is consistent with the argument that an increase in E increases the wave number k and reduces the de Broglie wavelength λ.

In the lowest energy level the ψ curve is always without a node, the next level always has one node, the third two nodes, etc. but the zeros will not be quite equally spaced and the ψ amplitude will not be uniform across the well. In particular it will increase near the potential walls as the particle is slowed down to give a higher probability of the particle being found there. Where $E < V$ the ratio

$$\frac{\partial^2\psi/\partial x^2}{\psi}$$

will be positive and the ψ curve must keep its convex side towards the axis as in exponential curves. The classical boundary $E = V$ must always mark the division where the character of the ψ curve changes from one form to the other and the two parts of the curve will only match for certain values of E.

The Harmonic Oscillator

As a final example to illustrate the fitting of ψ curves into a potential well we shall consider the potential curve $V = \frac{1}{2}sx^2$ of the harmonic oscillator in Figure 13.11. The calculation of the ψ curves is too complicated for this chapter but their essential features confirm what we may expect from our earlier examples. Moreover, by purely classical arguments we shall obtain a very good approximation to the wave mechanical results.

In 1901 Planck had postulated that the energy of such an oscillator could have the values $E = nh\nu$ where n was an integer and ν was the frequency. Schrödinger was able to derive this result in 1926 but one essential difference arises from the Uncertainty Principle which requires a minimum energy level or *zero point energy* of $\frac{1}{2}h\nu$.

For a classical oscillator the minimum energy $E = 0$, point 0 in Figure 13.11 gives the precise and simultaneous values $x = 0$ and $p = 0$; that is, a zero oscillation. The Uncertainty Principle forbids this. If a_0 is the smallest amplitude of the oscillator compatible with the Uncertainty Principle, then

$$a_0 \sim \tfrac{1}{2}\Delta x$$

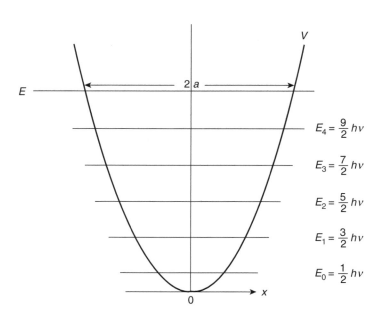

Figure 13.11 Potential energy curve V of a harmonic oscillator with allowed energy levels $E_n = (n + \frac{1}{2})h\nu$. The energy E (with oscillator amplitude a) is shown in the text to define an average value of the de Broglie wavelength $\lambda = h/(\frac{4}{3}mE)^{1/2}$

If p_0 is the maximum momentum of the oscillator with amplitude a_0 it may be either in the positive or negative direction so

$$p_0 \sim \tfrac{1}{2}\Delta p$$

The energy of a classical oscillator is given by

$$E = \tfrac{1}{2}m\omega^2 a_0^2 = \tfrac{1}{2}\omega(a_0)(m\omega a_0) = \tfrac{1}{2}\omega a_0 p_0$$
$$\approx \tfrac{1}{8}\omega\Delta x\Delta p \approx \tfrac{1}{8}h\omega \approx \tfrac{1}{2}\hbar\omega = \tfrac{1}{2}h\nu$$

All other energy levels will therefore take integral steps of $h\nu$ above this zero point energy.
Let us consider the energy level of the oscillator which has an amplitude a so that

$$E = \frac{p^2}{2m} + V = \frac{p^2}{2m} + \frac{1}{2}sx^2 = \frac{1}{2}sa^2 = \frac{1}{2}m\omega^2 a^2$$

so that

$$2a = \frac{2}{\omega}\sqrt{\frac{2E}{m}}$$

The value of the kinetic energy of the oscillator averaged over the distance $2a$ between $\pm a$ may be written

$$\frac{\int_{-a}^{a} p^2/2m\,\mathrm{d}x}{\int_{-a}^{a} \mathrm{d}x} = \frac{1}{2a}\int_{-a}^{a}\left(E - \frac{1}{2}m\omega^2 x^2\right)\mathrm{d}x = E - \frac{1}{6}m\omega^2 a^2 = \frac{2}{3}E$$

because

$$E = \tfrac{1}{2}m\omega^2 a^2$$

Thus, the average value of the kinetic energy

$$\frac{p^2}{2m} = \frac{2}{3}E$$

giving

$$p = \frac{h}{\lambda} = \sqrt{\frac{4mE}{3}}$$

This gives an average value for the de Broglie wavelength of

$$\lambda = \frac{h}{\sqrt{\dfrac{4mE}{3}}}$$

and we expect n half wavelengths to fit into the length $2a$ at energy E where

$$2a = \frac{2}{\omega}\sqrt{\frac{2E}{m}}$$

Thus

$$n\frac{\lambda}{2} = \frac{nh}{2\sqrt{4mE/3}} = \frac{2}{\omega}\sqrt{\frac{2E}{m}}$$

Writing $\omega = 2\pi\nu$ we have

$$E = \frac{\pi}{4}\sqrt{\frac{3}{2}}nh\nu = 0.96\,nh\nu$$

which is a fairly close approximation to $nh\nu$. The correct result, however, must take into account the zero point energy of $\frac{1}{2}h\nu$ and the energy levels are given by

$$E = (n + \tfrac{1}{2})h\nu, \quad n = 0, 1, 2, 3, \text{etc.}$$

The ψ curves for the first four energy levels are plotted in Figure 13.12 together with those for $|\psi|^2$.

We see that whilst a classical oscillator may never exceed its maximum amplitude a particle obeying a wave mechanical description has a finite probability of being found beyond this limit.

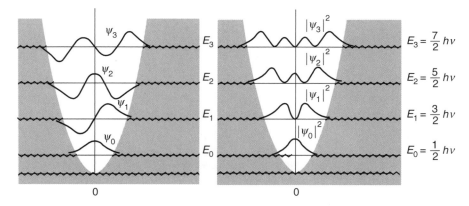

Figure 13.12 Wave functions $\psi(x)$ and probability densities $|\psi(x)|^2$ for the first four energy levels of the harmonic oscillator

(Problems 13.18, 13.19)

Electron Waves in a Solid

Bloch Functions and the Kronig–Penney Model

When electrons move through a solid, e.g. a metal, they meet a series of potential barriers generated by the atoms or ions located at the centre of the valleys between successive barriers. Figure 13.13 shows such a one-dimensional lattice array of ions. The electron wave function is derived via Bloch functions and the electron behaviour is demonstrated using the Kronig–Penney Model which replaces Figure 13.13 in the first instance with a periodic series of potential wells of finite depth as shown in Figure 13.14. An exact but unwieldy solution can be found for the situation described by Figure 13.14, but Kronig and Penney, by deepening the wells and reducing their separation, were able to show how the electrons behaved and to demonstrate the restrictions imposed on their motion.

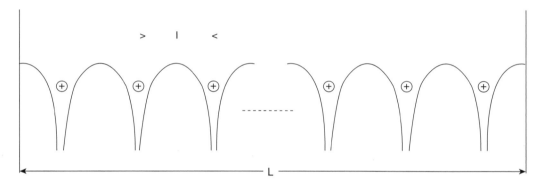

Figure 13.13 A one-dimensonal periodic array of potential barriers formed by ions or atoms located along a crystal lattice

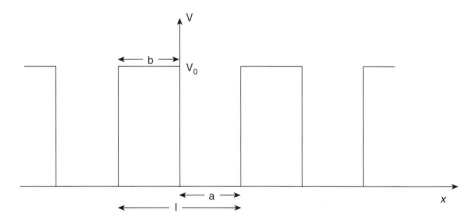

Figure 13.14 A series of finite potential wells used by Kronig and Penney as a first approximation of Figure 13.13

In Figure 13.14 the space between the potential wells is a, the well thickness is b and its height is V_0. The problem is similar to that described on p. 435 where the total energy of the electron is $E - V_0$ so the wave equation is

$$\frac{\partial^2 \psi}{\partial x^2} + \frac{2m}{\hbar^2}(E - V_0)\psi = 0$$

Now, $V(x)$ is periodic so $V_x = V(x + l)$ where $l = a + b$. Evidently, since the probability of finding an electron at x or at $x + l$ is the same, we have

$$|\psi(x)|^2 = |\psi(x + l)|^2$$

Hence, we may write $\psi(x + l) = \gamma\psi(x)$ where $\gamma\gamma^* = |\gamma|^2 = 1$ (γ^* is the complex conjugate of γ).

At this stage we could write $\gamma = e^{ikx}$, but this does not define k well enough to satisfy the boundary conditions at each end of the crystal. For periodic functions the conventional method to meet the boundary conditions is to form a ring of circumference of length $L = Nl$ where L is the length of the crystal and N is the number of atoms along its length. Note that in Figure 13.13 the potential barriers at each end of the crystal add l to its length.

Proceeding along the crystal (or around the ring) we have

$$\psi(x + 2l) = \psi(x + l + l) = \gamma\psi(x + l) = \gamma^2(\psi)$$

or for r integral steps

$$\psi(x + rl) = \gamma^r\psi(x) \qquad r = (0, 1, 2, 3 \ldots N - 1)$$

Now $r = 0$ and $r = N$ are identical positions (one complete circuit of the ring), so

$$\psi(x + Nl) = \gamma^N\psi(x) = \psi(x)$$

that is

$$\gamma^N = 1$$

We may now write

$$\gamma = e^{i2\pi r/N} \qquad (r = 0, 1, 2, 3 \ldots)$$

so that

$$\psi(x + l) = \gamma\psi(x) = e^{i2\pi r/N}\psi(x)$$

The Bloch function $\mu_k(x)$ is defined by

$$\psi(x) = \mu_k(x)e^{ikx}$$

where

$$\mu_k(x) = \mu_k(x+l)$$

Here, $k = 2\pi r/lN$ and $\mu_k(x)$ has the periodicity of the potential. Since r changes by units as we move along the crystal each step of r/N (for N large) is so small that $k = 2\pi r/lN$ may be considered as varying continuously.

The Bloch functions satisfy all conditions because

$$\psi(x+l) = e^{ik(x+l)}\mu_k(x+l) = e^{ikl}e^{ikx}\mu_k(x) = e^{i\frac{2\pi r}{N}}\psi(x) = \gamma\psi(x)$$

The wave equations of Figure 13.14 are

$$\frac{\partial^2 \psi_1}{\partial x^2} + \alpha^2 \psi_1 = 0 \qquad 0 < x < a \tag{13.5}$$

and

$$\frac{\partial^2 \psi_2}{\partial x^2} - \beta^2 \psi_2 = 0 \qquad -b < x < 0 \tag{13.6}$$

where

$$\alpha^2 = \frac{2mE}{\hbar^2} \quad \text{and} \quad \beta^2 = \frac{2m}{\hbar^2}(V_0 - E)$$

with

$$V(x) = V(x+l) \quad \text{and } l = a + b$$

The Bloch function $\mu_k(x) = \mu_k(x+l)$ where $l = a+b$, so for $x = -b$ we have $\mu_x(a) = \mu_k(-b)$, which is evident from Figure 13.14.

Earlier examples in this chapter have shown that the boundary conditions require $\psi(x)$ and its first derivative to be continuous across any potential change.

Applying $\psi(x) = \mu_k(x)e^{ikx}$ to equations (13.5) and (13.6), we have

$$\mu_1(x) = Ae^{i(\alpha-k)x} + Be^{-i(\alpha+k)x} \qquad 0 < x < a$$
$$\mu_2(x) = Ce^{(\beta-ik)x} + De^{-(\beta+ik)x} \qquad -b < x < 0$$

so that the boundary conditions are

$$\mu_1(0) = \mu_2(0) \qquad \text{with} \qquad \left(\frac{\partial \mu_1}{\partial x}\right)_{x=0} = \left(\frac{\partial \mu_2}{\partial x}\right)_{x=0}$$

and

$$\mu_1(a) = \mu_2(-b) \qquad \text{with} \qquad \left(\frac{\partial \mu_1}{\partial x}\right)_{x=a} = \left(\frac{\partial \mu_2}{\partial x}\right)_{x=-b}$$

which give four homogenous equations.

Remember that

$$\mu(x) = \mu(x + l)$$

As with the rectangular well on p. 435 these boundary conditions determine the permitted values of E (via α and β). Here, the boundary conditions require either $A = B = C = D = 0$ or the determinant of their coefficients to be zero. Equating the determinant of the coefficients to zero gives the unwieldy expression

$$\frac{\beta^2 - \alpha^2}{2\alpha\beta} \sin \alpha a \sinh \beta b + \cos \alpha a \cosh \beta b = \cos k(a + b) \qquad (13.7)$$

Kronig and Penney simplified this equation by allowing V_0 to tend to infinity as b approached zero in such a way that $V_0 b$ remained constant. This has two important implications. First, the potential wells become very deep so that Figure 13.14 approximates Figure 13.13. Second, their separation is narrowed so that $l = a + b \approx a$ and we may rewrite equation(13.7) as

$$V_0 b \left(\frac{ma}{\hbar^2}\right) \frac{\sin \alpha a}{\alpha a} + \cos \alpha a = \cos ka \qquad (13.8)$$

The values of $\alpha = (2mE/\hbar^2)^{\frac{1}{2}}$ which satisfy this equation determine the permitted energy values and wave functions of the electrons.

Note that when $V_0 \to \infty$ equation (13.8) requires $\sin \alpha a = 0$ to remain valid, leaving

$$\alpha = \pm \frac{n\pi}{a} \qquad (n = 1, 2, 3 \ldots)$$

or

$$E = \frac{\pi^2 \hbar^2 n^2}{2ma^2}$$

which are the quantized energies of the tightly bound electron in the infinitely deep potential of p. 420.

At the other extreme when $V_0 = 0$ equation (13.8) gives

$$\alpha = k = \left(\frac{2mE}{\hbar^2}\right)^{\frac{1}{2}}$$

which allows E to take any positive value. This gives a free particle solution to the wave equation (graphed as the dotted parabola in Figure 13.16).

Between these two extreme values of V_0 the permitted values of the energy E are displayed on the graph in Figure 13.15 where the left-hand side of equation (13.8) is plotted against αa where αa is written w and $V_0 b(\frac{ma}{\hbar^2})$ is written K.

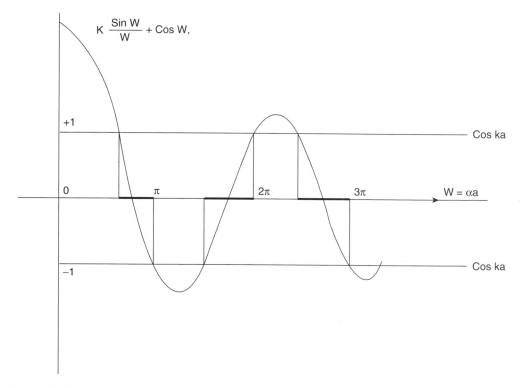

Figure 13.15 Allowed electron energy values are denoted by heavy horizontal lines which define the Brillouin zones. These occur when the left-hand side of equation (13.8) has values between ± 1. The curve is symmetric about the axis $w = 0$.

Now the limits of cos ka in equation (13.8) are ±1 and these determine the allowed values of $w = \alpha a$ indicated by the heavy horizontal line on the w or αa axis. These in turn denote the permitted ranges or bands of energy values which the electron may take. The bands increase with $w = \alpha a$ and between the bands are gaps where electron energies are forbidden. The limits of each energy band are defined by $\cos ka = \pm 1$ that is

$$ k = \pm \frac{n\pi}{a} \qquad (n = 1, 2, 3, \dots) $$

and the regions in k space defining the energy bands are known as Brillouin zones. The band for $n = 1$ is called the first Brillouin zone, $n = 2$ is the second Brillouin zone and so on. Figure 13.15 can be displayed as the energy E versus k graph in Figure 13.16 where the dotted parabola defines the free electron energy $E = \frac{\hbar^2}{2m} k^2$ and the heavy lines at the k boundaries denote the permitted electron energies in a given band. The cosine curves joining the zone boundaries are justified by Figure 5.15, which shows that no new information is gained by extending the k range beyond $-\pi/a \le k \le \pi/a$. This limited range of k values defines the reduced zone scheme.

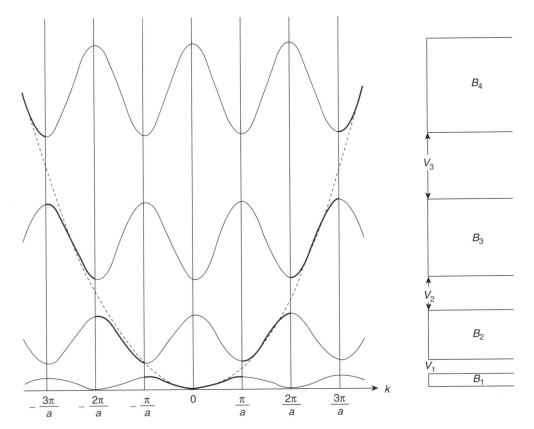

Figure 13.16 Figure 13.15 displayed as allowed electron energies versus k. The dotted parabola defines the free electron energy $E = \hbar^2 k^2 / 2m$ and the allowed energy bands are the Brillouin zones B_i. V_1, V_2, V_3 are the energy gaps between the zones. The cosine curves joining the zone boundaries are justified by Figure 5.15, i.e. all relevant information is contained in the region $\frac{-\pi}{a} \leq k \leq \frac{\pi}{a}$

The number of energy states (excluding spin) in each zone is determined by

$$k = \frac{2\pi r}{lN} = \frac{2\pi r}{aN} \qquad (r = 0, 1, 2, 3 \ldots N - 1)$$

for each k value represents an allowed energy state. Each value of r gives a different value of k; there are N such values. Hence, in this range

$$-\frac{\pi}{a} \leq k \leq \frac{\pi}{a} \text{ i.e. } \frac{2\pi}{a} = \frac{2\pi r}{Na} \qquad \text{where } a \approx l$$

the number of energy levels is equal to the number of atoms.

As $a + b = l \rightarrow \infty$ each band contracts to a single level which is N-fold degenerate since the electron can be bound to any one of the atoms. For finite values of l this degeneracy is removed and each discrete atomic level spreads into a band of N levels.

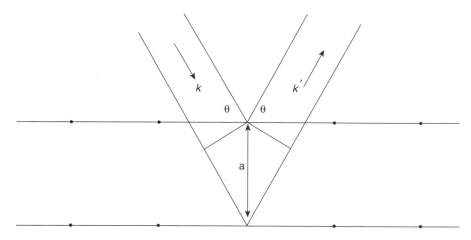

Figure 13.17 Elastic Bragg reflection occurs when electron waves are scattered by atoms in planes separated by a distance a. Principal maxima are formed when $2a\sin = n\lambda$

Only free electrons will escape interaction with the ions in the crystal lattice; almost free electrons will experience weak coupling to the lattice. Coupling which is strong enough to reflect electron waves may be seen in terms of Bragg reflection, Figure 13.17. Here, waves reflected by successive planes in a crystal which are separated by a distance a reinforce to give maxima on reflection when $2a\sin\theta = n\lambda$.

When $\theta = \pi/2$ and the coupling is strong enough the electron waves will be reflected from successive ions, Figure 13.18, giving a path difference of $2a$. Reflection maxima occur for

$$2a = \pm n\lambda = \pm n\frac{2\pi}{k}, \qquad \text{i.e. } k = \pm\frac{n\pi}{a}$$

Thus, Bragg reflections define the Brillouin zone boundaries.

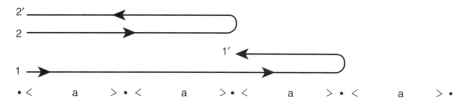

Figure 13.18 When $\theta = \pi/2$ in Figure 13.17 Bragg scattering by electron--ion interactions gives principal maxima when electron waves are reflected from ions separated by multiples of a. The condition $2a = n\lambda$ defines the Brillouin zone boundaries for $n = 1, 2, 3$, etc.

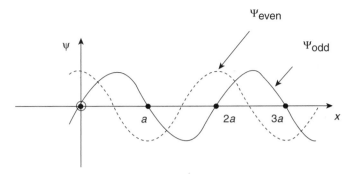

Figure 13.19 The wave function ψ (even) has an anti-node at an ion (atom) site. The anti-node for ψ (odd) is located midway between sites. This governs the energy of interaction, which is different for the two ψ values

Wave functions of electrons can be represented by travelling waves in both directions, i.e. by $e^{\pm ikx}$ and for $k = \pm n\pi/a$ standing waves will be formed by the sum or difference of, e.g.

$$e^{i\pi x/a} \quad \text{and} \quad e^{-i\pi x/a}$$

The sum of these terms creates

$$\psi_{\text{even}} = \cos\frac{\pi x}{a}$$

and their difference gives

$$\psi_{\text{odd}} = \sin\frac{\pi x}{a}$$

The energies associated with these two wave functions will differ when they interact with the ions. $\psi_{\text{even}} = \cos\pi x/a$ has anti-nodes (maxima) at the site of each ion so the electron–ion interaction is attractive and the energy corresponding to ψ_{even} is lowered. $\psi_{\text{odd}} = \sin\pi x/a$ has its anti-node midway between ion sites where the potential is repulsive, Figure 13.19. The calculation of these energy shifts requires knowledge of the effective potential, but it can be shown that for ψ_{even} the energy change at a given V_n in Figure 13.16, where V_n is the energy gap between bands, is $\Delta E = -\frac{1}{2}V_n$ and for ψ_{odd} the energy change is $\Delta E = \frac{1}{2}V_n$ (see Problem 13.21). Note that the band widths and gaps increase with n.

The band structure may also be demonstrated by considering the effect of tunnelling. Two widely separated equivalent potential wells may each contain a single electron occupying identical energy levels. When the potential well separation becomes small enough for the tunnelling of Figure 13.6 to be possible this symmetry is destroyed because the wave function of an electron spreads right across both wells and their separating potential barrier, Figure 13.20. There is a finite probability of finding an electron at any

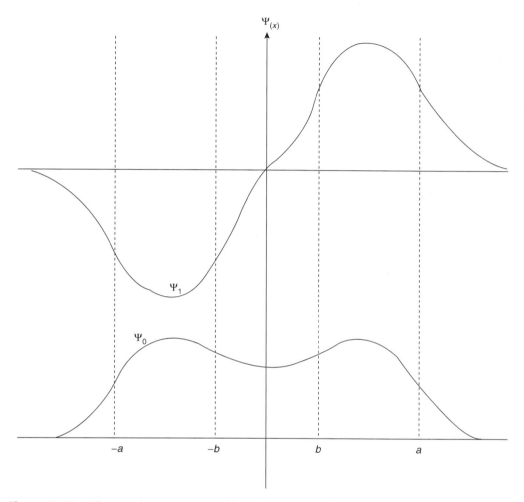

Figure 13.20 When an electron can tunnel between two potential wells (a, b) and − (a, b) it cannot exist in a single energy state. The higher of the two resulting energy states has a greater curvature

point x of its wave function $\psi(x)$ so the two electrons cannot occupy the same energy level and the single state splits into two. The lowest lying energy levels split into a narrow band of very closely spaced states since the barrier to tunnelling is very large for electrons in these levels. Higher energy levels have a wider spread and it is even possible for bands to overlap. The band structure helps to explain the difference between electrical conductors and insulators.

Once an energy level is occupied by an electron it cannot accept another electron. However, in a metal only the lower energy levels in a band or Brillouin zone are occupied and an applied electric field can accelerate electrons which move to occupy higher available energy states within the band. Insulators have completely filled energy bands so the electrons cannot move under the influence of an electric field – there are no empty neighbouring states.

However, a very strong electric field can cause an electron to jump from the top of a band across a gap to occupy an empty level immediately above the gap, so the insulator breaks down. A spark can jump across an air gap between two terminals; lightning is such a spark on a much larger scale. A semiconductor is basically an insulator with a very narrow forbidden gap where even a small energy change will switch the insulator into a conductor.

Phonons

Pages 135 and 162 showed that the elastic field in a crystal could sustain transverse and longitudinal modes of vibration along a chain of atoms acting as a series of coupled oscillators. In a normal mode of angular frequency ω_i every atom performed simple harmonic oscillations of ω_i. On p. 440 we saw that the energy of such oscillations at atomic and sub-atomic levels was quantized with values of $(n + 1/2)\hbar\omega$.

The concept of photons as quanta of energy $\hbar\omega$ associated with an electromagnetic field allows the analogy of *phonons* as quanta of energy associated with the elastic field. In a normal mode of angular frequency ω_i the energy of a phonon is $\hbar\omega_i$ so phonons can be seen as exciting a mode to an energy state $(n + \frac{1}{2})\hbar\omega_i$. When $n = 0$ the mode ω_i is left with the zero-point energy $\frac{1}{2}\hbar\omega_i$. A more detailed calculation of Debye's theory of specific heats (page 253) takes account of this quantization.

Normal modes are plane waves extending throughout the crystal and phonons are not localized particles. The uncertainty principle prevents an exact determination of a phonon position and it exists as a localized wave packet of combined modes with a small spread of frequency and wavelength and a group velocity $\frac{\partial \omega}{\partial k}$. The number of phonons, like that of photons, is not conserved. They are created and absorbed by collisions and, like photons, they obey Bose–Einstein statistics (appendix 1). However, unlike photons, they exist only within the crystal. They contribute to the crystal momentum but do not carry momentum. This is evident from Figure 5.15 where a lattice vibration has a wave number $k = k \pm \frac{m\pi}{a}$ $(m = 1, 2, 3, \ldots)$ so $\hbar k$ has no precise meaning. Indeed, when the mode oscillations are purely harmonic the equilibrium position is zero so phonon momentum is zero.

Phonon–phonon collisions are usually three-phonon processes in which both transverse and longitudinal waves are involved. They are characterized by energy conservation

$$\hbar\omega_1 = \hbar\omega_2 + \hbar\omega_3$$

and by phonon wave vector conservation

$$\mathbf{q}_1 = \mathbf{q}_2 + \mathbf{q}_3$$

A phonon of wave vector $\mathbf{q}_1$ can separate into two phonons with wave vectors $\mathbf{q}_2$ and $\mathbf{q}_3$. Alternatively, $\mathbf{q}_2$ can absorb $\mathbf{q}_3$ to form $\mathbf{q}_1$. Phonon–phonon collisions play a role in the thermal conductivity of a crystal; neutron interactions with the crystal lattice also involve the concept of phonons.

When particles, as waves, interact with crystal structures they create diffraction patterns when the particle wavelength is of the order of atomic separation within the crystal, typically $\sim 2 \times 10^{-10}$ m. The waves of X-rays striking a crystal create principal maxima on reflection to satisfy Braggs Law (p. 447) when the path difference

$$2a \sin \theta = n\lambda$$

where a is the separation between the reflecting (diffracting) planes. If k is normal to the particle wave fronts before striking the crystal and $\mathbf{k}'$ is normal to the wave front leaving the crystal the condition $|\mathbf{k}| = |\mathbf{k}'|$ defines the scattering as elastic, so Bragg scattering is elastic. Knowing the plane separation of a nickel crystal, determined by X-rays, Davisson and Germer were able to find the wavelength of electrons by Bragg elastic scattering (see Problem 13.20).

Neutrons with $\lambda \sim 2 \times 10^{-10}$ m have been used in non-elastic scattering experiments where $|\mathbf{k}| \neq |\mathbf{k}'|$ to probe the structure of crystals, that is, the atomic arrangements and separation. Where X-rays interact chiefly with electrons surrounding the nucleus of an atom, uncharged neutrons interact much more strongly with its nucleus; lattice vibrations are set up so phonons play a role in the scattering.

Non-elastic scattering may be seen in terms of Figure 13.21 where waves in the wave front normal to $\mathbf{k}$ are scattered by atoms 1 and 2 in a row where the atomic separation is a. The phase lag of the wave incident on atom 2 is $\frac{2\pi}{\lambda} a \sin \theta$ with respect to that striking atom 1, but after scattering it leads the wave scattered by atom 1 by a phase $\frac{2\pi}{\lambda'} a \sin \phi$. A diffraction maximum occurs when the phase difference

$$\frac{2\pi}{\lambda} a \sin \theta - \frac{2\pi}{\lambda'} a \sin \phi = ka \sin \theta - k'a \sin \phi = l2\pi \; (l = 1, 2, 3, \ldots)$$

i.e.

$$\mathbf{a}(\mathbf{k} - \mathbf{k}') = l2\pi$$

or

$$\mathbf{k} - \mathbf{k}' = l\frac{2\pi}{a}$$

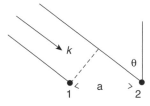 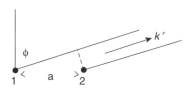

Figure 13.21 When electrons are scattered from atoms separation a, in the same plane, the scattering may be inelastic, i.e. $|\mathbf{k}| \neq |\mathbf{k}'|$. Here, the electron of wave number $k = 2\pi/\lambda$ strikes atom 1 ahead in phase of that striking atom 2 by $2\pi/\lambda\, a \sin\theta$, but after scattering it lags that from atom 2 by a phase difference $\frac{2\pi}{\lambda'} a \sin \phi$. Note that λ need not $= \lambda'$

Note that $\mathbf{k} - \mathbf{k}'$ is a vector in diffraction space and λ' need not equal λ. This is true for every row of lattice points in the x direction.

The expression $l2\pi/a$ represents a series of planes in $\mathbf{k}$ space with a separation $2\pi/a$. Crystal planes in a second dimension with separation b would form another series of planes $m\frac{2\pi}{b} (\text{m} = 1, 2, 3, \ldots)$ with separation $\frac{2\pi}{b}$ in $\mathbf{k}$ space having lines of intersection with the series l. A set of crystal planes in a third dimension with separation c would form a final set of planes $n\frac{2\pi}{c} (n = 1, 2, 3, \ldots)$ with separation $2\pi/c$ in $\mathbf{k}$ space. These three sets of planes would meet in points (l, m, n) in $\mathbf{k}$ space to form the *reciprocal lattice*. In three dimensions the diffracted vector $k - k'$ would end on a reciprocal lattice point l, m, n. There is no requirement for the directions a, b and c in the crystal to be mutually perpendicular, but a symmetry exists between the crystal lattice and its reciprocal in that planes in the one are perpendicular to rows of points in the other and the plane spacing in one is 2π times the reciprocal of the point spacing in the other.

When neutrons are diffracted from a crystal lattice in which a phonon of wave vector $\mathbf{q}$ and frequency ω is already excited, more than one diffraction maximum can appear. This first maximum will result from Bragg elastic scattering, i.e. $|\mathbf{k}| = |\mathbf{k}'|$.

A second maximum occurs in a vector direction

$$\mathbf{g} = \mathbf{k} - \mathbf{k}' + \mathbf{q}$$

or

$$\mathbf{k}' = \mathbf{k} + \mathbf{q} - \mathbf{g}$$

This suggests that a neutron of wave vector $\mathbf{k}$ has absorbed a phonon of wave vector $\mathbf{q}$ to become a neutron of wave vector $\mathbf{k}'$. In the scattering, because the neutron is initially outside the crystal, the crystal plus the phonon receives a momentum

$$\hbar(\mathbf{k} - \mathbf{k}') = \hbar(\mathbf{g} - \mathbf{q})$$

Conventionally, the momentum $\hbar\mathbf{g}$ is associated with the whole lattice while $\hbar\mathbf{q}$ (associated with the absorbed phonon) is known as the crystal or quasi-momentum of the phonon because it acts as a momentum when absorbed by the neutron.

In pure phonon–phonon collisions two processes may occur. The three phonons involved may begin and end in the same Brillouin Zone. this is called a normal process. In some cases, however, the third phonon may finish outside the Brillouin zone. This is known as the Umklapp process. This occurs when a phonon is Bragg reflected (at the edge of a Brillouin zone) at the same time as it absorbs another phonon. We know, however, that a phonon of wave vector $\mathbf{q}$ is identical with a phonon of wave vector $\mathbf{q} \pm \frac{2\pi}{a}$, so the third phonon may be considered as remaining within the Brillouin zone. Umklapp processes play a role in the thermal conductivity of a crystal in the following way.

When the crystal lattice vibrations are purely harmonic the separation between adjacent atoms during vibrations contributes an energy term $\propto (x_i - x_{i\pm1})^2$, where x_i is the displacement of an atom from its equilibrium position. In this case a phonon may travel along hundreds of atoms without hindrance. However, with increasing energy, i.e. temperature, vibrations become anharmonic and cubic terms replace the squared term above because separate normal modes become coupled. Effectively, a cubic term describes the emission of a phonon by another phonon or the decay of a phonon into two phonons and the energies of individual phonons are changed. The phonons constitute a gas where the phonons have approximately constant speed (unlike in a real gas), but have a larger number density and energy density at the hot end of the crystal. Heat flow is primarily by phonon flow with phonons being created at the hot end and destroyed at the cold end. The thermal resistance in an insulator is produced by collisions which reverse the group velocity of the phonons, and the Umklapp process involving high-energy phonons at Bragg reflection on the edge of the Brillouin zone is significant here.

(Problems 13.20, 13.21)

Problem 13.1

The energy of an electron mass m charge e circling a proton at radius r is

$$E = \frac{p^2}{2m} - \frac{e^2}{4\pi\varepsilon_0 r}$$

where p is its momentum.

Use Heisenberg's Uncertainty Principle in the form $\Delta p \Delta r \approx \hbar$ to show that the minimum energy (H_2 atom ground state) is

$$E_0 = \frac{-me^4}{8\varepsilon_0^2 h^2}$$

at a Bohr radius

$$r = \frac{\varepsilon_0 h^2}{\pi m e^2}$$

Problem 13.2

The observation of a particle annihilates its mass m and its rest mass energy is converted to radiation. Use the relations $\Delta p \, \Delta x \approx h$ and $E = pc$ for photons to show that the short wavelength limit on length measurement is the Compton wavelength

$$\lambda = \frac{h}{mc}$$

Show that this is 2.42×10^{-12} m for an electron.

Problem 13.3

When x and p vary simple harmonically it can be shown that the averaged values of the squares of the uncertainties satisfy the relation

$$(\overline{\Delta x^2})(\overline{\Delta p^2}) \approx \frac{\hbar^2}{4}$$

If the energy of a simple harmonic oscillation at frequency ω is written

$$E = \frac{p^2}{2m} + \frac{1}{2}m\omega^2 x^2$$

show that its minimum energy is $\frac{1}{2}h\nu$.

Problem 13.4

An electron of momentum p and wavelength $\lambda = h/p$ passes through a slit of width Δx. Its diffraction as a wave may be regarded in terms of a change of its momentum Δp in a direction parallel to the plane of the slit (its total momentum remaining constant). Show that the approximate position of the first minimum of the diffraction pattern is in accordance with Heisenberg's uncertainty principle. (Note that the variation of the intensity of the principal maximum in the pattern is a direct measure of the probability of the electron arriving at a point on the screen.)

Problem 13.5

A beam of electrons with a de Broglie wavelength of 10^{-5} m passes through a slit 10^{-4} m wide. Show that the angular spread due to diffraction is $5°47'$.

Problem 13.6

Show that the de Broglie wavelength of an electron accelerated across a potential difference V is given by

$$\lambda = h/(2m_e eV)^{1/2} = 1.29 \times 10^{-9} V^{-1/2}\,\text{m}$$

where V is measured in volts.

Problem 13.7

If atoms in a crystal are separated by 3×10^{-10} m (3 Å) show that an accelerating voltage of $\sim 3\,\text{kV}$ would be required to produce electrons diffracted by the crystal.

Problem 13.8

Electromagnetic radiation consists of photons of zero rest mass. Show that the average momentum per unit volume associated with an electromagnetic wave of electric field amplitude E_0 is given by

$$p = \frac{1}{2}\varepsilon_0 E_0^2/c$$

(Verify the dimensions of this relation.)

Problem 13.9

Show that the average momentum carried by an electromagnetic wave develops a radiation pressure

$$P = cp = \frac{1}{2}\varepsilon_0 E_0^2$$

when the wave is normally incident on a *perfect absorber* and a pressure

$$P = 2cp = \varepsilon_0 E_0^2$$

when the wave is normally incident on a *perfect reflector*. (Radiation incident from all directions within a solid angle of 2π will introduce a factor of $1/3$ in the expressions above.)

Problem 13.10

If the radiation energy from the sun incident upon the perfectly absorbing surface of the earth is $1.4\ \mathrm{kW\ m^{-2}}$ and the radiation comes from all directions within a solid angle of 2π show that the radiation pressure is about 10^{-11} of the atmospheric pressure.

Problem 13.11

In a carbon molecule the two atoms oscillate with a frequency of 6.43×10^{-11} Hz. Show that the zero point energy is 1.34×10^{-3} eV ($1\,\mathrm{eV} = 1.6 \times 10^{-19}$ J).

Problem 13.12

A particle of mass m moves in an infinitely deep square well potential of width $2a$ defined by

$$\begin{aligned} V(x) &= 0 \quad & -a \le x \le +a \\ V(x) &= \infty & |x| > a \end{aligned}$$

If it is described by the wave function

$$\begin{aligned} \psi(x) &= \frac{1}{\sqrt{a}}\left(1 - \frac{\pi^2 x^2}{8a^2}\right) \quad & \text{for } |x| \le a \\ &= 0 & |x| > a \end{aligned}$$

show by calculating $\int_{-a}^{a} |\psi(x)|^2\,\mathrm{d}x$ that the probability of finding it in the box is 0.96.

Show that in its normalized ground state, it is represented by $\psi(x) = (1/\sqrt{a})\cos(\pi x/2a)$ and expand this in powers of $\pi x/2a$ to compare it with the wave function above.

Problem 13.13

Show that the normalization constant for the wave function

$$\psi(xyz) = A \sin\frac{n_1 \pi x}{a} \sin\frac{n_2 \pi y}{b} \sin\frac{n_3 \pi z}{c}$$

describing an electron in a volume abc at the bottom of a deep potential well is equal to $(8/abc)^{1/2}$.

Problem 13.14

A total of N electrons occupy a volume V in a solid at a very low temperature between the energy levels 0 to E_F the Fermi energy.

Show that their total energy

$$\begin{aligned} U &= \int E\,\mathrm{d}n = \int_0^{E_F} E \frac{\mathrm{d}n}{\mathrm{d}E}\,\mathrm{d}E \\ &= \frac{3}{5} N E_F \end{aligned}$$

giving an average energy per electron of $\frac{3}{5}E_F$.

Problem 13.15

Copper has one conduction electron per atom, a density of 9 and an atomic weight of 64. Show that n_0, the number of free electrons per unit volume is $\approx 8 \times 10^{28}$ m^{-3} and that the value of its Fermi energy level is about 7 eV (1 eV $= 1.6 \times 10^{-19}$ J).

Problem 13.16

The probability of a particle of mass m penetrating a distance x into a classically forbidden region is proportional to e$^{-2\alpha x}$ where

$$\alpha^2 = 2m(V - E)/\hbar^2$$

If x is 2×10^{-10} m (2 Å) and $(V - E)$ is 1 eV (1.6×10^{-19} J) show that

$$\mathrm{e}^{-2\alpha x} = 0.1 \text{ for an electron}$$
$$= 10^{-43} \text{ for a proton}$$

Problem 13.17

A particle of total energy E travels in a positive x direction in a region where the potential energy $V = 0$. The potential suddenly drops to a very large negative value. Show that, quantum mechanically, the amplitude of the reflected wave tends to unity and that of the transmitted wave to zero. Note that this implies non-classical total reflection.

Problem 13.18

Show that Schrödinger's equation for a one dimensional simple harmonic oscillator of frequency ω is given by

$$\frac{\mathrm{d}^2\psi}{\mathrm{d}x^2} + \frac{2m}{\hbar^2}\left[E - \frac{1}{2}m\omega^2x^2\right]\psi = 0$$

and verify that if $a^2 = m\omega/\hbar$ then

$$\psi_0(x) = (a/\sqrt{\pi})^{1/2}\,\mathrm{e}^{-a^2x^2/2}$$

and

$$\psi_1(x) = (a/2\sqrt{\pi})^{1/2}2ax\,\mathrm{e}^{-a^2x^2/2}$$

are respectively the normalized wave functions for $E_0 = \frac{1}{2}\hbar\omega$ (zero point energy) and $E_1 = \frac{3}{2}\hbar\omega$.

Problem 13.19

The normalized wave function for a one-dimensional harmonic oscillator with energy $E_n = (n + \frac{1}{2})\hbar\omega$ is

$$\psi_n = N_nH_n(ax)\,\mathrm{e}^{-a^2x^2/2},$$

where

$$N_n = (a/\pi^{1/2}2^nn!)^{1/2}$$
$$a^2 = m\omega/\hbar$$

and

$$H(y) = (-1)^n e^{y^2} \frac{d^n}{dy^n} e^{-y^2}$$

Verify that $\psi_0(x)$ and $\psi_1(x)$ of Problem 13.18 satisfy the expression for ψ_n and calculate $\psi_2(x)$ and $\psi_3(x)$.

Problem 13.20

Davisson and Germer (1927) fired electrons with an energy of 54 eV at a nickel crystal which had an atomic plane separation of 0.91×10^{-10} m (0.91 Å). Bragg reflection gave a diffraction maximum at $65°$. Calculate the reflected electron momentum p and the kinetic energy to show that the difference between the incident and scattered kinetic energies was within 3.9%.

Problem 13.21

The perturbed energies of ψ (odd) and ψ (even) due to electron–ion interactions are given by

$$\Delta E = \frac{\int \psi^* V \psi \, dx}{\int \psi^* \psi \, dx} \qquad \text{where } \psi^* \text{ is the complex conjugate of } \psi$$

If the zero of energy is taken as the mean value of the potential then the potential may be written as a Fourier series in the form

$$V = - \sum_{n=1}^{\infty} V_n \cos 2\pi n x / a$$

where the V_n are the potential gaps in Figure 13.16. They are positive numbers for a potential with strong negative peaks at the lattice sites. For travelling waves $\psi = e^{\pm ikx}$ so $\psi^* \psi = 1$, which gives $\Delta E = 0$ in the above expression except for $\psi = \sin kx$ or $\cos kx$ when $k = n\pi/a$ where a is the periodicity of the lattice.

Show that for $\psi = \sin ka$

$$\Delta E = - \sum_{n=1}^{\infty} \frac{\int \sin^2 kx V_n \cos \frac{2\pi nx}{a} \, dx}{\int \sin^2 kx \, dx}$$

$$= \frac{1}{2} V_n \quad \text{for} \quad k = n\pi/a$$

Show that $\psi = \cos kx$ in the above expression gives $\Delta E = -\frac{1}{2} V_n$ for $k = n\pi/a$

Summary of Important Results

De Broglie Wavelength $\lambda = h/p$

Heisenberg's Uncertainty Principle (Bandwidth Theorem)

$$\Delta x \, \Delta p \approx h$$

$$\Delta E \, \Delta t \approx h$$

determines zero point energy.

Schrödinger's time independent wave equation

$$\frac{d^2\psi(x)}{dx^2} + \frac{2m(E-V)}{\hbar^2}\psi(x) = 0$$

$$\psi(x) = A\,e^{ikx} + B\,e^{-ikx},$$

where

$$k^2 = \frac{2m(E-V)}{\hbar^2} \quad E > V$$

$$\psi(x) = C\,e^{\alpha x} + D\,e^{-\alpha x},$$

where

$$\alpha^2 = \frac{2m(V-E)}{\hbar^2} \quad V > E$$

Probability per unit length of finding a particle at x

$$P(x) = |\psi(x)|^2$$

Normalization

$$\int |\psi(xyz)|^2\,dx\,dy\,dz = 1$$

all space

Harmonic oscillator

Energy levels $E_n = (n + \frac{1}{2})h\nu$

14

Non-linear Oscillations and Chaos

The oscillations discussed in this book so far have all been restricted in amplitude to those which satisfy the equation of motion where the restoring force is a linear function of the displacement. This restriction was emphasized in Chapter 1 and from time to time its limiting influence has required further discussion; for example, in Chapter 6 on acoustic waves in a fluid. We now discuss some of the consequences when this restriction is lifted.

We begin with simple examples in mechanical, solid state and electrical oscillators. More complicated behaviour associated with chaos in these oscillators is also examined together with the appearance of chaos in biological and fluid mechanical systems.

Free Vibrations of an Anharmonic Oscillator – Large Amplitude Motion of a Simple Pendulum

In Figure 1.1 the equation of motion of the simple pendulum was written in terms of its angular displacement as

$$\frac{\mathrm{d}^2\theta}{\mathrm{d}t^2} + \omega_0^2\theta = 0$$

where $\omega_0^2 = g/l$. Here, an approximation was made by writing θ for $\sin\theta$; the equation is valid for oscillation amplitudes within this limit. When $\theta \geq 7°$ however, this validity is lost and we must consider the more complicated equation

$$\frac{\mathrm{d}^2\theta}{\mathrm{d}t^2} + \omega_0^2\sin\theta = 0$$

Multiplying this equation by $2\mathrm{d}\theta/\mathrm{d}t$ and integrating with respect to t gives $(\mathrm{d}\theta/\mathrm{d}t)^2 = 2\omega_0^2\cos\theta + A$, where A is the constant of integration. The velocity $\mathrm{d}\theta/\mathrm{d}t$ is zero at the maximum angular displacement $\theta = \theta_0$, giving $A = -2\omega_0^2\cos\theta_0$ so that

$$\frac{\mathrm{d}\theta}{\mathrm{d}t} = \omega_0[2(\cos\theta - \cos\theta_0)]^{1/2}$$

The Physics of Vibrations and Waves, 6th Edition   H. J. Pain
© 2005 John Wiley & Sons, Ltd

or, upon integrating,

$$\omega_0 t = \int \frac{\mathrm{d}\theta}{\{2[\cos\theta - \cos\theta_0]\}^{1/2}}$$

If $\theta = 0$ at time $t = 0$ and T is the new period of oscillation, then $\theta = \theta_0$ at $t = T/4$, and using half-angles we obtain

$$\omega_0 \frac{T}{4} = \int_0^{\theta_0} \frac{\mathrm{d}\theta}{2[\sin^2\theta_0/2 - \sin^2\theta/2]^{1/2}}$$

If we now express θ as a fraction of θ_0 by writing $\sin\theta/2 = \sin(\theta_0/2)\sin\phi$, where, of course, $-1 < \sin\phi < 1$, we have

$$\tfrac{1}{2}(\cos\theta/2)\delta\theta = (\sin\theta_0/2)\cos\phi\,\delta\phi$$

giving

$$\frac{\pi}{2}\frac{T}{T_0} = \int_0^{\pi/2} \frac{\mathrm{d}\phi}{[1 - (\sin^2\theta_0/2)\sin^2\phi]^{1/2}}$$

where $T_0 = 2\pi/\omega_0$.

Expansion and integration gives

$$T = T_0(1 + \tfrac{1}{4}\sin^2\theta_0/2 + \tfrac{9}{64}\sin^4\theta_0/2 + \cdots$$

or approximately

$$T = T_0(1 + \tfrac{1}{4}\sin^2\theta_0/2)$$

(Problem 14.1)

Forced Oscillations -- Non-linear Restoring Force

When an oscillating force is driving an undamped oscillator the equation of motion for such a system is given by

$$m\ddot{x} + s(x) = F_0\cos\omega t$$

where $s(x)$ is a non-linear function of x, which may be expressed in polynomial form:

$$s(x) = s_1 x + s_2 x^2 + s_3 x^3 \ldots$$

where the coefficients are constant. In many practical examples $s(x) = s_1 x + s_3 x^3$, where the cubic term ensures that the restoring force $s(x)$ has the same value for positive and negative displacements, so that the vibrations are symmetric about $x = 0$. When s_1 and s_3 are both positive the restoring force for a given displacement is greater than in the linear case and, if supplied by a spring, this case defines the spring as 'hard'. If s_3 is negative the

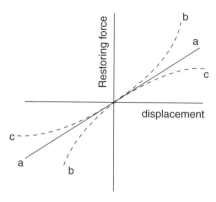

Figure 14.1 Oscillator displacement versus restoring force for (a) linear restoring force, (b) non-linear 'hard' spring, and (c) non-linear 'soft' spring

restoring force is less than in the linear case and the spring is 'soft'. In Figure 14.1 the variation of restoring force is shown with displacement for s_3 zero (linear), s_3 positive (hard) and s_3 negative (soft). We see therefore that the large amplitude vibrations of the pendulum of the previous section are soft-spring controlled because

$$\sin\theta \approx \theta - \tfrac{1}{3}\theta^3$$

Figure 14.2 shows a mass m attached to points D and D′, a vertical distance $2a$ apart, by two light elastic strings of constant stiffness s and subjected to a horizontal driving force $F_0\cos\omega t$. At zero displacement the tension in the strings is T_0 and at a displacement x (not limited in value) the tension is $T = T_0 + s(L - a)$ where L is the stretched string length.

The equation of motion (neglecting gravity) is

$$m\ddot{x} = -2T\sin\theta + F_0\cos\omega t$$
$$= -2[T_0 + s(L - a)]\frac{x}{L} + F_0\cos\omega t$$

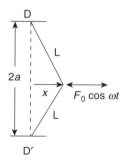

Figure 14.2 A mass m supported by elastic strings between two points D and D′ vertically separated by a distance $2a$ and subjected to a lateral force $F_0\cos\omega t$

Inserting the value

$$L = a\left[1 + \left(\frac{x}{a}\right)^2\right]^{1/2}$$

and expanding this expression in powers of x/a, we obtain by neglecting terms smaller than $(x/a)^3$

$$m\ddot{x} = -\frac{2T_0}{a}x - \frac{(sa - T_0)}{a^3}x^3 + F_0\cos\omega t$$

which we may write

$$\ddot{x} + s_1 x + s_3 x^3 = \frac{F_0}{m}\cos\omega t$$

where

$$s_1 = \frac{2T_0}{ma} \quad \text{and} \quad s_3 = \frac{sa - T_0}{ma^3}$$

If s_3 is small we assume (as a first approximation) the solution $x_1 = A\cos\omega t$, which yields from the equation of motion

$$\ddot{x}_1 = -s_1 A\cos\omega t - s_3 A^3\cos^3\omega t + \frac{F_0}{m}\cos\omega t$$

Since $\cos^3\omega t = \frac{3}{4}\cos\omega t + \frac{1}{4}\cos 3\omega t$, this becomes

$$\ddot{x}_1 = -(s_1 A + \tfrac{3}{4}s_3 A^3 - F_0/m)\cos\omega t - \tfrac{1}{4}s_3 A^3\cos 3\omega t$$

Integrating twice, where the constants become zero from initial boundary conditions, gives as a second approximation to the equation

$$\ddot{x} + s_1 x + s_3 x^3 = \frac{F_0}{m}\cos\omega t$$

the solution

$$x_2 = \frac{1}{\omega^2}\left(s_1 A + \frac{3}{4}s_3 A^3 - \frac{F_0}{m}\right)\cos\omega t + \frac{s_3 A^3}{36\omega^2}\cos 3\omega t$$

Thus, for s_3 small we have a value of ω appropriate to a given amplitude A, and we can plot a graph of amplitude versus driving frequency. Note that we have a third harmonic. We see that for a system with a non-linear restoring force resonance does not exist in the same way as in the linear case. In the example above, even when no damping is present, the amplitude will not increase without limit for a driving force of a given frequency, for if ω is the natural frequency at low amplitude it is no longer the natural frequency at high amplitude. For s_3 positive (hard spring) the natural frequency increases with increasing amplitude and the

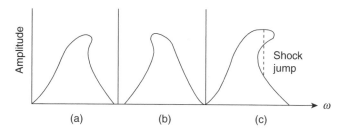

Figure 14.3 Response curves of amplitude versus frequency for oscillators having (a) a 'hard' spring restoring force, and (b) a 'soft' spring restoring force. In the extreme case (c) the tilt of the maximum is sufficient to allow multi-valued amplitudes at a given frequency and 'shock jumps' may occur (See Figure 15.1 for comparable behaviour in a high amplitude sound wave.)

amplitude versus frequency curve has a tilted maximum (Figure 14.3a). For a soft spring, s_3 is negative and the behaviour follows Figure 14.3b. It is possible for the tilt to become so pronounced (Figure 14.3c) that the amplitude is not single valued for a given ω and shock jumps in amplitude may occur (see the next chapter on the development of a shock front in a high amplitude acoustic wave).

(Problems 14.2, 14.3)

Thermal Expansion of a Crystal

Chapter 1 showed that the curve of potential energy versus displacement for a linear oscillator was parabolic. Small departures from this curve are consistent with anharmonic oscillations. Consider the potential energy curve for a pair of neighbouring ions of opposite charge $\pm e$ in a crystal lattice such as that of KCl. If r is the separation of the ions the mutual potential energy is given by

$$V(r) = \frac{\alpha e^2}{r} + \frac{\beta}{r^p}$$

where α and β are positive constants and $p = 9$. This is plotted in Figure 14.4, which shows that the potential energy curve is no longer parabolic. The first term of $V(r)$ is the energy due to Coulomb attraction; the second is that of a repulsive force. The value of α depends upon the presence of neighbouring ions and is about 0.3. The constant β can be found in terms of α and the equilibrium separation r_0 because, in equilibrium,

$$\left(\frac{dV}{dr}\right)_{r=r_0} = \frac{\alpha e^2}{r_0^2} - \frac{p\beta}{r_0^{p+1}} = 0$$

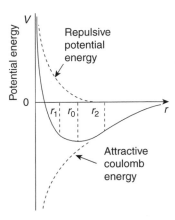

Figure 14.4 Non-parabolic curve of mutual potential energy between oppositely charged ions in the lattice of an ionic crystal (NaCl or KCl). The combination of repulsive and attractive forces yields an equilibrium separation r_0. Very small energy increments give harmonic motion about r_0 but oscillations at higher energies are anharmonic, leading to thermal expansion of the crystal

giving

$$\beta = \frac{\alpha e^2 r_0^{p-1}}{p}$$

X-ray diffraction from such crystals gives $r_0 = 3.12\,\text{Å}$ for KCl, so that β may be found numerically.

To consider small displacements from the equilibrium value r_0 let us expand $V(r)$ about $r = r_0$ in a Taylor series to give

$$V(r) = V(r_0) + x\left(\frac{\mathrm{d}V}{\mathrm{d}r}\right)_{r_0} + \frac{x^2}{2!}\left(\frac{\mathrm{d}^2V}{\mathrm{d}r^2}\right)_{r_0} + \frac{x^3}{3!}\left(\frac{\mathrm{d}^3V}{\mathrm{d}r^3}\right)_{r_0}$$

where $x = r - r_0$. Since $(\mathrm{d}V/\mathrm{d}r)_{r_0} = 0$, we may write

$$V(r) - V(r_0) = V(x) = A\frac{x^2}{2!} + \frac{Bx^3}{3!}$$

The quantity $Ax^2/2$ is the quadratic term familiar in the linear oscillator, so that for very small disturbances the bottom of the potential energy curve is parabolic, and a small gain in energy causes the ion pair to oscillate symmetrically about $r = r_0$. An increase in the ion pair energy involves the second term $Bx^3/6$, and oscillations are no longer symmetric about r_0, because $|r_2 - r_0| > |r_1 - r_0|$ in Figure 14.4. Hence the time average for $r - r_0$ is not zero as it is for a linear oscillator, and this time average $r_t > r_0$. If all ion pairs acquire this amount of energy, for example by heating, the crystal expands. We may consider the force between the two ions as

$$F = -\frac{\mathrm{d}V}{\mathrm{d}x} = -Ax - \frac{Bx^2}{2}$$

and note that the quadratic term here is responsible for the lack of symmetry in the motion. If it were a cubic term as in the previous example the symmetry of motion about r_0 would still occur. The coefficient A in the force equation is the force constant in the discussion on crystals in Chapters 5 and 6 and leads directly to Young's modulus. The coefficient B gives information on the coefficient of thermal expansion of the crystal.

(Problems 14.4, 14.5)

Non-linear Effects in Electrical Devices

A feature of the non-linearity in the mechanical devices discussed earlier was the introduction of harmonics of the fundamental frequency of the driving force. It is comparatively simple to avoid these effects of non-linearity in electronic systems by choosing a small linear portion of the operating characteristic and amplifying the response in stages. In an electromechanical device such as a piezoelectric crystal linearity is again achieved by restricting all oscillations to small amplitudes and amplifying the response. In electroacoustic devices such as microphones and loudspeakers the introduction of harmonics often leads to severe distortion. In the loudspeaker of Figure 14.5 even if a pure sinusoidal wave is delivered to the speech coil it is difficult to provide a mechanical suspension for the cone which has a linear response. The cone acts as a piston radiating acoustic power, and limitation of amplitude together with inevitable mismatching of acoustic impedances reduces the efficiency of transforming electrical into acoustic power to less than 10%. Fortunately the ear is a sensitive device.

Non-linear electrical oscillators are, however, often used, and Figure 14.6a shows a 'relaxation oscillator' circuit where a capacitance is discharged very rapidly through a gaseous conductor such as a hydrogen tube. E is the constant charging potential and i is the instantaneous value of the current which charges the capacitor through the resistor R to a potential V_s, the striking potential, at which the gas in the tube is ionized. The tube

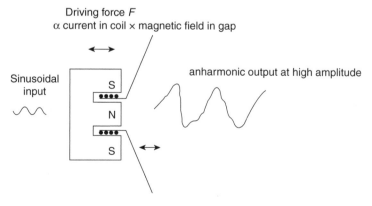

Figure 14.5 A pure sinusoidal wave input to an electroacoustical device such as a loudspeaker will lead to distorted sound output if the cone suspension has a non-linear stiffness at high amplitudes

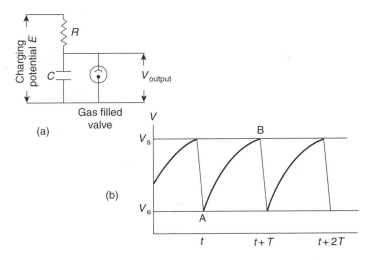

Figure 14.6 Electrical circuit of a non-linear 'relaxation oscillator'. A capacitance C is charged through a resistance R to a potential $V_s < E$, at which the gas-filled valve strikes and rapidly discharges the condenser to an extinction potential V_e, when the valve ceases to conduct and the cycle is repeated

becomes highly conducting and discharges the capacitance in a negligibly short time to V_e, the extinction potential, at which the tube ceases to conduct. The capacitance charges again to V_s and the cycle is repeated. The variation of voltage across the capacitance with time is shown in Figure 14.6b. Assume that at point A and time t the capacitance has just discharged. If current i_0 is flowing at time $t = 0$ then

$$V_e = E - i_0 R e^{-t/RC}$$

The capacitance charges to the potential V_s in a time τ so that

$$V_s = E - i_0 R e^{-(t+\tau)/RC}$$

giving

$$
\begin{aligned}
V_s - V_e &= i_0 R \left(e^{-t/RC} - e^{-(t+\tau)/RC} \right) \\
&= i_0 R e^{-t/RC} [1 - e^{-\tau/RC}] \\
&= (E - V_e)[1 - e^{-\tau/RC}]
\end{aligned}
$$

giving

$$e^{-\tau/RC} = \frac{E - V_s}{E - V_e}$$

or

$$\tau = RC \left[\log_e \left(\frac{E - V_e}{E - V_s} \right) \right]$$

The period of oscillation is therefore directly proportional to the charging time constant *RC*.

A more sophisticated circuit produces a linear charging system with a very short discharge time so that the exponential voltage output becomes linear and gives a 'sawtooth' waveform. From Chapter 10 we know that this periodic function contains many harmonics. A sawtooth voltage output applied to the time base of an oscilloscope produces a linear sweep of the spot across the tube.

Electrical Relaxation Oscillators

Van der Pol and Chaos (1926–1927)

The work of Van der Pol continues to attract the attention of research workers in chaos chiefly because of an equation he derived at that time. His relaxation oscillator was a multivibrator, a two stage resistance-capacity coupled amplifier with the output of the second triode fed back as input to the grid of the first. His analysis used the mechanical form of the damped simple harmonic equation with a negative resistance term which increased the amplitude, thus

$$\ddot{x} - \alpha x + \omega^2 x = 0$$

with a solution

$$x = C e^{+\alpha t/2} \sin\left[(\omega^2 - \alpha^2/4)t + \phi\right]$$

for $\alpha > 0$ and $\alpha^2/4 < \omega^2$.

He restricted the unlimited growth of x by replacing α with $\alpha - 3\gamma/x^2$ where γ is a constant, writing $\omega t = t'$ and $x = (\alpha/3\gamma)^{1/2} v$ to give his equation the form

$$\ddot{v} - \varepsilon(1 - v^2)\dot{v} + v = 0$$

where $\varepsilon = \alpha/\omega$ and $\dot{v} = \mathrm{d}v/\mathrm{d}t'$.

It is this equation with a forcing term $A \sin \omega_0 t$ on the right hand side which is known as Van der Pol's equation and which has formed the basis of a number of studies in chaos, one of which we shall meet later. Van der Pol found that as ε increased his oscillator gradually assumed the period $\tau = RC$ with the output for $\varepsilon = 10$ shown in Figure 14.7 (Van der Pol, 1926).

Even more interesting from the viewpoint of chaos was the oscillator by which he could produce subharmonics of its natural frequency. Such a phenomenon, period doubling, tripling, etc. is now recognized as an early sign of chaos, indeed Li and Yorke (1975) have published a paper entitled 'Period 3 implies Chaos'.

Van der Pol's period doubling circuit is shown in Figure 14.8. With $E_0 = 0$ and $C = 10^{-3} \mu F$ the relaxation frequency of the system was 10^3 cycles. Setting $E_0 \sin \omega t$ at $7.5 \sin 2\pi 10^3 t$ he was able, by increasing C through the range 5–$40 \times 10^{-3} \mu F$ to produce subharmonics $\omega/2, \omega/3 \ldots \omega/40 \ldots \omega/200$. He registered the output on a pair of loosely coupled telephone earpieces and his paper makes the interesting comment that 'often an irregular noise is heard in the telephone receivers before the frequency jumps, however this is a subsidiary phenomenon'. In fact, such *internally* generated noise accompanied by subharmonics is one of the early signs of chaos (Van der Pol and Van der Mark, 1927).

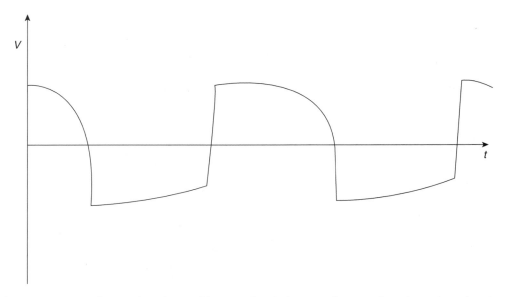

Figure 14.7   Non-linear relaxation oscillations of period $\tau = RC$ for an unforced Van der Pol system

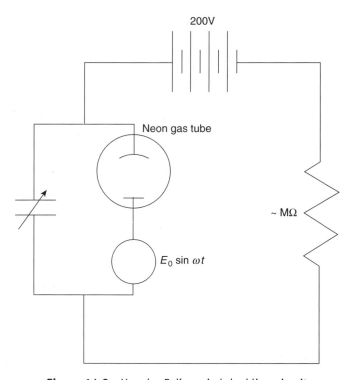

Figure 14.8   Van der Pol's period doubling circuit

Chaos in Population Biology

Chronological accounts of a modern research topic rarely present the most coherent picture. The significance of early developments is not always recognized until much later; indeed the first recorded strange or chaotic attractor, that of Lorenz in 1963, comes at the end of this account but only because of its level of sophistication. Even the simple example with which we begin was not fully explained when it first appeared.

Despite its simplicity the example of population biology reveals many of the characteristics displayed by chaotic systems. These are:

- The chaos is deterministic and not random; that is, the paths followed by trajectories are governed by solutions to given non-linear equations.

- Trajectories from closely neighbouring starting points diverge with time.

- Trajectories can, according to the conditions, finish on a stable point attractor, they can diverge to infinity from a repellor or at some stage they can orbit in what is known as a limit cycle.

- Such a limit cycle can develop an infinite series of period doubling; odd number periods may be generated, also completely aperiodic trajectories which still remain within a bounded region of space.

- With the appearance of chaotic motion the sharp definition of these frequencies is gradually overcome by a growing background of wide band noise which is *internally* generated.

A number of equations dealing with population biology has been widely studied but we consider the simplest, a quadratic equation discussed by May (1976) in a classic review.

This is known as the logistic map and is given by

$$x_{n+1} = 4\lambda x_n(1 - x_n)$$

where the subscripts refer to the year in which the population was measured and λ is a parameter. Restricting the values of x and λ to $0 < x < 1$ and $0 < \lambda < 1$ is a scaling device which keeps the dynamics within the limits of a diagram. Because it involves only the coordinate x this logistic equation is known as a one-dimensional map.

Much of the behaviour of populations under this quadratic rule is shown by the interaction of the parabola and the straight line bisector $x_{n+1} = x_n$ of gradient unity and this behaviour is divided into three distinct categories by the λ ranges $0 < \lambda < \frac{1}{4}, \frac{1}{4} < \lambda < \frac{3}{4}$ and $\frac{3}{4} < \lambda < 1$.

To illustrate the general use of the bisector consider what happens to a population with a constant reproduction rate; that is, the straight line $x_{n+1} = 4\lambda x_n$. Figure 14.9a shows the line for $\lambda > \frac{1}{4}$ compared with the bisector $x_{n+1} = x_n$. Taking x_0 as the starting value of the population gives x_1 on the $\lambda > \frac{1}{4}$ line which then projects horizontally to the same value (x_1) on the bisector. This gives the value x_1 on the base line which projects vertically to the $\lambda > \frac{1}{4}$ line to give x_2 and the process is repeated. Evidently for $\lambda > \frac{1}{4}$ the population

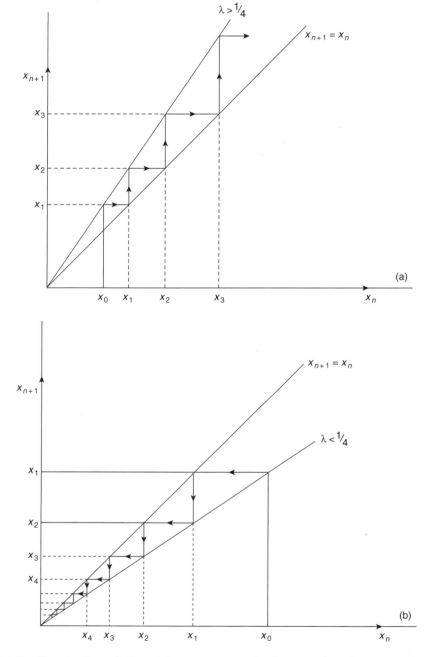

Figure 14.9   Change of population with constant reproduction rate given by $x_{n+1} = 4\lambda x_n$. (a) for $\lambda > \frac{1}{4}$ the population $\to \infty$ as the trajectories move away from the origin (a repellor). (b) For $\lambda < \frac{1}{4}$ the population is extinguished, all trajectories moving to the stable point attractor at zero. The initial population at x_0 gives x_1 on the $\lambda > \frac{1}{4}$ line which projects horizontally to the same value on the bisector $x_{n+1} = x_n$. The value x_1 projects vertically to x_2 on the $\lambda > \frac{1}{4}$ line and the process repeats itself. Similarly for $\lambda < \frac{1}{4}$

increases without limit, the trajectories move to infinity from a repellor. For $x < \frac{1}{4}$, Figure 14.9b, the same process of horizontal and vertical projection produces $x_1 < x_0$ and the population is extinguished, all trajectories moving to the stable point attractor at zero.

The method is equally applicable to the parabola

$$x_{n+1} = 4\lambda x_n (1 - x_n)$$

For $\lambda > \frac{1}{4}$ we have Figure 14.10 and where the curve and the bisector intersect we have $x_{n+1} = x_n$ corresponding to a fixed point in the iteration process. Writing this value as $x_{n+1} = x_n = x^*$ we find from $x^* = 4\lambda x^* (1 - x^*)$ the two roots $x^* = 0$ and $x^* = 1 - \frac{1}{4\lambda}$ each of which is a fixed point.

Restricting x and λ to the values between 0 and 1 gives for $\lambda < \frac{1}{4}$ only the value $x^* = 0$ but for $\frac{1}{4} < \lambda < 1$, x^* may take both values. If x^* is stable; that is, a fixed point to which the end points of all trajectories become infinitely close, Figure 14.10, it is a point attractor and this stability depends on the slope of the curve at x^*. We write $x_{n+1} = 4\lambda x_n (1 - x_n) = f(x_n)$ and if $-1 < f'(x) < 1$ at x^*, x^* is stable. When the slope $f'(x)$ equals -1 stability is lost and x^* bifurcates into two new values, each of which is stable. This is called a pitchfork bifurcation and is the origin of the period doubling sequence in the logistic map. Odd numbered periodic cycles arise at a later stage from bifurcations into pairs of new values, only one of each pair being stable. These are called tangent bifurcations.

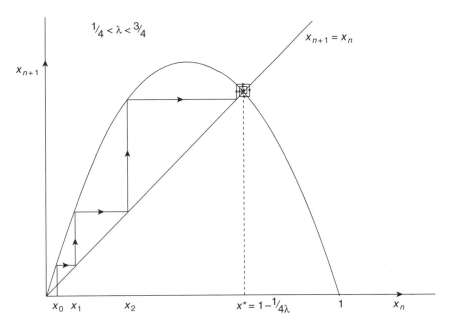

Figure 14.10 The logistic equation $x_{n+1} = 4\lambda x_n (1 - x_n)$ cut by the bisector $x_{n+1} = x_n$ at the points $x^* = 0$ and $x^* = 1 - \frac{1}{4\lambda}$. When $\frac{1}{4} < \lambda < \frac{3}{4}$ the latter value of x^* is a stable point attractor for all trajectories as shown

The dependence of stability upon $f'(x)$ at the fixed point x^* follows from Taylor's theorem for, with $x_{n+1} = f(x_n)$ and $x_n = x^* + \varepsilon_n$ where ε_n is a very small quantity, we have

$$x_{n+1} = f(x^* + \varepsilon_n) \approx f(x^*) + \varepsilon_n f'(x^*)$$
$$= x^* + \varepsilon_n f'(x^*)$$

because $x^* = f(x^*)$ at this fixed point x^*.

Now $x_{n+1} = x^* + \varepsilon_{n+1}$, giving $f'(x^*) = \varepsilon_{n+1}/\varepsilon_n$ and for $n \to \infty$, $\varepsilon_{n+1} \to 0$ only if $-1 < f'(x) < 1$.

Thus, $x^* = 0$ is a stable point attractor for all trajectories when $\lambda < \frac{1}{4}$ but becomes unstable at $\lambda = \frac{1}{4}$ while $x^* = 1 - \frac{1}{4\lambda}$ is a stable point attractor for all trajectories when $\frac{1}{4} < \lambda < \frac{3}{4}$. At $\lambda = \frac{3}{4}$ the slope of $f(x)$ at $x^* = 1 - \frac{1}{4\lambda}$ equals -1, stability is lost, x^* bifurcates and a stable oscillation between two new values x_1^* and x_2^* develops. We can see this by studying the behaviour of x_{n+2} versus x_n, obtained by a double application of the logistic equation.

We can express $x_{n+2} = f(x_{n+1}) = ff(x_n) = f^2(x_n)$ where the superscript defines the double application. A graph of $f^2(x)$, which is symmetric, is shown in Figure 14.11a where the central minimum decreases as λ increases. The bisector is now of course $x_{n+2} = x_n$ and, as shown, it cuts $f^2(x)$ at three fixed points. The value of λ is chosen so that x_1^* is near the minimum and x_2^* is near a maximum. The slope of $f^2(x)$ (written $f^{2\prime}(x)$) at x_1^* and x_2^* is therefore close to zero and x_1^* and x_2^* are stable fixed points of $f^2(x)$. It is at this value of $\lambda = \frac{3}{4}$ that period doubling begins.

The third fixed point x^* is clearly the original fixed point of $f(x)$. This follows from noting that the point $x^* = x_n = x_{n+1} = x_{n+2}$ falls on both $f(x)$ and $f^2(x)$ and on their respective bisectors. In addition, the stability behaviour of x^* is the same for $f(x)$ and $f^2(x)$. We can show this via the chain rule, for if

$$x_2 = f(x_1) = f^2(x_0) \quad \text{where} \quad x_1 = f(x_0)$$

then

$$f^{2\prime}(x) = f'(x_1) = \frac{d[f(x_1)]}{dx_1} \frac{dx_1}{dx} = \left[\frac{df(x_1)}{dx_1} \right] f'(x)$$

where all derivatives are evaluated at $x = x_0$. This result holds for higher values of the superscript n in $f^n(x)$.

Taking x_0 as the fixed point x^* then

$$x^* = x_0 = x_1 = x_2$$

and

$$f^{2\prime}(x^*) = f'(x^*)f'(x^*) = (f'(x^*))^2.$$

Thus, if x^* is stable (unstable) in $f(x)$ then it is stable (unstable) in $f^2(x)$.

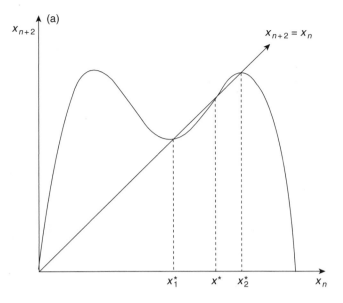

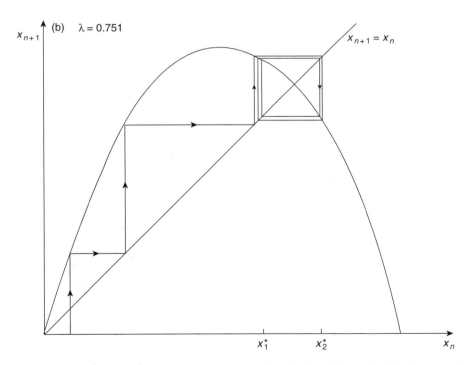

Figure 14.11 (a) x_1^* and x_2^* are two of the three fixed points formed by the intersection of $x_{n+2} = f^2(x_n)$ and its bisector $x_{n+2} = x_n$. The third fixed point is the original fixed point $x^* = 1 - \frac{1}{4\lambda}$ of $x_{n+1} = f(x_n)$. (b) When the value of λ is just greater than $\frac{3}{4}$ period doubling begins between two new fixed points x_1^* and x_2^*

The stable fixed points x_1^* and x_2^* of $f^2(x)$ for $\lambda > \frac{3}{4}$ are not fixed points of $f(x)$. Clearly, since these points lie on the bisector $x_{n+2} = x_n$, each will return to itself every second iteration. This can occur only when the expressions

$$x_1^* = f(x_2^*) \quad \text{and} \quad x_2^* = f(x_1^*)$$

jointly hold so a trajectory ends in the cycle $x_1^* \to x_2^* \to x_1^* \to x_2^*$, Figure 14.11(b).

(Problem 14.6)

In the same way that x_1^* and x_2^* became the two stable points at $\lambda_1 = \frac{3}{4}$ they will become simultaneously unstable for some larger value λ_2 when $f^{2\prime}(x^*) = -1$. At λ_2, x_1^* and x_2^* will each bifurcate to two stable points to give a stable 4-cycle period based on the stable fixed points of $f^4(x)$. As the period doubling sequence continues, via pitchfork bifurcations, the values $\lambda_1, \lambda_2, \lambda_3, \lambda_4 \ldots$ for the cycles $2, 2^2, 2^3, 2^n \ldots$ converge geometrically and Feigenbaum (1978) found that for this period doubling sequence the limit as $n \to \infty$ is given by

$$\delta_{n \to \infty} = \frac{\lambda_{n+1} - \lambda_n}{\lambda_{n+2} - \lambda_{n+1}} = 4.6692016$$

This result appears to be verified not only for the logistic map but for other non-linear equations with a single maximum and many experiments, computer simulated and otherwise, support Feigenbaum's result.

The value of λ at which the cycle $2^n (n \to \infty)$ is approached is given by $\lambda_\infty = 0.8925$. This is illustrated in Figure 14.12 where the successive bifurcations of 2^n cycles become

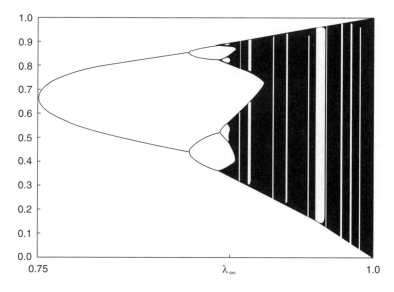

Figure 14.12 Bifurcations at period doubling for the logistic map begin at $\lambda = \frac{3}{4}$ and reach the limit 2^∞ at λ_∞. Between λ_∞ and $\lambda = 1$ chaotic behaviour is interspersed with regions or windows at which odd numbered cycles of period k and their harmonics $k2^n$ appear. Some cycles are aperiodic (Figure 14.13). (From Tabor, 1989)

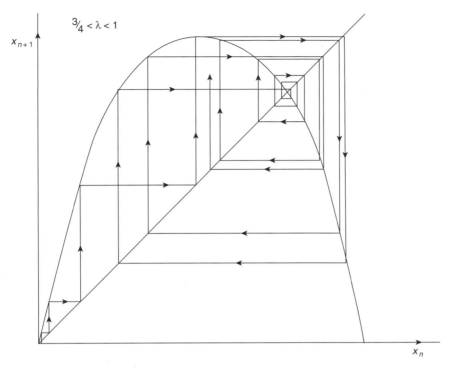

x_{n+1}

$3/4 < \lambda < 1$

x_n

Figure 14.13 An aperiodic cycle which remains bounded within the system for $\frac{3}{4} < \lambda < 1$

increasingly compressed in the λ space. Between the values of λ_∞ and $\lambda = 1$ a very rich behaviour is observed; there is an infinite number of different periodicities and an uncountable number of very long cycles of no measurable period but which remain bounded within the system (Figure 14.13).

The order in which these cycles appear has been successfully predicted by Metropolis *et al.* (1973). The first odd cycle appears at $\lambda = 0.9196$ and the first period 3 cycle appears at $\lambda = 0.9571$. This is an important cycle because of the paper by Li and Yorke entitled 'Period 3 implies Chaos' (Li and Yorke, 1975).

We can examine the origin of the first period 3 cycle in Figure 14.14(a). At some value λ^* the bisector $x_{n+3} = x_n$ is tangent to the curve $x_{n+3} = f(x_n)$ at the three fixed points x_1^*, x_2^*, x_3^*. The slope of $f^3(x_n)$ at these points must equal $+1$ and each of these three unstable fixed points bifurcates into a pair of which one is stable and the other is unstable. This is the tangent bifurcation. The period 3 cycle orbits between the three stable fixed points (one from each bifurcation) and we can follow the bifurcation process by increasing λ beyond λ^* by a small quantity. This heightens the maxima and deepens the minima so that the bisector now cuts $f^3(x_n)$ in pairs of points one on each side of the tangent position. A typical pair is shown in Figure 14.14(b) on a magnified scale. The tangent point T splits into points A and B each of which moves along the curve from T as λ increases. Point A moves from a gradient position of $+1$ around the curve maximum to a gradient position of less than 1 and forms the stable fixed point of the bifurcated pair. Point

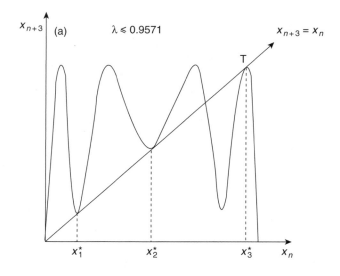

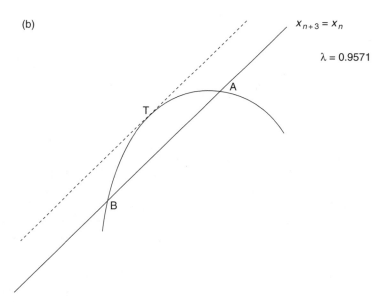

Figure 14.14 (a) The first period 3 cycle appears at $\lambda = 0.9571$. Just below this value of λ the bisector $x_{n+3} = x_n$ is tangent to $x_{n+3} = f^3(x_n)$ at three unstable fixed points (gradient $= +1$). A small increase of λ splits these points into pairs, one point of each pair becoming stable. (b) Magnification at tangent point T which splits into a pair A and B with a small increase in λ. At T the gradient is $+1$ (unstable), A is stable at a reduced gradient and B is unstable at an increased gradient

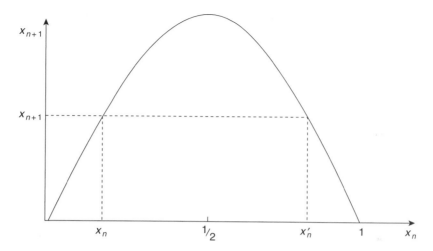

Figure 14.15 The one-dimensional logistic equation $x_{n+1} = 4\lambda x_n(1-x_n)$ is non-invertible because trajectories cannot be traced uniquely backwards to their origins. Each x_{n+1} can arise from two different values of x_n

B moves from T (gradient +1) along the curve to a steeper gradient position and remains unstable.

Thus, to quote May, 'the fundamental stable dynamical units are of basic period k which arise by tangent bifurcation along with their associated cascade of harmonics of periods $k2^n$ which arise by pitchfork bifurcation. The hierarchy of stable cycles of period 2^n (namely, $k = 1$) is merely a special case albeit a conspicuously important one'.

The one dimensional logistic map has one profound limitation. Figure 14.15 shows that it is symmetric about the point $x = \frac{1}{2}$ so that any x_{n+1} can arise from one of two different values x_n and x'_n. This fails an essential requirement in chaos theory, namely that all trajectories may be traced uniquely backwards in time to their origins. This property is known as 'invertibility' and clearly the logistic map is non-invertible.

Chaos in a Non-linear Electrical Oscillator

The development of the varactor has made it possible to display many features of the preceding section on a cathode ray oscilloscope in a first year university laboratory experiment. The varactor acts as a diode in the forward direction but behaves in the reverse direction as a variable non-linear capacitance in a series LCR circuit. Testa *et al.* (1982) confirmed not only many of the results above but, in addition, supported two predictions made by Feigenbaum (1979). These were

1. That bifurcation at period doubling follows a distinct procedure—as a 2^n cycle loses stability after 2^n iterations, a point of the attractor just misses duplicating itself with duplication occurring only after another 2^n iterations. Thus each element of the cycle splits into closely spaced pairs with 2^n iterations required to visit an element from its

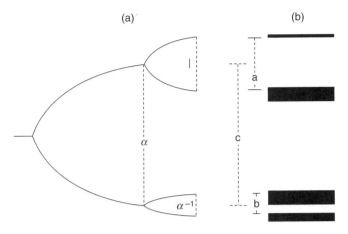

Figure 14.16 In the period doubling process the separation of adjacent elements in a pair is reduced by a universal factor α from one bifurcation to the next. For period doubling between 16 and 32 $\alpha = a/b = 2.35$ and $\alpha = c/a = 2.61$. Reproduced by permission of The American Physical Society from Testa *et al.* (1982)

    adjacent neighbour. From one bifurcation to the next, separation of adjacent elements in a pair is reduced by a universal factor $\alpha = 2.5029$ (Figure 14.16).

2. After a spectral component in the period doubling process has been generated, its amplitude remains approximately constant during further bifurcation and each new subharmonic of this frequency can be predicted as having its amplitude reduced by $10 \log_{10}\mu$ where

$$\mu = \frac{4\alpha}{(2 + \frac{1}{\alpha^2})^{1/2}} = 6.57$$

A typical varactor LCR circuit is shown in Figure 14.17 with the non-linear capacitance given by

$$C(V) = C_0/(1 + V_c/\beta)^\gamma$$

where V_c is the voltage across the varactor. In Testa's experiment $C_0 = 300\,\text{pF}$, $\beta = 0.6, \gamma = 0.5, L = 10\,\text{mH}$ and $R = 28\,\Omega$. For low values of V_0 this gave a high Q resonance circuit at a frequency of 93 kHz. With f fixed near the resonance frequency in the driving voltage $V_0 \sin 2\pi f t$, V_0 was varied and the varactor voltage $V_c(t)$ was measured. Testa *et al.* assumed that V_0 played the role of λ in the logistic equation and that V_c corresponded to x. A real time display on a double beam CRO of $V_c(t)$ and $V_0(t)$ clearly revealed threshold values of V_{0n} for bifurcations into subharmonics f/n where $n = 2, 4, 8, 16$. At $n = 4$ (not shown by Testa) this would appear as Figure 14.18.

    Figure 14.19 was obtained on the oscilloscope screen by Testa with a slow horizontal scan of V_0 versus the varactor voltage V_c which was magnified in selected steps of 10 mV. The numbers on the horizontal axis indicate the generation of particular periods and

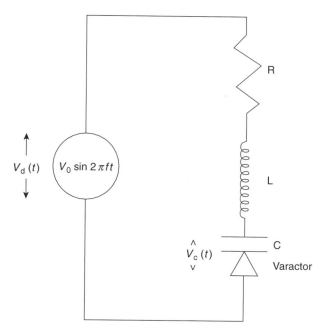

Figure 14.17 Non-linear LCR series circuit where the non-linear element is the varactor C which acts as a diode in the forward direction but becomes a variable non-linear capacitance in the reverse direction

bifurcations are clearly visible. The threshold values of V_0 for these periods are shown in Table 14.1. The first four threshold values V_{0n} gave

$$\delta_1 = \frac{V_{02} - V_{01}}{V_{03} - V_{02}} = 4.257 \pm 0.1$$

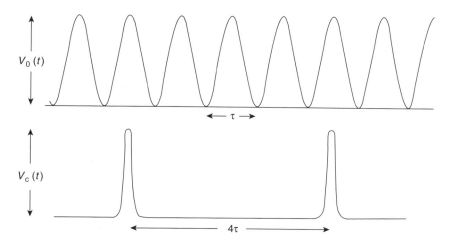

Figure 14.18 Double beam oscilloscope showing driving voltage $V_0(t)$ at frequency f and varactor voltage $V_c(t)$ at frequency $f/4$. Values of V_{0n} for appearance of f/n are given in Table 14.1

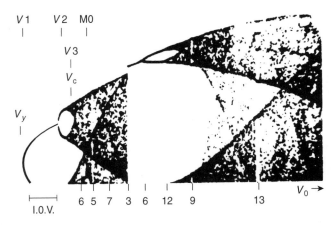

Figure 14.19 Slow horizontal scan of V_0 versus V_c. The numbers on the horizontal axis indicate the generation of particular periods. Bifurcations are clearly visible. Threshold values of V_0 for various periods are shown in Table 14.1. Reproduced by permission of The American Physical Society from Testa *et al.* (1982)

Table 14.1   Table of periods and the threshold values V_0 at which they appear

Period	Threshold V_0 (rms volts)	comments
2	0.639	
4	1.567	
8	1.785	Threshold for periodic bifurcation
16	1.836	
32	1.853	
Chaos	1.856	Onset of noise
12	1.901	Window
24	1.902	
6	2.073	Window
12	2.074	
5	2.353	Window
10	2.363	
7	2.693	Window
14	2.696	
3	3.081	
6	3.338	Wide Window
12	3.711	
24	3.821	
9	4.145	Window
18	4.154	

Reproduced by pemission of the American Physical Society from Testa *et al.* (1982)

and

$$\delta_2 = \frac{V_{03} - V_{02}}{V_{04} - V_{03}} = 4.275 \pm 0.1$$

in the Feigenbaum convergence series.

To test the first of Eigenbaum's predictions the values of c, a and b in Figure 14.16 were measured for the bifurcations between periods 16 and 32. These gave

$$\alpha = \frac{a}{b} = 2.35 \quad \text{and} \quad \alpha = \frac{c}{a} = 2.61$$

As periods doubled the power reduction in their frequency components was measured and the results were consistent with Feigenbaum's analysis.

Phase Space

One of the most vital concepts in the description of chaos is that of phase space. In one dimension, e.g. the logistic equation, trajectories can be followed without introducing it. In higher dimensions it is essential.

The idea of phase space has many applications in physical sciences. Students meet it initially in the Maxwell–Boltzmann statistical distribution where the question is asked: 'Given N gas particles at a temperature T occupying a volume V, what fraction of N will be found in the velocity range v to $v + dv$ in the small volume range dV?' We shall discuss this application to statistical distributions in an appendix at the end of the book.

The number of dimensions of phase space is determined by the number of coordinates required to define the complete physical state of the system. For each gas particle above we need six dimensions, three for the v_x, v_y, v_z components in velocity space and three for the $x\, y\, z$ components in the configuration space V.

Each point in phase space defines the complete physical state of the system (here a gas particle) and trajectories in phase space follow the physical development of the system.

When the energy of an ensemble of systems (particles) is conserved the phase space or volume associated with them remains constant, but if any energy is dissipated the phase volume contracts. This contraction generates a sub-space, there is a reduction in the number of coordinates required and their range is reduced.

Figures 14.20–14.23 show, in turn, the two dimensional phase space diagrams of different oscillators using the coordinates $\dot{x}$ and x.

1. A linear simple harmonic oscillator (Figure 14.20).

2. a damped simple harmonic oscillator (Figure 14.21).

3. an undamped non-linear oscillator formed by a pendulum supported on a light rigid rod (Figure 14.22)

4. (a) an undamped oscillator with a potential energy

$$V = -\tfrac{1}{2}ax^2 + \tfrac{1}{4}bx^4$$

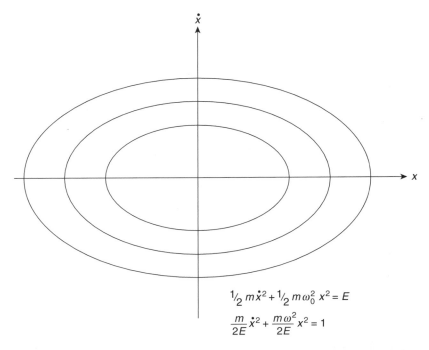

$$\tfrac{1}{2}\,m\dot{x}^2 + \tfrac{1}{2}\,m\omega_0^2\,x^2 = E$$

$$\frac{m}{2E}\dot{x}^2 + \frac{m\omega^2}{2E}x^2 = 1$$

Figure 14.20 Linear simple harmonic oscillator represented in the two dimensional phase space of $\dot{x}$ and x. Each ellipse corresponds to a curve of constant energy and encloses a constant area of phase space

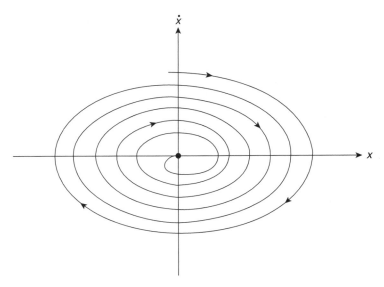

Figure 14.21 The energy loss per cycle in a damped simple harmonic oscillator is shown in its phase space diagram as a reduction of area with each cycle as its trajectory spirals to a stable point attractor at the origin

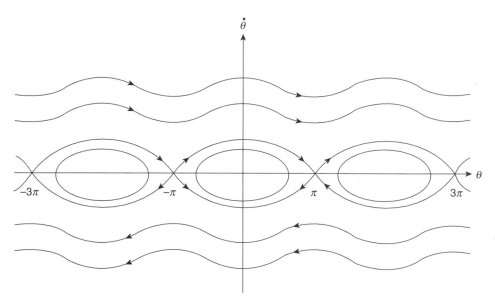

Figure 14.22 Phase portraits for a non-linear pendulum on a light rod. The closed curves represent energy values up to the limit $\dot{\theta} = 0$ at pendulum amplitude $\theta = \pm \pi$ ($\theta = 0$ is the hanging rest position). The open curves represent fast rotations with energy values large enough for $\dot{\theta} > 0$ at $\theta = (2n + 1)\pi$

(b) the oscillator of 4(a) now lightly damped (Figure 14.23).
The features of each will now be described, introducing ideas which are frequently met in chaotic systems.

1. The trajectory in $\dot{x}x$ phase space for a simple harmonic oscillator of constant energy is an ellipse of constant area. Its potential energy curve is the familiar parabola of p. 10.

2. For a lightly damped simple harmonic oscillator where energy is dissipated the phase space is an inward spiral on to the equilibrium zero position which is a stable point attractor. As energy is lost each orbit of the spiral encloses a smaller element of phase space than its predecessor, unlike (1).

3. Here we plot the phase portraits for a large range of pendulum energies E. The closed curves represent those energy values up to the limit where the pendulum (rigid rod) stands on its head with zero velocity and angular amplitude $\theta = \pm \pi$ measured from the hanging rest position. Higher E values have open curves because their rotations are fast enough to pass through the values of $\theta = (2n + 1)\pi$ with velocities $\dot{\theta} > \theta$. The largest closed curve has pointed ends, at maximum amplitude θ, because $\dot{\theta}$ is small for changes of θ in that range. Each interval of 2π along the horizontal axis represents a complete rotation.

    The curves passing through $\theta = \pm \pi$ evidently separate those energies capable of allowing complete rotations from those which cannot. Such a curve is called a separatrix and the points $\theta = \pm \pi$ are called saddle points.

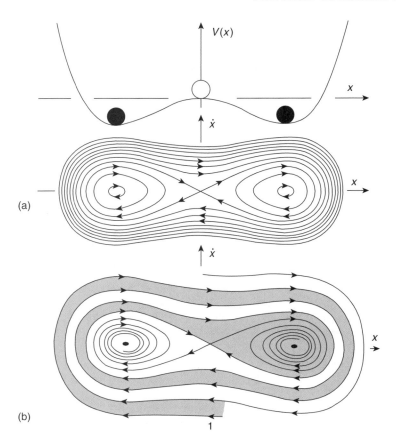

Figure 14.23   Potential energy curve $V = -\frac{1}{2}ax^2 + \frac{1}{4}bx^4$ with phase portraits for the damped and undamped oscillators. For the undamped oscillator energies $V(x) < 0$ restrict the motion to that potential well containing the $\dot{x}x$ starting position. (a) When the starting position is on the curve $V(x) > 0$ the trajectories cross the potential barrier repeatedly. (b) For the damped oscillator trajectories from a given range of $\dot{x}x$ starting positions will finish at the bottom of a particular potential well (indicated by the shaded region known as the basin of attraction). The other basin of attraction is unshaded. Reproduced by permission of John Wiley & Sons from Thompson and Stewart (1986)

4. The potential energy curve $V = -\frac{1}{2}ax^2 + \frac{1}{4}bx^4$ is drawn together with the phase portraits for the undamped and damped oscillators. For the undamped oscillator any starting position with total energy less than $V(x) = 0$ restricts the motion to one or other of the potential wells. For any starting position greater than $V(x) = 0$ the motion may cross the potential barrier repeatedly. The trajectory associated with motion starting from rest at any of the three $V(x) = 0$ positions is the separatrix through the saddle point.

   If the oscillator now has a small damping term $r\dot{x}$ the final rest position is determined exclusively by its starting values $\dot{x}$ and x. The saddle connection is broken and the two equilibrium states are now competing point attractors. Starting positions of $(\dot{x}, x)$ which lie in the dotted regions of the phase space generate trajectories which will come to

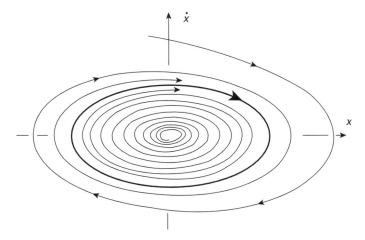

Figure 14.24 Repellor and limit cycle. Phase trajectories of an oscillator governed by the equation $m\ddot{x} - r\dot{x} + d\dot{x}^3 + sx = 0$. For x small and r positive, trajectories spiral outwards from the repellor at the origin. For large $\dot{x}$, the $d\dot{x}^3$ term dominates and trajectories spiral inwards. These effects balance at some boundary to form a stable limit cycle. Reproduced by permission of John Wiley & Sons from Thompson and Stewart (1986)

equilibrium in the dotted attractor spiralling to rest at the minimum of the right hand potential well. Similarly the clear region of phase space defines the starting positions and trajectories which will finish at the minimum of the left hand potential well. Each of these two phase space regions is called a *basin of attraction*.

Repellor and Limit Cycle

To illustrate the concepts of repellor and limit cycle in two dimensional phase space we consider the damped non-linear oscillator governed by the equation

$$m\ddot{x} - r\dot{x} + d\dot{x}^3 + sx = 0$$

When x is very small we can neglect the $d\dot{x}^3$ term and if r is positive we have negative damping giving outwardly spiralling trajectories from the central point which is therefore a repellor. For large values of $\dot{x}$, $d\dot{x}^3$ is the dominant term and the trajectories spiral inwards. These competing effects are balanced at some boundary to form a steady state oscillation in a stable limit cycle of fixed period, Figure 14.24.

The Torus in Three-dimensional $(\dot{x}, x, t)$ Phase Space

Extending the ideas about phase space let us consider the generation of a torus by following the trajectory of a particle (or system) subject to the influence of two perpendicular circular simple harmonic motions of angular frequencies ω_0 and ω_1, where ω_0 traces a circle in the azimuthal plane with a radius r_0 while ω_1 causes the particle to spiral on the surface of a torus of radius r_1 (Figure 14.25). A cross section of the torus will be a circle of radius r_1,

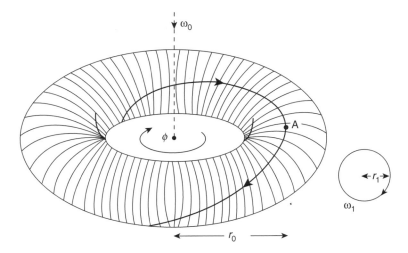

Figure 14.25 Torus in $(\dot{x}xt)$ phase space generated by a system subject to the influence of two perpendicular circular simple harmonic motions. The trajectory of the system spirals on the torus surface

and the particle will register some point on the circumference of the circle each time it passes the cross section. If $\omega_1 = \omega_0$ this point will be identical for each period $\tau_0 = 2\pi/\omega_0$. However, if $\omega_1 \neq \omega_0$ the particle will arrive at different points on the circle circumference after each interval τ_0; for example, if $\omega_1 = 3\omega_0/4$, the particle will travel only $\frac{3}{4}$ of the circumference for each τ_0 and will register the points A B C D of Figure 14.26 in that order.

Such a cross section is called a Poincaré section in phase space and is a vital tool in describing the multiple excursions of trajectories in phase space associated with chaos. It is always taken at some fixed interval of the system such as τ_0, a typical example, as we shall see, is the period of the force driving an oscillator displaying chaotic motion.

The Poincaré section for a simple harmonic oscillator taken in the upper half plane containing the $\dot{x}$ axis but normal to the x axis consists of only one spot at the maximum value of $\dot{x}$ as the system passes through this position at intervals of τ_0. A similar section for the damped oscillator will register a series of points between $\dot{x}$ maximum and the origin as the trajectory spirals inwards.

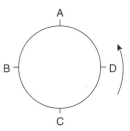

Figure 14.26 When $\omega_1 = 3\omega_0/4$ in Figure 14.25, the system will register the points ABCD in that order at a given cross section. This is an example of a Poincaré section in phase space

If the motion associated with ω_1 is not circular the surface of the torus will be distorted. We shall see that it can be pulled out, crinkled and folded back on itself so that its Poincaré section will assume remarkable shapes. When the repeated excursions of trajectories are located on such a surface it is called a manifold. The final state of such a distorted surface represents the reduced phase space which follows the dissipation of energy. Within this space is located the attractor to which the orbiting trajectories are bound.

Chaotic Response of a Forced Non-linear Mechanical Oscillator

Fifty years ago no engineer calculating the forced vibrations of a beam via the equation

$$\ddot{x} + k\dot{x} + x^3 = B\cos t$$

could have foreseen the complexity of response which computer simulated solutions have uncovered. Ueda (1980) has found no fewer than 21 distinct regions of behaviour using a range of B values (0–25) and k values (0–0.8) where the units are unspecified. Five of these 21 regions display chaos, the others contain a variety of different attractors. Thompson and Stewart (1986) have chosen particular B and k values from Ueda to illustrate many basic features of chaotic oscillators and the use of Poincaré sections to identify them. Even with the same B and k values the long term behaviour of the oscillator is found to depend critically upon the starting values of $\dot{x}$ and x and Figure 14.27 shows the phase trajectories and wave forms of five stable periodic motions around attractors for $B = 0.2$ and $k = 0.08$ where the letter A denotes the starting point in each case.

We have already noted that one sign of impending chaos in a system is the divergence with time of phase trajectories from almost identical starting positions even though their behaviour is determined by the same equation. For a forced damped oscillator we saw on p. 58 that this behaviour consists of two terms, a transient which decays with time leaving the steady state component.

One of Ueda's chaotic regimens lies in the B range (6–8) and the k range (0.03–0.1) and Thompson and Stewart chose $B = 7.5$ and $k = 0.05$ for their illustration. Figure 14.28 shows phase trajectories of the oscillator for two almost identical starting positions labelled A and a of $(\dot{x}, x)$. Because the vibration waveform of the oscillator is so irregular there is only one way of registering the passage of time on this two-dimensional phase diagram and that is by marking off the constant period τ_0 associated with $\cos t$ of the driving force. This gives points B and b and the trajectory divergence is already evident. This divergence may be traced over many periods of τ_0 and is found to be exponential with time. We can associate the points B and b and their successors after each interval of τ_0 with the formation of our Poincaré section of the torus on p. 489. Figure 14.29 shows the history of the single phase trajectory which started at A marked off in alphabetical order over the first nine periods of τ_0. Note that each letter represents a maximum of the driving force $B\cos t$ and that all letters fall on the right hand side of the $\dot{x}$ axis, that is x positive.

Tracing this complicated trajectory on the three-dimensional $(\dot{x}xt)$ phase surface of the torus would separate that is time resolve, the apparent trajectory crossing points in the two-dimensional picture. If now only the Poincaré section points A, B, C, D, etc. are plotted

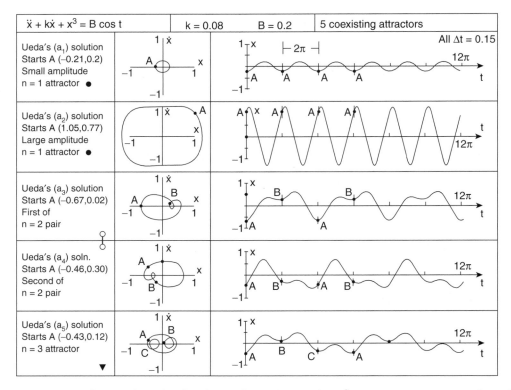

Figure 14.27 Phase trajectories for the oscillator $\ddot{x} + 0.08\dot{x} + x^3 = 0.2\cos t$ are seen to depend critically upon the starting values of $\dot{x}$ and x. The letter A denotes each starting position. Reproduced by permission of John Wiley & Sons from Thompson and Stewart (1986)

over a very large number of intervals of τ_0 they build up a shape of which that shown in Figure 14.30 is typical.

Irrespective of any starting position or of the size and duration of any transients all long term, steady state, Poincaré section points eventually settle to contribute to this pattern. It bears the signature of a chaotic attractor for high resolution displays a fine structure known as fractal. It is an example of the stretching and folding of an ensemble of steady state trajectories in phase space during which the trajectories become thoroughly mixed; that is, change from one set of close neighbours to another. The important point is that despite mixing, the trajectories retain their distinct identities and never merge; their time histories are invertible.

A Brief Review

We now review briefly the discussion so far in order to present a clearer picture of what we shall expect to identify in following sections.

We saw on p. 474 how chaos could be approached via period doubling but that the symmetry of the population biology equation created an ambiguity on the route to chaos, so

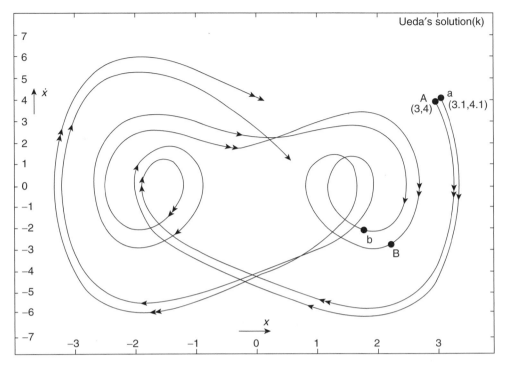

Figure 14.28 Two phase trajectories from almost identical starting positions A and a for the oscillator $\ddot{x} + 0.05\dot{x} + x^3 = 7.5\cos t$. After one period of the driving force the trajectories have diverged respectively to B and b. Reproduced by permission of John Wiley & Sons from Thompson and Stewart (1986)

that no final point on a trajectory could be uniquely time reversed back to its origin. This essential time reversal arises from the continuity of unique solutions to the non-linear equations governing the system. The solution at a given time defines the complete state of the system and occupies a point in phase space so that, with time, the trajectory traces a line in phase space. However, trajectories with close origins in a chaotic attractor system diverge exponentially with time while the energy dissipation always associated with chaotic attractors requires the phase space volume to contract. To reconcile these contradictory features, phase space of at least three dimensions is required and the problem is resolved essentially through stretching and folding this phase space. The distortion of phase space on a torus surface is an example of this.

To illustrate this process of stretching and folding, which we shall discuss later in more detail, we may consider two trajectories, originally close neighbours, which diverge as they spiral outwards on a plane (Figure 14.31) leaving the plane only to fold over by attraction and return back to the centre of the spiral. The divergence; that is, the sensitivity to initial conditions results from the stretching process and the folding comes from the attraction. The uniqueness of the trajectories in phase space ensures that they remain distinct, that they never merge, no matter how complex the phase space structure becomes. This complexity

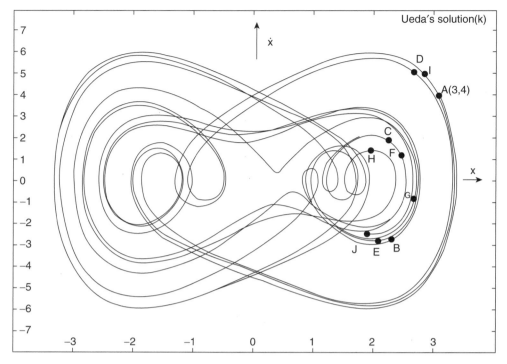

Figure 14.29 A single phase trajectory traced over the first nine periods of the driving force in Figure 14.28. In three dimensional phase space the apparent crossing points would be separated by time resolution. Reproduced by permission of John Wiley & Sons from Thompson and Stewart (1986)

is revealed by the fractal nature of the highly resolved Poincaré section of the chaotic attractor in Figure 14.30.

We now explain what is meant by fractal structure and discuss how theories of phase space distortion or mapping produce it.

Fractals

In topology a curve has a dimension of one and a surface a dimension of two. There are higher integral dimensions. The word 'fractal' was coined by Mandelbrot in 1975 to express the idea of a 'shape' with a non-integral dimension. He has since published books on the subject containing many beautiful computer generated patterns. The essential feature of all these fractal patterns is that they are self similar which means that, irrespective of scale, they retain the same geometric appearance. A well known example is the Koch snowflake.

Koch Snowflake

Figure 14.32 shows an equilateral triangle of side length $3l$. On the central section of each side is placed a similar triangle of side l and the process is repeated indefinitely to produce

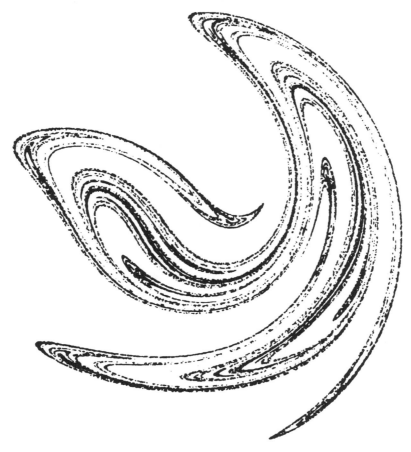

Figure 14.30 Poincaré section for an oscillator similar to that of Figures 14.28 and 14.29. High resolution displays a fractal fine structure. Reprinted with permission from 'Steady motions exhibited by Duffing's equation: A picture book of regular and chaotic motions', by Yoshisuke Ueda, published in *New Approaches to Nonlinear Problems in Dynamics*, pp. 311--322. Copyright 1980 by the Society for Industrial and Applied Mathematics, Philadelphia, Pennsylvania. All rights reserved

a curve of infinite length $(3l \times \frac{4}{3} \times \frac{4}{3} \ldots)$ but which encloses a finite area less than that of the circle surrounding the original triangle.

Mandelbrot was first led to the idea of fractals by studying noise on a transmission line. He found that the pattern or the distribution of the noise remained the same whether taken over a period of an hour, a minute or a second; that is, self similarity prevailed. He identified the pattern as belonging to a Cantor set which dates from the nineteenth century and which G. D. Birkhoff had suggested in the 1920s might be significant in dynamical systems.

Cantor Set

The Cantor set (Figure 14.33) is constructed by removing the centre part l of a line of length $3l$ and repeating the process indefinitely. We define the total set of points lying on

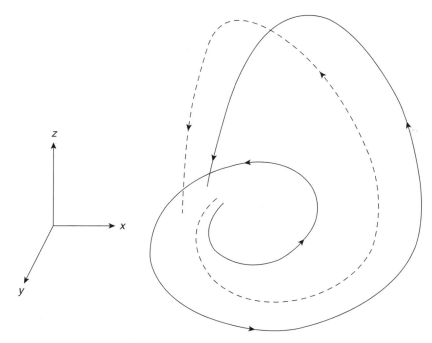

Figure 14.31 Trajectories around a chaotic attractor diverge yet remain within a bounded region. This is achieved by the stretching and folding of phase space

the line segment l to be some function $f(l)$ and assume this total set to be preserved so that

$$f(3l) = 2f(l)$$

If then $f(l)$ is considered to vary as some power δ of l so that $f(l) \sim l^\delta$ we have $f(3l) = 2f(l)$ giving $(3l)^\delta = 2l^\delta$ so that $3^\delta = 2$ and $\delta = \log 2/\log 3 = 0.6309$. This is the non-integral fractal dimension of the Cantor set.

Figure 14.32 The Koch snowflake has a fractal non-integral dimension. The final pattern has infinite length but encloses a finite area less than that of the circle surrounding the original triangle

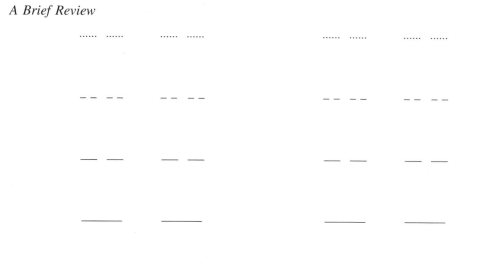

Figure 14.33 A Cantor set has a fractal non-integral dimension and is produced by removing the central third of a line and repeating the process indefinitely with the remaining segments. Poincaré sections of chaotic attractors have a Cantor set-like structure

(Problem 14.7)

The importance of the Cantor set is that the highly resolved Poincaré section of a chaotic attractor such as that on p. 493 reveals a Cantor set-like structure. It results from stretching the phase space and folding it closely into layers. It is the signature of a chaotic attractor and we now look at how this may be achieved.

Smale Horseshoe

The mathematical process which describes the stretching and folding of phase space is called mapping and a number of such maps have now been devised to produce this effect, e.g. the Smale horseshoe (Smale, 1963).

In this example (Figure 14.34) a square is taken, stretched to double its length while its width is reduced to form a rectangle of area less than the square. The square may be taken as a cross section of a particular volume of phase space containing an ensemble or collection of trajectories the ends of which are shown as dots within the square. The reduction of area in the stretching process is equivalent to reducing the phase space by energy dissipation; at the same time it separates trajectories from their neighbours. The rectangle is then folded over into a horseshoe, the stretching and folding process is now repeated with the horseshoe again and again, so that successive cross sections reveal a Cantor set-like structure. The relative positions of the original trajectories are completely changed in this process.

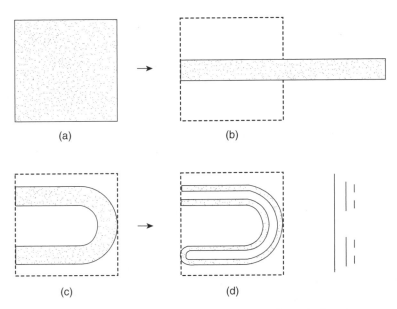

Figure 14.34 The Smale horseshoe takes a square cross section of phase space containing an ensemble of trajectories (dotted ends), stretches the square to a rectangle of reduced area and folds the rectangle into a horseshoe. The process is repeated continuously with successive cross sections revealing a Cantor-set-like structure. The relative positions of the trajectories are changed in the process as the trajectories are mixed

Chaos in Fluids

Turbulence in fluids is the most widely observed of all chaotic motions. Fast flowing water from a tap or around a blunt obstacle loses its low speed coherence and flow symmetry. A satisfactory description of the behaviour is made more difficult because:

- The theory of the liquid state is less well developed than that of gases and liquids.

- Experimental methods have until recently used probes which disturb the state of the system being measured.

The second of these difficulties has now been overcome by the development of laser-Doppler techniques combining the holographic system (p. 404) with the Doppler effect (p. 141).

Typically, a laser beam of frequency ν_0 and wavelength λ_0 is split so that one half acts as a reference beam while the other is focused on a small fluid element (~ 0.1 mm diameter) moving with a velocity u. This beam is scattered through an angle θ with a frequency ν_s. The relationship between ν_0 and ν_s is shown in Figure 14.35b. In Figure 14.35a the scattered beam joins the reference beam which is now modulated to give a component at the detector of the Doppler shift frequency $\nu_D = \nu_s - \nu_0$. If $\mathbf{k}_0$ and $\mathbf{k}_s$ are the wave number vectors associated respectively with ν_0 and ν_s then the component of the velocity u parallel

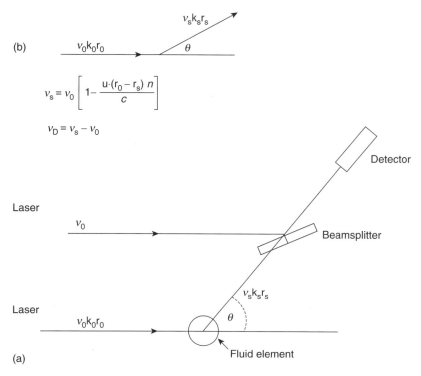

Figure 14.35 (a) Scheme of the laser-Doppler technique for velocity measurements in a fluid. (b) The vector relationship between the scattered frequency ν_s, the incident laser frequency ν_0 and the fluid velocity u; $\mathbf{r}$ is a unit vector, n the refractive index of the fluid and c the velocity of light. The Doppler shift frequency is $\nu_D = \nu_s - \nu_0$

to the vector $\mathbf{k} = \mathbf{k}_0 - \mathbf{k}_s$ depends only upon λ_0, $\sin \theta/2$ and ν_D. Velocities in the range of 10^{-6} to $10^3\,\mathrm{ms}^{-1}$ are capable of being measured by this system.

The frequency ν_D is so much greater than the frequencies associated with the fluid motion that the measured $u(t)$ gives an instantaneous velocity value. Continuous records of $u(t)$ over long periods may be Fourier analysed to show sharply defined frequency components when the flow is periodic with the appearance of broad band noise when the flow becomes chaotic.

Chaos in fluids has been studied chiefly in two systems:

1. Couette flow where the appropriate parameter is the dimensionless Reynolds number.

2. Rayleigh–Bénard convection where the parameter is the dimensionless Rayleigh number. This system is the model used by Lorenz in finding the original strange attractor.

Couette Flow

This flow was completely defined in the classic paper of G. I. Taylor (1923). In its simplest form it is produced in a fluid contained in the gap between two concentric cylinders with radii differing by about a centimetre. One of the cylinders is fixed while the others rotates with an angular velocity ω although sometimes both cylinders may rotate with different angular velocities. The outer cylinder is usually glass, allowing observation of the flow. At low speeds of angular rotation the flow is symmetric in the azimuthal direction (Figure 14.36a).

For flow in one dimension the relevant equation would read

$$\rho u_x \frac{\partial u_x}{\partial x} = \frac{-\partial p}{\partial x} + \frac{\mu \partial^2 u_x}{\partial x^2}$$

where ρ is the fluid density, u_x is the velocity in the x direction, p is the pressure and μ is the fluid viscosity. Each term in the equation has the dimensions of force per unit volume; the left hand side term may be considered as an inertial force and the last term may be seen as the viscous force. Flow symmetry depends on the relative strengths of these forces and the Reynolds number is written dimensionally as

$$Re = \frac{\text{inertial force}}{\text{viscous force}} = \frac{\rho u^2}{L} \frac{L^2}{\mu u} = \frac{uL}{\eta}$$

where $\eta = \mu/\rho$ is the kinematic viscosity and u and L are a characteristic velocity and length of the system.

For Couette flow

$$Re = \frac{r_i \omega d}{\eta}$$

where r_i is the radius of the inner cylinder and d is the width of the cylindrical gap.

For slow speeds, that is low Re, any departure from symmetry is overcome by the viscous force restoring the system to equilibrium but as Re increases with increasing ω, the inertial effects of any departure from symmetric flow may be too great for the restoring viscous force and purely azimuthal Couette flow is lost.

This loss of symmetry for high Re first shows itself as a series of vortices around each azimuthal flow line, so that fluid elements follow a spiral path in the azimuthal direction (Figure 14.36b). These vortices, called Taylor cells, are seen to arise as follows.

An elemental toroid of the fluid initially at radius r_1, circulating at angular velocity ω_{r_1} is displaced to radius r_2. If its angular momentum is conserved we have

$$\omega_{r_1} r_1^2 = \omega'_{r_1} r_2^2$$

where ω'_{r_1} is its new angular velocity. Its centrifugal force will exceed that of the fluid originally at r_2 circulating with angular velocity ω_{r_2} if

$$\left| \omega'_{r_1} \right| > \left| \omega_{r_2} \right|$$

Hence an instability develops if $\left| \omega_{r_1} r_1 \right|^2 > \left| \omega_{r_2} r_2 \right|^2$ for $r_2 > r_1$; that is, if

$$\frac{\mathrm{d}}{\mathrm{d}r} \left| \omega r^2 \right| < 0$$

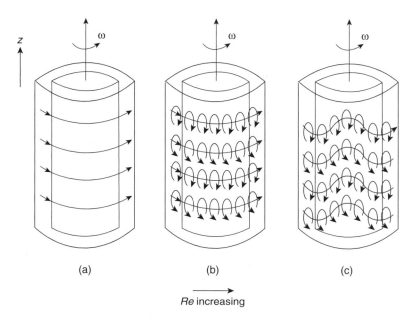

Figure 14.36 In Couette flow a liquid is contained in the gap between two concentric cylinders one of which has an angular velocity ω with respect to the other. At low Reynolds number *Re* the flow is azimuthal as in (a). As *Re* increases flow symmetry is lost and vortices develop (b). A further increase of *Re* develops transverse waves along the lines of vortices (c)

This is known as the Rayleigh criterion for the instability of Couette flow.

When the inner layers of the fluid are moving more rapidly than the outer layers they tend to move outwards because the centrifugal force is greater than the pressure holding them in place. A whole layer cannot move out uniformly because the outer layers are in the way so it breaks into cells which circulate.

The rotational motion of a fluid element in a Taylor cell appears as a periodic velocity variation in the z direction of Figure 14.36. Increasing *Re* that is the angular velocity of the cylinder, now causes harmonic oscillations of the vortices in the z direction as transverse waves travel around the azimuthal torus (Figure 14.36c). The frequency of these waves will be registered via the velocity measurements and as *Re* increases still more, other frequencies are generated and broad band noise begins to dominate with the appearance of chaos (Figure 14.37).

Rayleigh–Bénard Convection

In this process heat provides the energy driving asymmetries in the flow. The incompressible fluid is contained between two horizontal plates about a centimetre apart, the lower of which is heated. For a small constant temperature difference between the plates the thermal conductivity and viscosity of the fluid ensure that the heat is conducted upwards in an orderly fashion (Figure 14.38a). When the temperature gradient is too steep the effect of these forces in maintaining equilibrium is overcome, flow symmetry is

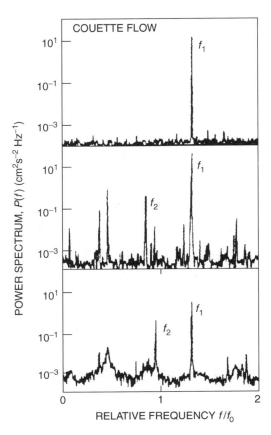

Figure 14.37 The number of frequencies of the waves in Figure 14.36c increases with *Re* but broad band noise begins to dominate with the appearance of chaos in the bottom figure. Reproduced by permission of the American Institute of Physics from Swinney and Gollub (1978)

lost and convective rolls in both clockwise and anti-clockwise directions can develop (Figure 14.38b).

This occurs at some critical value of the Rayleigh parameter which we derive from the relevant equations. These are, in the positive *z* direction

$$\rho u_z \frac{\partial u_z}{\partial z} = -\frac{\partial p}{\partial z} + \rho\mu \frac{\partial^2 u_z}{\partial z^2} - \rho g \alpha \Delta T$$

$$u_z \frac{\mathrm{d}T}{\mathrm{d}z} = K \frac{\mathrm{d}^2 T}{\mathrm{d}z^2}$$

In the last term of the first equation *g* is the acceleration due to gravity, α is the thermal expansion coefficient and ΔT is the constant temperature difference between the plates. This term is the buoyancy force which drives the warmer, less dense, liquid upwards. In the second equation *K* is the thermal diffusivity (p. 190) and equals $k/\rho C_\mathrm{p}$ where *k* is the thermal conductivity and C_p is the specific heat at constant pressure.

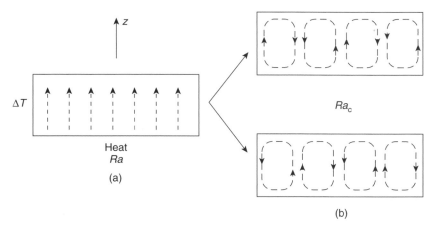

Figure 14.38 (a) at low Rayleigh numbers *Ra* fluid in a Rayleigh--Bénard cell conducts heat away from the base in a symmetric fashion. At some critical value Ra_c flow symmetry is lost (b) and convective rolls develop in clockwise or anti-clockwise directions

In the first equation the buoyancy force responsible for upward motion is opposed by the viscous term. If the strength of these forces is comparable, a low pressure gradient in the fluid will keep the inertial force on the left hand side low enough for the flow to remain symmetric.

Comparable values of the buoyancy and viscous terms will give

$$\mu\frac{\partial^2 u_z}{\partial z^2} \approx g\alpha\Delta T$$

to yield some characteristic velocity

$$U \sim \frac{g\alpha\Delta T L^2}{\mu} \tag{14.1}$$

where L, a characteristic length, is usually the depth of the liquid.

The second equation determines the temperature distribution and the ratio

$$\frac{u_z \mathrm{d}T/\mathrm{d}z}{K\mathrm{d}^2T/\mathrm{d}z^2} \sim \frac{UL}{K} \tag{14.2}$$

tells us that for K large enough the thermal conductivity will distribute the heat rapidly enough for the symmetric conduction process to prevail. Combining (14.1) and (14.2) using the common factor U gives the Rayleigh number

$$Ra = \frac{g\alpha\Delta T L^3}{\mu K}$$

When the Rayleigh number is small enough, μ and K govern the conduction process. At some critical Rayleigh number Ra_c convective fluid motion driven by ΔT replaces pure

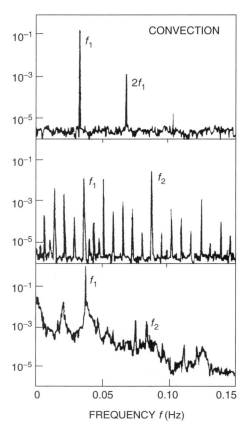

Figure 14.39 The development of frequencies in the velocity flow spectrum at the critical Rayleigh number Ra_c with the onset of noise as chaos sets in (bottom figure). Reproduced by permission of the American Institute of Physics from Swinney and Gollub (1978)

heat conduction, instabilities develop and the flow becomes asymmetric. At the critical value Ra_c convective rolls in the right or left handed direction begin to show, with a single frequency and its harmonics appearing in the velocity flow spectrum. Increasing Ra beyond Ra_c introduces further frequency components which are followed by the onset of noise as chaos sets in (Figure 14.39).

The Strange Attractor of Lorenz

Lorenz (1963) used the Rayleigh–Bénard process as the basis of his model of atmospheric convection in assessing the possibility of long range weather forecasting. The physical model is so restricted that it yields only the most rudimentary information about weather patterns, enough however to show that long range forecasting is not feasible because phase trajectories starting from almost identical positions diverge after a relatively short time.

The two-dimensional convection rolls which appear in the rectangular cross section of Figure (14.38b) when $Ra > Ra_c$ can be described by two velocity components together

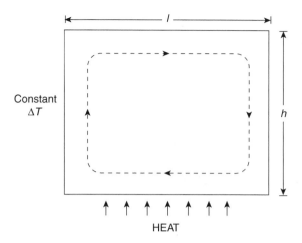

Figure 14.40 The first mode $X(t)$ in the Lorenz equations gives a single convective roll, clockwise for X positive, anti-clockwise for X negative. Warm rising fluid in this mode indicates where X and Y have the same sign. The ratio h/l determines the geometric factor b in the Lorenz equations

with the deviation of the temperature from the linear conduction profile of low Ra. These three quantities, two of velocity and one of temperature, were expanded in two-dimensional Fourier series with terms (modes) of the form $A_{ij}(t)\sin k_i x \sin k_j z$ (p. 248) where the time dependence now appears in the amplitude coefficient. These expansions were used in the hydrodynamic equations of the last section to produce an infinite set of ordinary differential equations, but Lorenz reduced this number to three by considering only the first three modes of the Fourier expansion.

The first mode $X(t)$ determined by the velocity components gives a single convective roll filling the rectangular cell (Figure 14.40). The second mode $Y(t)$ describes the temperature differences between ascending and descending currents in the convective roll and the third mode $Z(t)$ represents the departure from linearity of the vertical temperature profile.

Each mode is a phase space coordinate and the modes XYZ represent the physical state of the system at a given time.

The Lorenz equations take the form

$$\dot{X} = \sigma(Y - X)$$
$$\dot{Y} = rX - Y - XZ$$
$$\dot{Z} = XY - bZ$$

where σ is the ratio of the fluid viscosity to its thermal conductivity, r is the ratio Ra/Ra_c and b is a geometric factor governed by the ratio h/l (height/length) of the cell in Figure 14.40. Lorenz took $\sigma = 10$ (the approximate value for water) and $b = 8/3$.

To show that the volume of phase space containing the trajectories decreased with time, Lorenz used a transport theorem of fluid dynamics relating the space rate of change of vectors describing a flow integrated over a volume V to the time rate of change of the same volume. The vector in phase space may be written as $F(\dot{X}, \dot{Y}, \dot{Z})$ to give

$$\frac{\mathrm{d}}{\mathrm{d}t} V(t) = \int_V \operatorname{div} F \, \mathrm{d}V$$

Div F from p. 203 is given by

$$\frac{\partial \dot{X}}{\partial X} + \frac{\partial \dot{Y}}{\partial Y} + \frac{\partial \dot{Z}}{\partial Z}$$

with a value of $-(\sigma + b + 1) = -13.67$ in Lorenz's equations, so $\mathrm{d}V(t)/\mathrm{d}t$ is negative.

This reduction in phase space volume indicates that the trajectories will eventually be confined to some limiting manifold.

The overall behaviour of the system can be conveniently divided into various ranges of the value of $r = Ra/Ra_c$.

When

$$\dot{X} = \dot{Y} = \dot{Z} = 0$$

there are three solutions to the Lorenz equations. These are

(1) $X = Y = Z = 0$

(2) $X = Y = +[b(r - 1)]^{1/2} : Z = (r - 1)$

(3) $X = Y = -[b(r - 1)]^{1/2} : Z = (r - 1)$

When $r < 1$ solution (1) corresponds to a steady process of pure conduction with no convection, typical behaviour for small ΔT. Solutions (2) and (3) correspond to states of steady convection which exist only when $r > 1$.

If there is now a small perturbation from the condition $\dot{X} = \dot{Y} = \dot{Z} = 0$ the behaviour of (1) remains stable as pure conduction for $r < 1$, trajectories moving to the origin $X = Y = Z = 0$ as a point attractor. As r increases beyond unity, steady convection will give way to the right and left handed convective rolls of solutions (2) and (3) which now correspond to separate stable attractors each with its own basin of attraction and set of spiralling trajectories.

At $r \approx 13.9$ the separation between the basins of attraction is lost and trajectories move between (2) and (3) before settling on one or the other. At $r \approx 24.7$ (2) and (3) lose their stability as limit cycles and beyond this value of r the trajectories form two connecting bands, one centred on (2), the other on (3). (2) and (3) are now chaotic attractors with trajectories orbiting aperiodically around one before switching to the other.

Problem 14.1
If the period of a pendulum with large amplitude oscillations is given by

$$T = T_0 \left(1 + \frac{1}{4} \sin^2 \frac{\theta_0}{2} \right)$$

where T_0 is the period for small amplitude oscillations and θ_0 is the oscillation amplitude, show that for θ_0 not exceeding $30°$, T and T_0 differ by only 2% and for $\theta_0 = 90°$ the difference is 12%.

Problem 14.2

The equation of motion of a free undamped non-linear oscillator is given by

$$m\ddot{x} = -f(x)$$

Show that for an amplitude x_0 its period

$$T_0 = 4\sqrt{\frac{m}{2}} \int_0^{x_0} \frac{dx}{[F(x_0) - F(x)]^{1/2}}, \quad \text{where} \quad F(x_0) = \int_0^{x_0} f(x)\, dx$$

Problem 14.3

The equation of motion of a forced undamped non-linear oscillator of unit mass is given by

$$\ddot{x} = s(x) = F_0 \cos \omega t$$

Writing $s(x) = s_1 x + s_3 x^3$, where s_1 and s_3 are constant, choose the variable $\omega t = \phi$, and for $s_3 \ll s_1$ assume a solution

$$x = \sum_{n=1}^{\infty} \left(a_n \cos \frac{n}{3}\phi + b_n \sin \frac{n}{3}\phi \right)$$

to show that all the sine terms and the even numbered cosine terms are zero, leaving the fundamental frequency term and its third harmonic as the significant terms in the solution.

Problem 14.4

If the mutual interionic potential in a crystal is given by

$$V = -V_0 \left[2\left(\frac{r_0}{r}\right)^6 - \left(\frac{r_0}{r}\right)^{12} \right]$$

where r_0 is the equilibrium value of the ion separation r, show by expanding V about V_0 that the ions have small harmonic oscillations at a frequency given by $w^2 \approx 72\, V_0/mr_0^2$, where m is the reduced mass.

Problem 14.5

The potential energy of an oscillator is given by

$$V(x) = \tfrac{1}{2}kx^2 - \tfrac{1}{3}ax^3$$

where a is positive and $\ll k$.

Assume a solution $x = A \cos \omega t + B \sin 2\omega t + x_1$ to show that this is a good approximation at $w_0^2 = w^2 = k/m$ if $x_1 = \alpha A^2/2w_0^2$ and $B = -\alpha A^2/6w_0^2$, where $\alpha = a/m$.

Problem 14.6

Prove that when $\lambda > 0.75$ in Figure 14.11 then the slopes of $f^2(x)$ at x_1^* and x_2^* are the same.

Non-linear Oscillations and Chaos

Problem 14.7
Use the arguments in the paragraph on the Cantor set (p. 495) to show that the Koch snowflake has a fractal dimension of 1.2618.

Recommended Further Reading

Non-linear Dynamics and Chaos by Thompson, J. M. T. and Stewart, H. B., Wiley, New York (1986).

References

Feigenbaum, M. J. (1978), *Physica*, **7D**, 16.
Feigenbaum, M. J. (1979), *Phys. Lett.*, **74A**, 375.
Li, T. Y. and Yorke, J. A. (1975), *Am. Math. Monthly*, **82**, 985.
Lorenz, E. N. (1963), *J. Atmos. Sci.*, **20**, 130.
May, R. M. (1976), *Nature*, **261**, 459.
Metropolis, N., Stein, M. L. and Stein, P. R. (1973), *J. Combin. Theory*, *A*, **15**, 25.
Smale, S. (1963), *Differential and Combinatorial Topology* (ed. S. Cairns), Princeton University Press, Princeton, NJ.
Swinney, H. L. and Gollub, J. P. (1978), *Physics Today*, August, 41.
Tabor, M. (1989), *Chaos and Integrability in Non-linear Dynamics*, Wiley, New York.
Taylor, G. I. (1923), *Phil. Trans. Roy. Soc. (London)*, *A*, **223**, 289.
Testa, J., Perez, J. and Jeffries, C. (1982), *Phys. Rev. Lett.*, **48**, 714.
Thompson, J. M. T. and Stewart, H. B. (1986), *Non-linear Dynamics and Chaos*, Wiley, New York.
Ueda Y. (1980) *New Approaches to Non-linear Dynamics* (ed. P. J. Holmes), S. I. A. M., Philadelphia, p. 311.
Van der Pol, B. (1926), *Phil. Mag.*, (7)2, 978.
Van der Pol, B. and Van der Mark, J. (1927), *Nature*, **120**, 363.

15

Non-linear Waves, Shocks and Solitons

Non-linear Effects in Acoustic Waves

The linearity of the longitudinal acoustic waves discussed in Chapter 6 required the assumption of a constant bulk modulus

$$B = -\frac{\mathrm{d}P}{\mathrm{d}V/V}$$

If the amplitude of the sound wave is too large this assumption is no longer valid and the wave propagation assumes a new form. A given mass of gas undergoing an adiabatic change obeys the relation

$$\frac{P}{P_0} = \left(\frac{V_0}{V}\right)^{\gamma} = \left[\frac{V_0}{V_0(1+\delta)}\right]^{\gamma}$$

in the notation of Chapter 6, so that

$$\frac{\partial P}{\partial x} = \frac{\partial p}{\partial x} = -\gamma P_0 (1+\delta)^{-(\gamma+1)} \frac{\partial^2 \eta}{\partial x^2}$$

since $\delta = \partial\eta/\partial x$.

Since $(1+\delta)(1+s) = 1$, we may write

$$\frac{\partial p}{\partial x} = -\gamma P_0 (1+s)^{\gamma+1} \frac{\partial^2 \eta}{\partial x^2}$$

The Physics of Vibrations and Waves, 6th Edition   H. J. Pain
© 2005 John Wiley & Sons, Ltd

Pressure

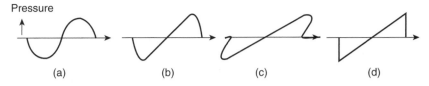

| (a) | (b) | (c) | (d) |

Figure 15.1 The local sound velocity in a high amplitude acoustic wave (a) is pressure and density dependent. The wave distorts with time (b) as the crest overtakes the lower density regions. The extreme situation of (c) is prevented by entropy-producing mechanisms and the wave stabilises to an *N* type shock-wave (d) with a sharp leading edge

and from Newton's second law we have

$$\frac{\partial p}{\partial x} = -\rho_0 \frac{\partial^2 \eta}{\partial t^2}$$

so that

$$\frac{\partial^2 \eta}{\partial t^2} = c_0^2 (1 + s)^{\gamma+1} \frac{\partial^2 \eta}{\partial x^2}, \quad \text{where} \quad c_0^2 = \frac{\gamma P_0}{\rho_0} \tag{15.1}$$

Physically this implies that the local velocity of sound, $c_0(1 + s)^{(\gamma+1)/2}$, depends upon the condensation s, so that in a finite amplitude sound wave regions of higher density and pressure will have a greater sound velocity, and local disturbances in these parts of the wave will overtake those where the values of density pressure and temperature are lower.

A single sine wave of high amplitude can be formed by a close fitting piston in a tube which is pushed forward rapidly and then returned to its original position. Figure 15.1a shows the original shape of such a wave and 15.1b shows the distortion which follows as it propagates down the tube. If the distortion continued the wave form would eventually appear as in Figure 15.1c, where analytical solutions for pressure, density and temperature would be multi valued, as in the case of the non-linear oscillator of Figure 14.3c. Before this situation is reached, however, the wave form stabilizes into that of Figure 15.1d, where at the vertical 'shock front' the rapid changes of particle density, velocity and temperature produce the dissipating processes of diffusion, viscosity and thermal conductivity. The velocity of this 'shock front' is always greater than the velocity of sound in the gas into which it is moving, and across the 'shock front' there is always an increase in entropy. The competing effects of dissipation and non-linearity produce a stable front as long as the wave retains sufficient energy. The *N*-type wave of Figure 15.1d occurs naturally in explosions (in spherical dimensions) where a blast is often followed by a rarefaction.

The growth of a shock front may also be seen as an extension of the Doppler effect (p. 141), where the velocity of the moving source is now greater than that of the signal. In Figure 15.2a as an aircraft moves from S to S′ in a time *t* the air around it is displaced and the disturbance moves away with the local velocity of sound v_S. The circles show the positions at time *t* of the sound wave fronts generated at various points along the path of the aircraft but if the speed of the aircraft u is greater than the velocity of sound v_S regions of high density and pressure will develop, notably at the edges of the aircraft structure and

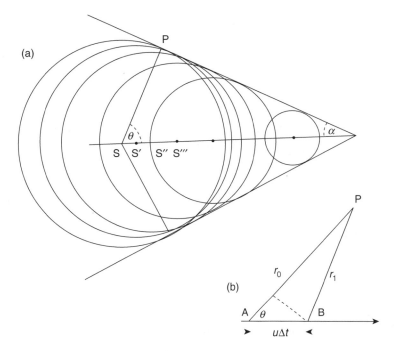

Figure 15.2 (a) The circles are the wavefronts generated at points S along the path of the aircraft, velocity $u > v_S$ the velocity of sound. Wavefronts superpose on the surface of the Mach Cone (typical point P) of half angle $\alpha = \sin^{-1} v_S/u$ to form a shock front. (b) At point P sound waves arrive simultaneously from positions A and B along the aircraft path when $(u/v_S) \cos\theta = 1$. $(\theta + \alpha = 90°)$

along the conical surface tangent to the successive wavefronts which are generated at a speed greater than sound and which build up to a high amplitude to form a shock. The cone, whose axis is the aircraft path, has half angle α where

$$\sin\alpha = \frac{v_S}{u}$$

It is known as the 'Mach Cone' and when it reaches the ground a 'supersonic bang' is heard.

The growth of the shock at the surface of the cone may be seen by considering the sound waves in Figure 15.2(b) generated at points A (time t_A) and B (time t_B) along the path of the aircraft, which travels the distance $AB = x = u\Delta t$ in the time interval $\Delta t = t_B - t_A$. The sound waves from A will travel the distance r_0 to reach the point P at a time

$$t_0 = t_A + \frac{r_0}{v_S}$$

Those from B will travel the distance r_1 to P to arrive at a time

$$t_1 = t_B + \frac{r_1}{v_S}$$

If x is small relative to r_0 and r_1, we see that

$$r_1 - r_0 \approx x\cos\theta = u\Delta t \cos\theta$$

so the time interval

$$t_1 - t_0 = t_B - t_A + \frac{(r_1 - r_0)}{v_S}$$

$$= \Delta t - \frac{u\Delta t \cos\theta}{v_S} = \Delta t\left(1 - \frac{u\cos\theta}{v_S}\right)$$

For the aircraft speed $u < v_S$, $t_1 - t_0$ is always positive and the sound waves arrive at P in the order in which they were generated.

For $u > v_S$ this time sequence depends on θ and when

$$\frac{u}{v_S}\cos\theta = 1$$

$t_1 = t_0$ and the sound waves arrive simultaneously at P to build up a shock.

Now the angles θ and α are complementary so the condition

$$\cos\theta = \frac{v_S}{u}$$

defines

$$\sin\alpha = \frac{v_S}{u}$$

so that all points P lie on the surface of the Mach Cone.

A similar situation may arise when a charged particle q emitting electromagnetic waves moves in a medium of refractive index greater than unity with a velocity v_q which may be greater than that of the phase velocity v of the electromagnetic waves in the medium ($v < c$). A Mach Cone for electromagnetic waves is formed with a half angle α where

$$\sin\alpha = \frac{v}{v_q}$$

And the resulting 'shock wave' is called Cerenkov radiation. Measuring the effective direction of propagation of the Cerenkov radiation is one way of finding the velocity of the charged particle.

Shock Front Thickness

The extent of the region over which the gas properties change, the shock front thickness, may be only a few mean free paths in a monatomic gas because only a few collisions between atoms are necessary to exchange the energy required to raise them from the equilibrium conditions ahead of the shock to those behind it. In a polyatomic gas the collisions are effective in producing a rapid increase in translational and rotational mode energies, but vibrational modes take much longer to reach their new equilibrium, so that the shock front thickness is very much greater.

Within the shock front thickness the state of the gas is not easily found, but the state of the gas on one side of the shock may be calculated from the state of the gas on the other side by means of the conservation equations of mass, momentum and energy.

Equations of Conservation

In a laboratory, shock waves are produced in a tube which is divided by a diaphragm into a short high-pressure section and a much longer low-pressure section. When the diaphragm bursts the expanding high pressure gas behaves as a very fast low-inertia piston which compresses the low pressure gas on the other side of the diaphragm and drives a shock wave down the tube. The profile of this shock wave is the step function shown as the dotted line in Figure 15.3, and the gas into which the shock is propagating is considered to be at rest. This simplifies the analysis, for we can consider the situation in Figure 15.3 as it appears to an observer O travelling with the shock front velocity u_1 into the stationary gas. The shock front is located within the region bounded by the surfaces A and B of unit area, each of which remains fixed with respect to the observer. The stationary gas which moves through the shock front from surface B acquires a flow velocity $u < u_1$ and a velocity relative to the shock front of $u_2 = u_1 - u$. From the observer's viewpoint the quantity of gas flowing into unit area of the region AB per unit time is $\rho_1 u_1$, where ρ_1 is the density of

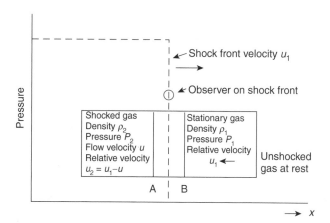

Figure 15.3 The pressure 'step profile' of a shock wave developed in a shock tube is shown by the dotted line. The plane cross-sections at A and B remain fixed with respect to the observer O moving with the shock front at velocity u_1 into unshocked gas at rest of pressure p_1 and density ρ_1. The shocked gas has a pressure p_2, a density ρ_2 and a velocity u, with a relative velocity $u_2 = u_1 - u$ with respect to the shock front. The states of the gas at A and B are related by the conservation equations of mass, momentum and energy across the shock front. Experimental measurement of the shock velocity u_1 is sufficient to determine the unknown parameters if the stationary gas parameters are known

the gas ahead of the shock. The quantity leaving unit area of AB per unit time is $\rho_2(u_1 - u) = \rho_2 u_2$, where ρ_2 is the density of the shocked gas.

Conservation of mass yields $\rho_1 u_1 = \rho_2 u_2 = m$ (a constant mass). The force per unit area acting across the region AB is $p_2 - p_1$, which equals the rate of change of momentum of the gas within the unit element, which is $m(u_1 - u_2)$. The conservation of momentum is therefore given by

$$p_1 + \rho_1 u_1^2 = p_2 + \rho_2 u_2^2.$$

The work done on unit area of the region per unit time is $p_1 u_1 - p_2 u_2$, and this equals the rate of increase of the kinetic and internal energy of the gas passing through unit area of the shock wave.

The difference

$$p_1 u_1 - p_2 u_2 = \frac{p_1}{\rho_1} m - \frac{p_2}{\rho_2} m$$

so that if the internal energy per unit mass of the gas is written $e(p, \rho)$, then the equation of conservation of energy per unit mass becomes

$$\frac{1}{2} u_1^2 + e_1 + \frac{p_1}{\rho_1} = \frac{1}{2} u_2^2 + e_2 + \frac{p_2}{\rho_2}$$

where for an ideal gas $p/\rho = RT$ and $e = c_v T = (1/\gamma - 1)p/\rho$, where T is the absolute temperature, c_v is the specific heat per gram at constant volume and $\gamma = c_p/c_v$, where c_p is the specific heat per gram at constant pressure.

These three conservation equations

$$\rho_1 u_1 = \rho_2 u_2 = m \quad \text{(mass)}$$
$$p_1 + \rho_1 u_1^2 = p_2 + \rho_2 u_2^2 \quad \text{(momentum)}$$

and

$$\frac{1}{2} u_1^2 + e_1 + \frac{p_1}{\rho_1} = \frac{1}{2} u_2^2 + e_2 + \frac{p_2}{\rho_2} \quad \text{(energy)}$$

together with the internal energy relation $e(p, \rho)$ completely define the properties of an ideal gas behind a shock wave in terms of the stationary gas ahead of it.

In an experiment the properties of the gas ahead of the shock are usually known, leaving five unknowns in the four equations, which are the shock front velocity u_1, the density of the shocked gas ρ_2, the relative flow velocity behind the shock u_2, the shocked gas pressure p_2 and its internal energy e_2. In practice the shock front velocity u_1 is measured and the other four properties may then be calculated.

Mach Number

A significant parameter in shock wave theory is the Mach number. It is a local parameter defined as the ratio of the flow velocity to the local velocity of sound. The Mach number of the shock front is therefore $M_s = u_1/c_1$, where u_1 is the velocity of the shock front propagating into a gas whose velocity of sound is c_1.

The Mach number of the gas flow behind the shock front is defined as $M_f = u/c_2$, where u is the flow velocity of the gas behind the shock front ($u < u_1$) and c_2 is the local velocity of sound behind the shock front. There is always an increase of temperature across the shock front, so that $c_2 > c_1$ and $M_s > M_f$. The physical significance of the Mach number is seen by writing $M^2 = u^2/c^2$, which indicates the ratio of the kinetic flow energy, $\frac{1}{2}u^2 \text{ mol}^{-1}$, to the thermal energy, $c^2 = \gamma RT \text{ mol}^{-1}$. The higher the proportion of the total gas energy to be found as kinetic energy of flow the greater is the Mach number.

Ratios of Gas Properties Across a Shock Front

A shock wave may be defined in terms of the shock Mach number M_s, the density or compression ratio across the shock front $\beta = \rho_2/\rho_1$, the temperature ratio across the shock T_2/T_1 and the compression ratio or shock strength $y = p_2/p_1$.

Given the shock strength, $y = p_2/p_1$, the conservation equations are easily solved to yield

$$M_s = \frac{u_1}{c_1} = \left(\frac{y + \alpha}{1 + \alpha}\right)^{1/2}$$

where

$$\alpha = \frac{\gamma - 1}{\gamma + 1}$$

$$\beta = \frac{\rho_2}{\rho_1} = \frac{\alpha + y}{1 + \alpha y}$$

and

$$\frac{T_2}{T_1} = y\left(\frac{1 + \alpha y}{\alpha + y}\right)$$

Alternatively these may be written in terms of the experimentally measured parameter M_s as

$$\frac{p_2}{p_1} = y = M_s^2(1 + \alpha) - \alpha$$

$$\frac{\rho_2}{\rho_1} = \beta = \frac{M_s^2}{1 - \alpha + \alpha M_s^2}$$

and

$$\frac{T_2}{T_1} = \frac{[\alpha(M_s^2 - 1) + M_s^2][\alpha(M_s^2 - 1) + 1]}{M_s^2}$$

For weak shocks (where p_2/p_1 is just greater than 1) β, T_2/T_1 and M_s are also just greater than unity, and the shock wave moves with the speed of sound.

Strong Shocks

The ratio $p_2/p_1 \gg 1$ defines a strong shock, in which case

$$M_s^2 \to \frac{(\gamma + 1)}{2\gamma} y$$

and

$$\beta = \frac{\rho_2}{\rho_1} \to \left(\frac{\gamma + 1}{\gamma - 1}\right)$$

a limit of 6 for air and 4 for a monatomic gas for a constant γ. The flow velocity

$$u = u_1 - u_2 \to \frac{2u_1}{(\gamma + 1)}$$

and the temperature ration

$$\frac{T_2}{T_1} = \left(\frac{c_2}{c_1}\right)^2 \to \frac{(\gamma - 1)}{(\gamma + 1)} y$$

The temperature increase across strong shocks is of great experimental interest. The physical reason for this increase may be seen by rewriting the equation of energy conservation as $\frac{1}{2}u_1^2 + h_1 = \frac{1}{2}u_2^2 + h_2$, where $h = (e + p/\rho)$ is the total heat energy or enthalpy per unit mass. For strong shocks $h_2 \gg h_1$ of the cold stationary gas and $u_1 \gg u_2$, so that the energy equation reduces to $h_2 \approx \frac{1}{2}u_1^2$, which states that the relative kinetic energy of a stationary gas element just ahead of the shock front is converted into thermal energy when the shock wave moves over that element. The energy of the gas which has been subjected to a very strong shock wave is almost equally divided between its kinetic energy and its thermal or internal energy. This may be shown by considering the initial values of the internal energy e_1 and pressure p_1 of the cold stationary gas to be negligible quantities in the conservation equations, giving the kinetic energy per unit mass behind the shock as

$$\tfrac{1}{2}u^2 = \tfrac{1}{2}(u_1 - u_2)^2 = e_2$$

the internal energy per unit mass of the shocked gas.

In principle, the temperature behind very strong shock waves should reach millions of degrees. In practice, real gas effects prevent this. In a monatomic gas high translational energies increase the temperature until ionization occurs and this process then absorbs energy which otherwise would increase the temperature still further. In a polyatomic gas the total energy is divided amongst the various modes (translational, rotational and vibrational) and the temperatures reached are much lower than in the case of the monatomic gas. The reduction of γ due to these processes is significant, since with increasing ionization $\gamma \to 1$, and the temperature ratio depends upon the factor $(\gamma - 1)/(\gamma + 1)$ which becomes very small.

(Problems 15.1, 15.2, 15.3, 15.4, 15.5, 15.6)

Solitons

We have seen that a pulse, limited in space, is also limited in time. Fourier analysis shows that a pulse is the superposition of a large number of components with different frequencies and that the high frequency components contribute to the vertical edges of the pulse Figure 10.3. The superposition of these components changes as phase differences develop; different frequencies will have different phase velocities and the pulse disperses.

It is surprising, therefore, that high amplitude solitary waves or solitons are known to exist. The first recorded observation of a soliton is that of Scott–Russel (1844) who saw a single wave about 40 cm high travelling along a canal in Scotland. Rayleigh (1876) developed an expression for the shape of this soliton based on the hydrodynamics of waves in shallow water.

That expression, the bell-shaped Figure 15.4 is given by

$$\eta = a\,\text{sech}^2\alpha(x - x_0)$$

where

$$\alpha = \frac{1}{2}\sqrt{\frac{3a}{h^2(h + a)}}.$$

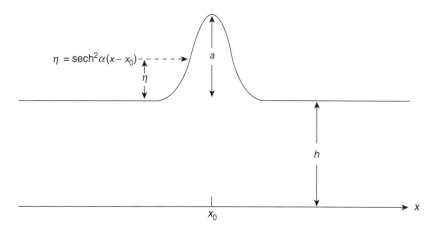

Figure 15.4 The solitary wave (soliton) on a shallow canal seen by Scott–Russel (1844) was described as a sech^2 bell-shaped function by Rayleigh (1876). The canal depth is *h*, the soliton amplitude is *a* and η measures a displacement on the soliton curve. The soliton is centred at x_0 and α is a function of *a* and *h*

η, a, h and x_0 are all shown in Figure 15.4. The coordinate x_0 about which the static figure is centred is replaced by ct when the soliton is moving; c is the soliton velocity and t is the time. We shall see that c is related to the height of the soliton. Larger amplitude solitons move faster.

Further sightings of solitons on Dutch canals led to a thorough discussion of waves with finite amplitude in shallow water by Korteweg and de Vries (1895). Their equation describing soliton behaviour is known as the KdV equation and is now taken as the basis of soliton theory. We shall not pursue the relevant fluid dynamics necessary to obtain the KdV equation but we shall obtain its mathematical form by a method which may lack formal rigour but which provides a good working model. It also emphasizes the physical characteristics which produce a soliton.

The underlying physics of solitons is the competition between two processes. One of these causes a high amplitude or non-linear wave to break; we have seen this in the formation of a shock wave in Figure 15.1c. This results from the increased phase velocities of the high amplitude non-linear components of the wave.

In a soliton this is opposed by the dispersion of the wave components in such a way that a stable profile is maintained.

We shall derive the form of the KdV equation and then discuss the following topics:

- Solitons, Schrödinger's equation and elementary particles.

- Solitons in optical fibres. Telecommunications..

A list of references is given at the end of the chapter.

Non-Linearity

Equation (15.1) shows that the higher amplitude components of an acoustic wave propagate with a phase velocity

$$v = \frac{\partial x}{\partial t} = c_0(1 + s)^{\gamma + 1/2}$$

where c_0 is the phase velocity of a small amplitude linear wave and s, the condensation, is a measure of the compression in the wave.

We may expand this, to a first order, to give

$$v = \frac{\partial x}{\partial t} = c_0\left(1 + \frac{\gamma + 1}{2} s \ldots\right) \tag{15.2}$$

In a linear, low-amplitude, right-going wave we have

$$\eta = \eta_m \, e^{i(\omega t - kx)}$$

So, denoting $\partial\eta/\partial t$ as η_t and $\partial\eta/\partial x$ as η_x we have

$$\eta_t/\eta_x = \frac{-\omega}{k} = -c_0$$

or

$$\eta_t + c_0\eta_x = 0 \tag{15.3}$$

Throughout this chapter we shall indicate partial differentiation with respect to a variable by writing that variable as a subscript. Thus, $\eta_t = \partial\eta/\partial t$; $\eta_x = \partial\eta/\partial x$; $\eta_{tt} = \partial^2\eta/\partial t^2$ and $\eta_{xx} = \partial^2\eta/\partial x^2$. Replacing c_0 in equation (15.3) by v in equation (15.2) gives

$$\eta_t + c_0\left[1 + \left(\frac{\gamma+1}{2}\right)s\right]\eta_x = 0$$

which, because $s = k\eta$ is in phase with η_t (Figure 6.2), becomes

$$\eta_t + c_0\left[1 + \left(\frac{\gamma+1}{2}\right)k\eta\right]\eta_x = \eta_t + c_0\eta_x + c_0\left(\frac{\gamma+1}{2}\right)k\eta\eta_x = 0 \tag{15.4}$$

We are interested in non-linear effects and after removing the linear contribution of equation (15.4) we are left with the non-linear expression

$$\eta_t + b\eta\eta_x = 0 \tag{15.5}$$

where

$$b = c_0\left(\frac{\gamma+1}{2}\right)k$$

Equation (15.5) provides the first two terms of the KdV equation. We now consider the third, the dispersion term, which competes with the non-linear $b\eta\eta_x$ term.

Dispersion and the Form of the KdV Equation   A typical dispersion equation is that for transverse and longitudinal waves in a periodic structure given by equation (5.12) as

$$v = \frac{\omega}{k} = c_0\left(\frac{\sin ka/2}{ka/2}\right)$$

where k is the wave number and a is the particle separation. For small k, long λ, we may expand the sine term to give

$$v = \frac{\omega}{k} = \frac{c_0}{ka/2}\left[\frac{ka}{2} - \left(\frac{ka}{2}\right)^3 \cdots +\right]$$

or

$$\omega = c_0 k \left[1 - \left(\frac{ka}{2} \right)^2 \right] = c_0 k - dk^3 \qquad (15.5a)$$

where

$$d = c_0 a^2 / 4$$

Writing a linear wave in the form

$$\eta = \eta_m \, e^{i(\omega t - kx)}$$

gives

$$\eta_t = i\omega\eta, \quad \eta_x = -ik\eta \quad \text{and} \quad \eta_{xxx} = ik^3\eta$$

which, with equation (15.5a), gives

$$\eta_t + c_0 \eta_x + d\eta_{xxx} = 0$$

Again, the contribution $\eta_t + c_0 \eta_x$ applies only to linear waves and replacing this for non-linear waves by equation 15.5

$$\eta_t + b\eta\eta_x$$

gives

$$\eta_t + b\eta\eta_x + d\eta_{xxx} = 0 \qquad (15.6)$$

where b and d are constant coefficients. This is the form of the KdV equation which describes soliton behaviour. The coefficients b and d depend upon the particular soliton under discussion.

We gain an insight into the effect of the dispersion term by considering the following. Let us write a right-going linear wave in the form

$$\eta = \eta_m \, e^{i(\omega t - kx)} = \eta_m \, e^{ik(c_0 t - x)}$$

where

$$\omega = c_0 k$$

The effect of dispersion, from the previous section, changes $\omega = c_0 k$ to

$$\omega = c_0 k \left[1 - \left(\frac{ka}{2} \right)^2 \right]$$

so we have

$$\eta = \eta_m \exp\left(ik \left[c_0 \left\{ 1 - \left(\frac{ka}{2} \right)^2 \right\} t - x \right] \right)$$

and dispersion has the effect of shifting the wave. Note that in this case of normal dispersion the shift retards the higher k, shorter wavelength terms.

Mathematically, this dispersive shift is used to offset the steepening, wave breaking effects of non-linearity. The technique, known as a Gardner–Morikawa transformation, is to choose a coordinate system which moves with the velocity c_0, the pulse rides on this moving coordinate so that dispersion relative to c_0 is much reduced. In addition, because any dispersive change is now so much slower, a much longer time scale $\tau > t$ is chosen and the final aim is to show that changes in the soliton profile are negligible in the τ time scale.

The Elements of the KdV Equation Although we derived the form of the KdV equation using the amplitude η, the equation is most often written in terms of a quantity u which may represent any property of the wave which varies with distance and time.

In their paper 'The Discovery of the Soliton' (1965) Zabusky and Kruskal used the equation in the form

$$u_t + uu_x + \delta^2 u_{xxx} = 0 \tag{15.7}$$

where $\delta \ll 1$.

Their experiment was made by computer simulation. In the absence of the third dispersive term the non-linear equation

$$u_t + uu_x = 0 \tag{15.8}$$

describes the development of the shock wave of Figure 15.1. The positive pulses of Figure 15.1a, b and c are superposed in Figure 15.5 with u plotted against x. It is evident that u_t increases with higher values of u and equation (15.8) retains a single valued solution only as long as the gradient u_x of the leading edge becomes increasingly negative as the pulse steepens.

Now equation (15.8) is satisfied by any function $u = f(x - ut)$—see Problem 15.7—and

$$u_x = (1 - u_x t)f' \tag{15.9}$$

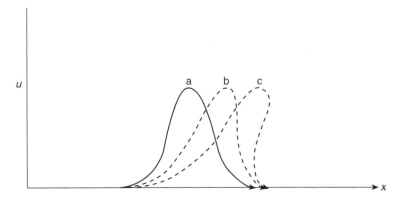

Figure 15.5 Figs. 15.1 (a), (b) and (c) superimposed to show breaking of a non-linear wave

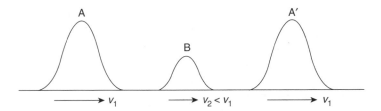

Figure 15.6 The velocity of a soliton increases with its magnitude and solitons are transparent in mutual collisions, each retaining its own identity. A large soliton *A* overtakes a smaller soliton *B* to emerge as *A'* with *B* unaffected

where

$$f' = \partial f / \partial (x - ut)$$

Taking the pulse profile at $t = 0$ as $u = f(x) = \cos \pi x$ equation (15.9) shows that $u_x = -\infty$ at $u \approx 0$ (the foot of the pulse) when $x = 0.5$ and $t = 1/\pi$. At this point the wave becomes infinitely steep and breaks. This behaviour was observed by Zabusky and Kruskal. When Zabusky and Kruskal added the third dispersion term in their computer experiment to give the KdV equation

$$u_t + uu_x + \delta^2 u_{xxx} = 0$$

they found that after a time $t = 1/\pi$ the solution broke into a train of solitary waves (solitons) of successively larger amplitudes with the larger waves travelling faster than the smaller ones. Even more important from the point of view of optical solitons, after one soliton had overtaken another, each soliton retained its unique identity (Figure 15.6). Solitons are transparent to each other and are unaffected by mutual collisions.

(Problems 15.7, 15.8)

Two Important Forms of the KdV Equation

1. The KdV equation for shallow water waves may be written in the form

$$u_t + 6uu_x + u_{xxx} = 0 \tag{15.10}$$

    with a solution

$$u(x, t) = 2\alpha^2 \text{sech}^2 \alpha(x - ct)$$

$$= 2\frac{\partial^2}{\partial x^2} \log \left[1 + e^{2\alpha(x-ct)} \right]$$

    or

$$u(x, t) = 2\frac{\partial^2}{\partial x^2} \log \left[1 + e^{-2\alpha(x-ct)} \right]$$

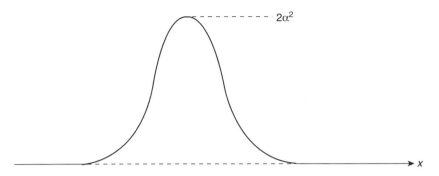

Figure 15.7 The KdV equation $u_t + 6uu_x + u_{xxx} = 0$ has a soliton solution $u(x, t) = 2\alpha^2 \operatorname{sech}^2 \alpha(x - ct)$ with a maximum value of $2\alpha^2$

Note that the exponents in the log solutions may be positive or negative.

The sech^2 form of the solution may be seen to fit equation (15.10) with a soliton velocity $c = 4\alpha^2$ (twice the maximum value of u) by showing that

$$u_t = 2\alpha uc \tanh \phi, \quad \text{where } \phi = \alpha(x - ct)$$
$$uu_x = -2\alpha u^2 \tanh \phi$$

and

$$u_{xxx} = -8\alpha^3 u \tanh \phi + 12\alpha u^2 \tanh \phi$$

The sech^2 shape of the soliton is shown in Figure 15.7. Its peak value is

$$u = 2\alpha^2$$

(Problems 15.9, 15.10)

2. The second important form of the KdV equation is

$$u_t - 6uu_x + u_{xxx} = 0 \tag{15.11}$$

(the shallow water wave form with a negative second term). This has a time independent soliton solution of

$$u(x) = -2\alpha^2 \operatorname{sech}^2(x - x_0)$$

where x_0 locates the centre of the soliton. This solution may be shown to satisfy equation (15.11) by calculating u_x and u_{xxx} as for equation (15.10).

A graph of this soliton, Figure 15.8, shows its minimum to have a value of $-2\alpha^2$. Its importance is its connection with Schrödinger's equation, which we now discuss.

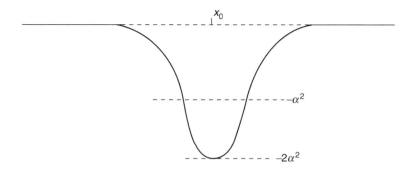

Figure 15.8   The KdV equation $u_t - 6uu_x + u_{xxx} = 0$ has a time independent solution $u(x) = -2\alpha^2 \operatorname{sech}^2 \alpha(x - x_0)$ with a minimum value of $-2\alpha^2$. This equation is related via Miura's transformation to Schrödinger's equation which has an eigenvalue of $\lambda = -\alpha^2$

(Problem 15.11)

Solitons, Schrödinger's Equation and Elementary Particles

In 1968, Miura found a remarkable connection between equation (15.11) and the equation

$$v_t + 6v^2 v + v_{xxx} = 0 \tag{15.12}$$

which itself has a soliton solution.
Miura showed that if $v^2 + v_x = u$ then

$$\left(\frac{\partial}{\partial x} + 2v \right) (v_t - 6v^2 v_x + v_{xxx}) = u_t - 6uu_x + u_{xxx} \tag{15.13}$$

(Problem 15.12)

So if v satisfies equation (15.12) with the sign of its second term changed, then u satisfies equation (15.11). Now Miura's transformation with

$$v^2 + v_x = u(x) \quad \text{and} \quad v = \psi_x / \psi$$

yields

$$\psi_{xx} - u(x)\,\psi = 0 \tag{15.14}$$

(Problem 15.13)

If $u(x)$ is now transformed to $u(x) - \lambda$, where λ is a constant, then equation (15.14) becomes Schrödinger's equation

$$\psi_{xx} + (\lambda - u(x))\,\psi = 0$$

with λ as an eigenvalue.

So Miura's transformation has related the KdV equation

$$u_t - 6uu_x + u_{xxx} = 0 \qquad (15.11)$$

to Schrödinger's equation

$$\psi_{xx} + (\lambda - u(x))\,\psi = 0 \qquad (15.15)$$

Using the soliton solution

$$u = -2\alpha^2 \operatorname{sech}^2 \alpha(x - x_0)$$

of equation (15.11) we can show that the wave function

$$\psi = A \operatorname{sech} \alpha(x - x_0), \quad \text{where } A \text{ is a constant} \qquad (15.16)$$

satisfies equation (15.15) when the eigenvalue $\lambda = -\alpha^2$ which is half the value of the minimum of the soliton with which it is associated (Figure 15.8) (See Gardner *et al.*, 1967).

(Problems 15.14, 15.15, 15.16)

Since λ is negative this represents a bound state in wave mechanics.

Other values of $\lambda > 0$ may be associated with solitons but these are not bound states and are related to progressive waves.

The fact that solitons may be associated with Schrödinger's equation and retain their unique identities in mutual collisions has led physicists to postulate that solitons may appear as massive elementary particles much heavier than the proton.

Solitons may enter particle physics in another way, confined not only in space but in time. In this case they are called instantons. Instantons have already been used to explain a pattern of particle masses which had posed a long-standing puzzle.

There are four ways of making quark–antiquark mesons from light quarks. Three of these mesons have been known for many years: the negative, positive and neutral pi mesons (pions) with masses equivalent to about 140 MeV (an electron equivalent mass is $\sim 0.5\,\text{MeV}$).

The fourth meson has never been found but the eta meson has all the required properties except its mass which is about 550 MeV. Instantons explain this mass anomaly—they appear as energy excitations, located in space, in the field which binds the quarks together. They change the mass distribution among the mesons because they affect the various quark combinations in different ways (see Rebbi, 1979).

Optical Solitons

At the time of this writing the most practical use of solitons is in telecommunications. Optical fibres act as wave guides to microwaves and higher frequency electromagnetic waves and optical solitons are able to carry information along single mode silica fibres at multigigabit rates for distances greater than 9000 km, the width of the Pacific Ocean, with a bit error rate (BER) $< 10^{-9}$, the international standard. Modern fibres have a very low loss rate of < 1 dB km^{-1} and an effective area of $\sim 30\,\mu\text{m}^2$. The electrical power involved is very low and a total optical system is feasible including the amplifiers spaced along the cable. This permits a simpler, faster and more easily maintained system than that using conventional electronics. Research on optical solitons is world-wide but, for the English reader, the work of Linn Mollenauer and his colleagues at the A. T. & T. Bell Labs, New Jersey is the most accessible (see references).

Optical solitons have the normal sech^2 intensity profile and their amplitudes are given by sech wave function solutions to a non-linear Schrödinger equation (see Appendix, p. 555).

As with all solitons, optical solitons are produced by a balance between the competing effects of dispersion and non-linearity but the non-linearity of optical fibres is a very special case which contributes in a remarkable way to the maintenance of the soliton profile.

The Kerr Optical Effect and Self-phase Modulation   In some materials, including silica fibres, the index of refraction for light of a given wavelength varies with the intensity of the light. This is the Kerr optical effect, which is expressed by

$$n - n_0 = n_2 I$$

where n is the index of refraction for a light wave of intensity I (large enough for non-linearity), n_0 is the refractive index for a low amplitude wave of the same frequency and n_2 is a constant equal to $3.2 \times 10^{-16}\,\text{cm}^2\,\text{W}^{-1}$. The value of n_2 is small but the area of a single mode optical fibre $\sim 10^{-6}\,\text{cm}^2$, so we must think in terms of megawatts per square metre. Moreover, the effects of non-linearity build up over fibre distances of many kilometres.

Since $n_2 I$ is positive we have

$$n - n_0 = c\left(\frac{1}{v} - \frac{1}{v_0}\right) > 0$$

so the phase velocity v of a high amplitude wave is less than v_0, the phase velocity of a low amplitude linear wave of the same wavelength.

At a given wavelength this creates a phase retardation between the two amplitudes of

$$\Delta\phi = \frac{2\pi}{\lambda} L n_2 I$$

over a length L of the fibre. This phase retardation is obviously greater for the short wavelength high frequency components of the pulse, Figure 15.9, than for the lower frequencies and so in the high intensity central section of the pulse the higher frequencies are shifted towards the tail of the pulse while the lower frequencies advance to the front.

This process is opposed by the dispersive properties of the fibre because at the wavelength at which the solitons are centred; that is, $\lambda \sim 1.5\,\mu\text{m}$ (1500 nm) the dispersion is negative (anomalous) so that $\partial v_g/\partial\lambda < 0$, where v_g is the group velocity.

Negative dispersion advances the trailing higher frequencies and retards the lower frequencies, both in a direction towards the centre of the pulse, so the pulse sharpens

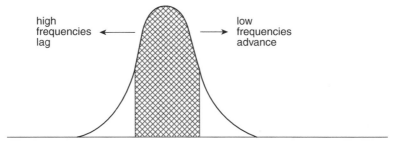

Figure 15.9 In the Kerr optical effect the velocity of light at a given wavelength depends upon its intensity. The high frequencies in the high intensity region of a soliton travelling in an optical fibre suffer a phase retardation; the low frequencies are advanced

towards a soliton sech^2 shape, Figure 15.10, and in a loss-free perfect silica fibre the soliton would maintain this shape indefinitely. In practice, the wavelength $\lambda \sim 1.5\,\mu\text{m}$ is close to the minimum of the loss versus wavelength behaviour of the fibre, which accounts for low loss fibres of < 1 dB km^{-1}. Optical amplifiers, which we shall discuss shortly, maintain the shape of the soliton over very long distances but even without amplification a soliton can travel several hundred kilometres along the fibre without changing its amplitude or shape.

This distance is called the soliton period, Figure 15.11, and is given by

$$z_0 = 0.322\,\frac{\pi^2 c\tau^2}{\lambda_{\text{vac}}^2 D} = 0.39\,\frac{\tau^2}{D} \quad \text{at} \quad \lambda \sim 1.55\,\mu\text{m}$$

where c is the velocity of light in free space, λ_{vac} is the wavelength in free space, τ is the full width at half the maximum value of the soliton and D is the group velocity dispersion parameter of the fibre; that is, the change in pulse delay with change in wavelength per unit of fibre length.

The units of τ are picoseconds and experimental solitons are produced in the range 1–50 ps. The units of D are picoseconds per nanometre per kilometre and experimental values of D are $\sim 10\,\text{ps}\,\text{nm}^{-1}\,\text{km}^{-1}$. At $D \sim 1\,\text{ps}\,\text{nm}^{-1}\,\text{km}^{-1}$ a 50 ps pulse has a soliton period $z_0 \approx 930\,\text{km}$.

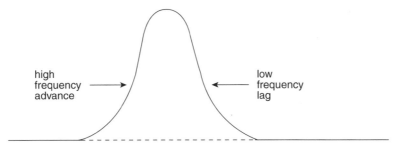

Figure 15.10 The effects of Figure 15.9 are reversed by the negative (anomalous) dispersion of the optical fibre at the wavelength on which the soliton is centred. This sharpens the soliton pulse

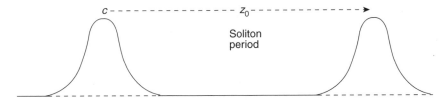

Figure 15.11 A soliton can travel several hundred kilometres in an optical fibre without being degraded in any way. This distance z_0, is called the soliton period

Experimental Aspects Experimentally, the solitons are produced by a mode locked laser with an additional fibre arm in the feedback loop. As the laser builds up from noise the initially broad pulses are considerably narrowed by passing through the fibre arm and then reinjected back into the laser cavity, forcing the laser itself to produce narrower pulses. This process is repeated until the pulses become solitons and are ready for injection, via coupling, into the transmission system. The laboratory cable is a fibre spool $\sim 75\,\mathrm{km}$ long and the solitons are recirculated through this loop to travel distances $> 10\,000\,\mathrm{km}$ if required.

A typical laser soliton source produces pulses of $\sim 50\,\mathrm{ps}$ with a power $\sim 0.5\,\mathrm{mW}$ at a repetition rate of 2.5 GHz.

The Raman Effect This plays a very important role in optical soliton transmission. It arises when molecules in a material absorb radiation and it involves the vibrational and sometimes the rotational energy levels of the molecules. Figure 15.12 shows the vibrational

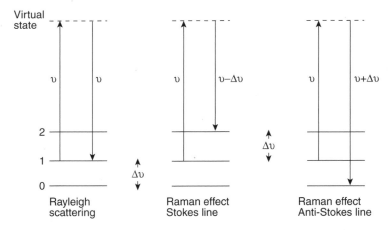

Figure 15.12 The Raman effect can degrade a soliton by transferring energy from its higher frequency to its lower frequency components. Vibrational energy levels in the optical fibre absorb higher frequency radiation ν from the soliton which reabsorbs it at a lower frequency $\nu - \Delta\nu$ (Stokes line). There are three possible processes. In Rayleigh scattering a photon returns to its original vibrational energy level, the Raman effect provides a frequency change $\Delta\nu = \pm 1$, where $\Delta\nu$ is the frequency interval between vibrational energy levels

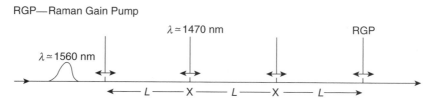

Figure 15.13 The transmission line acts as its own distributed amplifier when solitons accept higher energy photons via the Raman effect from optical pumps located at short intervals (distances $\ll z_0$, the soliton period). Excessive noise production is avoided by frequent low gain amplification (see Mollenauer *et al.*, 1986)

energy levels in a molecule with 0 as the ground state. Suppose initially that the molecule is in the energy level 1 and absorbs a photon of frequency ν which raises it to an excited level which may not be a stationary state. If the photon drops back to its original level the re-radiated photon of frequency ν is called Rayleigh scattering. However, selection rules also allow vibrational level changes $\Delta\nu = \pm 1$, where $\Delta\nu$ is the vibrational energy level interval, so the photon may drop back into level 2 or 0. The re-radiated or scattered photon will then appear at the frequencies $\nu - \Delta\nu$ (Stokes line) or $\nu + \Delta\nu$ (anti-Stokes line).

The Raman effect can 'degrade' a single soliton via a process known as the 'self-frequency shift'. Here the vibrational levels of the silica fibre molecules absorb energy from the higher frequencies in the soliton pulse and the scattered radiation acts as a Raman pump for the lower frequencies in the pulse because the fibre provides a Raman acceptance band over a broad frequency spectrum.

Indeed, although a power of 0.5 mW provides a stable single soliton, early experiments showed that solitons with powers >1 W suffered from 'self-frequency shift' to such an extent that the soliton initially narrowed but then formed smaller satellite solitons.

The Raman Effect and Optical Amplification   Solitons can gain energy via the Raman effect as well as lose it and this is the basis of amplification along an optical transmission line. One method results in the line acting as its own distributed amplifier. Laser pumps coupled into the line at regular intervals maintain the shape of a soliton by feeding in a frequency higher than that of the soliton, the energy difference being very close to the broad peak of the Raman gain band of the silica fibre. In Figure 15.13 the soliton wavelength is $\lambda = 1.5\,\mu\text{m}$ and the lasers pump energy at $\lambda \approx 1.4\,\mu\text{m}$. The pumps can also inject radiation in the counter-propagating direction, which helps to average out any effect of pump fluctuations; the penetration of the amplifying beam along the fibre is also enhanced. The intervals between the laser pumps are $\sim 30\,\text{km}$ which is a small fraction of the soliton period z_0 ($\sim$ several hundred kilometres). In this way, the gain per interval is kept low enough to avoid excessive amplification of noise.

A second method, Figure 15.14 uses lumped amplifiers in the form of short lengths $\sim 3\,\text{m}$ of optically pumped fibres doped with a rare earth such as Erbium. Again, the interval between these lumped amplifiers is $\ll z_0$ the soliton period to keep the noise amplification low. The lumped amplifiers are energized by laser diode chips and for an input of $\sim 10\,\text{mW}$ a gain of 30–40 dB is obtained at the useful wavelengths. The power of

A—Erbium-doped amplifying fiber
OP—optical pump ($\lambda = 1480$ nm)
C—coils of transmission line

$\lambda = 1532$ nm   OP   A   25 km   A   25 km   A   OP   25 km

C   C   C

Figure 15.14 Solitons are now maintained by lumped amplifiers in the form of ≈ 3 m lengths of optically pumped fibres doped with the rare earth Erbium separating 25 km lengths of transmission line. The interval between the low gain amplifiers $\ll z_0$ (the soliton period) to avoid noise amplification

these amplifiers is useful in multiplexing, the subject of the next section (see Desurvire, 1992).

Multiplexing This refers to the possibility of sending more than one channel of information down a single fibre. In current transmission systems, non-linear interaction causes severe interchannel interference but solitons are transparent to each other. They are unaffected by collisions and do not interfere with each other.

In multiplexing, two channels along a single fibre are provided by solitons which are polarized in planes perpendicular to each other.

Even more channels are possible with wavelength division multiplexing. Solitons of different wavelengths have different velocities and analysis shows that in a system using a chain of lumped amplifiers, adjacent WDM (wavelength division multiplexed) solitons interact just as in a lossless fibre so long as the collision length (twice the length of a soliton) is two or three times the amplifier spacing (Figure 15.15).

This implies that several multigigabit per second WDM channels spanning a wavelength separation of 1 or 2 nm may be used in a single fibre.

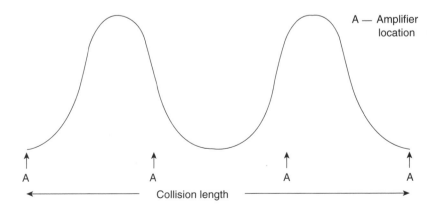

A — Amplifier location

A   A   A   A

Collision length

Figure 15.15 Wavelength division multiplexing is possible with solitons of different wavelengths and velocities. These solitons do not interfere with each other so long as the collision length (twice a soliton length) is two or three times the lumped amplifier spacing (see Mollenauer *et al.*, 1990)

In a conventional transmission line each channel must be isolated at the regenerative amplifiers and separately processed but one amplifier can handle all soliton channels and Erbium-doped amplifiers are powerful enough to do this.

Random Noise Effects and the Frequency Sliding Guiding Filter

There are two main sources of error which affect an optical soliton transmission system: fluctuations of pulse energy and arrival time at the receiver. Spontaneous emission (noise) always accompanies coherent Raman gain and at each amplifier, amplified spontaneous emission (ASE noise) is added to a soliton which can change its energy and its central frequency in a random way. The change of energy may affect the amplitude of a soliton and the accumulated effect may reduce a soliton to such an extent that its intended arrival as a ONE in the bit system is registered as a ZERO. Alternatively, amplified noise may register a ONE in a ZERO space. This contributes to the bit error rate (BER) which must be kept below the international standard of $< 10^{-9}$.

The ASE change in the frequency of the soliton changes its velocity and therefore affects its arrival time, throwing the pulse out of its proper time slot.

Amplitude and time jitter may be reduced by narrowing the bandwidth of the transmission line (Mollenauer, 1994), using a narrow band filter at each amplifier. Each filter is a low-finesse Fabry–Perot etalon (p. 343), centred on the true frequency peak of the soliton (Figure 15.16). A soliton whose frequency has been shifted from the filter peak suffers a loss across the spectrum provided by the filter. This, together with the non-linear effect which generates new frequencies, pushes the soliton back towards the filter peak. In this way, the noise-induced frequency shift is returned to zero rather than being maintained as it would in a broad-band transmission line.

Amplitude jitter is damped because a pulse with excess energy will narrow in time and broaden in spectrum more than the average and will suffer a greater loss at each filter. However, the soliton loss at each filter must be replaced at each amplifier by an excess gain with a resulting growth in noise.

Mollenauer *et al.* (1994), found that even when the soliton source laser was not tuned exactly to the filter peak frequency, the soliton was still guided rapidly on to the filter peak. The filter peak frequencies were therefore gradually slid with distance so that the soliton frequency followed the filters while the noise remained in its original frequency band and

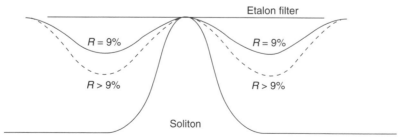

Figure 15.16 Noise effects in an optical transmission line are reduced using a narrow band Fabry--Perot etalon filter at each amplifier. The low finesse, $R \sim 9\%$, of fixed frequency filters can be increased, $R > 9\%$, if the frequency of the filters is gradually shifted with distance along the line. The soliton frequency has no difficulty in adjusting itself to this shift and noise is progressively reduced (see Mollenauer *et al.*, 1994)

its growth was inhibited. This noise reduction allowed the etalon filters to be strengthened to a higher finesse. Experiments with a soliton pulse width of $\tau \sim 16\,\mathrm{ps}$, $D \sim 0.5\,\mathrm{ps\,nm^{-1}\,km^{-1}}$, amplifier spacing $= 26\,\mathrm{km}$ with one filter per amplifier, and a frequency sliding rate of $7\,\mathrm{GHz}\ 10^{-3}$ km gave a net frequency shift over $9000\,\mathrm{km}$ (trans-Pacific distance) of a few soliton bandwidths, i.e. $0.5\,\mathrm{nm}$ at $\lambda = 1557\,\mathrm{nm}$. Such a series of sliding frequency etalon filters can operate over a range of wavelengths wide enough to allow several channels of wavelength division multiplexing.

Problem 15.1

The properties of a stationary gas at temperature T_0 in a large reservoir are defined by c_0, the velocity of sound, $h_0 = c_p T_0$, the enthalpy per unit mass, and γ, the constant value of the specific heat ratio. If a ruptured diaphragm allows the gas to flow along a tube with velocity u, use the equation of conservation of energy to prove that

$$\frac{c_0^2}{\gamma - 1} = \frac{\gamma + 1}{2(\gamma - 1)} c^{*2}$$

where c^* is the velocity at which the flow velocity equals the local sound velocity.

Hence show that if $u_1/c^* = M^*$ and $u_1/c_1 = M_s$, then

$$M^{*2} = \frac{(\gamma + 1)M_s^2}{(\gamma - 1)M_s^2 + 2}$$

Problem 15.2

Using a coordinate system which moves with a shock front of velocity u_1, show from the conservation equations that c^* in Problem 15.1 is given by

$$c^{*2} = u_1 u_2$$

where u_2 is the relative flow velocity behind the shock front.

Problem 15.3

Use the conservation equations to prove that the pressure ratio across a shock front in a gas of constant γ is given by

$$\frac{p_2}{p_1} = \frac{\beta - \alpha}{1 - \beta\alpha}$$

where $\beta = \rho_2/\rho_1$, the density ratio, and $\alpha = (\gamma - 1)/(\gamma + 1)$.

Problem 15.4

Use the results of Problems 15.1 and 15.2 with the equation of momentum conservation to prove that the shock front Mach number is given by

$$M_s = \frac{u_1}{c_1} = \sqrt{\frac{y + \alpha}{1 + \alpha}}$$

where $y = p_2/p_1$, the pressure ratio across the shock and $\alpha = (\gamma - 1)/(\gamma + 1)$. Hence show that the flow velocity behind the shock is given by

$$u = \frac{c_1(1 - \alpha)(y - 1)}{\sqrt{1(1 + \alpha)(y + \alpha)}}$$

Problem 15.5

The diagrams show (a) a shock wave of pressure p_2 and flow velocity u propagating into a stationary gas, pressure p_1, and (b) after reflexion at a rigid wall the reflected wave of pressure p_3 moving back into the gas behind the incident shock still at pressure p_2. Use the result at the end of Problem 15.4 to show that the flow velocity u_r behind the reflected wave is given by

$$\frac{u_r}{c_2} = \frac{(1-\alpha)(p_3/p_2 - 1)}{\sqrt{(1+\alpha)(p_3/p_2 + \alpha)}}$$

and since $u + u_r = 0$ at the rigid wall, use this result together with the ratio for $c_2/c_1 = (T_2/T_1)^{1/2}$ to prove that

$$\frac{p_3}{p_2} = \frac{(2\alpha + 1)y - \alpha}{\alpha y + 1}$$

where $y = p_2/p_1$ and $\alpha = (\gamma - 1)/(\gamma + 1)$.

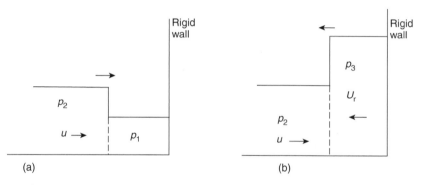

(a) (b)

Problem 15.6

Use Problem 15.5 to prove that the ratio

$$\frac{p_3 - p_1}{p_2 - p_1} \rightarrow 2 + \frac{1}{\alpha}$$

in the limit of very strong shocks. (Note that this value is 8 for $\gamma = 1.4$ and 6 for $\gamma = 5/3$, compared with the normal acoustic pressure jump of 2 upon reflexion.)

Problem 15.7

Equation (15.9) evaluates u_x for $u = f(x - ut)$. Obtain u_t in a similar way and use this with equation (15.9) to prove equation (15.8).

Problem 15.8

Burger's equation $u_t + uu_x - \nu u_{xx} = 0$ where $\nu > 0$ is a special case. It has a second-order dispersion term and is directly integrable. Show that $u = -2\nu\psi_x/\psi$ transforms Burger's equation into the diffusion equation

$$\frac{\partial^2 \psi}{\partial t^2} = \nu \frac{\partial^2 \psi}{\partial x^2}$$

For fluids, ν is a measure of viscosity which dissipates excess momentum in non-linear waves.

Problem 15.9
Show that $u(x,t) = 2\alpha^2 \operatorname{sech}^2 \alpha(x - ct)$ is a soliton solution of the KdV equation $u_t + 6uu_x + u_{xxx} = 0$ after calculating u_t, u_x and u_{xxx} as shown in the text.

Problem 15.10
For small values of q, $\log(1 + q) \approx q$. Show that values of $u(x,t)$ near the base of Figure 15.5(a) where $uu_x \approx 0$ may be written

$$u(x,t) \approx 2 \frac{\partial^2}{\partial x^2} e^{-2\alpha(x-ct)}$$

and that this satisfies the dispersion equation $u_t + u_{xxx} = 0$ if $c = 4\alpha^2$.

Problem 15.11
Use the method of Problem 15.9 to show that $u(x) = -2\alpha^2 \operatorname{sech}^2 \alpha(x - x_0)$ is a solution of the KdV equation $u_t - 6uu_x + u_{xxx} = 0$.

Problem 15.12
Prove equation (15.13) if $u = v^2 + v_x$.

Problem 15.13
Verify equation (15.14) for $u(x) = v_x + v^2$ and $v = \psi_x/\psi$.

Problem 15.14
Show that the wave function $\psi = A \operatorname{sech} \alpha(x - x_0)$ where A is a constant satisfies Schrödinger's equation (15.15) when $\lambda = -\alpha^2$.

Problem 15.15
KdV equations are invariant to a Galilean transformation. Show that the transformations $u \to u - \lambda$ where λ is constant together with $x \to x + 6\lambda t$ returns $u_t + 6uu_x + u_{xxx} = 0$ to its original form.

Problem 15.16
At time $t = 0$ a high amplitude signal has a profile $y = a \sin \pi x$ with $\partial y/\partial t = 0$. Thereafter, it propagates according to the non-linear wave equation

$$\frac{\partial^2 y}{\partial t^2} = c_0^2 \left(1 + \varepsilon \frac{\partial y}{\partial x} \right) \frac{\partial^2 y}{\partial x^2}$$

where ε is a small positive constant.

Show that the time required for the leading edge of a positive signal to become infinitely steep is given by

$$t = 4/c_0 \varepsilon a \pi^2$$

Hint: Rayleigh's method (Rayleigh, *Theory of Sound*, Vol. 2, Dover Press p. 35), shows the required time to be the reciprocal of the maximum value of $|du/dx|$ where du is the relative phase velocity between two points on the leading edge of a positive signal separated by a horizontal distance dx. Note that waves propagate in the positive and negative x-directions.

Bibliography

Solitons and Non-Linear Wave Equations by Dodd, K. R. *et al*. Academic Press, New York (1983).
Non Linear Waves, Solitons and Chaos by Infeld, E. and Rowlands, G., Cambridge University Press, Cambridge (1990).
Waves Called Solitons, Concepts and Experiments by Remoissenet, M., Springer Verlag, Berlin (1994).
Non-Linear Physics, Vol. 4, Contemporary Concepts in Physics by Sagdeev, R. Z., Usikov, D. A. and Zaslavsky G. H., Harwood Academic Press, Chur, Switzerland (1990).
Non-Linear Waves and Solitons by Toda, M., Kluwer Academic Publishers, Dordvecht (1989).

References

Desurvire, E. (1992), *Scientific American* (January).
Gardner, C. S., Greene, J. M., Kruskal, M. D. and Miura, R. (1967), *Phys. Rev. Lett.*, **19**, 1095.
Korteweg, D. J. and de Vries, G. (1895), *Phil. Mag.*, **39**, 422.
Miura, R. (1968), *J. Math. Phys.*, **9**, 1202.
Rayleigh, Lord, (1876), *Phil. Mag.*, (5), 1, 257.
Rebbi, C. (1979), *Scientific American* (February).
Sagdeev, R. Z. and Kennel, C. F. (1991), *Scientific American* (April).
Scott–Russel, J. (1844), *Proc. Roy. Soc. Edinburgh*, 319.
Zabusky, N. and Kruskal, M. D. (1965), *Phys. Rev. Lett.*, **15**, 240.

Selected references in chronological order from:
Mollenauer L. *et al.*: *Phys. Rev. Lett.*, **45**, 1095 (1980).
Fibreoptic Technology, April (1982).
Optics Lett., **9**, 13 (1984).
Phil. Trans. Royal Soc., London, **A 315**, 437 (1985).
Optic News, **12**, 42 (1986).
J. I. E. E., Journal of Quantum Electronics, **22**, 157 (1986).
Optics Lett., **13**, 675 (1988).
Phys. World, 29, September (1984).
Optics Lett., **15**, 1203 (1990).
J. Lightwave Technol., **9**, 3 (1991).
Laser Focus World, 159, November (1991).
Optics Lett., **17**, 1575 (1992).
Optics and Photonic News, **15**, April (1994).

Appendix 1: Normal Modes, Phase Space and Statistical Physics

The last line of the introduction to the first edition states that 'it is the wide validity of relatively few principles which this book seeks to demonstrate'. Here we apply that concept to the relationship between normal modes which feature in most of the book, phase space of chapter 14, and statistical physics.

Firstly, we wish to show that the expression for the number of normal modes *per unit volume* in the frequency range ν to $\nu + d\nu$ given on p. 253 as

$$dn = \frac{4\pi\nu^2 d\nu}{c^3}$$

is nothing more than the number of 'cells' of phase space *per unit volume* in the same range ν to $\nu + d\nu$ available to particles in a statistical distribution.

Moreover, we can easily convert this expression in the frequency ν to one in the velocity v, the momentum $p = mv$ or the energy E.

The particle may be a molecule in the classical Maxwell–Boltzmann distribution (M–B), a fermion of half integral spin in the quantum Fermi–Dirac distribution (F–D) or a boson or any other particle of integral spin in the quantum Bose–Einstein distribution (B–E). Bosons are the messengers of the force fields in physics, e.g. the photon in the electromagnetic field.

We shall see that each of these distributions is nothing more than the statement that

$$n_i = g_i \times \text{probable occupation of the phase space cell.}$$

Here n_i is a number of particles in the distribution and g_i is our expression $4\pi\nu^2\, d\nu/c^3$ (or its equivalent).

The expression for g_i is common to all three types of distribution but the occupation factor or relative probability of occupation depends on the way in which the particles are allowed to distribute themselves.

Firstly, let us examine the various equivalent forms of g_i. We write

$$g_i = g_i(\nu)\, d\nu = 4\pi\nu^2\, d\nu/c^3$$

The Physics of Vibrations and Waves, 6th Edition   H. J. Pain
© 2005 John Wiley & Sons, Ltd

as the number of phase space cells per unit volume in the frequency range ν to $\nu + d\nu$. For a quantum particle (p. 415) the momentum $p = \hbar k = h\nu/c$ where h is Planck's constant, k is the particle wave number $= 2\pi/\lambda$ and c is the velocity of light, so

$$g_i = g_i(p)\,dp = 4\pi p^2 dp/h^3$$

is the number of phase space cells per unit volume in the momentum range p to $p + dp$. Note that $4\pi p^2 dp$ is the volume of the shell in momentum space between spheres of radius p and $p + dp$.

All particles in statistical distributions are required to be free particles, that is having only kinetic energy with no potential energy interaction terms.

Thus, the energy of a particle $E = \frac{1}{2}mv^2 = p^2/2m$ where $p = mv$, m is the particle mass and v is its velocity. Now

$$p^2\,dp = (2m^3)^{1/2}E^{1/2}\,dE = m^2v^2m\,dv = m^3v^2\,dv$$

so

$$g_i = g_i(E)\,dE = 4\pi(2m^3)^{1/2}E^{1/2}\,dE/h^3$$

is the number of phase space cells *per unit volume* in the energy range E to $E + dE$ and

$$g_i = g_i(v)\,dv = 4\pi m^3v^2\,dv/h^3$$

is the number of phase space cells *per unit volume* in the velocity range v to $v + dv$.

Although we used the phase space of $\dot{x}$ or v with x in our discussion of chaos, the phase space of mv or p with x is much more commonly used in physics. The phase space of (p, x) reveals the significance of h^3 in the denominators of g_i. Consider the expression

$$4\pi p^2\,dpV/h^3$$

where V is the total volume (not the unit volume) so that the numerator expresses the phase space over the momentum range p to $p + dp$ and the volume $V = xyz$ of the system.

Heisenberg's Uncertainty Principle, p. 416, tells us that $\Delta x\Delta p \sim h$, so we may write $(\Delta x\Delta p_x)(\Delta y\Delta p_y)(\Delta z\Delta p_z)$ as h^3; that is, the 'volume' of a cell in (p, V) phase space. This volume is the smallest acceptable volume which a particle may occupy for it defines the volume associated with a particle as

$$\left(\frac{h}{\Delta p}\right)^3 \approx \lambda_{DB}^3$$

where λ_{DB} is the de Broglie wavelength of the particle (p. 412).

So g_i measures the number of phase space cells each of 'volume' h^3 *per unit volume* in the range p to $p + dp$. Each of these cells may or may not be occupied by a particle.

We now examine what we mean by a statistical distribution in order to find the probable occupation of a cell. This occupation factor is different for each of the three distributions M–B, F–D and B–E.

We consider a system, say a gas, of N particles occupying a volume V and having a total internal energy E. The macroscopic parameters E, V, N define a *macrostate*. The energy E may be partitioned in many different ways among the N particles subject only to the restrictions that $E = \sum n_i \varepsilon_i$ and $N = \sum n_i$ remain constant where ε_i represents the energy levels available to the particles. The probability of a system being found in a particular partition is proportional to W the number of ways of distributing the energy among the particles to achieve that partition.

Each different way is called a *microstate* and each *microstate* has *a priori* the same probability. Each microstate contributes to the statistical weight of a partition so that the particular partition reached by the greatest number of ways has the greatest statistical weight and is therefore the most probable. The most probable partition with W (maximum) defines the equilibrium of the macrostate and is written Ω (*EVN*).

It is here that we relate Ω (*EVN*) to the concept of entropy S. Entropy is a measure of the disorder of a system which increases as the system tends to equilibrium. At constant temperature and volume the internal energy E of the system may be written

$$E = F + TS$$

where T is the temperature, S is the entropy and the product TS is a measure of the energy of the system locked in the disorder amongst the particles and not available for work. F is defined as the Helmholtz free energy and measures the work which can be done by the system at constant temperature. At best, in an ideal reversible thermodynamic process the disorder energy TS remains constant, but in a natural or thermodynamically irreversible process TS increases at the expense of F as E remains constant.

An isolated system in equilibrium with the most probable partition of its energy among its particles represents a maximum of its entropy S and Boltzmann related S and Ω through his expression $S = k \log \Omega$ where k is Boltzmann's constant. Fluctuations from the equilibrium position are very small indeed and $\log \Omega$ is a very sharply defined function.

Calculating the value of W the statistical weight of a partition in order to find W (maximum) $= \Omega$ (*EVN*) for each of the three distributions is a mathematical exercise which is straightforward and a little tedious but which fails to reveal the underlying physics.

We shall make these calculations at the end of this appendix but we adopt the procedure of quoting the results below together with the forms in which we usually meet them. This will raise questions the answers to which are not evident in the mathematical derivation (Table A1.1).

For all three distributions the particles are identical and indistinguishable, the total energy E and number of particles N are constant. There are no restrictions on the number of particles having a particular energy in the M–B and B–E distributions but in the F–D distribution, Pauli's exclusion principle allows only one fermion per energy level (or two if we include spin).

Note firstly that the occupation factor or relative probability of occupation for each distribution includes the term $e^{\alpha + \beta \varepsilon_i}$, where α and β arise as multipliers in the mathematical derivation. The index of the exponential requires β to be the inverse of an energy and the relevant term in the normal form of the Fermi–Dirac distribution suggests that α is the ratio of two energies.

Table A1.1   The mathematical derivation for each statistical distribution in the left hand column is compared with its more familiar form on the right

$n_i = g_i \times$ occupation factor	Normal form
M–B $\quad n_i = g_i \times \dfrac{1}{e^{\alpha + \beta \varepsilon_i}}$	$\dfrac{n}{N} = \dfrac{4\pi p^2 \mathrm{d}p}{(2\pi mkT)^{3/2}}\, e^{-p^2/2mkT}$
$\qquad = g_i\, e^{-\alpha - \beta \varepsilon_i}$	$\hfill (p = mv)$
F–D $\quad n_i = g_i \times \dfrac{1}{e^{\alpha + \beta \varepsilon_i} + 1}$	$n(E)\,\mathrm{d}E = \dfrac{2.4\pi V (2m^3)^{1/2} E^{1/2}}{h^3} \times \dfrac{1}{e^{(\varepsilon_i - \varepsilon_F)/kT} + 1}$
B–E	$n(\nu)\,\mathrm{d}\nu\, h\nu = E(\nu)\,\mathrm{d}\nu$
$\qquad n_i = g_i \times \dfrac{1}{e^{\alpha + \beta \varepsilon_i} - 1}$	$= \dfrac{2.4\pi \nu^2 \,\mathrm{d}\nu\, h\nu}{c^3} \times \dfrac{1}{e^{h\nu/kT} - 1}$

Planck's radiation law

In comparing the two columns of the table several questions arise:

1. Is $\beta = 1/kT$?

2. What has happened to the α term in the normal form of M–B?

3. What is the physical significance of the α term?

4. What has happened to the α term in Planck's radiation law?

In question 1 let us integrate by parts the expression

$$\int e^{-\beta \varepsilon}\,\mathrm{d}p = [p e^{-\beta \varepsilon}]_{p=-\infty}^{p=+\infty} + \beta \int p \frac{\partial \varepsilon}{\partial p}\, e^{-\beta \varepsilon}\,\mathrm{d}p$$

where

$$\varepsilon = p^2/2m$$

For $\varepsilon \to \infty$ as $p \to \pm\infty$ the first term on the right hand side equals zero, leaving

$$\frac{1}{\beta} = \frac{\displaystyle \int p \frac{\partial \varepsilon}{\partial p}\, e^{-\beta \varepsilon}\,\mathrm{d}p}{\displaystyle \int e^{-\beta \varepsilon}\,\mathrm{d}p} = \overline{p \frac{\partial \varepsilon}{\partial p}}$$

the average value of

$$p \frac{\partial \varepsilon}{\partial p}$$

From the equipartition of energy

$$p\frac{\partial \varepsilon}{\partial p} = \frac{\overline{p^2}}{m} = kT = \frac{1}{\beta}$$

where kT is the average energy per particle.

In question 2 we note that the term $e^{-\alpha}$ in M–B has been replaced by $N/(2\pi mkT)^{3/2}$ and that h^3 has been lost from the denominator of $g_i(p)\,dp$. To explain this and its consequences let us write not n *per unit volume* but n_p in the range p to $p + dp$ over all $V = xyz$ as

$$n_p = \frac{V4\pi p^2 \, dp \, e^{-p^2/2mkT}}{h^3}$$

Then

$$N = \sum n_p = V \int_0^\infty \frac{4\pi p^2 \, dp \, e^{-p^2/2mkT}}{h^3}$$

where the standard definite integral is well known to have a value of $(2\pi mkT)^{3/2}$.

Thus

$$N = V(2\pi mkT)^{3/2}/h^3$$

Now the average particle momentum $\bar{p} = m\bar{v}$ where $\frac{1}{2}m\bar{v}^2 = kT$ ($\bar{v}$ is the most probable velocity).

Hence

$$(2\pi mkT)^{3/2} \approx \bar{p}^3$$

Thus, $(V/N)\bar{p}^3$ replaces $e^\alpha h^3$ and

$$e^\alpha = \frac{V\bar{p}^3}{Nh^3} = \frac{V}{N}\frac{1}{\lambda_{\mathrm{DB}}^3}$$

$$= \frac{\text{Volume available to each particle}}{\text{Volume associated with the thermal de Broglie wavelength of the particle}}$$

The value of $e^\alpha = 0.026\,m^{3/2}T^{5/2}$ at a pressure of one atmosphere, where m is measured in a.m.u. ($O^{16} = 16$).

For air at STP $e^\alpha \approx 10^6$ so for the Maxwell–Boltzmann distribution

$$\frac{g_i}{n_i} = e^{\alpha + \beta \varepsilon_i} \approx 10^6 \, e^{\varepsilon_i/kT} \gg 1$$

This states that there are many more states or cells available for occupation than there are particles to fill them, so the probable occupation of each cell is very small. This defines a classical distribution.

For the Bose–Einstein gas He^4 at 4 K and one atmosphere pressure e$^\alpha \sim 7.5$ so the gas is not safely classical.

Although it is not strictly applicable, for electrons in a metal at 300 K, e$^\alpha \sim 10^{-4}$ so the classical description for the Fermi–Dirac case is totally invalid.

A distribution which is not classical is said to be degenerate. Note that for high enough energies (temperatures) all three distributions become classical.

Before we examine the origin of α and its physical meaning let us note that a factor 2 appears in both the F–D and B–E distributions where each particle has two spin states for each energy level which must be accounted for. In Planck's radiation law these spin states are equivalent to the polarization states of electromagnetic waves. Note also in Planck's law that $E(\nu)\,\mathrm{d}\nu$, the energy per unit volume in the frequency range ν to $\nu + \mathrm{d}\nu$, is $n(\nu)\,\mathrm{d}\nu\, h\nu$ where $h\nu$ is the photon energy.

Turning to question (iii) on the significance of α we again use the expression $S = k \log \Omega$ or $\Omega = \mathrm{e}^{S/k}$. Consider a system in contact with a large reservoir at constant temperature, Figure A1.1, able to exchange both energy and particles with the reservoir. The combination of reservoir and system is isolated and its energy E, volume V and total number of particles N are all fixed and constant.

We ask 'What is the probability of finding the *system* in a particular microstate with n_j particles having *total* energy ε_j?' This will be proportional to the number of microstates in the reservoir after n_j and ε_j are supplied to the system.

The entropy equation with subscript R for reservoir becomes

$$S_\mathrm{R}(E - \varepsilon_j, N - n_j) = S_\mathrm{R}(E, N) - \varepsilon_j \left(\frac{\partial S}{\partial E} \right)_{NV} - n_j \left(\frac{\partial S}{\partial N} \right)_{EV}$$

where we neglect higher terms in the expansion.

Elementary thermodynamics shows that

$$\left(\frac{\partial S}{\partial E} \right)_{NV} = \frac{1}{T} \quad \text{and} \quad \left(\frac{\partial S}{\partial N} \right)_{EV} = \frac{-\mu}{T}$$

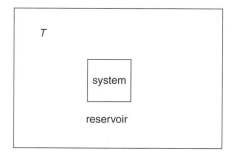

Figure A1.1   When a system, surrounded by a large reservoir with constant N, V and E receives n_j particles and total energy ε_j from the reservoir, the entropy change of the reservoir is $\Delta S = (n_j \mu - \varepsilon_j)/T$ where μ is the chemical potential

where μ is called the chemical potential. The chemical potential μ is the free energy per particle lost when the entropy S is increased in the relation $E = F + TS$ where E is constant. Thus, the entropy change may be written

$$\Delta S = S_R(E - \varepsilon_j, N - n_j) - S_R(E, N) = -\frac{\varepsilon_j}{T} + \frac{n_j\mu}{T}$$

Because the statistical weight Ω *(EVN)* represents the probability of a partition, the probability of the combination of two partitions may be written as the product of their statistical weights so we have

$$\Omega(E - \varepsilon_j, N - n_j) = \Omega(E, N)\,\Omega(\varepsilon_j, n_j)$$
$$= \Omega(E, N)\,e^{\Delta S/k}$$
$$= \Omega(E, N)\,e^{(n_j\mu - \varepsilon_j)/kT}$$

In order to show the relation between α and $-\mu/kT$, we take as an example a system of four fermions available to occupy any of four single particle energy states ε_1, ε_2, ε_3, ε_4 (Table A1.2). The particles and energies are supplied by the reservoir and each energy level may be filled or empty. The numbers of possible microstates of the system using 0, 1, 2, 3 or 4 particles are shown below together with their relative probabilities.

For any microstate in which a particular energy level is filled we can find another which differs only in having that energy level empty.

Table A1.2 Distribution of four fermions among four single particle energy states with numbers of possible microstates and their relative probabilities

	No particles	One particle	Two particles	Three particles	Four particles
Number of microstates	1	4	6	4	1
Energy level ε_4	0	0	0	0	1
Energy level ε_3	0	0	0	1	1
Energy level ε_2	0	0	1	1	1
Energy level ε_1	0	1	1	1	1
	$n_j = 0$ $\varepsilon_j = 0$	$n_j = 1$ $\varepsilon_j = \varepsilon_1$	$n_j = 2$ $\varepsilon_j = \varepsilon_1 + \varepsilon_2$	$n_j = 3$ $\varepsilon_j = \varepsilon_1 + \varepsilon_2 + \varepsilon_3$	$n_j = 4$ $\varepsilon_j = \varepsilon_1 + \varepsilon_2 + \varepsilon_3 + \varepsilon_4$
Relative probability of microstate	$e^{(0-0)/kT}$	$e^{(\mu - \varepsilon_1)/kT}$	$e^{[2\mu - (\varepsilon_1 + \varepsilon_2)]/kT}$	$e^{[3\mu - (\varepsilon_1 + \varepsilon_2 + \varepsilon_3)]/kT}$	$e^{[4\mu - (\varepsilon_1 + \varepsilon_2 + \varepsilon_3 + \varepsilon_4)]/kT}$

Thus, for example

$$\frac{\text{Relative probability of finding } \varepsilon_3 \text{ filled}}{\text{Relative probability of finding } \varepsilon_3 \text{ empty}} = \frac{p}{1-p}$$

$$= \frac{e^{[3\mu-(\varepsilon_1+\varepsilon_2+\varepsilon_3)]/kT}}{e^{[2\mu-(\varepsilon_1+\varepsilon_2)]/kT}}$$

$$= e^{(\mu-\varepsilon_3)/kT}$$

More generally

$$\frac{p}{1-p} = e^{(\mu-\varepsilon_i)/kT}$$

so

$$p = \frac{1}{e^{(\varepsilon_i-\mu)/kT}+1} = \overline{n}_i$$

where $n_i = g_i\overline{n}_i$ and $\overline{n}_i$ or the relative probability is the average occupation of a cell.

This is the Fermi–Dirac occupation factor and we can identify $\alpha = -\mu/kT$ (the ratio of two energies) where μ is the chemical potential. For the Fermi–Dirac distribution $\overline{n}_i \leq 1$ and Figure A1.2 shows $\overline{n}_i$ versus ε for electrons in a metal at $T = 0\,\text{K}$.

Each energy level is occupied by one electron until the top energy level ε_F the Fermi energy level is reached. At $T = 0\,\text{K}$ the electron with ε_F is the only one capable of moving to change the entropy of the system and we identify its free energy with that of the chemical potential μ. Note that, at ε_F for $T > 0$, $\overline{n}_i = \frac{1}{2}$ and this is indicated by the dotted curve at ε_F in the $\overline{n}_i$ versus ε graph.

We may apply a similar procedure to particles obeying Bose–Einstein statistics where there is no restriction on the number of particles n_i in the energy level ε_i. If n_i can take any value, three identical bosons available to three energy levels $(\varepsilon_1, \varepsilon_2, \varepsilon_3)$ can form the

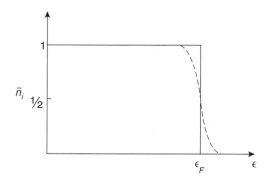

Figure A1.2 Occupation number $\overline{n}_i$ versus energy ε for electrons in a metal at $T = 0\,\text{K}$ (solid line). A slight increase in T permits the electrons near ε_F to move to higher energy levels (dotted curve)

microstates (3, 0, 0) (0, 3, 0) (0, 0, 3) (2, 1, 0)   (0, 2, 1) (1, 0, 2) (0, 1, 2) (2, 0, 1) (1, 2, 0) (1, 1, 1). The energy of each microstate is given by $\varepsilon_j = \sum n_i \varepsilon_i$ with $n_j = \sum n_i$. Suppose, as before, a large reservoir at temperature T surrounds a system to which it can supply particles and energy.

We consider a particular microstate of the system with $n_1, n_2, n_3 \ldots n_i$ particles in the various energy levels to have a *probability p when $n_i = 0$.*

If the system now takes n_i particles each of energy ε_i from the reservoir the probability of the microstate (now with $n_i \neq 0$) is given by

$$p\, e^{n_i(\mu - \varepsilon_i)/kT} = p\, e^{n_i x}$$

where $x = (\mu - \varepsilon_i)/kT$.

The total probability for the microstate with $n_i = 0, 1, 2, 3, \ldots$ is

$$1 = \sum_{n_i = 0}^{n_i = \infty} p\, e^{n_i x} = \frac{p}{(1 - e^x)}$$

because $\sum e^{n_i x}$ is a geometric progression.

Hence

$$p = (1 - e^x)$$

The average value

$$\bar{n}_i = \sum_{n_i = 0}^{n_i = \infty} n_i p\, e^{n_i x}$$

But

$$\sum n_i e^{n_i x} = \frac{\mathrm{d}}{\mathrm{d}x} \sum e^{n_i x} = \frac{\mathrm{d}}{\mathrm{d}x} \frac{1}{(1 - e^x)} = \frac{e^x}{(1 - e^x)^2}$$

Therefore

$$\bar{n}_i = \frac{p\, e^x}{(1 - e^x)^2} = \frac{(1 - e^x)e^x}{(1 - e^x)^2} = \frac{e^x}{(1 - e^x)}$$

$$= \frac{1}{e^{-x} - 1} = \frac{1}{e^{(\varepsilon_i - \mu)/kT} - 1}$$

The general expression for the Bose–Einstein distribution is therefore

$$n_i = g_i \bar{n}_i = g_i \times \frac{1}{e^{(\varepsilon_i - \mu)/kT} - 1}$$

Finally we discuss the absence of α or $-\mu/kT$ in Planck's radiation law, noting that this is a special case and that $-\mu/kT$ is retained in other applications of Bose–Einstein statistics.

Black body radiation is an equilibrium process, so that the system or cavity of a box of photons is in equilibrium with the reservoir at temperature T, the entropy S is a maximum and this process results from the continual emission and absorption of photons by the walls of the cavity. The number of photons in the cavity is not conserved, the energy requirement could be satisfied by a few high energy photons in the γ-ray region or by many photons in the low energy infrared frequencies. This means that the occupation numbers are not subject to the constraint which specifies the total number of particles in the gas.

Since N is not fixed, the entropy S of the reservoir is not affected by the n_j photons in the exponent $n_j\mu$ of the occupation factor for a given microstate; n_j has no role and $n_j\mu = 0$ giving $\mu = 0$.

The graph of the entropy S versus N, the total number of particles, gives low S values, that is few microstates or particle arrangements at low N (γ-rays) and also at high N (infrared) photons.

A typical microstate for γ-rays occupying the energy levels ε_i would read

$$n_1 = 0, \quad n_2 = 0, \quad n_3 = 0 \quad \text{with} \quad n_{\to\infty} \neq 0$$

and for infrared photons a typical microstate would read

$$n_1 \neq 0 \quad n_2 = 0 \quad n_3 = 0$$

Both of these are extremely unlikely and would contribute to partitions of low statistical weight.

At equilibrium the maximum of the S versus N curve occurs at that value of N providing the greatest number of microstates and here

$$\left(\frac{\partial S}{\partial N}\right)_{EV} = \frac{-\mu}{T} = 0$$

again giving $\mu = 0$.

Mathematical Derivation of the Statistical Distributions

The particles are identical but distinguishable by labels. All energy ε states are equally accessible and have the same *a priori* probability of being occupied. The statistical weight or probability of a particular partition is proportional to the number of different ways of distributing particles to obtain that partition.

Maxwell–Boltzmann Statistics

We start by filling the ε_1 states with n_1 particles from the constant total of N particles. We can do this in

$$\frac{N!}{n_1!(N - n_1)!}$$

different and distinguishable ways.

We now fill the ε_2 state with n_2 particles from the $N - n_1$ remaining particles. This gives

$$(N - n_1)!/n_2!(N - n_1 - n_2)!$$

different and distinguishable ways.

Proceeding in this way for all remaining energy states we have

$$W = \frac{N!}{n_1! n_2! n_3! \ldots}$$

as the number of different and distinguishable ways of choosing $n_1, n_2, n_3, \ldots$ from the N particles. Particles with the same ε_i may have g_i differing amounts of angular momentum, etc. This will give g_i cells associated with ε_i in each of which a particle with ε_i may be located. If g_i is the probability of having one particle in the ε_i range of cells then $g_i \times g_i = g_i^2$ is the probability of two particles in that range and $g_i^{n_i}$ is the probability of n_i particles with ε_i being in that range.

Hence the total number of different distinguishable ways is

$$W = \frac{N! g_1^{n_1} g_2^{n_2} g_3^{n_3} \cdots}{n_1! n_2! n_3! \ldots}$$

The particles are distinguished by labels and if we now remove the labels and the condition of distinguishable particles, we cannot recognize the difference in the partition when particles are exchanged. Therefore all $N!$ permutations among the particles occupying the different states give the same partition with the total number of ways

$$W = \frac{g_1^{n_1} g_2^{n_2} g_3^{n_3} \cdots}{n_1! n_2! n_3! \ldots}$$

We now maximize $\log W$ with the constraints that

1. The number of particles $N = \sum n_i = $ constant so that $dN = \sum dn_i = 0$.

2. The energy $E = \sum n_i \varepsilon_i = $ constant so that $dE = \sum \varepsilon_i dn_i = 0$.

$$\log W = \sum_i (n_i \log g_i - \log n_i!)$$

where for large n_i Stirling's formula gives

$$\log n_i! = n_i \log n_i - n_i$$

Hence

$$\log W = \sum n_i \log \frac{g_i}{n_i} + \sum n_i$$

and

$$d\left(\log W\right) = \sum d n_i \log\left(\frac{g_i}{n_i}\right) + \sum n_i\, d\log\left(\frac{g_i}{n_i}\right) + \sum d n_i$$

$$= \sum d n_i \log\left(\frac{g_i}{n_i}\right) - \sum n_i \frac{d n_i}{n_i}$$

$$\text{(because } g_i \text{ is constant and } \sum d n_i = 0)$$

$$= \sum d n_i \log\left(\frac{g_i}{n_i}\right).$$

If $\sum d n_i = 0$ then $-\alpha \sum d n_i = 0$ and
if $\sum \varepsilon_i\, d n_i = 0$ then $-\beta \sum \varepsilon_i\, d n_i = 0$
where α and β are called Lagrange multipliers.

Adding these constraint conditions to d(log W) gives

$$d\left(\log W\right) = \sum d n_i \left(\log\left(\frac{g_i}{n_i}\right) - \alpha - \beta\varepsilon_i\right)$$

Maximizing W gives d(log W)$=0$ which, since all the coefficients $d n_i$ are arbitrary and independent, leaves

$$\log\left(\frac{g_i}{n_i}\right) - \alpha - \beta\varepsilon_i = 0$$

for each n_i.

At W_{max} we have therefore

$$n_i = g_i \times \frac{1}{e^{\alpha + \beta\varepsilon_i}}$$

Fermi–Dirac Statistics

We begin again with labelled identical particles. Here the Pauli exclusion principle operates and no two particles may occupy the same state. The g_i are quantum states, e.g. spin gives a factor 2 to each g_i. Also g_i gives the maximum number of particles with ε_i so $n_i \le g_i$.

To fill the ε_i states with n_i particles we put one particle in a g_i cell and the next particle in any of the $(g_i - 1)$ remaining cells. We can do this in $g_i(g_i - 1)$ ways so the total number of ways of filling the states of energy ε_i with n_i particles is

$$g_i(g_i - 1)\ldots(g_i - n_i + 1)$$

$$= \frac{g_i!}{(g_i - n_i)!}$$

If now the labels are removed and the particles become indistinguishable we reduce the total of different distinguishable arrangements to $g_i!/n_1!(g_i - n_i)!$.

Applying this to all g_i gives the total number of different distinguishable ways as

$$W = \frac{g_1!}{n_1!(g_1 - n_1)!} \frac{g_2!}{n_2!(g_2 - n_2)!} \frac{g_3!}{n_3!(g_3 - n_3)!}$$

Maximizing $\log W$ with $\sum \mathrm{d}n_i = \sum \varepsilon_i \mathrm{d}n_i = 0$ we proceed as with the Maxwell–Boltzmann example to obtain for W (max) the condition that

$$\log\left(\frac{g_i}{n_i} - 1\right) - \alpha - \beta\varepsilon_i = 0$$

to give

$$n_i = g_i \times \frac{1}{\mathrm{e}^{\alpha + \beta\varepsilon_i} + 1}$$

Bose–Einstein Statistics

Here there is no exclusion principle and we begin again with labelled identical particles.

The number of distinguishable arrangements of n_i particles in the g_i cells of energy ε_i equals the number of ways of putting n_i objects in g_i boxes with any number allowed in a box. This means putting n_i particles in a row separated by $g_i - 1$ walls so that the number of ways is the number of permutations of $(n_i + g_i - 1)$ objects, i.e. particles and walls. This gives $(n_i + g_i - 1)!$ ways. If we now remove the particle labels to make them indistinguishable we reduce the number of ways by a factor of $n!$ to give $(n_i + g_i - 1)!/n_i!$ ways.

However, all permutations of the $g_i - 1$ dividing walls among the n_i particles give the same physical state, so the number of different distinguishable ways is given by $(n_i + g_i - 1)!/n_i!(g_i - 1)!$ and for all particles we have the number of ways

$$W = \frac{(n_1 - g_1 - 1)}{n_1!(g_1 - 1)} \frac{(n_2 + g_2 - 1)}{n_2!(g_2 - 1)} \cdots \text{etc.}$$

Maximizing $\log W$ as for the other two distributions gives $\mathrm{d}(\log W) = 0$ when

$$\log\left(\frac{g_i}{n_i} + 1\right) - \alpha - \beta\varepsilon_i = 0$$

that is, when

$$n_i = g_i \frac{1}{\mathrm{e}^{\alpha + \beta\varepsilon_i} - 1}$$

Table A1.3 Waves incident normally on a plane boundary between media of characteristic impedances Z_1 and Z_2

			Amplitude Coefficients				
Wave type	Impedance Z $+$ve for wave in $+$ve direction $-$ for wave in $-$ve direction	Boundary conditions		$\dfrac{Reflected_r}{Incident_i} = \dfrac{Z_1-Z_2}{Z_1+Z_2}$	$\dfrac{Transmitted_t}{Incident_i} = \dfrac{2Z_1}{Z_1+Z_2}$	$\dfrac{Reflected_r}{Incident_i} = \dfrac{Z_2-Z_1}{Z_1+Z_2}$	$\dfrac{Transmitted_t}{Incident_i} = \dfrac{2Z_2}{Z_1+Z_2}$
Transverse on string (p. . . .)	$\dfrac{-T(\partial y/\partial x)}{\dot y} = \rho c = (T\rho)^{1/2}$	$y_i + y_r = y_t$ or $\dot y_i + \dot y_r = \dot y_t$ $T\left[\dfrac{\partial y_i}{\partial x} + \dfrac{\partial y_r}{\partial x} = \dfrac{\partial y_t}{\partial x}\right]$	y and $\dot y$			$-T\dfrac{\partial y}{\partial x}$	
Longitudinal acoustic (p. . . .)	$\dfrac{\rho}{\dot\eta} = \rho_0 c = (B_a\rho)^{1/2}$	$\dot\eta_i + \dot\eta_r = \dot\eta_t$ $p_i + p_r = p_t$	η and $\dot\eta$			p	
Voltage and current on transmission line (p. . . .)	$\dfrac{V}{I} = \sqrt{\dfrac{L}{C}}$	$I_i + I_r = I_t$ $V_i + V_r = V_t$	I			V	
Electro-magnetic (p. . . .)	$\dfrac{E}{H} = \sqrt{\dfrac{\mu}{\varepsilon}}$	$H_i + H_r = H_t$ $E_i + E_r = E_t$	H			E	

All waves $\dfrac{\text{Reflected intensity}}{\text{Incident intensity}} = \left(\dfrac{Z_1-Z_2}{Z_1+Z_2}\right)^2$ $\dfrac{\text{Transmitted intensity}}{\text{Incident intensity}} = \dfrac{4Z_1Z_2}{(Z_1+Z_2)^2}$

546

Appendix 2: Kirchhoff's Integral Theorem

Kirchhoff's Integral Theorem is valid for any solution E of the scalar time independent Helmholtz equation (3), p. 187, that is

$$\frac{\partial^2 \mathbf{E}}{\partial x^2} + k^2 \mathbf{E} = 0$$

For the radial direction r in a spherical coordinate system this becomes

$$\frac{\partial^2 \mathbf{E}}{\partial r^2} + \frac{2}{r} \frac{\partial \mathbf{E}}{\partial r} = 0$$

which is satisfied by

$$\mathbf{E} = \frac{E_0}{r} \, e^{ikr}$$

where E_0/r is the amplitude at a distance r from the origin O of a spherical electromagnetic wave. We note that the amplitude of such a wave decays as $1/r$ where r is the distance from O.

Kirchhoff's Theorem states that the complex amplitude $\mathbf{E}_P$ at a point P is related to the complex amplitude $\mathbf{E}$ on a surface S enclosing P by

$$\mathbf{E}_P = \frac{1}{4\pi} \iint_S \left(\mathbf{E} \frac{\partial}{\partial n} \frac{e^{ikR}}{R} - \frac{e^{ikR}}{R} \frac{\partial \mathbf{E}}{\partial n} \right) \mathrm{d}S$$

where R is the distance from P to the surface element $\mathrm{d}S$ and $\mathbf{n}$ is the direction normal to $\mathrm{d}S$ (Figure A2.1).

If r is the distance from O to $\mathrm{d}S$, then

$$\mathbf{E} = \frac{E_0}{r} \, e^{ikr}$$

The Physics of Vibrations and Waves, 6th Edition   H. J. Pain
© 2005 John Wiley & Sons, Ltd

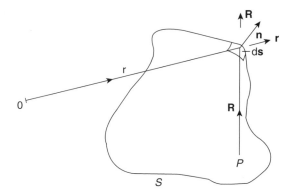

Figure A2.1   O is the origin of an electromagnetic wave. Kirchoff's Theorem relates its complex amplitude $\mathbf{E}_P$ at a point P to the complex amplitude $\mathbf{E}$ on a surface S enclosing P

and

$$\frac{\partial \mathbf{E}}{\partial n} = \frac{E_0}{r}\, e^{ikr}\left(ik - \frac{1}{r}\right)\cos\left(\mathbf{n}, \mathbf{r}\right)$$

The term $(ik - 1/r)$ shows that inside S there is a phase shift of $\pi/2$ rad and an amplitude factor $1/r$. However, for $r = m\lambda$, where m is large, then

$$k = \frac{2\pi}{\lambda} \gg \frac{1}{r} = \frac{1}{m\lambda}$$

so that $1/r$ may be neglected for distances much greater than λ.
    Similar arguments hold for

$$\frac{\partial}{\partial n}\frac{e^{ikR}}{R}$$

Thus, if P and O are many wavelengths from S, Kirchhoff's integral becomes

$$\mathbf{E}_P = \frac{-i}{\lambda}\iint E_0\, \frac{e^{ik(r+R)}}{rR}\, \frac{(\cos\mathbf{n}, \mathbf{R} - \cos\mathbf{n}, \mathbf{r})}{2}\, d\mathbf{S}$$

where the cosine terms generate an inclination factor $K(\chi)$ and $\cos(\mathbf{n}, \mathbf{R}) = \cos\chi$.
    The problem of showing that Huygens wavelets on an unobstructed wavefront do not propagate backwards reduces to that of demonstrating that $K(\chi)$ can be zero. This occurs where

$$\cos\left(\mathbf{n}, R\right) = \cos\pi = -1$$

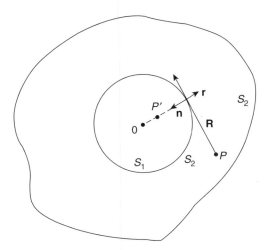

Figure A2.2 When P' is located on $\mathbf{r}$ within the surface of the spherical wavefront S_1, situated within S, $\mathbf{E}_{P'}$ is reduced to zero proving that Huygens wavelets do not propagate backwards

and

$$\cos\left(\mathbf{n}, \mathbf{r}\right) = \cos \pi = -1$$

This is achieved in the following way.

The surface S designated S_2 now encloses a spherical wavefront surface S_1 centre O. S_1 and S_2 are said to be doubly connected and the surface integral now includes S_1 and S_2 (Figure A2.2). At S_1 the normal $\mathbf{n}$ to dS on S_2 now points towards O and if the outer surface of S_2 is allowed to expand to infinity its contribution to the integral becomes zero. This leaves only the integral over the surface where S_1 and S_2 coincide. The singularity E_0/r at O is excluded from the integral.

If P is now located on $\mathbf{r}$, at P', that is in the direction of backward propagation of Huygens wavelets, then

$$\cos\left(\mathbf{n}, \mathbf{R}\right) = \cos \pi = -1$$

and

$$\cos\left(\mathbf{n}, \mathbf{r}\right) = \cos \pi = -1$$

$K(\chi)$ is then equal to zero. Any other position for P gives

$$K(\chi) = \frac{\cos \chi - \cos\left(\mathbf{n}, \mathbf{r}\right)}{2} = \frac{1 + \cos \chi}{2}$$

Appendix 3:
Non-Linear Schrödinger Equation

This equation describes phenomena in non-linear media with strong dispersion. It appears in several forms. For optical soliton purposes, Mollenauer *et al.* (1982) derive it from the equation

$$i\left(\frac{\partial u}{\partial z} + k_1 \frac{\partial u}{\partial t}\right) = \frac{-k_2}{2}\frac{\partial^2 u}{\partial t^2} + \gamma|u|^2 u \tag{A3.1}$$

where

$$k_1 = \frac{\partial k}{\partial \omega}, \qquad k_2 = \frac{\partial^2 k}{\partial \omega^2}, \quad \text{and} \quad \gamma = \frac{1}{2}k_0\frac{n_2}{n_0}$$

n_2 and n_0 appear in the Kerr Optical Equation $n - n_0 = n_2 I$.

Equation (A3.1) is satisfied by a pulse of the form

$$E(z,t) = u(z,t)\,e^{i(\omega_0 t - k_0 z)}$$

Using the transformation of Mollenauer *et al.* (1980), (A3.1) assumes the dimensionless form

$$-i\frac{\partial v}{\partial \xi} = \frac{1}{2}\frac{\partial^2 v}{\partial s^2} + |v|^2 v \tag{A3.2}$$

which has a soliton solution $u(\xi, s) = \text{sech}\,(s)e^{i\xi/2}$ where

$$s = T^{-1}(t - k_1 z) \qquad \xi = |k_2|T^{-2}z$$

and

$$v = T\left(\frac{\gamma}{|k_2|}\right)^{1/2} u$$

where T is a measure of the width of the input optical pulse.

The first term on the right hand of equation (A3.2) describes the effects of dispersion which may be seen as the kinetic energy term in the linear Schrödinger equation, while the second term corresponds to the energy of a self-trapping potential proportional to $|u|^2$ arising from the non-linear refractive index which may be interpreted in probability terms.

The Physics of Vibrations and Waves, 6th Edition   H. J. Pain
© 2005 John Wiley & Sons, Ltd

Index